RADIONUCLIDES: A TOOL FOR OCEANOGRAPHY

Proceedings of an International Symposium jointly organized by the 'Société Française pour l'Energie Nucléaire' (SFEN) and 'l'Institut National des Techniques de la Mer' (INTECHMER–CNAM) at Cherbourg, France, 1–5 June 1987.

The Symposium was subsidized by

Commission des Communautés Européennes (CCE)
Directorate of Fisheries Research (MAFF)
Centre National de la Recherche Scientifique (CNRS)
Commissariat à l'Energie Atomique (CEA)
Institut Français de Recherche pour l'Exploitation de la Mer (IFREMER)
Compagnie Générale des Matières Nucléaires (COGEMA)
Electricité de France (EDF)
Conseil Général de la Manche
Communauté Urbaine de Cherbourg
Conseils Municipaux des Villes de Cherbourg et Tourlaville
Syndicat Mixte des Après grands Chantiers du Cotentin (SMACC)

RADIONUCLIDES: A TOOL FOR OCEANOGRAPHY

Edited by

J. C. GUARY

Conservatoire National des Arts et Métiers, Institut National des Techniques de la Mer (INTECHMER), Cherbourg, France

P. GUEGUENIAT

Commissariat à l'Energie Atomique, Laboratoire de Radioécologie Marine, IPSN, La Hague, France

and

R. J. PENTREATH

Ministry of Agriculture, Fisheries and Food, Directorate of Fisheries Research, Lowestoft, UK

ELSEVIER APPLIED SCIENCE
LONDON and NEW YORK

ELSEVIER SCIENCE PUBLISHERS LTD
Crown House, Linton Road, Barking, Essex IG11 8JU, England

Sole Distributor in the USA and Canada
ELSEVIER SCIENCE PUBLISHING CO., INC
52 Vanderbilt Avenue, New York, NY 10017, USA

WITH 50 TABLES AND 203 ILLUSTRATIONS

British Library Cataloguing in Publication Data

Radionuclides: a tool for oceanography.
1. Oceanography. Use of tracers
I. Guary, J. C. II. Gueguenait, P.
III. Pentreath, R. J.
551.46′0028

ISBN 1-85166-298-7

Library of Congress Cataloging-in-Publication Data applied for

Printed in Northern Ireland by The Universities Press (Belfast) Ltd.

PREFACE

The Cherbourg Symposium on the subject of "RADIONUCLIDES: A TOOL FOR OCEANOGRAPHY" may be considered an important landmark in the evolution of French oceanography. It was given a wider context by the strong international participation, particularly from the European countries.

On reflection, after the symposium, it seems that a new stage has been reached which has profound implications for the future: it concerns the complementary contributions which can be made to oceanography by the study of the behaviour and distributions of radionuclides in the marine environment ("radionuclide oceanography") and by the techniques of remote sensing from orbiting satellites ("satellite oceanography"). Developed at much the same time, these two approaches permit the application of concepts and techniques which are among the most elaborate recently conceived in science. They have in common indispensible, multiple and complex connections with the major traditional fields of science: physics, chemistry, biology and geology, without which they remain non-productive or, at best, debatable in their conclusions.

Remote sensing from space examines the dynamic phenomena in the surface layer of the ocean and also penetrates deeper to provide data concerning the structure of the oceanic crust; the domain reserved for "radionuclide oceanography", which distinguishes it from its "satellite" homologue, is the mass of the ocean waters from the surface to the greatest depths and includes the unconsolidated sedimentary deposits.

The half lives of the radionuclides and the precise knowledge of the time and place of their introduction into the oceanic system which is often available, provide a time-frame for the study of processes in three dimensions. The only absolutely conservative tracers available to physical oceanographers remain "temperature and salinity"; yet more and more the distributions of the radionuclides are not only considered as circumstantial evidence by the physicists. On the contrary, the information allows tight constraints to be applied for the validation of models of the oceanic circulation.

In the field of geology, radionuclides play a major role in the study, with a previously unattainable spatio-temporal resolution, of the succession of climatic and tectonic events which have shaped the planetary environment and of which the sedimentary strata have preserved a record. In geology, connections must be made between the data acquired from satellite observations concerning the deep substrate and its dynamics and the isotopic dating information obtained from the covering of mobile sediment which accompanies the basement rock in its slow migration. A new field of geochemistry emerges from this confrontation.

In another connection, the elucidation of the evolution of the distributions of the many minor constituents in the water mass permits a novel approach to marine chemistry, in imposing, for example, precise time constants on the interactions between the physical forces detected by satellites and the behaviour of the biomass. Here again, one can glimpse the importance of a transfer of information coming from space-based observations, from the wind field to the colour of the sea, to marine radionuclide biogeochemistry. These interactions are indispensible, otherwise it is difficult to see how the problems posed by the parameterization of the fluxes of matter in the ocean may be resolved.

That is not all that can be said about radioactive tracers: in addition to the requirements set out above - the interactions with the basic sciences - there is an idea which must be kept carefully in mind, that of a necessary dynamic equilibrium between the process which is being investigated and the tracer which is being used as a probe. There exists meanwhile a recent technique called the "inverse method" which allows this requirement to be relaxed to a great extent, and personally, I regret that this symposium was not able to consider this question. Similarly, a little time could have been given to a comparison of the technical requirements and relative merits as tracers of the radionuclides and the novel chemicals such as the freons.

Faced with a large number of possible approaches, the symposium organizers had to make difficult choices. The main themes included (see list) cover the major areas of work and the organizers may be complimented on it.

It is necessary to remember that this was one of the first symposia to cover the whole field of "radionuclide oceanography".

This symposium has marked equally the emergence of new technologies: thus mass spectrometry has become a necessary complement to the detection of nuclear radiation at high sensitivity, particularly when it is combined with isotope acceleration.

The published proceedings of the symposium fall into two parts. The first comprises the high quality papers published in full after evaluation by referees. The second gathers together the abstracts of the remaining oral presentations and the posters, of which a full account is not provided here, although they generally have been, or will be, published in the wider scientific literature. Thus the totality of the contributions to the symposium are represented in these proceedings which provide as complete a record as possible.

I would like, before concluding this preface, to pay tribute to the memory of Alan Preston who contributed much to the organization and conduct of this symposium. Alan Preston always knew how to reconcile scientific rigour, ecological necessity and the economic constraints of the energy policy of his country. He remains an example to those among us who are confronted with these problems.

Finally, I wish to reiterate the urgent necessity for an evaluation to be made of the comparative prospects of "satellite oceanography" and "radionuclide oceanography". There are many who are looking forward to, or are already working towards, this end. A new symposium must be organized to bring together these two great scientific advances in our understanding of the marine environment.

Roger CHESSELET

THEMES

- ORIGIN AND USES OF RADIONUCLIDES IN THE MARINE ENVIRONMENT
- THE STUDY OF LARGE SCALE OCEANOGRAPHIC PROCESSES USING NATURAL RADIONUCLIDES
- THE NATURAL RADIONUCLIDES AS TRACERS OF SCAVENGING AND PARTICULATE TRANSPORT PROCESSES IN THE OPEN OCEAN AND COASTAL WATERS
- NATURAL RADIONUCLIDES AND MARINE SEDIMENTATION
- CHERNOBYL FALLOUT: AN OCEANOGRAPHIC TRACER
- THE USE OF RADIONUCLIDES DISCHARGED INTO THE SEA FOR THE STUDY OF WATER MOVEMENT IN COASTAL AND OPEN OCEAN SYSTEMS
- ARTIFICIAL RADIONUCLIDES AND SEDIMENTARY PROCESSES IN ESTUARINE AND COASTAL SYSTEMS
- THE DEVELOPMENT AND APPLICATION OF MODELS IN RESPECT OF RADIONUCLIDES IN THE MARINE ENVIRONMENT

PREFACE

Le Colloque de Cherbourg consacré aux "RADIONUCLEIDES : OUTIL OCEANOGRAPHIQUE", peut être considéré comme une étape importante dans l'évolution de l'océanographie française d'aujourd'hui. Une forte participation internationale, notamment européenne, lui a donné sa véritable dimension.

A la réflexion, après ce Colloque, il me semble qu'une nouvelle étape devra être franchie car elle est profondément porteuse d'avenir : il s'agit de la conjugaison de l'Océanographie nucléaire et de l'Océanographie spatiale. Nées toutes deux à peu près au même moment, elles mettent en oeuvre des concepts et des techniques parmi les plus élaborés que la science ait récemment conçus. Elles ont en commun des croisements obligés, multiples et complexes avec de grandes disciplines traditionnelles : physique, chimie, biologie, géologie, sans lesquelles elles demeurent inopérantes et parfois même contestables dans leurs interprétations.

L'Océanographie spatiale interroge les phénomènes dynamiques à la surface de l'océan et permet également de pénétrer profondément dans la structure de la croûte océanique ; le domaine d'intervention privilégié de l'Océanographie nucléaire, qui la distingue de son homologue "spatiale", est bien entendu la masse des eaux marines, depuis cette extrême surface jusqu'aux grands fonds, y compris la couverture sédimentaire encore peu compactée.

Les périodes de la désintégration des radionucléides, la connaissance souvent très précise du mode et du moment de leurs injections dans le système océanique, donnent accès à une chronologie des processus dans un système à trois dimensions. Les seuls traceurs absolument conservatifs que requiert l'océanographie physique demeurent "température et salinité" ; cependant de plus en plus les distributions des radionucléides ne sont plus considérées seulement comme des appoints de circonstance par les physiciens. Au contraire, elles permettent maintenant d'introduire dans les modèles de circulation océanique des contraintes très précieuses pour leur validation.

Dans le domaine de la géologie, les marqueurs radioactifs ont pris une place prépondérante en permettant de discerner, avec une résolution spatio-temporelle encore jamais atteinte, la succession des évènements climatiques et tectoniques qui ont affecté l'environnement planétaire et dont les édifices sédimentaires ont conservé la mémoire. En géologie des relais doivent être pris entre les données acquises par les satellites sur le substrat profond et sa dynamique, et les données de datations nucléaires de la couverture du sédiment meuble qui l'accompagne dans sa lente migration. Un domaine nouveau de la géochimie émerge de cette confrontation.

Par ailleurs, le suivi de l'évolution des distributions dans la masse d'eau d'une multitude de constituants mineurs, permet une approche nouvelle de la chimie marine, en imposant, par exemple, des constantes de temps précises aux interactions entre les forçages de type physique vus par les satellites et l'action de la biomasse. Là encore on peut entrevoir l'importance d'un transfert d'information venant du spatial, depuis les champs de vents jusqu'à la couleur de la mer, vers la biogéochimie nucléaire marine. Ces interactions sont à ce point indispensables, que l'on ne voit pas comment le problème posé par la paramétrisation des flux de matière dans l'océan pourrait être résolu autrement.

On ne peut pas tout faire dire aux traceurs radioactifs ; en plus des exigences énoncées ci-dessus - les croisements avec les disciplines de base - il est une notion qu'il faut garder soigneusement à l'esprit, celle d'un nécessaire équilibre dynamique entre les phénomènes que l'on veut tracer et le traceur que l'on s'est choisi. Il existe cependant une méthode récente dite méthode inverse qui permet quasiment de se soustraire à cette exigence et personnellement je regrette que ce Colloque n'ait pas pu faire le point sur cette méthode. De même, un peu de temps aurait pu être consacré à une comparaison, à une mise en perspective des performances et des exigences technologiques des traceurs radioactifs par rapport à de nouveaux traceurs chimiques tels que les fréons.

Face au grand nombre d'approches possibles, les organisateurs du Colloque ont du faire des choix difficiles (cf liste des thèmes en encadré). Les thèmes retenus recouvrent bien les axes majeurs de notre recherche et il faut les en féliciter.

Il faut rappeler qu'il s'agissait de l'un des premiers Colloques qui couvrait tout le champ de l'Océanographie nucléaire.

Ce Colloque a marqué également l'émergence de technologies récentes ; ainsi la spectrométrie de masse devient le complément quasi obligé de la détection du rayonnement radioactif à haute sensibilité, notamment quand la spectrométrie de masse est associée à un accélérateur d'isotopes.

Les Actes publiés se décomposent en deux parties. L'une comprend des articles de haut niveau publiés in extenso après avoir fait l'objet d'une évaluation par un comité de lecture. L'autre regroupe les résumés de toutes les communications présentées. Elles ne sont pas l'objet d'articles mais, pour la plupart, ont été ou seront publiées dans des grandes revues scientifiques. Nous avons voulu ainsi que l'ensemble des contributions figure dans ces Actes, afin que l'information soit aussi complète que possible.

Je voudrais, avant de conclure cette préface, rendre hommage à la mémoire d'Allan Preston qui a été un des promoteurs essentiels de ce Colloque. Allan Preston a toujours su concilier une parfaite rigueur scientifique, les exigences écologiques et les impératifs économiques de la politique énergétique de son pays. Il demeure un exemple pour beaucoup d'entre nous confrontés à ces problèmes.

Je souhaiterais enfin insister à nouveau sur la nécessité urgente d'une prospective comparée entre l'océanographie spatiale et l'océanographie nucléaire. Bien des chercheurs la pressentent ou la vivent. Un nouveau colloque devrait être organisé pour réunir ces deux grandes avancées scientifiques de notre connaissance récente du domaine de la mer.

Roger CHESSELET

THEMES

- ORIGINE ET APPLICATION DES RADIONUCLEIDES DANS LE MILIEU MARIN.
- ETUDE DES PROCESSUS OCEANOGRAPHIQUES A GRANDE ECHELLE A L'AIDE DES RADIONUCLEIDES NATURELS.
- LES RADIONUCLEIDES NATURELS, TRACEURS DES PROCESSUS D'ENTRAINEMENT ET DE TRANSPORT SOUS FORME PARTICULAIRE DANS L'OCEAN ET LE SYSTEME COTIER.
- LES RADIONUCLEIDES NATURELS ET LES PROCESSUS SEDIMENTAIRES MARINS.
- LA RETOMBEE DE TCHNERNOBYL : UN TRACEUR OCEANOGRAPHIQUE.
- UTILISATION DES RADIO-NUCLEIDES REJETES EN MER DANS L'ETUDE DES MOUVEMENTS DES MASSES D'EAU DANS LES SYSTEMES COTIERS ET OCEANIQUES.
- LES RADIONUCLEIDES ARTICIFICIELS ET LES PROCESSUS SEDIMENTAIRES COTIERS ET ESTUARIENS.
- DEVELOPPEMENT ET APPLICATIONS DE MODELES EN RELATION AVEC LES RADIONUCLEIDES DANS LE MILIEU MARIN.

CONTENTS

Preface v

Dedication. xxi

SESSION 1: ORIGIN AND APPLICATIONS OF RADIONUCLIDES IN THE MARINE ENVIRONMENT

The use of natural radionuclides in oceanography: an overview . . . 1
R. Chesselet and C. Lalou

Sources of artificial radionuclides in the marine environment 12
R. J. Pentreath

SESSION 2: STUDY OF LARGE-SCALE OCEANOGRAPHIC PROCESSES USING NATURALLY OCCURRING RADIONUCLIDES

(*Chairman*: K. COCHRAN)

Use of tritium and helium-3 for the study of oceanographic processes. An example: the northeastern Atlantic ocean 35
C. Andrié and L. Merlivat

Mesure du couple tritium/helium océanique par spectrometrie de masse 45
P. Jean-Baptiste, C. Andrié et M. Lelu

Les isotopes du beryllium dans le milieu marin et quelques applications 55
F. Yiou, G. M. Raisbeck et D. Bourlès

Determination of carbon-14 by accelerator mass-spectrometry: oceanographic applications 64
J. C. Duplessy, E. Bard, M. Arnold and P. Maurice

Ra-226 and Ba in the north east Atlantic deep and bottom water. Limitations of Ra-226 as a time tracer 74
M. Rhein and W. Roether

Determination of thorium in seawater by neutron activation analysis and mass spectrometry 84
Chih-An Huh

Sorbents for the extraction of radionuclides from natural waters . . . 92
R. M. Gershey and D. R. Green

SESSION 3: NATURALLY OCCURRING RADIONUCLIDES AS TRACERS OF SCAVENGING AND PARTICULATE TRANSPORT PROCESSES IN OPEN AND COASTAL ENVIRONMENTS

(*Chairman*: C. LALOU)

Marine scavenging of Th and Pa—evidence from mid-ocean ridge, continental margin and open ocean sediments 101
G. B. Shimmield and N. B. Price

Effect of natural colloidal matter on the equilibrium adsorption of thorium in seawater 111
Sherry E. H. Niven and Robert M. Moore

Temporal changes of Th-234 concentrations and fluxes in the northwestern Mediterranean 121
P. Buat-Ménard, H. V. Nguyen, J. L. Reyss, S. Schmidt, Y. Yokoyama, J. La Rosa, S. Heussner and S. W. Fowler

Scavenging and bioturbation in the Irish sea from measurements of $^{234}Th/^{238}U$ and $^{210}Pb/^{226}Ra$ disequilibria 131
P. J. Kershaw, P. A. Gurbutt, A. K. Young and D. J. Allington

$^{234}Th/^{238}U$ disequilibrium to estimate the mean residence time of thorium isotopes in a coastal marine environment 143
A. Battaglia, W. Martinotti and G. Queirazza

SESSION 4: NATURALLY OCCURRING RADIONUCLIDES IN MARINE SEDIMENTS AND OVERLYING WATERS

(*Chairman*: R. D. CHERRY)

Sources and sinks of lead-210 in Puget sound 153
A. E. Nevissi

The geochemistry of uranium in marine sediments 162
Christina Barnes and J. K. Cochran

Uranium in the marine environment. A geochemical approach to its hydrologic and sedimentary cycle 171
Renata Djogić, G. Kniewald and M. Branica

Some aspects of the marine geochemistry of uranium 183
J. Toole, M. S. Baxter and J. Thomson

Natural radioactivity of heavy mineral sands: a tool for coastal sedimentology? 195
R. J. de Meijer, L. W. Put, R. D. Schuiling, J. de Reus and J. Wiersma

SESSION 5: CHERNOBYL FALLOUT AS AN OCEANOGRAPHIC TRACER

(*Chairman*: A. WALTON)

Characteristics of Chernobyl fallout in the southern Black Sea 204
H. D. Livingston, K. O. Buesseler, E. Izdar and T. Konuk

Spatial and temporal variations in the levels of cesium in the north western Mediterranean seawater (1985–1986) 217
D. Calmet, J. M. Fernandez, P. Maunier and Y. Baron

The contamination of the North sea and Baltic sea by the Chernobyl fallout 227
H. Nies and Ch. Wedekind

Cs-137 and other radionuclides in the benthic fauna in the Baltic sea before and after the Chernobyl accident 240
P-O. Agnedal

SESSION 6: USE OF ARTIFICIAL RADIONUCLIDES IN THE STUDY OF WATER MOVEMENTS IN COASTAL AND OPEN OCEAN SYSTEMS

(*Chairman*: A. PRESTON)

Artificial radioactivity in the northeast Atlantic 250
H. Nies

Utilisation de radionucléides artificiels (^{125}Sb–^{137}Cs–^{134}Cs) pour l'observation (1983–1986) des déplacements de masses d'eau en Manche 260
P. Guegueniat, R. Gandon, Y. Baron, J. C. Salomon, J. Pentreath, J. M. Brylinski et L. Cabioch

Determination of distribution processes, transport routes and transport times in the North sea and the northern north Atlantic using artificial radionuclides as tracers 271
H. Kautsky

Sellafield radiocaesium as a tracer of water movement in the Scottish coastal zone 281
P. E. Bradley, B. E. Economides, M. S. Baxter and D. J. Ellet

The application of seaweeds as bioindicators for radioactive pollution in the Channel and southern North sea 294
A. Aarkrog, H. Dahlgaard, S. Duniec, P. Guegueniat and E. Holm

Fucus vesiculosus as an indicator for caesium isotopes in Irish coastal waters 304
I. R. McAulay and D. Pollard

Répartition de deux traceurs radioactifs (^{106}Ru–Rh, ^{60}Co) chez deux espèces indicatrices (*Fucus serratus*, L., *Mytilus edulis*, L.) le long du littoral français de la Manche 312
P. Germain, Y. Baron, M. Masson et D. Calmet

SESSION 7: ARTIFICIAL RADIONUCLIDES AND SEDIMENTARY PROCESSES IN COASTAL WATERS AND ESTUARIES

(*Chairman*: M. S. BAXTER)

The behaviour of dissolved plutonium in the Esk estuary, U.K. . . . 321
M. Kelly, S. Mudge, J. Hamilton-Taylor and K. Bradshaw

Mixing process in near-shore marine sediments as inferred from the distributions of radionuclides discharged into the northeast Irish sea from BNFL, Sellafield 331
D. S. Woodhead

Radionuclide distributions in the surface sediments of Loch Etive . . 341
T. M. Williams, A. B. MacKenzie, R. D. Scott, N. B. Price and I. M. Ridgway

Chemical partitioning of plutonium and americium in sediments from the Thule region (Greenland) 351
E. Holm, J. Gastaud, R. Oregioni, A. Aarkrog, H. Dahlgaard and J. N. Smith

SESSION 8: DEVELOPMENT AND APPLICATION OF MODELS TO DESCRIBE THE BEHAVIOUR OF RADIONUCLIDES IN THE MARINE ENVIRONMENT

(*Chairman*: R. J. PENTREATH)

Polonium-210 in selected categories of marine organisms: interpretation of the data on the basis of an unstructured marine food web model 362
R. D. Cherry and M. Heyraud

Modèle mathématique du transport des radionucléides sur le plateau continental nord-européen 373
S. Djenidi, J. C. J. Nihoul et A. Garnier

A Lagrangian model for long term tidally induced transport and mixing. Verification by artificial radionuclide concentrations 384
J. C. Salomon, P. Guegueniat, A. Orbi and Y. Baron

Modelling the distribution of soluble and particle-adsorbed radionuclides in the Irish sea 395
P. A. Gurbutt, P. J. Kershaw and J. A. Durance

High-level radioactive waste disposal into the seabed: investigation of the sediment-barrier efficiency by means of analytical models 407
D. Boust, M. Poulin and R. Chesselet

ROLE OF THE CEC IN THE DEVELOPMENT OF PROGRAMMES TO INVESTIGATE THE BEHAVIOUR OF ARTIFICIAL RADIONUCLIDES IN THE MARINE ENVIRONMENT

Description des résultats obtenus en matière d'océanographie dans le cadre du programme radioprotection de la CCE 416
G. Desmet et C. Myttenaere

LIST OF PAPERS AND POSTERS PUBLISHED IN THE FORM OF ABSTRACTS

^{10}Be in the water column of the equatorial Atlantic 427
M. Segl, A. Mangini, G. Bonani, H. J. Hofmann, M. Suter and W. Wölfli

Vertical flux of radionuclides in marine systems mediated by biogenic debris 428
N. S. Fisher, J. K. Cochran and S. Krishnaswami

^{10}Be fluxes from sediment traps 429
G. M. Raisbeck, F. Yiou, S. Honjo, J. Klein and R. Middleton

Natural radionuclides in two anoxic marine environments 430
J. F. Todd, R. J. Elsinger and W. S. Moore

Processes involved in long-term behaviour of americium in deep-sea sediments investigated by rare earth elements analogy 431
D. Boust, C. Lambert, R. Chesselet and J. L. Joron

^{10}Be/^{9}Be as a potential paleoceanographic tracer (*Poster*) 432
D. Bourlès, G. M. Raisbeck and F. Yiou

Mixing rates on the continental slope derived from radium measurements. (*Poster*) 432
P. A. Gurbutt

Le krypton-85 océanique: technique de mesure et intérêt en océanographie (*Poster*) 433
P. Jean-Baptiste, J. Radwan et J. F. Ternon

Radionuclides in submarine hydrothermal systems (*Poster*) 434
D. Kadko

Radiocarbon dating of Irish Sea sediments (*Poster*) 435
P. J. Kershaw

Major ions as master parameters in radionuclide partition between sediments and seawater (*Poster*) 436
A. M. Hansen, L. D. Mee and J. O. Leckie

Particle associated ^{10}Be in the ocean (*Poster*) 437
F. Yiou, G. M. Raisbeck, D. Bourlès and M. Bacon

Teneurs en ^{99}Tc chez *Fucus sp.* et dans les eaux de mer sur le littoral de la Manche: synthèse (*Poster*) 438
M. Masson, F. Patti et D. Calmet

The marine dispersion of radionuclides from Winfrith (*Poster*) 439
S. E. Nowell

Processus sédimentaire et niveaux d'activité du ^{137}Cs dans les sédiments de Méditerranée nord occidentale (*Poster*) 440
D. Calmet et J. M. Fernandez

Determination of recent sedimentation rate in the Tyrrhenian sea using 239, 240-Pu and 210-Pb (*Poster*) 441
R. Delfanti, C. Papucci and G. Paganin

Distributions of 239, 240-Pu and 14-C in north-east Atlantic sediments (waste disposal and NOAMP areas) (*Poster*) 442
C. Papucci and R. Delfanti

Sellafield waste radionuclides in the intertidal sediments of Skyreburn Bay, south west Scotland (*Poster*) 443
A. B. MacKenzie and R. D. Scott

Pathways of removal of Chernobyl-induced radionuclides from northwestern Mediterranean waters 444
S. W. Fowler, P. Buat-Ménard, S. Ballestra, Y. Yokoyama and H. V. Nguyen

Sellafield tracers in the Norwegian sea and Arctic ocean 445
J. N. Smith, K. M. Ellis and E. P. Jones

Radioactive tracers from the European nuclear industry in Fram Strait and Greenland Sea oceanography 446
H. Dahlgaard, A. Aarkrog and E. Holm

L'utilisation des traceurs radioactifs artificiels et des capteurs radiométriques en sédimentologie dynamique 447
A. Caillot

Origin and fate of artificial radionuclides in the Rhone delta, France . 448
J. M. Martin and A. Thomas

Accumulation du ^{137}Cs dans les sédiments estuariens et littoraux de la Manche 449
P. Walker, P. Geuguéniat, J. P. Auffret et Y. Baron

Author index 451

List of participants 453

Dedication

ALAN PRESTON

The sudden and totally unexpected death of Alan Preston on the 10th January 1988 was a great shock to everyone. Alan had been greatly interested in the objectives and purpose of the Symposium on Radioactivity and Oceanography, held in Cherbourg 2 to 5 June 1987, for it brought together two features of marine scientific endeavour in which he had maintained a life-time's interest, and in the development of which he had played a significant role. The first of these was the subject matter of the symposium: the application of studies made in relation to radioactive waste disposal to the subject of oceanography as a whole, together with the necessity to understand the cycling of naturally-occurring radionuclides in the oceans in order to ensure that such wastes could be disposed of in a safe and acceptable manner. The second was the occasion: the bringing together of scientists from all over the world, from many disciplines, in order to celebrate the opening of a laboratory designed to train students in the techniques required to support research into the marine environment. The strong Anglo-French element of the meeting, and the attendance of so many young scientists, was a particular satisfaction. It is thus both fitting and proper that his interest, energy, and many contributions devoted to the subject of radionuclides in the marine environment should be recognised by the dedication of this volume to him.

Alan Preston was born on 23 May 1929. He was educated at Newcastle High School in North Staffordshire and subsequently at Reading University where he took a special Honours degree in marine zoology. He then joined the staff of the Ministry of Agriculture, Fisheries and Food (MAFF) Laboratory at Lowestoft, UK, in 1951 but departed in 1953 to do his national service, being commissioned to the Royal Corps of Signals. On his return to MAFF he was involved with the research work at the Fisheries Radiobiological Laboratory (FRL), work which covered a considerable range of subjects including radiobiology, radiochemistry and radiological protection. During the late 1950s and throughout the 1960s the marine environment received a varied input of artificially-produced radionuclides, initially from the atmospheric testing of nuclear weapons and subsequently from the authorised discharges of waste from the nuclear industries. There were many schools of thought at that time on how best to manage such discharges; Alan and his team largely pioneered what has become the standard method, the critical pathway approach, involving the painstaking identification of critical groups via habit surveys, plus detailed analyses for the occurrence of radionuclides in a wide range of environmental materials. After becoming head of FRL in 1965, Alan succeeded in building a system of discharge control laid squarely on the basis of authorisation, inspection, monitoring and dose assessment. It is the one still in use today. He was also conscious of the necessity to base such a system on a sound understanding of the behaviour of radionuclides in the environment, and was personally involved in such research.

In 1972 his appointment to Deputy Director resulted in his responsibilities being broadened to cover the whole of marine pollution. This enabled him to further his objective of bringing common sense to bear upon our use and management of the seas in an equitable way with regard to the disposal of any waste. This ever-broadening experience resulted in him being appointed Director in 1980, since when, although being responsible for all aspects of fisheries' research, he never lost interest in his original subject of radioactivity in the marine environment.

At all these levels Alan had played a large role internationally and his influence can truly be said to have been world-wide. He was made an Honorary Professor of the University of East Anglia in 1981 and Buckland Foundation Professor in the same year. At the time of his death he was a Governor and Council Member of the Marine Biological Association of the UK, and Vice-President of the International Council for the Exploration of the Seas.

Alan was always interested in all that went on around him and had developed an enormous circle of friends, both at home and abroad. He and his wife Trixie, together with their four children, had played host to innumerable visitors to the laboratories at Lowestoft, and did so with great pleasure. All who knew him join together in offering his family their love and deepest sympathy. Although nothing can compensate for their loss, they may draw some comfort from the fact that he left behind him a great legacy of learning and inspiration. In summing up the Symposium, the contents of which he had listened to with great attention, Alan concluded that it was the first significant step in bringing two very important marine science disciplines closer together. He expressed the hope that such an inter-disciplinary and international approach would continue; it is for the readers of this book to ensure that it will.

May he R.I.P.

R J Pentreath

THE USE OF NATURAL RADIONUCLIDES IN OCEANOGRAPHY: AN OVERVIEW

R. Chesselet and C. Lalou
Centre des Faibles Radioactivités
Laboratoire mixte CNRS-CEA
91190 Gif sur Yvette, France

ABSTRACT

After a brief review of the origin of the principal natural radionuclides used in oceanography, it will be shown that their use depends on their half-life and, above all, of their biogeochemical behaviour.

Following this behaviour, the problems can be divided in:

- Large scale oceanic circulation, implying the use of a gaseous nuclide with a half-life long enough: C-14;
- Air-sea exchanges, using gaseous and soluble nuclides with shorter half-life: H-3, Rn-222.
- Internal geochemistry of the ocean, using the fact that in the U and Th series, there is a succession of soluble and insoluble elements. These insoluble/soluble properties of the elements are used as tracers of the particulate matter, to evaluate exchange rates, and possibly to predict the fate of potential metallic pollutants.

Going deeper, it will be shown that some elements can diffuse from the sediment and help to calculate diffusion and mixing rates at the sediment-water interface.

- Sedimentary processes, in which the biological cycle of C-14 is used to date the different levels of deep sea cores, the reactive radionuclides (Th-230, Pa-231, Be-10, Al-26) are used to calculate sedimentation rates, and Pb-210 to calculate bioturbation coefficients.

The coastal sedimentary dynamics is possible to study with short-lived radionuclides as Pb-210, Th-234, Th-228, Ra-226...

Other less used nuclides may eventually be used, e.g. K-40 to date by the K/Ar method volcanic ash layers that are reference layers in the sedimentary sequence, Sr isotopes that can provide information about the origin of the detrital component of the sediment.

- Geochemical processes within the ocean: essentially by the study of the ages and growth rates of polymetallic concretions, using Th-230, Pa-231, Be-10, Al-26...
- Hydrothermal processes, by dating either Mn oxides or sulfides associated to the black smokers, or by following the He-3/He-4 anomaly.

Finally, the use of natural radionuclides to study global changes in the oceans, due to climatic variations by dating fossil sea-levels or precise climatic events will be mentionned.

INTRODUCTION

In this introductory paper, we will briefly give the general laws of the repartitioning of natural radionuclides in the oceans, from which follows their use. Thereafter, some examples of their applications will be given, especially concerning the problems which are not treated in the following papers.

As often in research, at the beginning, the new methods seem to work easily with some simple assumptions, but when going more precisely in the subjects, the assumptions are revealed to be too simplistic, and finer studies become necessary. It seems that for the natural radionuclides as tracers of oceanic process, we are now in this second phase where the crude assumptions about their geochemistry must be revised. This can be clearly seen, by reading the titles of the papers given at this conference which concern either technological improvements or new studies on the real geochemistry of the elements in the water column which govern their potential use as tracers in the oceans.

In the first part of this paper, we will give, in increasing order of importance, some fundamental ideas about the origin, half-life and behaviour of the main natural radionuclides used in oceanography. Thereafter, some simple examples of application in the different domains will be given as well as some possible future applications.

THE NATURAL RADIONUCLIDES

A) Their origin.

The natural radionuclides used in oceanography have essentially two origins: i) cosmonuclides which are formed in the high atmosphere by the interaction of cosmic rays with the elements present in the atmosphere, the most generally used are: Carbon 14, Tritium, Berylium 10 and 7, Aluminium 26; ii) primitive radionuclides which have been formed during the nucleosynthesis and which have a half life long enough to persist on earth, their radioactive decay gives rise to shorter half life products they are Uranium 238 and 235, Thorium 232 with their decay series.

As a general rule, cosmonuclides will arrive to the surface of the oceans through the atmosphere while primitive radionuclides which are stocked in the continental rocks will arrive, due to weathering, through the rivers.

B) Their half-live.

It is evident that the half-life of any radiotracer is of importance because one of the main uses of these radiotracers is to introduce the time factor. For this the half life of the element must be of the same order as the time span covered by the studied phenomenon. The cosmonuclides present a large spectrum of half life, from some years for tritium to more than one million year for Be-10. The use of primitive radionuclides will be dependent of the half life of the daughter products since U-238, U-235 and Th-232 have too long a half life to be directly used, but in each series, radionuclides with half-life from days to several hundred thousands of years are continuously formed (radionuclides with shorter half lifes, from seconds to hours, are not useful).

C) Their biogeochemical behaviour.

This is the most important characteristic that defines the potential use of radionuclides since it may be a "La Pallisse truth" to say that to use an element it is necessary that it is present...

Figure 1 summarizes the different possible behaviours of radionuclides in the ocean which are the result of the physico-chemical condition of the oceanic environment. When an element enters into the

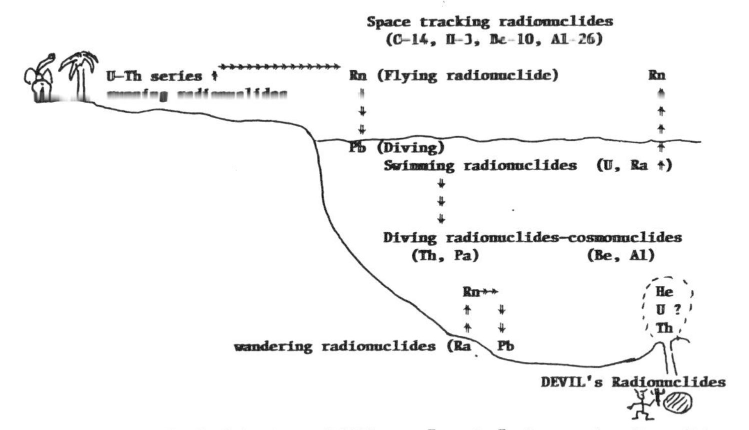

Figure 1: The behaviour of different "species" of natural radionuclides used in oceanography.

ocean, whatever it is cosmogenic or "running", it may follow two principal behaviours: swimming or diving. The swimming radionuclides are those which, being soluble in the ocean follow the water masses and then may be used to quantify their movements, the principal are: C-14, H-3, U. The diving radionuclides are those which, being highly reactive in the marine environment are scavenged by the particulate matter in suspension or taken in the biological cycle, so that they go relatively rapidily to the depth of the oceans, the main diving radio/cosmonuclides are Th, Pa, Be, Al... An important fact is that in the U an Th series, the decay products are different elements, so that they have different behaviours (Fig.2), hence

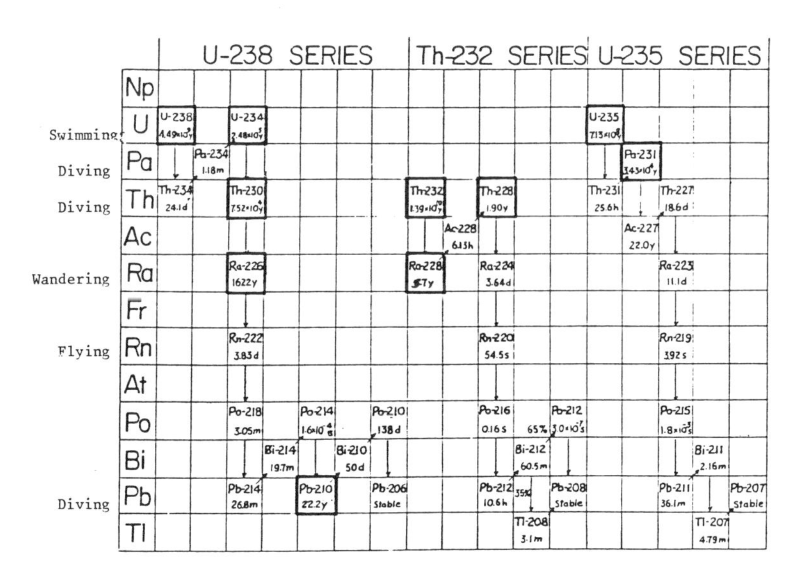

Figure 2: Natural radionuclides series with their behaviours

disequilibria are created in these series, leading to new possibilities due either to an eventual reequilibration of a daughter product with its progenitor once in a solid phase or to a decrease of an excess of the insoluble nuclide.

Two secondary behaviours may be noted at the two main interfaces, the sea-air interface and the sediment water interface. In the first, we have a category which is the flying nuclides, the best exemple is Rn. Rn is a gas produced from Ra which is a swimming nuclide. But on the continent, Rn is able to escape from the rocks and so to mark the continental air mass in the marine atmosphere, and to give Pb (a diving radionuclide) which will participate to the mixing processes in the shallow coastal sedimentation; it is also able to escape from the upper layers of the ocean, allowing air-sea exchange to be quantified. At the sediment water interface, we find the wandering radionuclides: radium present in the interstitial waters of sediment decays to Rn which may escape and here also produce Pb, which in turn comes back to the sediment.

A relatively newly discovered phenomenon in the deep of the ocean is due to the Devil who drives primitive radionuclides out from the magmatic rocks through the high temperature hydrothermal systems. We will see that their localized input may allow these hydrothermal phenomena and their impigement on the oceanic chemistry to be quantified.

USE OF NATURAL RADIONUCLIDES IN OCEANOGRAPHY

In the following paragraphs we will attempt to give some flashes of the possible uses of natural radionuclides. We will use some pioneering studies rather than presently developing works because, first, they show more clearly the philosophy and, second, the latter will be presented in the papers given in this meeting, so we will not trespass on the other participant's field.

A) Physical oceanography.

In general, in physical oceanography, the radionuclide studies have been essential either to help in establishing models (1,2) or to constraint preexisting models (3). The studies need swimming radionuclides and have principally been made with C-14 for relatively long term ocean circulation and with H-3 for short term circulation.

Before using C-14, it is necessary to consider some important periods in its cycle. If men would have not existed, C-14 activity would have stayed constant in all carbon reservoirs, including the most important, the ocean. The first man's modification to the natural carbon cycle is known as Suess'effect (4). Due to the industrial development, the increasing combustion of fossil fuels (dead carbon) lead to a decline of its natural activity that might be used to study the sea-air exchanges at the scale of some decades... but before this can be made, the inverse effect has been introduced by the nuclear tests in the atmosphere that gave rise to a drastic increase of C-14 activity (100%) during the 1960's (Figure 3). The maximum of the increase was around 1964 and when the atmospheric nuclear tests stopped, the C-14 activity in the air decreased. This decrease is not a radioactive decrease, but reflects the absorption of atmospheric CO_2 in the oceans. This is presently used to calculate the rate of this absorption.

i) C-14 as a tracer of water masses movements in an ocean. At the scale of an ocean, C-14 has been used to evaluate the age of North Atlantic Deep Water (NADW). The first attempt was made by Broecker et al., as early as 1960 (5), giving an age of about 700 years, making the hypothesis that the age may be deduced from the differences in ΔC-14 of surface waters to deep waters ($-28°/_{\circ\circ}$ for surface waters in central

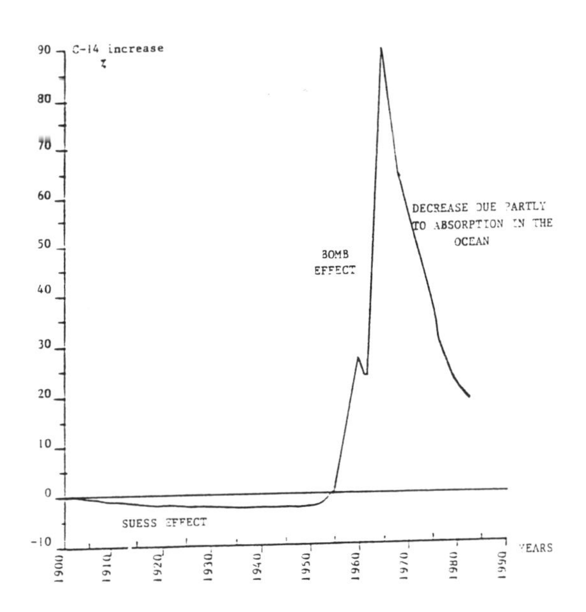

Figure 3: Δ C-14 activity in the atmosphere.

Atlantic to $-105^{\circ}/_{\circ\circ}$ for NADW). But later on, the results of GEOSECS program have shown that the Northern waters feeding the NADW do not have the same ΔC-14 as the surface waters of the central Atlantic and that a contribution of the Antarctic bottom water of 11% must be considered, so that the reference value became $-80^{\circ}/_{\circ\circ}$ (6), giving an age of about 230 years. Lastly, Broecker (7) reassessed the proportion of antarctic water contribution to NADW which leads to and age of about 100 years.

ii) C-14 as a tracer of the world ocean. At the scale of the global ocean, the GEOSECS programm is one of the master pieces of chemical oceanography. Numerous C-14 profiles have been made in all the oceans (8,9). All the measured profiles show a decrease of ΔC-14 from the surface values (corrected for artificial C-14 contribution) to the deep waters in all oceans, this difference increases from the North Atlantic Deep Waters ($\sim-100^{\circ}/_{\circ\circ}$) to the Antarctic ocean ($\sim-170^{\circ}/_{\circ\circ}$), to the Equatorial Indian ocean ($\sim-180^{\circ}/_{\circ\circ}$) to be the lowest value ($\sim-250^{\circ}/_{\circ\circ}$) in the deep North Pacific waters. This implies that the turnover time of the ocean is around 1,000 years.

iii) Short term oceanic circulation using tritium. Tritium (H-3) is a cosmonuclide having an half life of 12.5 years. Its activity is conventionaly given in Tritium Units (1TU = 1 H-3 atom per 10^{18} H atoms). Natural tritium has been completely overhelmed by tritium formed during nuclear tests (by a factor of several hundreds). This artificial tritium entered the surface waters of the oceans and, due to its short half life is already decreasing. It is used either like a dye, marking the water masses or, associated to its decay product, He-3, as a clock. Examples of this use will be given by Andrié and Merlivat (10).

iv) Air-Sea exchanges.

At a global scale, C-14 may also be used to measure the rates of air-sea exchange. Ideed, one of the main problems for the end of the century will be to evaluate the capability of the ocean to absorb the excess CO_2 released in the atmosphere by industries. This is not yet done, but studies on the rate of penetration of artificial C-14 in the oceans will be presented.

A more local tracer of air-sea exchange with shorter half life is the Ra-226-Rn-222 ratio. In the ocean, under the thermocline, Rn is at equilibrium with its father Ra-226; above the thermocline, in the well mixed layer, their is a deficit of Rn-222 relative to Ra-226. Ra-226 is in solution in the sea water, Rn 222 being a gas is able in this turbulent layer to diffuse towards the air. The total deficit of Rn is a measure of the rate of exchange. It has been shown that it is proportional to the wind velocity (11). The same profiles allow the calculation, using a gas exchange model (12),of the thickness of the stagnant film supposed to exist at the air water interface. From a large number of profiles obtained in GEOSECS program, broad trends have been shown: the exchange rate increases from Equatorial zones to North and South temperate zones and is maximum in Antarctica.

B) Sedimentation and geochemistry.

In this field, the potential tracers are essentially the diving nuclides. The different sedimentary and geochemucal problems that may be studied with the natural radionuclides are summarized in figure 4.

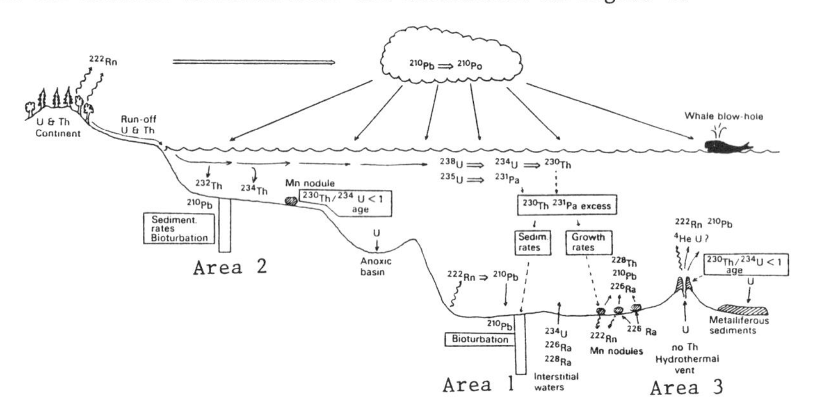

Figure 4: Problems that may be solved using natural radionuclides

On this figure the behaviour of the elements and their potential uses are given and we have schematically separated 3 areas: area 1 is the deep area, area 2 is the coastal area and area 3 is the Devil's realm. We will briefly describe these 3 areas.

1) In area 1, the main phenomenon is the sedimentation. The natural radionuclides may be used to study the sedimentation rates, or to date some levels in the sediment cores. Figure 5 gives a schematic diagramm of what may be done on an ideal core. The diving radionuclides, Pa-231, Th-230, Al-26, Be-10 fall to the sediment, the decrease of the "excess" of activity for the first two or of their activity for the two others with depth in the core, that is with time, allows sedimentation rates to be calculated. The half life of the elements, from which depends the time span covered is from about 30,000 years to 1.5 Million years so it would provide a true continuum if Al-26 were more easy to measure. Now with the new method of accelerator mass spectrometry (AMS) Be-10 becomes relatively easy to measure, but not yet Al-26. Another way to study these cores is to try to date some levels. For this, when the sediment contains a biogenic carbonaceous component (Foraminifera or Coccoliths), it may be dated to

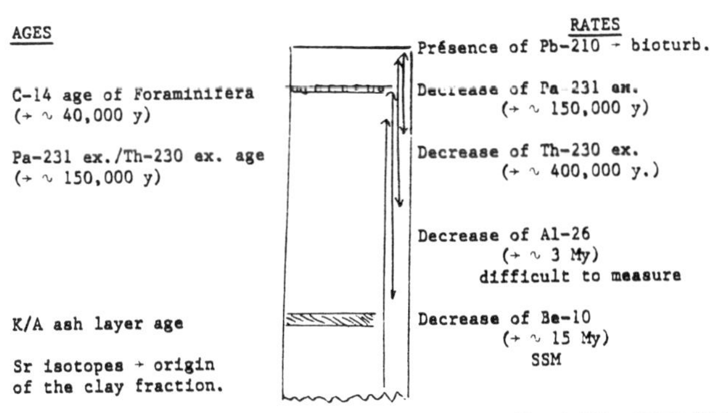

Figure 5: Different methods to measure ages or sedimentation rates in an idealized core.

about 40,000 years with C-14. Here also the AMS technic is useful, allowing work on very minute quantities of material, giving rise to an extremely high resolution in time. Theoretically, another dating tool would be the use of the ratio Th-230/Pa-231, because the production ratio is known so the increase of this ratio from about 11, may be translated in ages. We will see in the following that there are some problems... If by chance a volcanic ash layer is present in the core, it may be used as a marker for all the oceanic basin which has been covered by those ashes and may be dated by K/A method. Always on the same core, the analysis of the isotopic ratios of Sr nuclides may be used to find the origin of the detritic clay fraction by comparison with the one of the rocks from the surrounding continents (13).

On top of the deep sea sediments, geochemical phenomena have led to the formation of polymetallic concretions. Applying the same methods as the one used to measure the sedimentation rates, very slow rates, of the order of mm per million years have been found. This raised a problem: how can these formations stay on top of sediments accumulating 1000 times faster. For Lalou et al., (14) there is an artefact and the growth rates are not realistic, others, like Krishnaswami and Cochran (15) after a very detailed study of all nuclides in Mn nodules conclude, essentially from the difference in Th-230 and Pa-231 activity in the upper and lower surfaces of nodules, that they are kept on the surface by a rolling mechanism and these differences of activity may be used to calculate the time ellapsed since the nodule has attained its present orientation. The question is still open.

At the sediment-sea water interface, other problems may be studied using the wandering nuclides. As in the air-sea interface, at this interface, radium in solution in the interstitial water of the sediments decays giving Rn-222 which is able to diffuse from the sediment to the sea water. Attempts have been made to use this radon to measure vertical eddy diffusivity, which lead to the conclusion that there is an inverse relationship with the density gradient. This hypothesis was found too simplistic. For exemple, the importance of horizontal motions was neglected... This Rn-222, while in the water gives Pb-210, this nuclide being a diving nuclide comes back to the surface of the sediment. Due to its very short half life (22 years), it cannot survive at depth, but in fact it is often found at some cm depth in the cores. This is due to the gardening of the surface of the sediment by the deep sea fauna, and the profile of distribution of Pb-210 may be used to calculate bioturbation coefficients. This presence of Pb-210 at the surface of deep-sea sediments is often hidden by the fact that the very surface of deep-sea cores is seldom sufficiently well preserved, but, for exemple, in a core sampled from a submersible, with great care in the Famous area, Nozaki et al. (16) measuring together sedimentation rates by C-14 and Pb-210 have been able to calculate mixing coefficients of $0.10.10^{-8}$ cm^2/s.

2) In area 2 or coastal environment the problems are partly the same as the ones of deep-sea sedimentation, but the rates and times are shorter so that the diving radionuclides with shorter half-life are used, e.g. Th-234, daughter product of U-238. With a half life of 24.1 day, it is used to study very rapid phenomena. With this nuclide, it has been shown (17) that in coastal sediments, Th-234 is scavenged more efficiently by fine particles than by coarse ones. Since these particles may be resuspended and recirculated during storms it allows a prediction of the fate of polluants having the same behaviour as Th-234 in this unstable environment. Anomalous repartition of Th-234 may also be used for bioturbation measurements in these areas of high and intense productivity. Lead 210 is another nuclide of interest in coastal environment with its 22.3 years half life. It is formed in the atmosphere through a flying nuclide, Rn-222 issuing from the continental rocks, and is introduced in the sea by dry or wet fallout. It is used in the same manner as Th-234, but its longer half life enables longer chronological sequences to be established.

Until now, the uses of radionuclides as tracers of the sedimentology or geochemistry of deep sea sedimentation as well as of coastal sedimentation, have considered the water mass of the oceans as a black box. The different behaviours between swimming and diving radionuclides being deduced more from theoretical chemical considerations than from in situ measurements. But as the studies are more and more numerous and precise, problems arise. One of the main problem in deep sea sedimentation is that if we try to establish a budget for Th-230 or Pa-231, it does not work. Theoretically, uranium concentration in the oceans being constant around 3 ppb, U-238/U-235 ratio being also constant at the time scale under consideration, as well as the U-234/U-238 ratio (1.14 ± 0.02), and knowing the half life of the nuclides, one calculates a Th-230 flux of 2.2 dpm/cm^2/1000 y./1000 m depth and a production ratio Th-230/Pa-231 = 10.7. So in a given core, the total Th-230 or Pa-231 (Σ Th-230 or Pa-231) activity in excess per surface unit must reflect the equilibrium between the production rate in the water column overlying the core and the decay of these nuclides. In fact it is not the case, generally the value in sediments is higher than the theoretical one for high sedimentation rates and lower for slow sedimentation rates, so it seems linked to the flux of particulate matter acting as scavengers of the ocean in the water column. Moreover, even though it was postulated that Th-230 and Pa-231, both diving nuclides, have the same behaviour, it is frequent that the ratio Th-230/Pa-231 at the top of the sedimentary column or in the uppermost layer of Mn deposits is different from the theoretical production ratio. The difficulty in measuring the concentrations of these nuclides in the water column is the reason why, until recently, this water column was considered as a black box. The aformentionned problems of flux and the development of new techniques now allow us to open the black box and to study the behaviour of the elements in situ and not just theoretically. The technical support is essentially the development of sediment traps and of large volume filtration devices which allow to obtain a quantity of particulate matter sufficient for the measurement of Th or Pa isotopes (18), as well as of Be-10 (19). For this last nuclide, as we have seen before, the use of accelerator mass spectrometry has also been essential to open the field. It will be shown in the following papers in this book that, at least for natural radionuclides they are essentially devoted to such studies: the speciation of the radionuclides in the water column, their relations with the particulate matter, organic or mineral, or to their geochemical behaviour in special environments.

In addition to the interest of these studies for a better use in the knowledge of deep sedimentation, they are useful in understanding the pathways of stable trace metals known as potential pollutants of the oceans.

3) The third area is the Devil's realm. The Devil's history in the oceans began with helium. Two helium istotopes are present in the ocean, He-4 which is the result of the radioactive α decay of U and Th series and He-3 which is primordial. From these two different origins, it has been assumed that in deep sea, He-4 formed in the sediment may wander out regularily so that it must be in excess and the ratio He-4/He-3 must be higher in deep waters than in surficial waters (20). Measurements made somewhat later, in 1969 (21) have shown that in fact, in deep-sea, there is an excess of primordial He-3 which, later on, has been shown to be related to the heat flow issuing from the ridge crests (22), and then to the newly discovered hydrothermal phenomena. The helium anomaly may then be considered as a tracer of the hydrothermal impact and may be used to calculate the input of other metals to the ocean. This is not a restricted phenomenon having only a restricted impact since Lupton and Craig (22) have reported an excess of 50% for He-3 over the East Pacific Rise, with an extension of the plume 2000 km to the west of the ridge crest. These hydrothermal vents not only deliver to the oceans large quantities of elements in soluble form, but also form important deposits that may be dated by the use of natural radionuclides. It is important to evaluate the intensity and history of deep-sea hydrothermalism. Different methods may be applied, depending on the nature of the deposit and on the time span covered. For young sulfide deposits, as those recently found in the median valley of the oceanic ridges, the decrease of the ratio Pb-210/Pb may be used until about 100 years, provided that the initial ratio is known. This initial ratio may be measured in the particulate sulfides issuing from the black smokers, or can be calculated using a cross-checking with the ratio Th-228/Ra-228 in very young samples. For fossil deposits, either sulfide or manganese oxides, from about some thousands of years to about 300,000 years, the Th-230/U-234 ratio may be used making the assumption that U in solution is coprecipitated with the deposits while thorium, being depleted in sea-water, does not enter at the origin in the sample. Many examples of such datatings and their consequences are given in Lalou and Brichet (23).

C) Global Climatic Changes

Using the possibilities of datating given by the natural radionuclides, the study of global changes in the ocean is opened. During the Quaternary, climatic variations introduced a series of glacial and interglacial periods (24) which, in turn, were reflected by changes in the sea-level. These variations led to the edification of fossil beaches, including coral terraces which have been dated giving some precision to the climatic variations. From 0 to 30,000 years; C-14 is used; after, from ∿10,000 to 150,000 years, Pa-231/U-235 ratio is used, and to 350,000 years, Th-230/U-234. The two last methods are possible due to the fact that Th and Pa being diving nuclides, they are not available when the aragonitic crystals of corals formed while the swiming U may enter in the crystal lattice. The building up of equilibrium between Th (or Pa) and U is a good clock. These methods have been extensively used the last 20 years, and recently summarized by Moore (25). Other marine formations like phosphorites may also be dated by these methods (26).

ACKNOWLEDGMENTS

The authors want to acknowledge with thanks the organizers of the meeting, and specially Dr J.C. Guary. Thanks are also due to W.F. Fitzgerald who helped us to give its final english form to this paper. Financial support has been obtained through CNRS and CEA. This is C.F.R Contribution n°863.

REFERENCES

1. Craig, H., The natural distribution of radiocarbon and the exchange time of carbon dioxide between the atmosphere and sea. Tellus, 1957, 9, 1-17.

2. Suess, H.E., and Revelle, R., Carbon dioxide exchange between the atmosphere and ocean and the question of an increase of atmospheric CO_2 during the past decades. Tellus, 1957, 9, 18-27.

3. Munk, W.H., Abyssal recipes. Deep Sea Res., 1966, 13 707-730.

4. Suess, H.E., Radiocarbon concentration in modern wood. Science, 1955, 122 415-17.

5. Broecker, W.S.,Gerard, R., Ewing, M., and Heezen, B.C., Natural radiocarbon in the Atlantic ocean. J. Geophys. Res., 1960, 65, 2903-2931.

6. Stuiver, M., The C-14 distribution in west Atlantic abyssal waters. Earth Planet. Sci. Lett., 1976, 32, 322-330.

7. Broecker, W.S., A revised estimate for the radiocarbon age of North Atlantic Deep Water. J. Geophys. Res. 1979, 84 C6, 3218-3226.

8. Stuiver, M., and Ostlund, H.G., GEOSECS Atlantic Radiocarbon. Radiocarbon 1980, 22, 1-24.

9. Ostlund, H.G. and Stuiver, M., GEOSECS Pacific Radiocarbon. Radiocarbon, 1980, 22 25-53.

10. Andrié, C. and Merlivat, L., Utilisation du tritium et de l'hélium-3 en océanographie, contribution à l'étude de la circulation de la Méditerranée occidentale et de l'Atlantique Nord-Est. (This volume)

11. Bainbridge, A.E., Broecker, W.S., Horowitz, R.M., Li, Y.H., Mathieu, G., and Sarmiento, J., GEOSECS Atlantic and Pacific surface hydrography and Radon Atlas. GOG Publication, 1977, 120, SIO, La Jolla, Calif.

12. Broecker, W.S. and Peng, T.H., Tracers in the Sea. ELDIGIO Pub, 1982, 690 pp.

13. Grousset, F.E., and Chesselet, R., The Holocene sedimentary regim in the northern mid atlantic ridge region. Earth Planet. Sci.Lett., 1986, 78, 271-287.

14. Lalou, C., Brichet, E., and Bonté, Ph., Some new data on the genesis of manganese nodules. In: Geology and Geochemistry of Manganese, Pub.

House of the Hungarian Academy of science, 1976, III, 31-90.

15. Krishnaswami, S. and Cochran, J.K., Uranium and thorium series nuclides in oriented ferromanganese nodules: growth rates, turnover times and nuclide behavior., Earth Planet. Sci. Lett., 1978, 40, 45-62.

16. Nozaki, Y., Cochran, J.K., Turekian, K.K., and Keller, G., Radiocarbon and Pb-210 distribution in submersible-taken deep-sea cores from project Famous., Earth Planet. Sci. Lett., 1977, 34, 167-173.

17. Aller, R.C., Benninger, L.K. and Cochran, J.K., Tracking particle associated processes in nearshore environments by the use of Th-234/U-238 disequilibrium. Earth Planet. Sci. Lett., 1980, 47, 161-175.

18. Brewer, P.G., Nozaki, Y., Spencer, D.W. and Fleer, A.P., Sediment trap experiments in the deep North Atlantic: isotope and elemental fluxes, J. Mar. Res., 1980, 38, 703-728.

19. Raisbeck, G.M., Yiou, F., Honjo, S., Klein, J., and Middleton, R., Be-10 fluxes from sediment trap (This volume)

20. Suess, H.E., and Wanke, H., On the possibility of a helium flux through the ocean floor. In; Progress in Oceanography, 3, Pergamon Press, Oxford, 1965, 347-353.

21. Clarke, W.B., Beg, M.A. and Craig, H., Excess He-3 in the Atlantic. Earth Planet. Sci. Lett., 1969, 6, 213-220.

22. Jenkins, W.J., Edmond, J.M. and Corliss, J.B., Excess He-3 and He-4 in Galapagos submarine hydrothermal water. Nature, 1968, 272, 156-158.

23. Lalou, C. and Brichet, E., On the isotopic chronology of submarine hydrothermal deposits., Chemical Geology (Isotope Geoscience section), 1987, 65 in press.

24. Emiliani, C., Paleotemperature analysis of Carribean cores P 6304-8 and P 6304-9 and a generalized temperature curve for the past 425,000 years. J. Geol., 1966, 61, 171-183.

25. Moore, W.S., Late pleistocene sea level history. In: Uranium series disequilibrium, Applications to environmental problems, M. Ivanovitch, R. Harmon, Eds. 1982, Clarendon Press, Oxford, 481-496.

26. Veeh, H.H. and Burnett, W.C., Carbonate and phosphate sediments. In: Uranium series disequilibrium, Application to environmental problems, M. Ivanovitch R. Harmon Eds., 1982, Clarendon Press, Oxford, 171-183.

SOURCES OF ARTIFICIAL RADIONUCLIDES IN THE MARINE ENVIRONMENT

R. J. Pentreath
Ministry of Agriculture, Fisheries and Food
Directorate of Fisheries Research
Lowestoft, Suffolk NR33 OHT
UK

ABSTRACT

The marine environment has received man-made radionuclides from many sources: the detonation of nuclear weapons in the atmosphere, the localized inputs of low-level liquid wastes from the nuclear industries, the disposal of packaged wastes into the deep sea, lost military hardware and, most recently, from a power reactor accident. These various sources have, collectively, introduced a wide-range of nuclides both globally and locally that provide a means of studying the rates of oceanographic processes. In order to do so, however, it is necessary to consider the various origins of these nuclides because they usually have more than one source; fortunately sufficient is usually known of the isotopic composition to differentiate between them.

INTRODUCTION

The large-scale introduction of radionuclides into the marine environment arose initially from the detonation of nuclear weapons in the atmosphere in the 1940s; peak periods were from 1954 to 1958 and again from 1961 to 1962. A partial test-ban treaty on atmospheric testing was agreed between the USA, USSR and the UK in 1963 but, in late 1964, China carried out its first atmospheric test. France had tested weapons in North Africa during 1960 and 1961 and, in 1966, began atmospheric testing in the Pacific islands. And at least five tests, of relatively low yield, have been made underwater. Such inputs, with the exception of the last, have injected a wide range of radionuclides into the oceans via the atmosphere and indirectly via freshwater run-off from land. The quantities entering the oceans have thus varied in both space and time but by and large, for the longer-lived nuclides, have resulted in widespread, low-level contamination. The radionuclide content of this input has been augmented by plutonium from a satellite power source which burned up upon re-entry, some

46 km above the Mozambique channel, following a malfunction during launch in April 1964.

The second major source of input has been the deliberate controlled discharge of low-level liquid effluents from the nuclear power industry. These have occurred as relatively point-source inputs, the most important of which have been those from the Hanford plant via the Columbia River into the Pacific Ocean, from the BNFL Sellafield plant in the United Kingdom into the Irish Sea, and from the CEA plant situated at Cap de la Hague in France. In addition, numerous other localized low-level inputs occur from nuclear power stations situated on the coast.

A somewhat different source has been the disposal of low-level packaged wastes into the deep ocean. Such disposals have taken place in the Pacific and Atlantic Oceans at a number of sites, the best documented of which are those which have been made by European countries in the NE Atlantic Ocean. Yet another input has been that from the recent accident at Chernobyl. All of these sources can be used - and have been used - to study marine environmental processes. And such studies need to be made, so that the continued introduction of radionuclides into the marine environment can be based upon a sound scientific basis of their behaviour.

Many of the artificial radionuclides present in the oceans have arisen from more than one source. This is to be expected, because both nuclear weapons and the nuclear power industry make use of the fission of heavy atomic nuclei. The sources can usually be differentiated because their locations are known and because, quite frequently, the nuclides occur in different ratios. Thus it may be necessary to examine more than one nuclide in order to identify the major source. The chemical forms of a nuclide may also differ from one source to another, and in many cases such information is imprecisely known. The following is a brief review of the most useful nuclides which are present, and their principal sources. Their value as tracers arises primarily from the fact that many of them - ^{3}H, ^{99}Tc, ^{125}Sb - are largely soluble and thus conservative to water masses whereas others, such as the transuranium nuclides, are adsorbed onto particulate material and become incorporated into settled sediments. Nuclides such as ^{134}Cs and ^{137}Cs have been used for both purposes because, although the largest fraction at any one time remains in the water column, they do have a sufficiently high affinity for particulate materials to act as tracers of processes within the sea bed. Another value is that of their different half-lives, particularly of those nuclides of the same, or chemically-similar, elements. The ratios - or to be pedantic, the quotients - of different nuclides are frequently used. These are sometimes given with the longer-lived nuclide as the divisor so that the quotient arising from a source, at a specific moment of input, decreases with time in an exponential manner. But quotients from many of the sources are continually variable with time and, presentationally, it is usually more useful to discuss relative quotients when their values remain > 1. It is this convention which has been used in this paper.

ATMOSPHERIC FALLOUT

Radioactive fallout consists of the fission products arising from the fissile material used in the explosive device, of the nuclides arising from the neutron activation of its components, and of those arising from the neutron activation of surrounding materials. All of the primary materials and their immediate surroundings are initially in gaseous form but, within a few seconds, micro-particles are formed of iron and aluminium which have a tendency to incorporate those radionuclides which have oxides with high boiling points. The remainder of the material condenses into smaller particles. Such fractionation results in a 'local' fallout of the larger particles - plus entrained debris - which returns to earth within a few hundred km of the site of detonation, and an injection of smaller particles into the troposphere resulting in far-field contamination. Tropospheric fallout accounts for the majority of short-lived radionuclides deposited within a few months of an explosion, usually at approximately the same latitude. Particles injected into the stratosphere result in world-wide distribution although this, again, is primarily confined to the hemisphere in which the detonation took place. Such stratospheric fallout has been the origin of wide-spread contamination from longer-lived nuclides, primarily from the large-yield thermonuclear devices. The mean residence times of particles in the stratosphere depend upon altitude, latitude of injection, and the season in which the detonation occurred. Values for the lower stratosphere range from about 3 to 12 months in the polar regions to between 8 and 24 months in the equational regions. Exchange with the troposphere occurs at well defined areas - the subtropical tropopause gaps - where the downward transfer of material is at a maximum in late winter, giving rise to peaks of deposition from the troposphere to the ground during the spring. Deposition occurs either as dry or wet particulate matter, the latter being differentiated between that which results from droplet formation within clouds (rain out) and that which is scavenged by falling raindrops (wash out).

Explosive yield data are not available for each test and the quantities of radionuclides injected into the atmosphere have therefore been estimated by the direct measurement of certain ones, such as ^{90}Sr and ^{137}Cs, plus assumptions of the ratios of these nuclides to others. The $^{137}Cs/^{90}Sr$ quotient is assumed to be relatively constant, at 1.6. Fission product nuclides are produced in proportion to fission yields, and particular neutron activation products, such as ^{3}H and ^{14}C, are assumed to be produced in proportion to the fusion yields. Total quantities of the longer-lived nuclides produced by weapon's testing have been calculated by UNSCEAR[1], from which a selection of data is given in Table 1. Not included in the table is ^{14}C, of which some 220 PBq have been injected into the atmosphere up until 1980.

For present-day purposes, however, such information is of limited value because of the effect of decay. A more useful estimate is that of the actual quantities obtaining in the environment, and these are also given in the table, with a separate estimate of the quantities present in the North Atlantic Ocean[2]. By far the most abundant is ^{3}H, which arises primarily from the thermonuclear explosions in the 1960s.

The major tests have taken place in the northern hemisphere; the annual and cumulative depositions of ^{90}Sr for both hemispheres is shown in Figure 1(a) and (b). As can be seen, the quantities initially received in each hemisphere were quite different, but since the end of the 1960s the total inventory in both has decreased because the rate of introduction from the few tests carried out in recent years is less than the rate of loss due to radioactive decay. Not that these quantities necessarily remain in these areas. Deposition onto land will have been redistributed to some

Table 1. Introduction and inventories, to 1983, of a number of long-lived radionuclides arising from weapon's fallout and SNAP-9A[2]

Nuclide	Half-life (years)	Production (TBq)	Deposited inventory (TBq)		
			Northern hemisphere	Southern hemisphere	North Atlantic Ocean
^{3}H	$1.2\ 10^{1}$	$2.4\ 10^{8}$	$6.0\ 10^{7}$	$1.5\ 10^{7}$	$1.3\ 10^{7}$
^{90}Sr	$2.9\ 10^{1}$	$6.0\ 10^{5}$	$3.0\ 10^{5}$	$9.5\ 10^{4}$	$6.4\ 10^{4}$
^{137}Cs	$3.0\ 10^{1}$	$9.6\ 10^{5}$	$4.7\ 10^{5}$	$1.5\ 10^{5}$	$9.4\ 10^{4}$
^{238}Pu(a)	$8.8\ 10^{1}$	$3.3\ 10^{2}$	$3.2\ 10^{2}$	$4.8\ 10^{2}$	$6.4\ 10^{1}$
$^{239+240}Pu$	$2.4\ 10^{4}$(b)	$1.3\ 10^{4}$	$1.0\ 10^{4}$	$2.6\ 10^{3}$	$2.0\ 10^{3}$
^{241}Pu	$1.4\ 10^{1}$	$1.7\ 10^{5}$(c)	$5.0\ 10^{4}$	$1.2\ 10^{4}$	$1.0\ 10^{4}$
^{241}Am	$4.3\ 10^{2}$(d)	$5.5\ 10^{3}$	$2.8\ 10^{3}$	$7.0\ 10^{2}$	$5.6\ 10^{2}$

(a) includes SNAP; (b) ^{239}Pu; (c) assumed input in 1963; (d) from ultimate decay of ^{241}Pu.

extent by freshwater run-off, and deposition over the sea will have been considerably dispersed by advection from the latitudinal bands at which it arrived at the sea surface.

Two fallout nuclides, ^{3}H and ^{14}C, have been extensively used to study oceanographic processes - particularly in the GEOSECS programme from 1972 to 1977. At that time, about a decade after the large atmospheric inputs, it was found that the northern temperate oceans had far more ^{3}H at any given density level than their equivalent in the southern oceans, and thus reflected the different quantities introduced into the atmosphere of each hemisphere. But it was also shown that the ^{3}H in the northern waters had penetrated to comparatively much greater depths, when related to surface-water values, thus reflecting the areas of deep water formation. Indeed it was possible to trace water from the Denmark Straits down into the deep water of the western North Atlantic basin, and then along the margin of the North American continent as far south as Florida, a total distance of some 4800 km[3].

The injection of fallout radionuclides also permitted a re-evaluation of the estimated residence times of certain elements in the oceans. It was shown, for example, that ^{90}Sr and ^{137}Cs were present in deep sea sediments to the extent that their rate of removal from the overlying water column was orders of magnitude greater than had previously been supposed[4].

The transuranium nuclides have also been introduced in considerable quantities and their distribution, particularly in the Pacific Ocean, has been studied in some detail. The results of the GEOSECS programme indicated a sub-surface $^{239+240}Pu$ maximum concentration at a depth of some 500 m in the North Pacific Ocean and another, near-bottom, maximum concentration especially in Western, Central and Northern Pacific locations[5]. More recent studies[6] have shown that the western North Pacific, between 30°N and 40°N, must have received proportionately more 'close-in' fallout than those to the east. Such studies have also demonstrated the rapidity with which these nuclides - strongly associated with particulate materials - are transported to the deep sea bed, and the extent and rate to which they become incorporated into sediments. Pacific sediments taken at depths of

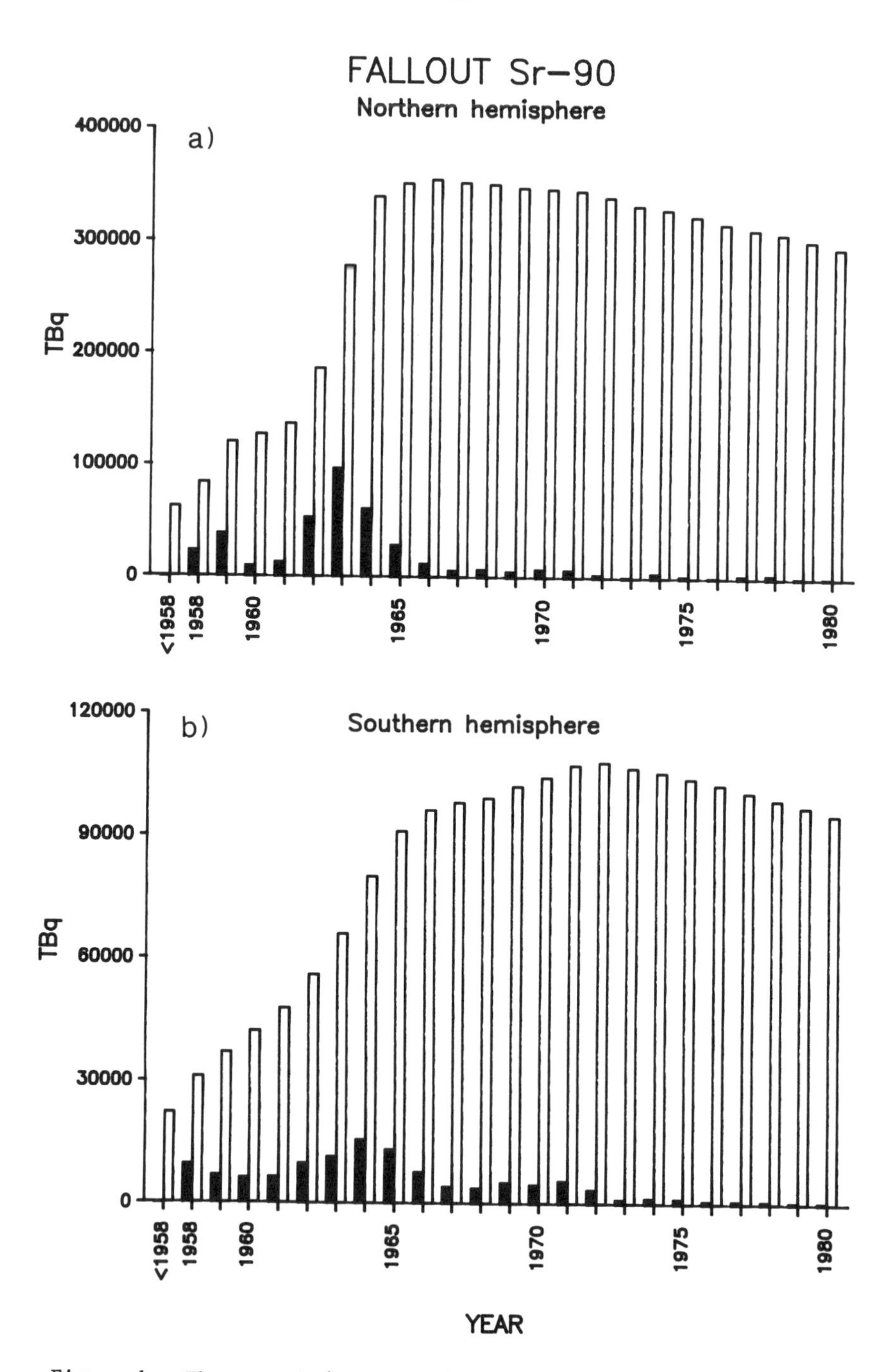

Figure 1. The annual (solid bar) and cumulative (open bar) deposition of ^{90}Sr from fallout in (a) the northern and (b) the southern hemispheres[1].

5.5 km clearly show $^{239+240}Pu$ incorporated some 30 to 50 cm into them, presumably by bioturbatory processes[6]. More recently, Livingston _et al._[7] have produced evidence to indicate that the high bottom-water concentrations arise from desorption of the Pu from bottom sediments.

Of particular value, over longer time-scales, will be the plutonium isotopes ratios. With a ^{238}Pu half-life of 87.7 years, the $^{239+240}Pu/^{238}Pu$ quotient will continue to change, providing a very useful dating technique. Its value, however, will depend on careful analyses of contemporary samples because the Pu α-isotope quotient is far from constant. The average quotient in weapon's fallout was at least 40, and may have been as high as 50, but the actual values are less than this, globally, because they have been affected by the input from the USA satellite SNAP 9A which burned up some 46 km above the Mozambique channel shortly after take off in April 1964. This contained plutonium metal, primarily ^{238}Pu, as a source of heat to provide power through thermo-electric converters. The SNAP (Systems for Nuclear Auxiliary Power) device contained about 0.63 PBq of ^{238}Pu and 0.48 TBq of ^{239}Pu. It is estimated that 73% was eventually deposited in the southern hemisphere, the ground-level period of maximum deposition being 1966-67; depletion from the atmosphere was essentially completed by about 1971[8]. The effect this had on the ratios of the cumulative deposition of Pu isotopes up until this time is given in Table 2.

Table 2. Estimated cumulative fallout Pu quotients, as of 1971, in relation to bands of latitude. The quotient due to weapon's fallout was assumed to be 41.7, the lower values resulting from SNAP-9A[2]

Latitude	$^{239+240}Pu/^{238}Pu$	Latitude	$^{239+240}Pu/^{238}Pu$
Northern hemisphere		Southern hemisphere	
90-80	50.0	0-10	17.5
80-70	40.0	10-20	4.5
70-60	25.0	20-30	4.9
60-50	29.4	30-40	5.7
50-40	27.8	40-50	4.5
40-30	27.0	50-60	4.1
30-20	28.6	60-70	4.2
20-10	26.3	70-80	3.3
10- 0	43.5	80-90	2.5

AUTHORIZED DISCHARGES INTO COASTAL WATERS

Although the inputs from weapon's testing have been considerable in quantity, their widespread dispersal has resulted in low-level ambient concentrations. In contrast, the controlled discharges of low-level liquid effluents from the nuclear industry into various coastal waters has resulted in higher, but localized, concentrations. As with weapon's fallout, it is of limited value to discuss the quantities of nuclides discharged without reference to their rates of decay and, indeed, rates of input.

The first major input of radionuclides into the environment began in 1944 - a year before the first weapon's explosion - with the start up of the plutonium production reactors at Hanford, on the Columbia River, in the USA. These reactors were cooled by drawing water in from the river, passing it once through the reactor, and returning it again downstream. The effluent thus contained large quantities of short-lived neutron activation product nuclides. At their peak period of use, between the mid-1950s and mid-1960s, about 1 TBq per day of mixed radionuclides entered the river.

The Hanford plant is situated some 560 km from the Pacific Ocean and thus the major nuclides to enter the Oregon and Washington coastal waters were the longer-lived ^{51}Cr, ^{65}Zn and ^{32}P. The plutonium reactors were shut down in 1971 and thus, apart from relatively small quantities of plutonium - ^{239}Pu, arising from the decay of ^{239}Np[9] - there is very little left of value in terms of an oceanographic tracer.

Not long after the Hanford plant began operating, and again initially for military purposes, two reactors were constructed in the UK on the shores of the Irish Sea. These were the Windscale piles 1 and 2 which operated up until 1957; they were air cooled and thus unconnected with freshwater or the sea. But the necessity to process the fuel after removal from the reactors resulted in discharges of low-level liquid effluent to sea. These discharges, carried out under authorization, began in 1952. The Sellafield site also possesses four Magnox reactors, and a prototype AGR - now being decomissioned. But its major function has been the reprocessing of spent fuel from the UK civil nuclear power programme, under the management of British Nuclear Fuels (BNFL) since 1971. The French, too, operate a fuel reprocessing plant (UP2) at Cap de la Hague in northern France, which discharges low-level liquid effluent into the English Channel. This facility, managed by the Campagnie Général des Matières Nucléaires (COGEMA), reprocesses both natural uranium and oxide fuels. Routine discharges have been made since 1966.

The quantities of a number of nuclides with half-lives $\geqslant$ 1 year discharged by these two plants are given in Table 3, together with estimates of the decay-corrected inventories obtaining in the environment up until 1986 for Sellafield, and until 1985 for Cap de la Hague. These values have been calculated for Sellafield from the annual discharge data given in BNFL annual reports and, prior to 1978, from NRPB-R171 Addendum[10]; and for La Hague, primarily from the data in Calmet & Guegueniat[11] plus more recent information (Guegueniat, pers. com.). The quantities and relative proportions of nuclides arising in the effluents depend on a number of factors: the nature of the fuel, its burn-up time, the nature and length of storage prior to reprocessing, the method of reprocessing, and the nature of effluent treatment plants. All of these are variable with time and each site - like any other large industrial chemical plant - continues to be developed to meet modern requirements of the industry and the constraints of authorizing government departments.

Table 3. Approximate introduction and inventories (TBq) of a number of radionuclides arising from Sellafield and La Hague. (Calculated largely from data reported in NRPB-R171 Addendum, 1986[10] and Calmet & Gueguenlat[11])

Radionuclide	Sellafield (to 1986)		La Hague (to 1985)	
	Total	Cumulative inventory	Total	Cumulative inventory
^{3}H	25230	16735	10190	8470
^{106}Ru	27545	285	4905	800
^{137}Cs	41080	33480	940	760
^{90}Sr	6120	4690	755	675
$^{239/240}Pu$	680	680	-	-
$^{238}Pu + ^{239/240}Pu$	-	-	3	3
^{241}Am	535	815+	-	-
^{125}Sb	170*	65*	1000	795
^{99}Tc	305*	305*	-	-

+Includes grow-in from ^{241}Pu.
*Since 1978.

It is instructive to consider some of these discharges in detail. The most soluble nuclide arising is ^{3}H, and its discharge from Sellafield is shown in Figure 2a. With a relatively constant rate of fuel throughput, and a 12 year half-life, the total environmental inventory has increased, but its rapid dispersion results in low ambient concentrations. The distribution of ^{3}H in the immediate vicinity of Sellafield has been studied by Hetherington & Robson[12]. A more versatile, and extensively used, tracer of water mass movement has been ^{137}Cs, together with the shorter-lived ^{134}Cs. The former isotope has been by far the dominant longer-lived nuclide discharged, although its total environmental inventory is now decreasing (Figure 2b). The burn-up time of fuel reprocessed at Sellafield has been relatively constant since the mid-1960s, and thus the annual - and cumulative - Cs isotope quotient (Figure 3a) is primarily a reflection of dwell time in the storage ponds. Discharges were at a peak during 1974 to 1978, owing to a cessation of reprocessing in 1974 which led to an increase in the corrosion of the magnox fuel cladding and a subsequent increase in the concentrations of nuclides in the cooling ponds. In 1976, following chemical treatment of the pond water, zeolite skips were used specifically to reduce the Cs isotope concentrations until a new treatment plant (SIXEP) came on stream in 1985. The $^{137}Cs/^{90}Sr$ quotient has also changed markedly over the last two decades (Figure 3b); it is considerably greater than that of fallout. The distribution of Sellafield-derived Cs has been the subject of numerous studies both within the Irish Sea (e.g. Jefferies _et al._[13]) and further afield (e.g. Kautsky[14]; Dahlgaard _et al._[15]).

A potentially useful long-term tracer of water movement is ^{99}Tc: it is highly soluble, and very long-lived. It is discharged from Sellafield periodically, arising from medium-active liquors which are stored prior to discharge to allow for the decay of other nuclides. Since 1978, when detailed data were first published by BNFL, some 300 TBq have been discharged. Fallout has contributed approximately 100 to 140 TBq globally[26] of ^{99}Tc and thus local inputs such as that from Sellafield are readily detectable.

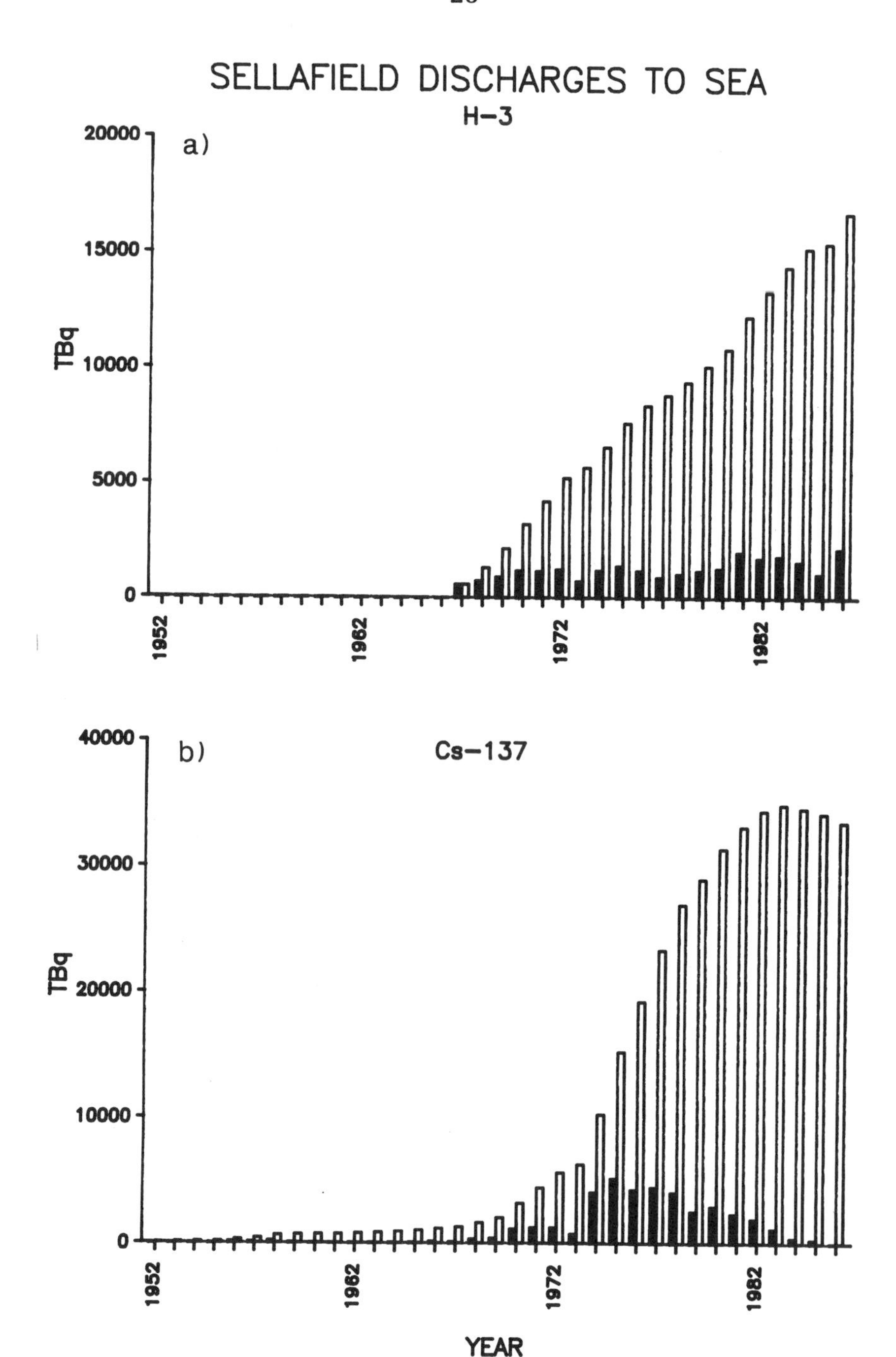

Figure 2. The annual (solid bar) and cumulative (open bar) discharges of (a) ^{3}H and (b) ^{137}Cs from Sellafield into the Irish Sea.

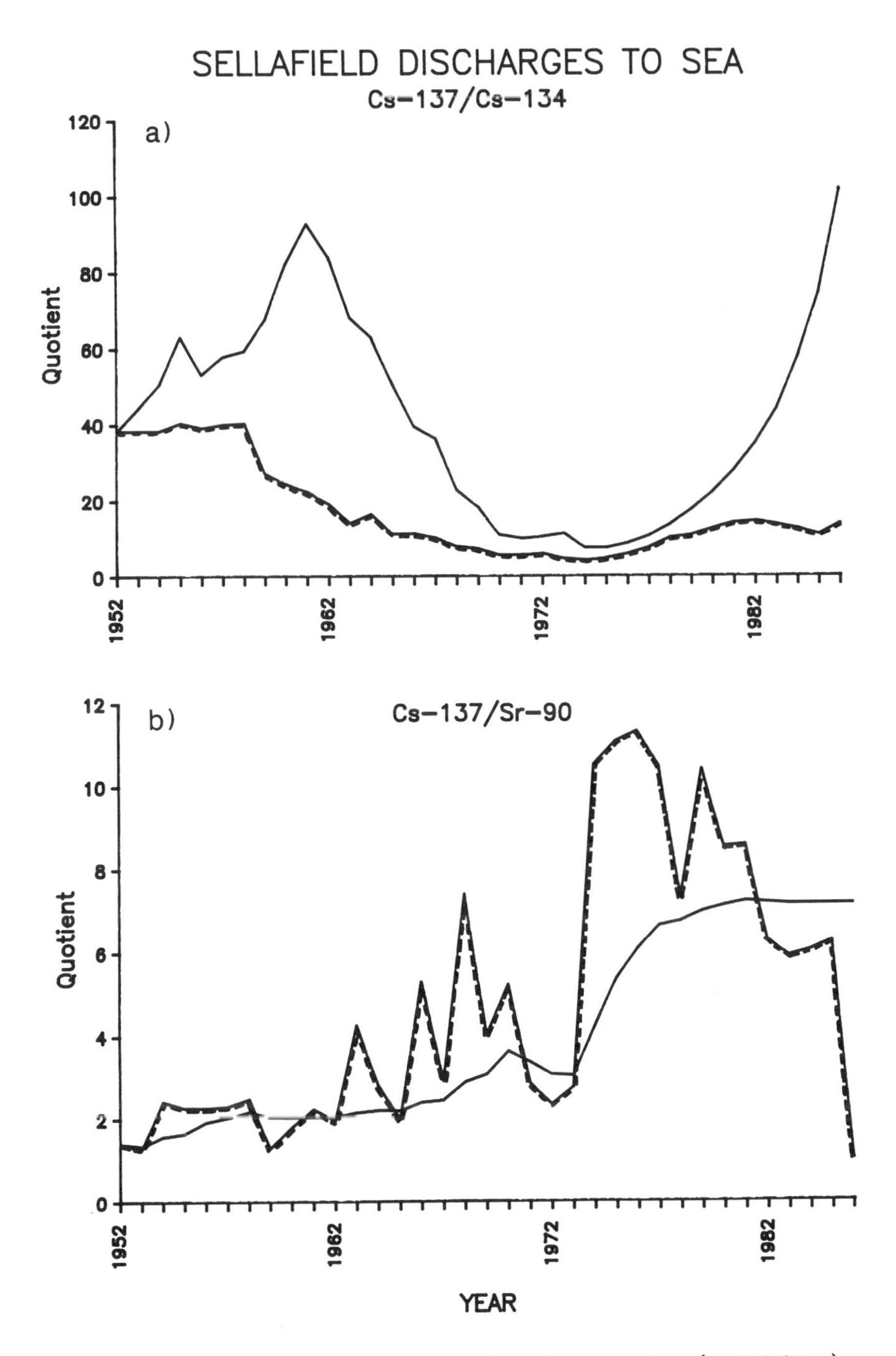

Figure 3. The annual (pecked line) and cumulative (solid line) quotients of (a) $^{137}/^{134}$Cs and (b) ^{137}Cs/^{90}Sr in the Sellafield liquid effluents.

Another of the elements with different isotopes is plutonium; it has been extensively studied, providing a variety of information on environmental processes. Its chemical form in the effluent, and in sea water, is reasonably well understood[16] and it is known to be highly associated with particulate material. The long-lived $^{239+240}Pu$ (Figure 4a) have been shown to be largely incorporated into the sea bed off Sellafield, together with another transuranium nuclide, ^{241}Am, which is not only discharged directly but also arises from the decay of ^{241}Pu (Figures 4b and 5a). The total environmental inventory of ^{241}Am - including that from ^{241}Pu decay - will continue to increase, reaching an asymptotic value in about 50 years time some 1.3 times that existing in 1987. In consequence, the $^{241}Am/^{239+240}Pu$ quotient will also change with time. The annual and cumulative values (Figure 5b) are of use because they indicate the limits which may be observed in near-shore settled sediments, depending on the extent to which all past discharges have been homogeneously mixed into them. Off the Sellafield coast it has been demonstrated that Pu concentrations can vary markedly between sediment cores at the same depth[17,18] and even in the same core[19], an effect resulting from bioturbation[20,21]. An identification of these processes has been greatly assisted by the use of the $^{239+240}Pu/^{238}Pu$ quotients (Figure 6a) which, prior to 1978, are not known precisely for the effluents, but have been inferred from a time-series of surface mud sediments in an estuary just south of the site. Also of value are the $^{137}Cs/^{239+240}Pu$ values (Figure 6b) because, although the bulk (90%) of the Cs discharged remains in the water column, the quantities discharged are such that both Cs nuclides are readily detectable in the sediments. Thus, from the short-lived nuclides such as $^{95}Zr/^{95}Nb$, through nuclides such as ^{144}Ce, ^{106}Ru, ^{60}Co and the Cs nuclides, to the transuranium series, a complete suite exists to examine the depth, rate, and nature of sediment mixing and transport. The most extensive use made of these tracers is the study by Hamilton and Clarke[22] of the Esk estuary. Many radionuclides were analysed, and the data used to model the sedimentary history of the area. This study also demonstrates the necessity to combine radiochemical data with more fundamental geological and ecological information in order to produce a sensible interpretation of past events. Other studies relating to the Sellafield discharges have been summarized by Pentreath[23].

A similar range of nuclides to that at Sellafield has been discharged from La Hague, although the quantities in the past (Table 3) have generally been much lower. The two plants have used similar reprocessing technologies but the effluent treatment plants have been different. Discharges of the Cs isotopes from La Hague are shown in Figure 7. The annual $^{137}Cs/^{134}Cs$ quotient has been between 4 and 6 in recent years, that of Sellafield between 10 and 14. The $^{137}Cs/^{90}Sr$ quotients are also different: less than 2 from La Hague (Figure 8a), about 5 to 10 from Sellafield over the last decade. Less detailed information has been published regarding the nature of the transuranium nuclides discharged from La Hague, but these have been substantially less than Sellafield and the $^{137}Cs/Pu\ \alpha$ quotient has thus been substantially higher (Figure 8b).

The fate and local dispersion of the La Hague discharges has been extensively studied, with a generally-similar behaviour of nuclides being observed, as reviewed by Calmet & Guegueniat[11] and Auffret <u>et al</u>[24]. One intriguing difference between the two sites, however, has been the quantities of ^{125}Sb discharged (Figure 9). This nuclide is highly soluble, lending itself to a tracer of water movement in the channel[25].

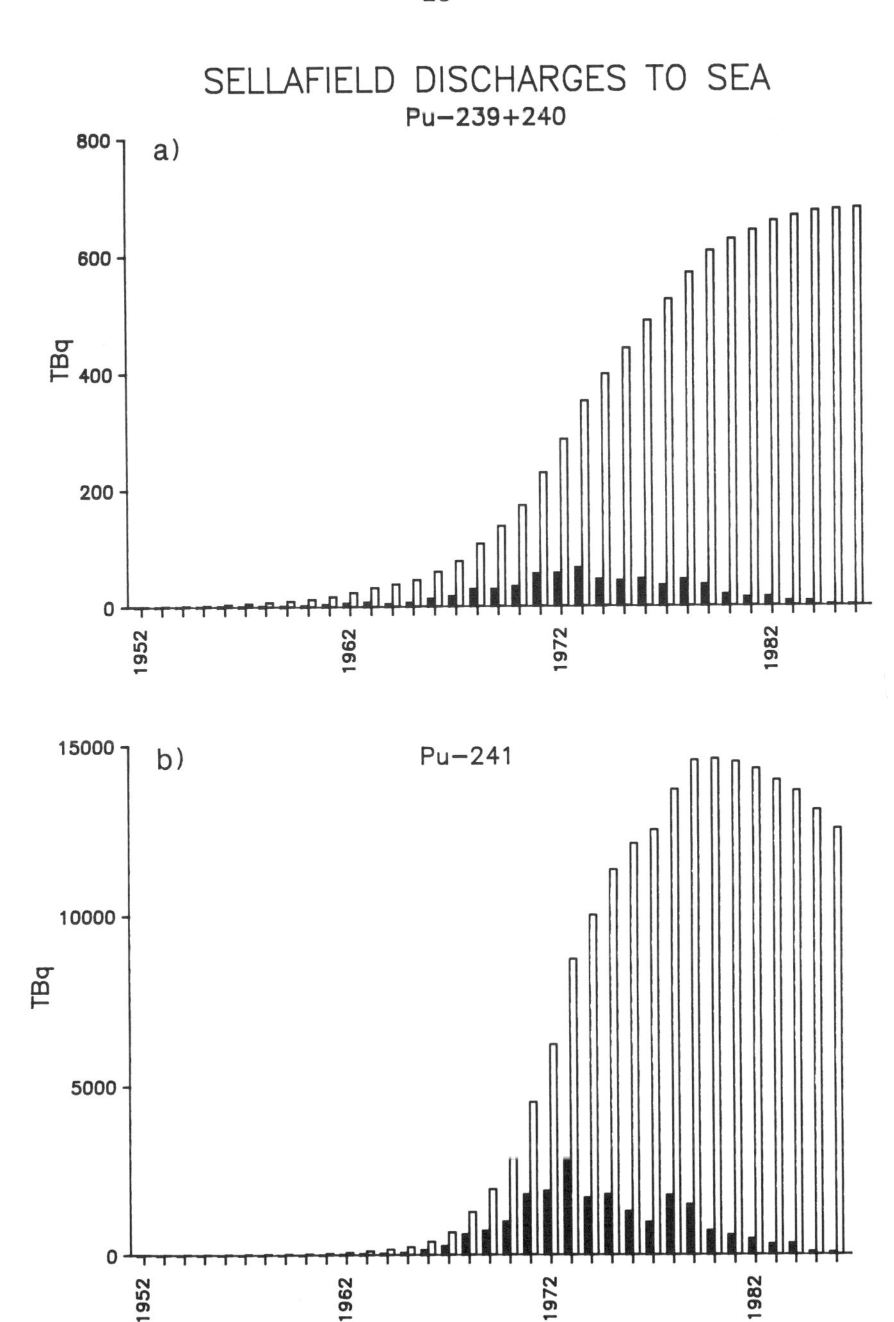

Figure 4. The annual (solid bar) and cumulative (open bar) discharges of (a) $^{239+240}Pu$ and (b) ^{241}Pu from Sellafield into the Irish Sea.

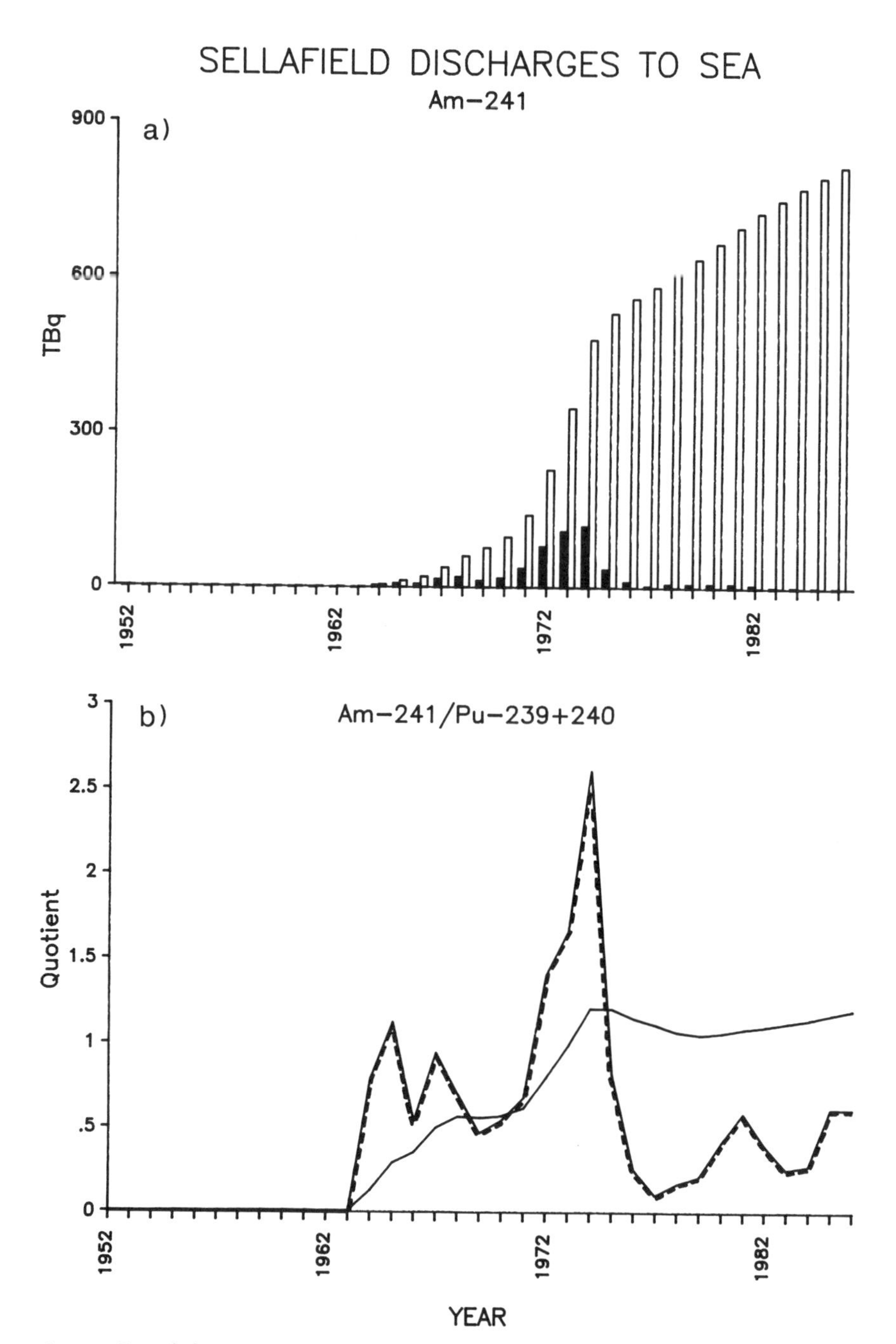

Figure 5. (a) The annual (solid bar) and cumulative (open bar) discharges to sea of ^{241}Am - the latter including grow-in from ^{241}Pu - from Sellafield and (b) the consequent annual (pecked line) and cumulative (solid line) $^{241}Am/^{239+240}Pu$ quotients.

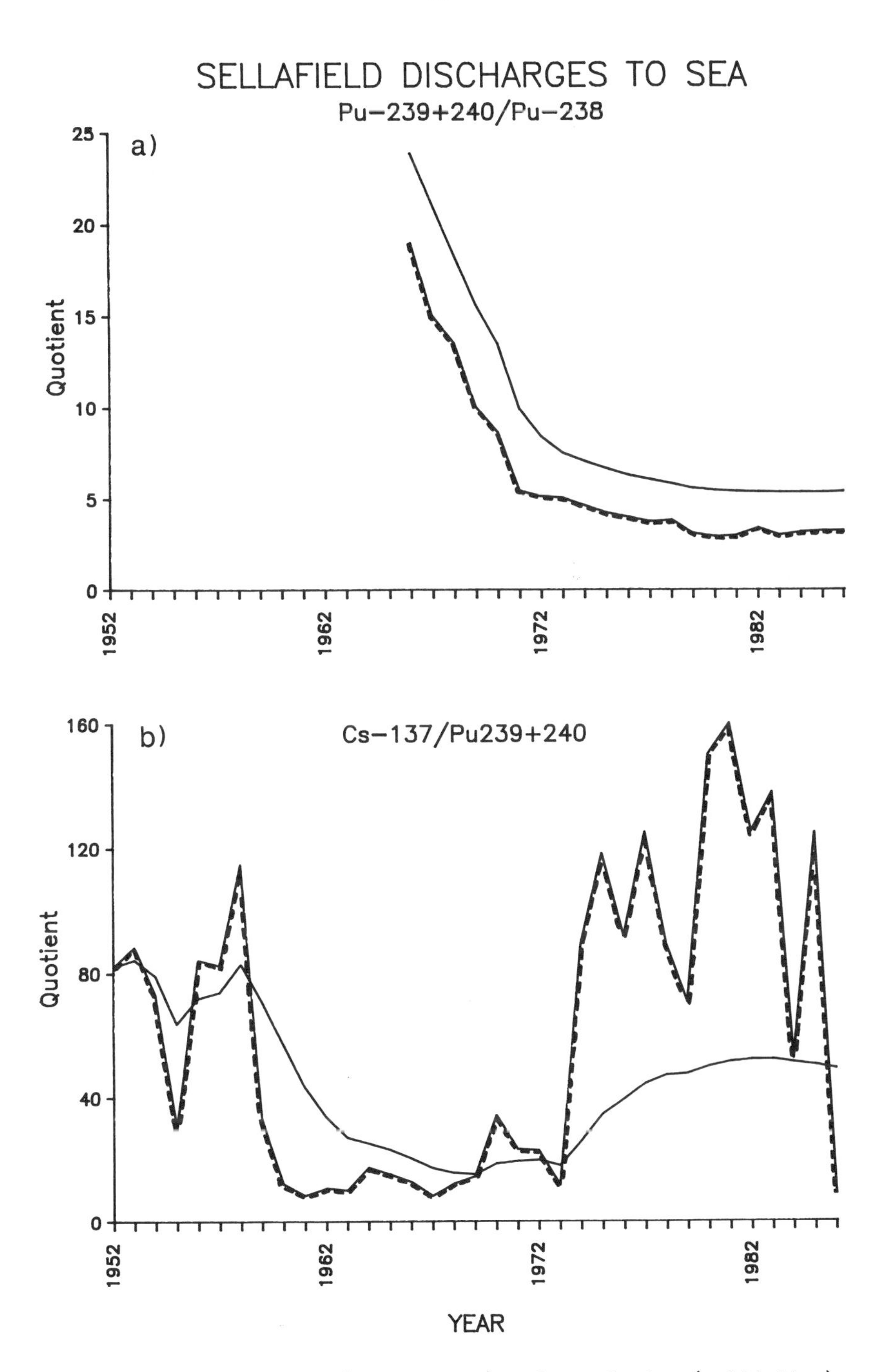

Figure 6. The annual (pecked line) and cumulative (solid line) quotients of (a) $^{239+240}Pu/^{238}Pu$ and (b) $^{137}Cs/^{239+240}Pu$ in the Sellafield liquid effluents.

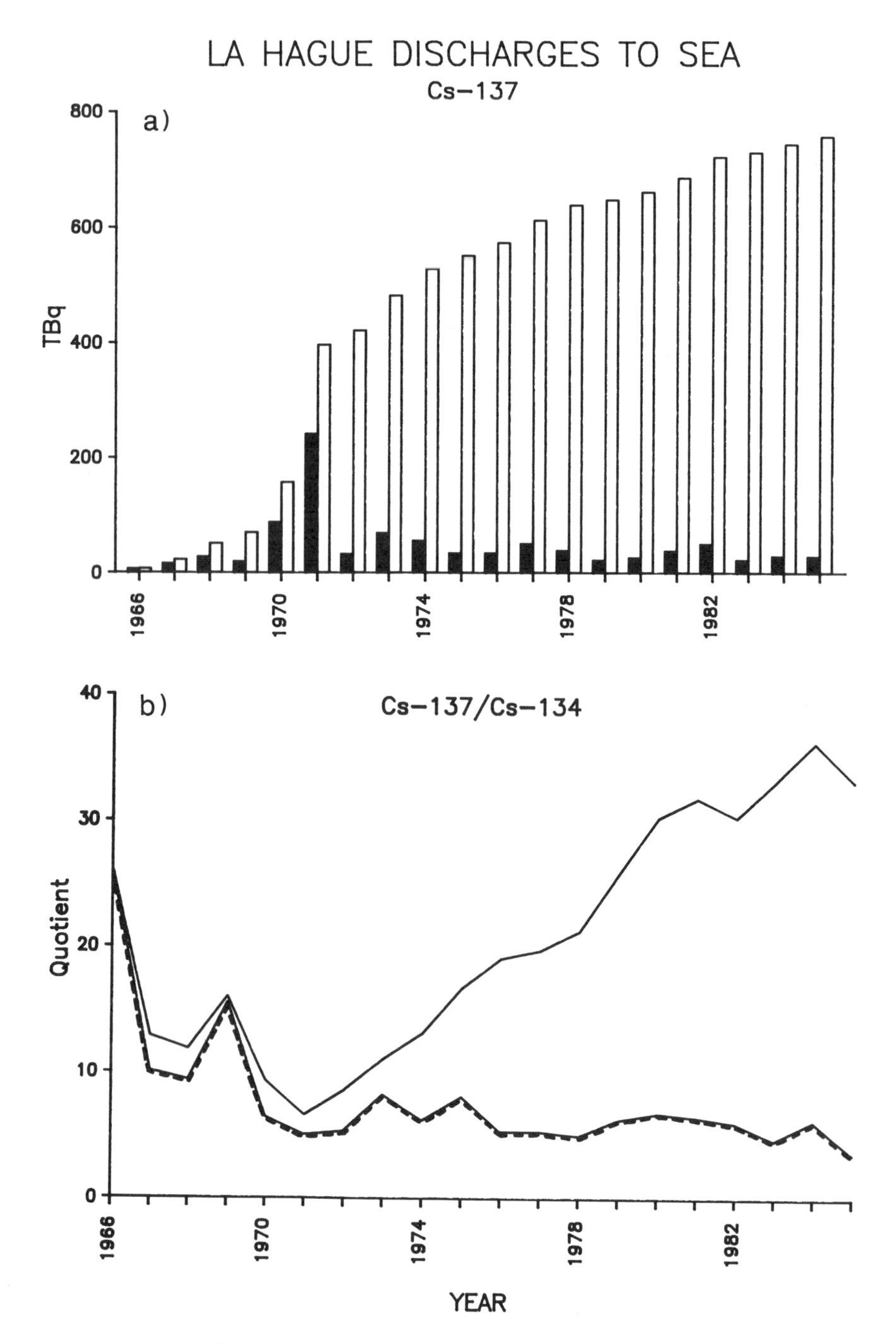

Figure 7. (a) The annual (solid bar) and cumulative (open bar) discharges of ^{137}Cs to sea from La Hague and (b) the annual (pecked line) and cumulative (solid line) $^{137}Cs/^{134}Cs$ quotients.

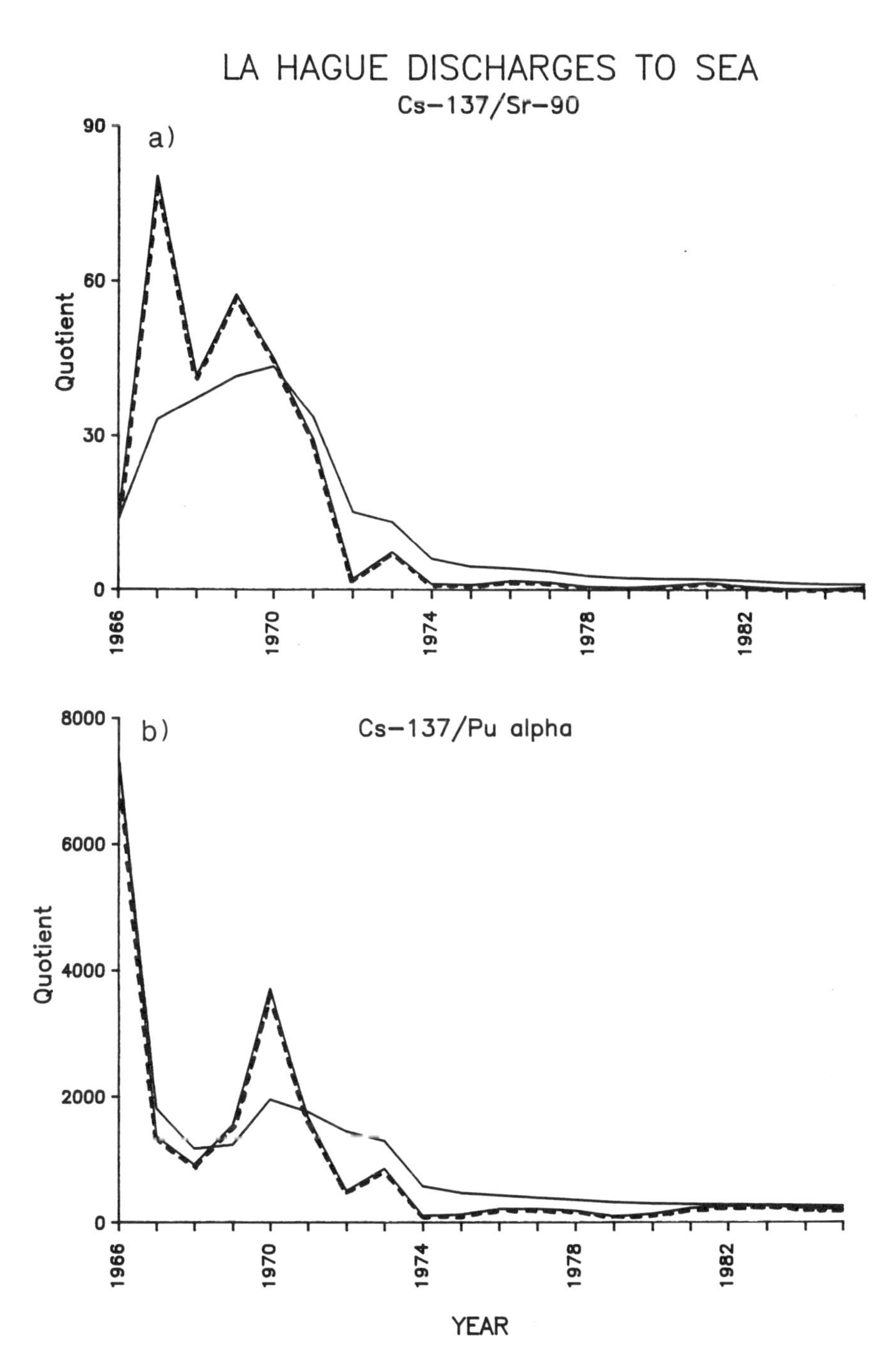

Figure 8. The annual (pecked line) and cumulative (solid line) quotients of $^{137}Cs/^{90}Sr$ and ^{137}Cs/Pu-alpha in the La Hague liquid effluents.

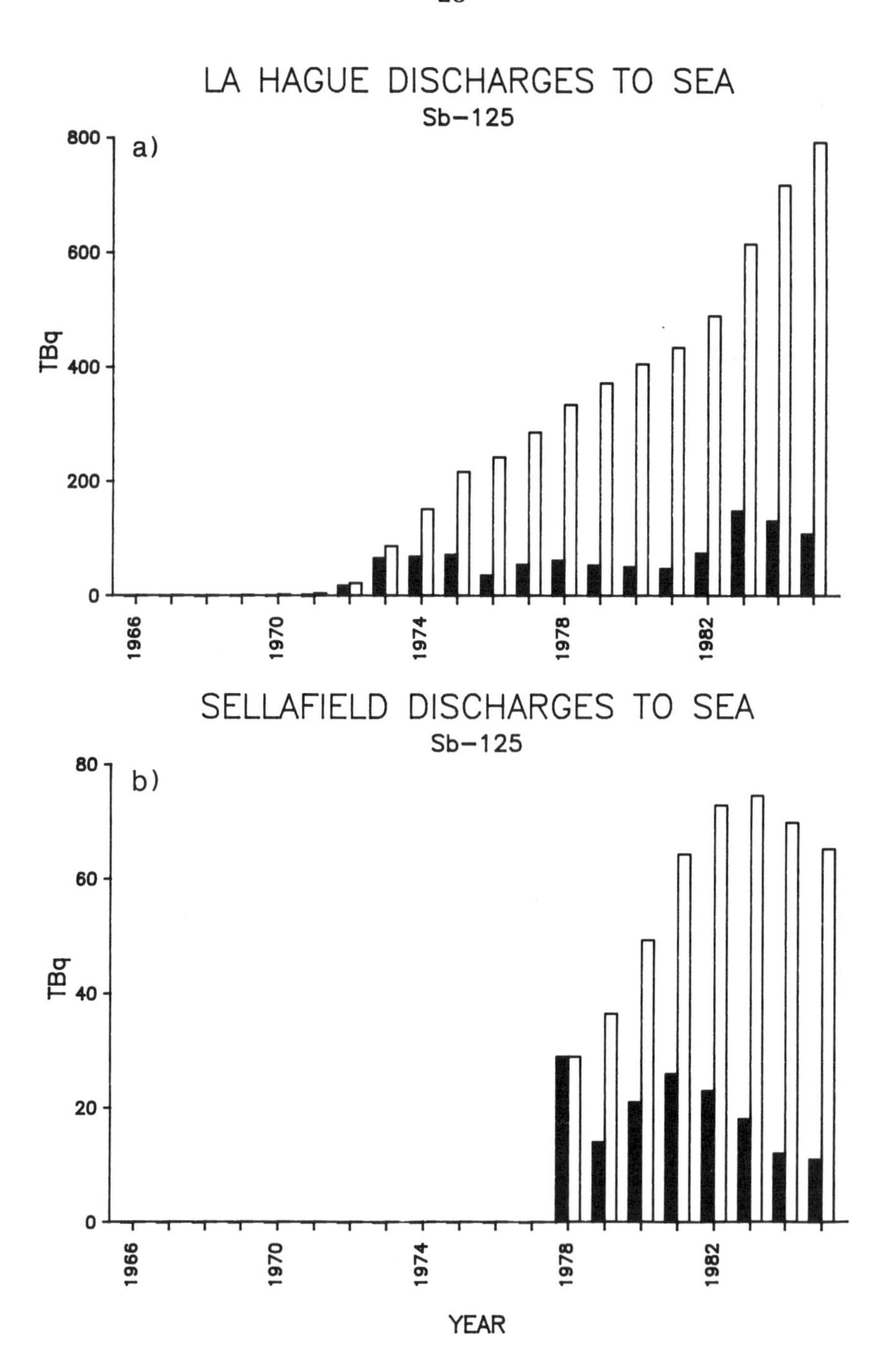

Figure 9. The annual (solid bar) and cumulative (open bar) discharges of ^{125}Sb from (a) La Hague and (b) Sellafield; data from the latter only being available since 1978.

A third source of interest is that of the reprocessing facility at Trombay, in India, but the quantities discharged are expressed in "^{90}Sr equivalent" units[27]. There are, however, intriguing differences between this site and those at Sellafield and La Hague, primarily due to the impact of the monsoon period (June to September) when land drainage and river run-off can be considerable, resulting in low salinities and very high sediment loads. And of equal value are the minor sites, such as effluents from power stations, which can provide useful data to examine local dispersion characteristics. Site-to-site comparisons may also reveal valuable information on the interaction between elements in different chemical forms and different receiving environments[19].

SEA DISPOSAL AND OTHER SOURCES

Solid packaged wastes have been dumped in the deep sea since 1947. The USA has used sites off their Pacific and Atlantic coasts, and a number of European countries have used sites in the NE Atlantic Ocean. The latter have been described and discussed in some considerable detail[2], and a further description is not warranted here; to date, a large mixture of radionuclides is estimated to have been dumped in them, none of which has been unequivocably identified in environmental samples. The most promising nuclide for detection is ^{3}H (Figure 10a and b), both because it represents a large fraction (23000 TBq) of the total quantity of nuclides dumped (about 42000 TBq) which, allowing for decay, amounts to an inventory of some 13200 TBq of ^{3}H as of 1987, and because of its high mobility and solubility. Samples in and around the NEA dump site, which is at about 4 km depth, have been taken by MAFF-DFR and are currently being analysed.

The major sites used by the USA have generally been in somewhat shallower waters (1 to 3 km depth). Samples of sediment taken close to some packages have indicated local contamination by nuclides such as those of Pu[28,29] but not, as yet, of the sites in general[30]. Two other Pacific sites which need to be mentioned are those used by Japan[31] and the Republic of Korea. The Japanese packages consisted primarily of scrap wastes from the manufacture of sealed radiation sources, mainly ^{60}Co (about 150 TBq), and the Korean wastes contained a range of short-lived nuclides such as ^{32}P, ^{59}Fe, ^{129}Te and ^{198}Au.

A somewhat unusual source of seabed release is that remaining from a crashed B-52 bomber off Thule in Greenland at a depth of some 200 m. This released about 1 TBq of $^{239+240}Pu$ into the sea, plus about 0.1 TBq of ^{241}Am arising from the decay of ^{241}Pu. The distribution of these nuclides has been studied in detail by Aarkrog and his colleagues[32,33]. Other seabed sources exist - such as lost nuclear submarines - but data are not readily available on their location, inventories, or resulting environmental concentrations.

CHERNOBYL

The accident to the USSR reactor Chernobyl Unit 4, in the spring of 1986, resulted in a release of a number of radionuclides into the atmosphere. At the time of the incident the total inventory of the core was approximately 4 x 10^{7} TBq[34]. Apart from the noble gases, which would have been completely expelled, it is estimated that about 10 to 20% of the volatile nuclides - those of I, Cs, Te - were expelled, plus some 3-6% of

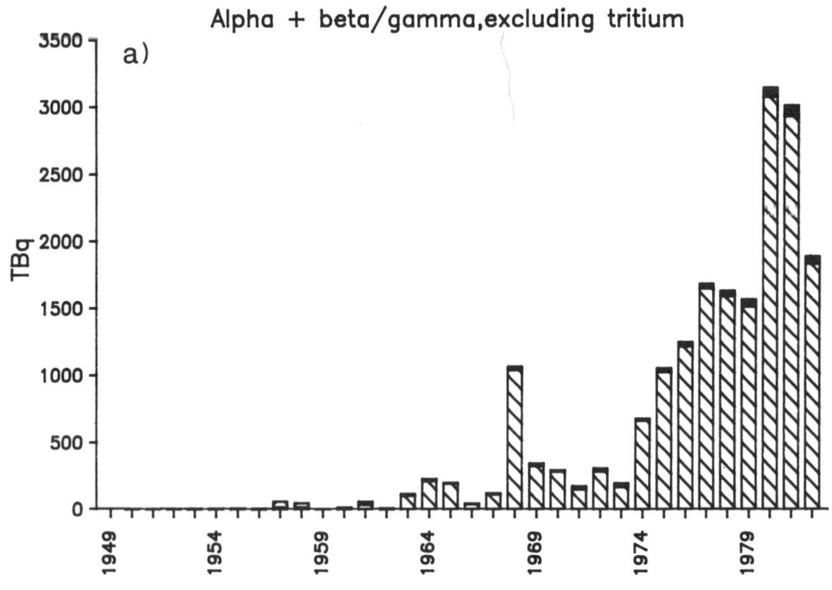

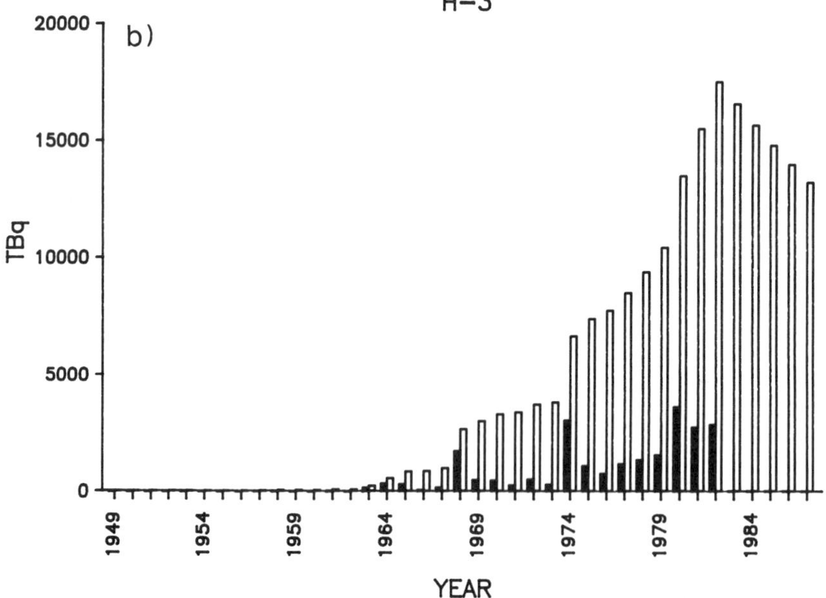

Figure 10. The quantities of (a) alpha-emitting (solid shading) and beta-emitting (cross-hatched shading) nuclides, excluding ^{3}H, and (b) an estimate of the annual (solid bar) and cumulative (open bar) quantities of ^{3}H which have been dumped by European countries in the NE Atlantic Ocean.

the more refractory elements such as Ba, Sr, Ce and Pu. The data given in Table 4 are for some of the nuclides of potential interest, although it is important to note that they are indeed estimates, with an error of the order of ± 50%. They were released over a period of about 10 days, but not at a constant rate. About 25% was released during the first day, after which the rate of release gradually declined until, five days later, it increased again. The more important feature, however, is the relative abundance of nuclides. From the data in Table 4 it can be seen that the quotients of interest are approximately as follows: $^{137}Cs/^{134}Cs$, 2; $^{137}Cs/^{90}Sr$, 5; $^{239+240}Pu/^{238}Pu$, 2; and $^{137}Cs/^{239+240}Pu$, 600.

Table 4. Approximate quantities of a number of nuclides released into the atmosphere as a result of the accident to the Chernobyl Unit 4 reactor, USSR, 26 April to 6 May 1986[34]

Nuclide	Quantity (TBq)	Nuclide	Quantity (TBq)
^{90}Sr	8000	^{238}Pu	30
^{95}Zr	140800	^{239}Pu	25.5
^{106}Ru	58000	^{240}Pu	36
^{134}Cs	19000	^{241}Pu	5100
^{137}Cs	37700	^{242}Cm	780
^{144}Ce	89600		

Much has already been published about the initial dispersion of the atmospheric release, which occurred largely at altitudes of less than 2 km[35]. It has resulted in contamination - by direct deposition and run-off - of a number of areas including the Caspian, Black, Mediterranean, Baltic, North and Irish Seas, and even the NE Atlantic Ocean to at least 70°N[36]. It will be interesting to compare the behaviour of nuclides arising from Chernobyl in these areas with earlier data on the same nuclides arising from weapon's fallout.

DISCUSSION

All of these artificial radionuclides in the oceans are potential tracers of environmental processes, but a degree of caution is warranted in their use. First of all it must be noted that, ideally, tracers need to be in a known chemical form: this is not always possible. Secondly, the occurrence of radioisotopes serves primarily to trace the cycling of stable isotopes of the same element. Their principal application, however, has been to study the rates of water movement, or of particles to which they become attached, and as such this is an indirect application. It depends on the simple assumption that certain nuclides are soluble and that others are adsorbed onto particles. In reality these assumptions are only crude approximations because the precise behaviour of nuclides depends on their chemical form at any one time and the materials with which they interact. Thus some nuclides can be rendered more soluble by changes in valence state or association with soluble organics; similarly nuclides can both adsorb onto and desorb from particles, depending on the ambient chemical environment. Thus it is important to gain as much information as possible on the chemical form of the nuclide being used as a tracer, and how changes in it can alter the interpretation of the data.

Specific nuclides can be used to study the rates of dispersion of water masses, the removal of particles from the water column, to provide labelled time horizons in settled sediments, and to estimate mixing rates within sediments in ways similar to those which have been used with naturally-occurring radionuclides. In fact not all of the potentially useful applications have been fully realized. For example, use is made of the disequilibrium between ^{234}Th and ^{238}U to study scavenging rates of the former from coastal waters; the ^{234}Th has a 24-day half-life. Use could be made of the ^{90}Sr/^{90}Y pair, the ^{90}Y having a 2.7 day half-life and an affinity for particles which is many orders of magnitude greater than that of ^{90}Sr. Differences between ^{241}Pu and ^{241}Am may provide information on relative mobilities in sediments in relation to organic complexation and longer-term diagenetic processes. Three Eu nuclides in effluents, ^{155}Eu, ^{154}Eu and ^{152}Eu have half-lives of 4.96, 8.8 and 13.3 years respectively, which could enable short-term events in sediments to be studied. There are no doubt many other examples of potential use. But the one factor which such tracer applications have in common is that such data should not be considered in isolation. It is essential that other relevant ecological, non-tracer, information be gathered and considered simultaneously so that conclusions are based on an overall evaluation; otherwise the radionuclide data can often be misinterpreted. The two approaches are, in any case, complementary and essential to one another. The distributions of radionuclides are an indication of the effect and rates of environmental processes. And it is essential to understand the effects and rates of such processes in order to predict the future distribution of radioactive wastes discharged or otherwise disposed of into the environment.

REFERENCES

1. United Nations Scientific Committee on the Effects of Atomic Radiation, 1982, UN, New York, 773 pp.

2. Nuclear Energy Agency, Review of the continued suitability of the dumping site for radioactive waste in the North-East Atlantic, OECD, Paris, 1985, 448 pp.

3. Broecker, W. S. and Peng, T.-H., Tracers in the Sea, Columbia University, 1982, 690 pp.

4. Noshkin, V. E. and Bowen, V. T., In Radioactive Contamination of the Marine Environment, IAEA-PUB-313, Vienna, 1973, pp.671-686.

5. Bowen, V. T., Noshkin, V. E., Livingston, H. D. and Volchok, H. L., Earth Planet. Sci. Lett., 1980, 49, 411-434.

6. Livingston, H. D., Fallout plutonium in western North Pacific sediments, IAEA-TECDOC-368, Vienna, 1986, pp.27-34.

7. Livingston, H. D., Mann, D. R., Casso, S. A., Schneider, D. L., Suprenant, L. D. and Bowen, V. T., J. Environ. Radioact., 1987, 5, 1-24.

8. Perkins, R. W. and Thomas, C. W., Worldwide fallout, In Transuranic Elements in the Environment, ed. W. C. Hanson, U.S. Dept. of Energy, 1980, pp.53-82.

9. Beasley, T. M., Science, 1981, 214, 913-915.

10. Stather, J. W., Dionian, J., Brown, J., Fell, T. P. and Muirhead, C. R., The risks of leukaemia and other cancers in Seascale from radiation exposure, National Radiological Protection Board, NRPB-R171, Addendum, 1986, 158 pp.

11. Calmet, D. and Gueguenìat, P., In Behaviour of Radionuclides Released into Coastal Waters, IAEA-TECDOC-329, Vienna, 1985, pp.111-144.

12. Hetherington, J. A. and Robson, J. C., In Behaviour of Tritium in the Environment, IAEA-PUB-498, 1979, pp.283-301.

13. Jefferies, D. F., Steele, A. K. and Preston, A., Deep Sea Res., 1982, 29(6A), 713-738.

14. Kautsky, H., Dt. hydrogr. Z., 1986, 39 (4), 139-159.

15. Dahlgaard, H., Aarkrog, A., Hallstadius, L., Holm, E. and Rioseco, J., Rapp. P.-v. Réun. Cons. int. Explor. Mer, 1986, 186, 70-79.

16. Pentreath, R. J., Harvey, B. R. and Lovett, M. B., In Speciation of Fission and Activation Products in the Environment, ed. R. A. Bulman and J. R. Cooper, Elsevier Applied Science Publishers, London, 1986, pp.312-325.

17. Hetherington, J. A., Mar. Sci. Commun., 1978, 4, 239-274.

18. Pentreath, R. J., Jefferies, D. F., Lovett, M. B. and Nelson, D. M., In Marine Radioecology, 3rd NEA Seminar, Tokyo, NEA/OECD, Paris, 1980, pp.203-221.

19. Pentreath, R. J., J. Cont. Shelf Res., 1987 (In press).

20. Kershaw, P J., Swift, D. J., Pentreath, R. J. and Lovett, M. B., Nature, 1983, 306, 774-775.

21. Kershaw, P. J., Swift, D. J., Pentreath, R. J. and Lovett, M. B., Sci. Tot. Environ., 1984, 40, 61-81.

22. Hamilton, E. I. and Clarke, K. R., Sci. Tot. Environ., 1984, 35, 325-386.

23. Pentreath, R. J., In Behaviour of Radionuclides Released into Coastal Waters, IAEA-TECDOC-329, Vienna, 1985, pp.67-110.

24. Auffret, J. P., Gueguenìat, P., Lepy, M. C., Patry, J. P. and Saur, H., In La Baie de Seine, IFREMER, Actes de Colloques, 4, 1986, pp.273-282.

25. Gueguenìat, P. and Le Hir, P., In Impacts of Radionuclide Releases into the Marine Environment, IAEA-PUB-565, Vienna, 1981, pp.481-499.

26. Schulte, E. H. and Scoppa, P., Sci. Tot. Environ., 1987, 64, pp.163-179.

27. Patel, P. and Patel, S., In Behaviour of Radionuclides Released into Coastal Waters, IAEA-TECDOC-329, Vienna, 1985, pp.145-182.

28. Schell, W. R. and Sugai, S., Hlth Phys., 1980, 39, 475-496.

29. Bowen, V. T. and Livingston, H. D., In Impacts of Radionuclide Releases into the Marine Environment, IAEA-PUB-565, Vienna, 1981, pp.33-63.

30. Noshkin, V. E., Wong, L. M., Jokela, T. A., Eagle, R. J. and Brunk, J. H., UCRL-52381, Lawrence Livermore National Lab., 1978, 17 pp.

31. Ichikawa, R., In Management of Low and Intermediate Level Radioactive Wastes, IAEA-PUB-264, Vienna, 1970, pp.91-99.

32. Aarkrog, A., Hlth Phys., 1977, 32, 271-284.

33. Aarkrog, A., Dahlgaard, H. and Nilsson, K., Hlth Phys., 1984, 46, 29-44.

34. International Atomic Energy Agency, Summary Report on the Post-Accident Review Meeting on the Chernobyl Accident, Safety Series No. 75, 1986, INSAG-1, Vienna, 106 pp.

35. Persson, C., Rodhe, H. and De Geer, L.-E., Ambio, 1987, 16, 20-31.

36. Mitchell, N. T. and Steele, A. K., J. Environ. Radioact., 1987 (In press).

USE OF TRITIUM AND HELIUM-3 FOR THE STUDY OF OCEANOGRAPHIC PROCESSES. AN EXAMPLE : THE NORTHEASTERN ATLANTIC OCEAN

C. Andrié and L. Merlivat

Laboratoire de Géochimie Isotopique
- LODYC (UA CNRS 1206) - CEA/IRDI/DESICP -
Département de Physico Chimie - CEN de Saclay
91191 GIF sur YVETTE cedex, FRANCE

ABSTRACT

These results relate to the TOPOGULF cruise during summer 1983. The sampled area is located near the Azores Islands on both sides of the Mid-Atlantic ridge. The tritium content of the surface waters reveals a strong front along the Azores current. At depth, the spatial distribution of tritium is studied along isopycnals. In the thermocline, it is essentially the anticyclonic gyre which is responsible for the northeast - southwest gradient in tritium concentration. Some information about the "age" of the water masses is given by the use of both tritium and Helium-3. In this way some areas where ventilation processes are active are identified. At middepth, a strong contrast exists between the young waters originating from the Labrador Sea and the more stagnant waters in the South East of the sampled area.

INTRODUCTION

In the form of HTO molecules tritium is an ideal tracer for the water masses circulation studies.

Large amounts of tritium entered the ocean from the atmosphere after the nuclear bomb testing starting in 1954. The larger inputs occured between 1963 and 1965. The time distribution of tritium concentration in surface water is relatively well documented (DREISIGACKER and ROETHER, 1978).

The transient nature of tritium and its half life (12.43 years) has already provided information about circulation processes in the North Eastern Atlantic ocean (SARMIENTO et al., 1982).

For the study of water mass transport and ventilation processes the simultaneous use of tritium and its radioactive daughter ^{3}He is specially adequate to deduce interior ocean travel times (JENKINS, 1980).

These two tracers have complementary boundary conditions at the ocean-atmosphere interface : this interface acts as a sink for the atmospheric tritium into the ocean and as a source for the oceanic gaseous helium-3 into the atmosphere. The "age" of a water mass, i.e the time elapsed since the water parcel has been isolated from the atmosphere can be determined from its helium-3 and tritium contents (JEAN-BAPTISTE et al, this issue).

The data are reported in $\delta^{3}He$ % for the ^{3}He excess (i.e. the anomaly of the isotopic ratio $^{3}He/^{4}He$ of the sample relative to the isotopic ratio of the atmosphere, expressed as a percentage) and in TU units for tritium (1 TU unit represents 1 tritium atom for 10^{18} hydrogen atoms). The age of each sample is calculated from $\tau = \lambda^{-1} \ln \left(1 + \frac{[^{3}HE]}{[^{3}H]}\right)$ where λ is the tritium decay constant equal to 0.0557 y^{-1} and $[^{3}He]$ and $[^{3}H]$ are the respective helium 3 and tritium contents of the sample (expressed in numbers of atoms).

FIELD WORK AND BACKGROUND

Figure 1 shows the stations sampled during the TOPOGULF cruise (July-August 1983). These are located in the area 24°N-40°N, 23°W-50°W near the Azores Islands along the Eastern and Western flanks of the Mid-Atlantic Ridge and perpendicular to it. The hydrological and nutrient data are given in the Data Report by the TOPOGULF Group (1986).

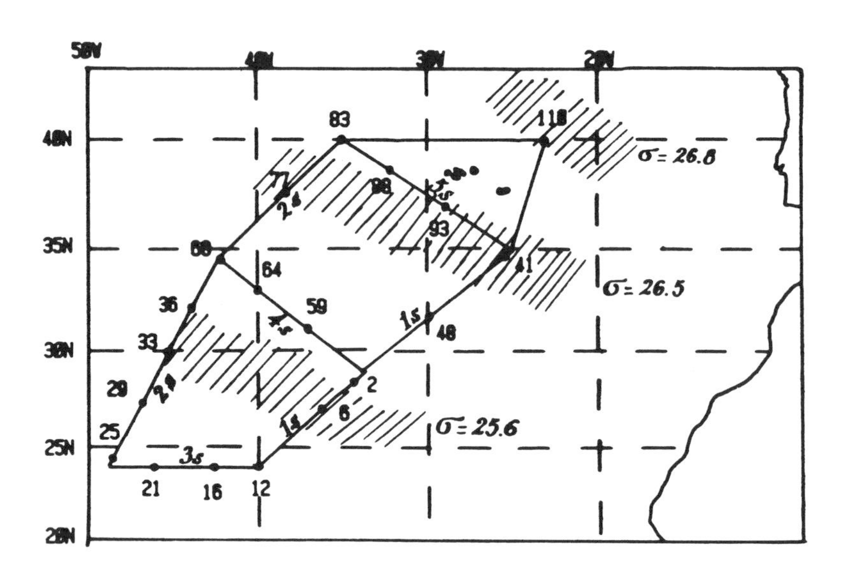

Figure 1 : Map of the TOPOGULF cruise (tracks 1s, 2s, 3s, 4s and 5s). The shaded are the approximate locations of the 25.6, 26.5 and 26.8 isopycnal levels.

Previous works relative to chemical or transient tracers distributions in the thermoclinal and middepth North Atlantic waters (JENKINS, 1980 ; SARMIENTO et al., 1982 ; KAWASE and SARMIENTO, 1985, 1986, THIELE et al., 1986) have shown that the tracers transport into subtropical anticyclonic gyre was essentially isopycnal.

We successively discuss the helium-3 versus density distributions along the five principal tracks of the cruise and their corresponding tritium and ^3H-^3He "age" distributions. In addition, salinity and oxygen data are used to identify some water masses along some specific isopycnal levels.

RESULTS AND DISCUSSION

Shaded areas in figure 1 correspond to the water outcropping location of the respective sigma-theta levels 25.6, 26.5 and 26.8 determined from historical wintertime temperature data from SARMIENTO et al., 1982.

In Figure 2 are given the respective helium-3 versus sigma-theta distributions of the tracks 1s and 2s on the Eastern and Western flanks of the MAR and on the tracks 3s, 4s and 5s perpendicular to the MAR (see Figure 1). On each Figure are drawn the respective sigma-theta levels 25.6, 26.5, 26.8. 27.1 and 27.4. The four first levels outcrop in winter in the subtropical anticyclonic gyre, the last one outcrops in the subarctic gyre. These distributions are the results of three processes : convection processes which are responsible for the helium-3 evasion, tritium penetration which generates helium-3, and the age of the water masses which increases the helium-3 regrowth by radioactive decay.

For the surface as well as for the first isopycnal level 25.6 the mean δ^3He values obtained for the samples over the whole TOPOGULF area are respectively - 1.21 % and - 1.44 %. These values agree with the previously observed excess ^3He in the ocean surface layer (FUCHS et al., 1987) as compared to the solubility equilibrium value of - 1.7 % from BENSON and KRAUSE, 1980.

Figure 3 gives the respective tritium and ^3He-^3H age versus sigma-theta distributions for the Western and Eastern tracks 2s and 1s.

We observe a very homogeneous tritium surface content in the area located South of 34°N (4.5 ± 0.15 TU81N). The tritium versus salinity diagram on Figure 4 shows that this homogeneity does not extend to salinities smaller than 36.4 ‰ which characterize the Azores front. North of the front we observe the greatest tritium concentration. This tritium increase with latitude is essentially the result of the latitudinal distribution of the tritium input function.

We report, for some isopycnal levels, informations relative to ventilation, transport and mixind using tritium and helium-3 data :

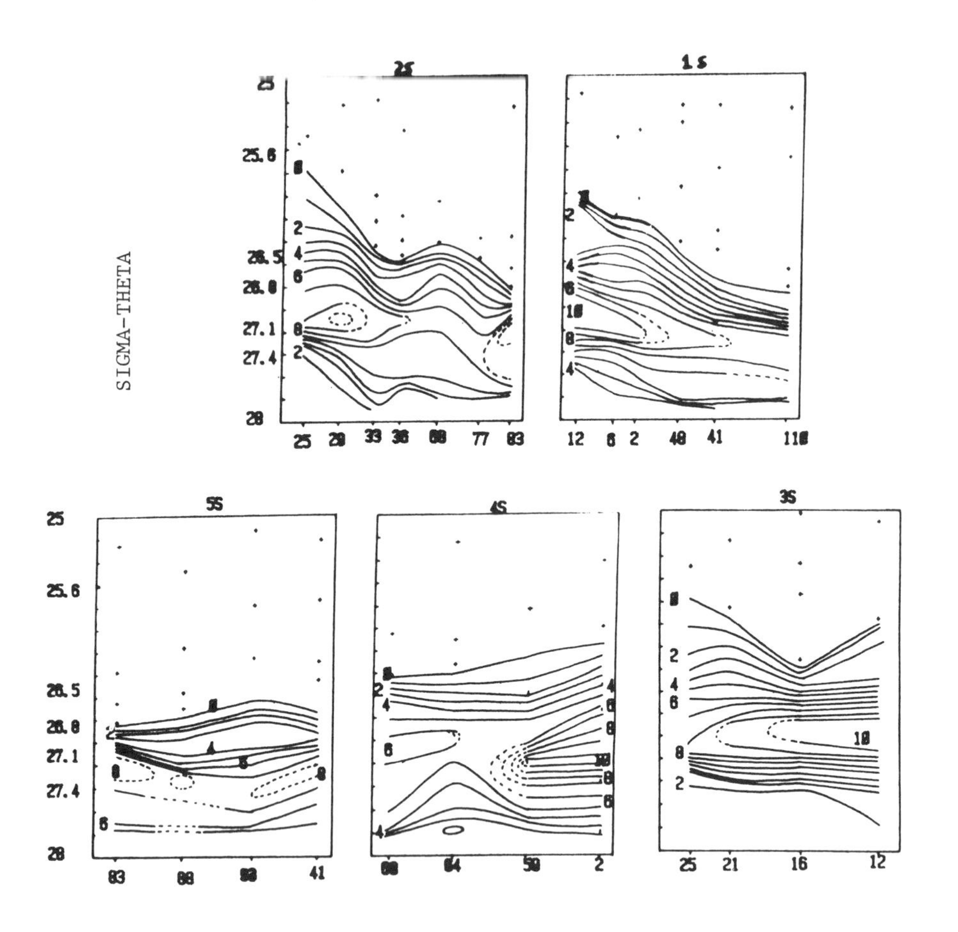

Figure 2 : δ^3He (%), valves plotted on diagrams of sigma-theta versus station number for the 1s, 2s, 3s 4s and 5s tracks in figure 1.

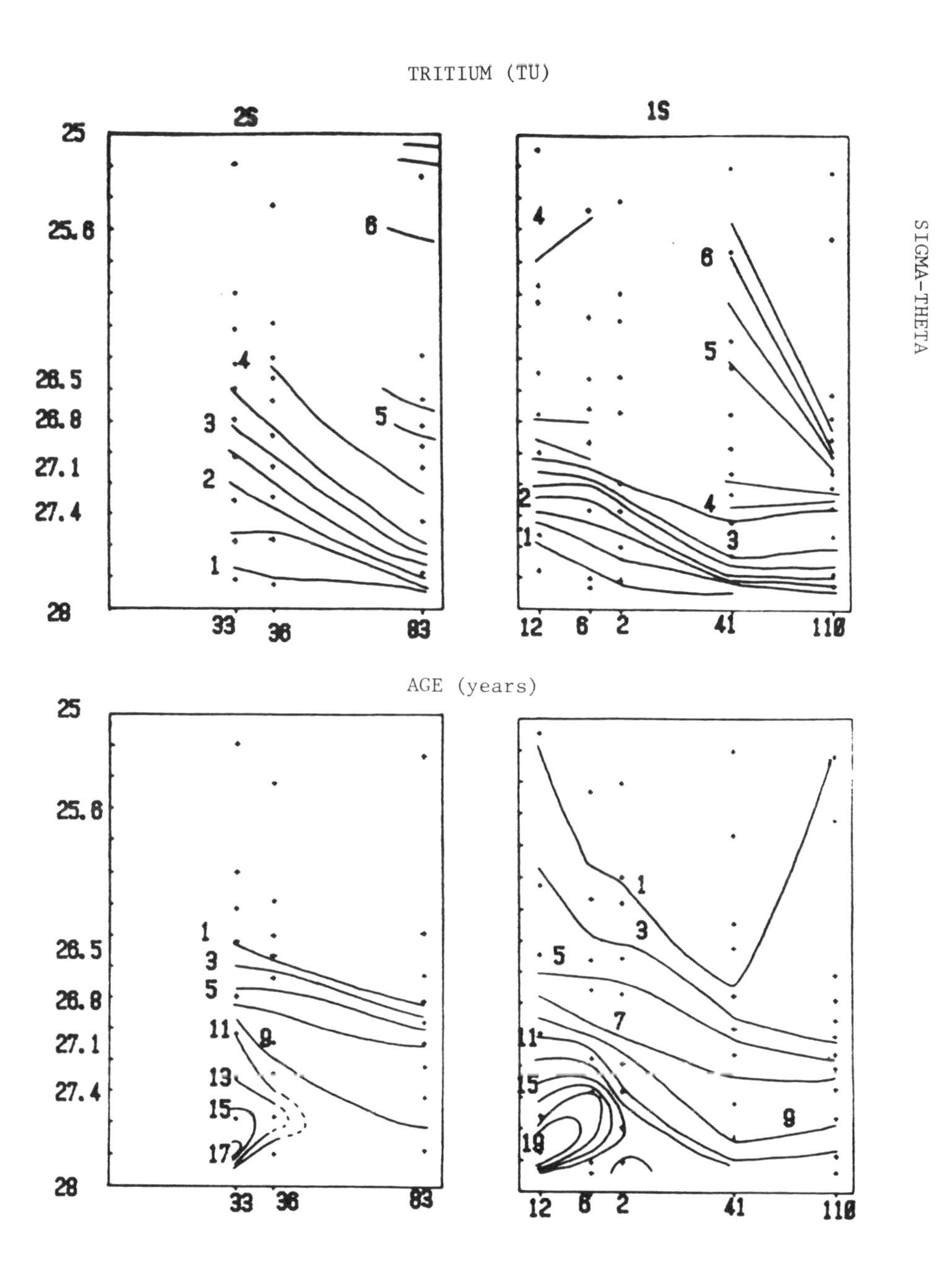

Figure 3 : Tritium concentrations and water mass "ages" plotted on diagrams of sigma-theta versus station number for the Western and Eastern tracks 2s and 1s.

On the 26.5 isopycnal level, we observe tritium concentrations in the South (Stations 12, 6, 2) higher than the corresponding tritium content of the surface waters. On figure 1, we can observe that the wintertime outcrop of this isopycnal takes place North of the Azores front : the tritium enrichment of the Southern stations on the 26.5 sigma-theta level is the result of the transport of tritium enriched waters from an area located to the north of 35°N by the anticyclonic gyre. On the same isopycnal, we observe for the station 33 and 36 relatively low tritium concentrations (Figure 3) correlated with negative δ^3He values (Figure 2) suggesting a remainder of surface convection. On the 26.5 isopycnal level these stations are more imprinted by the 18°C Mode Water than by the anticyclonic gyre. In the North we observe everywhere negative δ^3He values as the result of recent convection.

On the 26.8 isopycnal level in the North (Figure 5) we observe a subsurface maximum for the three stations 83, 93 and 110. This level is the first one for which the wintertime outcrop is located North of the 4°N latitude : this maximum corresponds to the enriched tritium surface waters of Northern origin.

The wintertime outcrop of the 27.4 isopycnal occurs North of the anticyclonic gyre. This can explain the second maxima observed for stations 83 and 93 (Figure 5). The tritium versus salinity and versus oxygen diagrams on Figure 6 indicates the tritium enriched 83, 88 and 93 are originating from the Labrador Sea water characterized by a high oxygen content. On the opposite, the Mediterranean input is well noticeable for stations 110 and 41 with a high salinity responsible for a tritium decrease. On this level, we observe the greatest North-South assymetry due to the imprint of the Antarctic Intermediate Water characterized by a low salinity and a low tritium content (see stations 6, 12 and 16 on Figure 6).

At intermediate depths (27.1 isopycnal level), the tritium data permit to quantify the transit time of the waters. The Northeast-Southwest assymetry of the tritium distributions (Figure 3) suggests that the gyre scale circulation has dominated the redistribution process during the twenty years duration of the transient. We can evaluate from the tritium content of respectively 4.6 and 3 TU for the stations 110 and 12 (Figure 3) a transit time of around 7.5 years assuming an isopycnal advection process. This leads to an advection rate of 1cm/s.

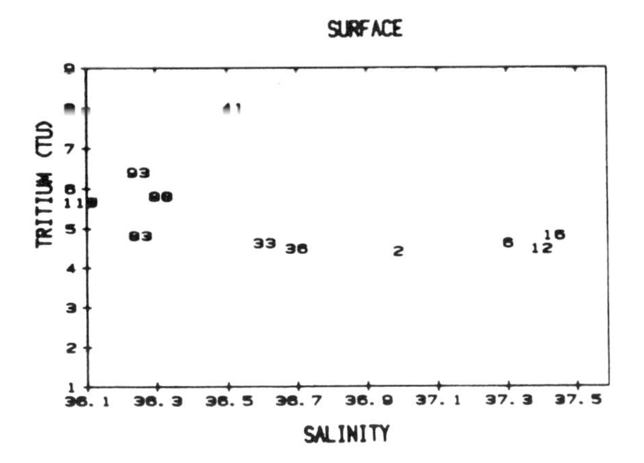

Figure 4 : Tritium versus salinity diagram for the surface waters.

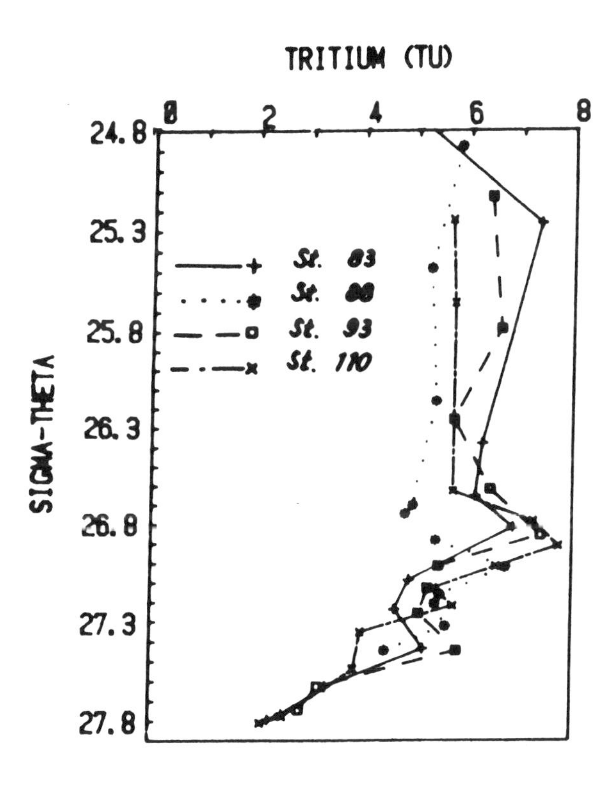

Figure 5 : Tritium versus sigma-theta distributions of the Northern stations.

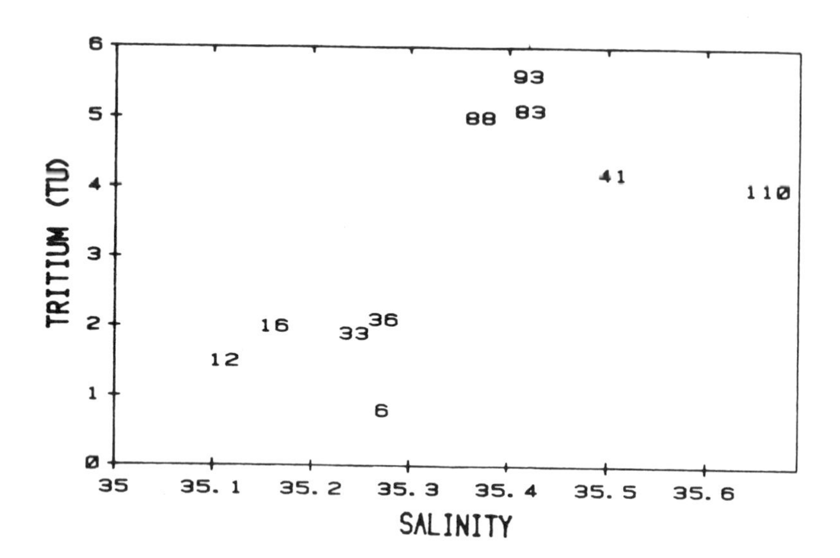

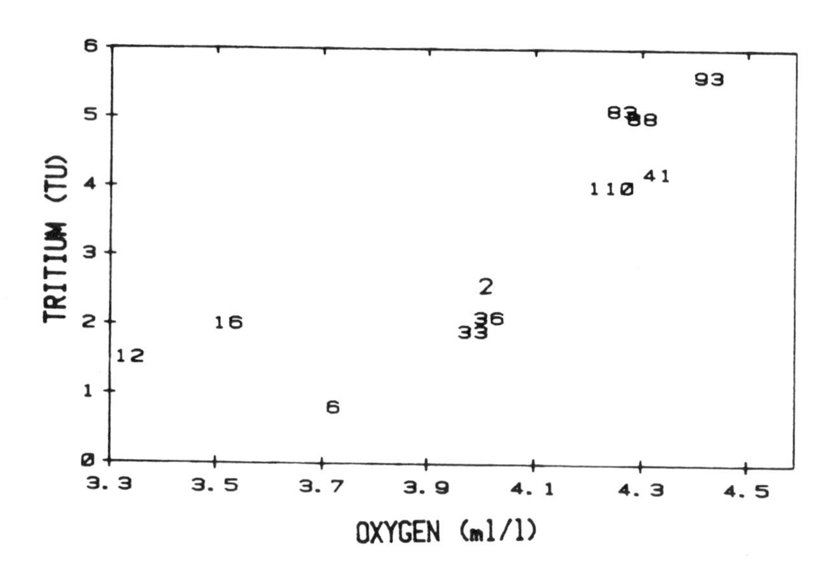

Figure 6 : Tritium versus salinity and tritium versus oxygen diagrams for the 27.4 isopycnal level. Numbers plotted are station numbers

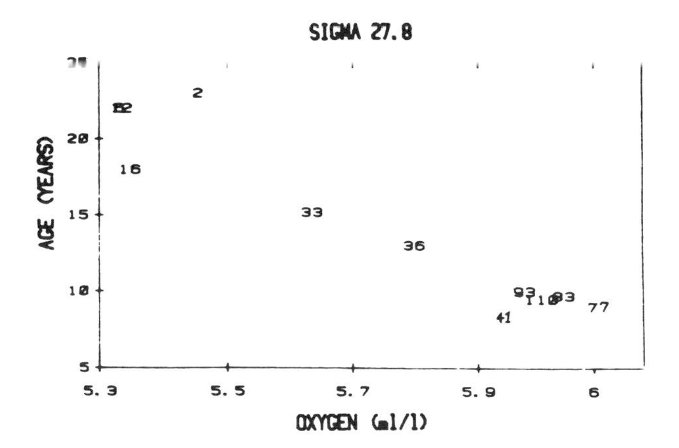

Figure 7 : Age versus oxygen diagram for the 27.8 isopycnal level. The young waters (age ≃ 10 years) of oxygen enriched content originate from the Labrador Sea. The old waters (stations 2, 6, 12, 16) of low oxygen are "stagnant" waters.

CONCLUSION

Finally, the simultaneous use of tritium and helium data gives important information relative to the thermocline ventilation.

The "age" of each sample, determined by its tritium and helium-3 contents, is an useful tool to characterize the ventilation degree of the waters.

For the waters located just above the 27.8 isopycnal level we identify (Figure 3) two kinds of water masses : young water masses about 10 years old in the North of the Eastern and Western sections, old water masses at least 20 years old in the South of the Eastern section.

The young waters are of Labrador origin while the old ones correspond to stagnant waters corresponding to the previously named "shaded area" water (Figure 7). Our evaluation of the age of the LSW in the Northern area (of around 10 years) agrees with the volumetric calculation of TALLEY and Mc CARTNEY (1982) giving a ventilation time for the LSW of around 9 years.

REFERENCES

BENSON B.B. and KRAUSE D. Jr (1980). Isotopic fractionation of helium during solution : a probe for the liquid state. J. Solut. Chem. 9 (12), 895-909.

DREISIGACKER E. and ROETHER W. (1978). Tritium and ^{90}Sr in North Altantic surface water. EARTH planet. Sci. Lett. 38, 301-312.

FUCHS G., ROETHER W. and SHLOSSER P. (1987). Excess ^{3}HE in the ocean surface layer. J. Geophys. Res. In press.

JEAN-BAPTISTE P., ANDRIE C. and LELU M. (1987). Mesure du couple tritium - helium océanique par spectrométrie de masse (this issue).

JENKINS W.J., (1980). Tritium and Helium-3 in the Sargasso, Mar. Res. 38 (3), 533-569.

KAWASE M. and SARMIENTO J.L. (1985). Nutrients in the Atlantic thermocline. J. Geophys. Res. 90 (C5), 8961-8979.

KAWASE M. and SARMIENTO J. L. (1986). Circulation and nutrients in middepth Atlantic waters. J. Geophys. Res 91 (C8), 9749-9770.

SARMIENTO J. L., ROOTH C.G.H. and ROETHER W. (1982). The North Atlantic tritium distribution in 1972. J. Geophys. Res. 87 (C10), 8047-8056.

TALLEY L. D. and Mc CARTNEY M. S. (1982). Distribution and circulation of Labrador Sea water. J. Physical Oceanogr. 12, 1189-1205.

THIELE G., ROETHER W., SCHLOSSER P., KUNTZ R., SIEDLER G. and STRAMMA L. (1986). Baroclinic flow and transient - tracer fields in the canary-cape-verde basin. J. Physical Oceanogr. 16, 814-826.

TOPOGULF Group (1986). TOPOGULF : A joint programme initiated by IFREMER, Brest (France), IFM, Kiel (W. Germany. Data Report Vol. 1, Institüt für Meereskunde, Kiel, 183 pp.

MESURE DU COUPLE TRITIUM/HELIUM OCEANIQUE PAR SPECTROMETRIE DE MASSE

P. Jean-Baptiste, C. Andrié et M. Lelu

Laboratoire de Géochimie Isotopique - LODYC
(UA CNRS 1206) - C.E.A. - IRDI/DESICP -
Département de Physico Chimie
91191 GIF sur YVETTE CEDEX, FRANCE

RESUME

Nous décrivons la technique expérimentale de mesure des isotopes de l'hélium (^3He et ^4He) dans l'eau de mer par spectrométrie de masse ainsi que la méthode de dosage du tritium océanique par analyse de l'hélium-3 de décroissance. Les protocoles analytiques sont reportés ainsi que la précision de chacune des méthodes. Enfin, l'intérêt de l'acquisition du couple de données hélium-3/tritium est explicité au-travers de la détermination de "l'âge" d'une masse d'eau.

ABSTRACT

We describe the experimental procedure of the mass spectrometric measurements of the dissolved helium isotopes (^3He and ^4He) in seawater and the analytical method of the oceanic tritium determination by the β-regrowth technique. The experimental procedures and the analytical accuracies of each method are reported. Finally, the interest of the helium-3/tritium pair is emphazised through the determination of a water mass "age".

INTRODUCTION

Le tritium et son descendant, l'hélium-3, connaissent une utilisation croissante en océanographie. De par leur caractère chimiquement inerte vis-à-vis du milieu océanique et de par l'aspect transitoire de l'injection du tritium dans l'océan, le couple ^3He-^3H constitue une source unique d'information par rapport aux traceurs à l'état stationnaire.

La spectrométrie de masse permet d'accéder à des mesures de tritium et d'hélium-3 suffisamment précises pour pouvoir évaluer "l'âge" d'une masse d'eau c'est-à-dire le temps écoulé depuis qu'elle a quitté l'interface océan-atmosphère.

Nous décrivons le principe de la méthode, les procédures analytiques relatives aux mesures du rapport isotopique $^3He/^4He$ et du tritium ainsi que la précision de la mesure et la reproductibilité des résultats obtenus.

Nous avons appliqué cette méthode en Mer Méditerranée Occidentale (PHYCEMED 1981) [5] ainsi qu'en Atlantique Nord-Est pendant la campagne TOPOGULF 1983 [3 et 4].

PRINCIPE DE LA METHODE

Le principe de la mesure du tritium (Fig. 1) consiste à stocker dans un container étanche une eau préalablement totalement dégazée et à mesurer après stockage par spectrométrie de masse la quantité d'hélium-3 produite par la décroissance du tritium d'après la réaction $^3H \rightarrow {}^3He^+ + e^-$. La technique utilisée est dérivée de celle mise au point par CLARKE et al [1] et JENKINS [2]. Les échantillons (40 cm^3 d'eau de mer) sont prélevés dans des tubes de cuivre en faisant circuler par gravité l'eau s'écoulant des bouteilles Niskin. Chaque extrêmité du tube est alors fermée par une pince de frigoriste. Le taux de fuite de ce système a été mesuré par ressuage d'hélium. Il est équivalent aux meilleures vannes métalliques à soufflet (quelques 10^{-9} atm.cm^3/s). L'étanchéïté totale vis-à-vis d'une éventuelle contamination des échantillons par échange isotopique avec la vapeur d'eau tritiée atmosphérique a également été vérifiée par des tests en milieu deutéré et mesure du rapport D/H de l'échantillon par spectrométrie de masse.

Dans la pratique, au laboratoire, le même échantillon permet dans un premier temps la mesure du rapport isotopique $^3He/^4He$ par extraction des gaz dissous et par la suite, celle du tritium.

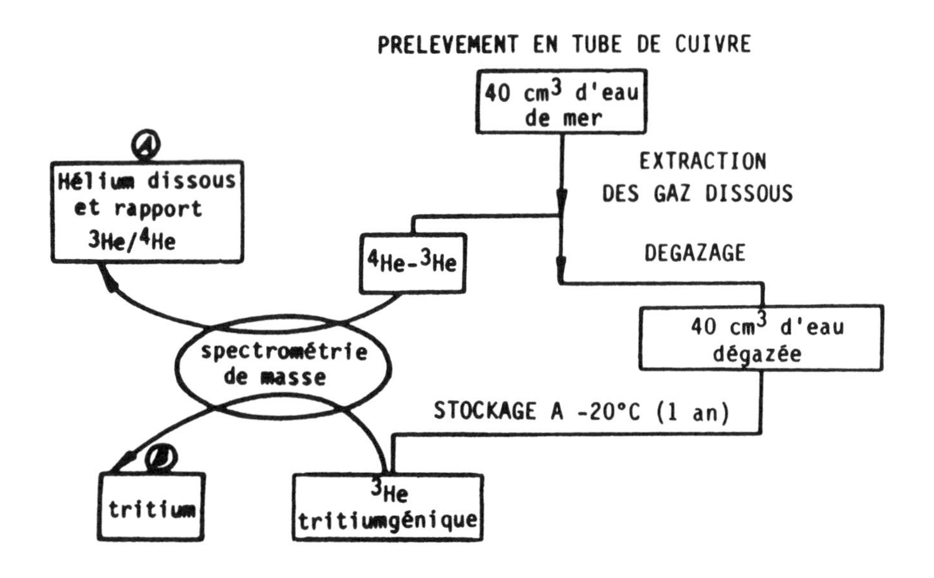

Fig. 1 : Principe de la méthode
(Analytical procedure)

EXTRACTION DES GAZ DISSOUS

La phase d'extraction de l'hélium s'effectue sur une ligne préalablement pompée à un vide secondaire (10^{-5}Torr) (Figure 2).

L'échantillon (tube de cuivre) est placé sur la ligne équipée d'un ballon en verre destiné à recueillir l'eau et d'une ampoule de verre contenant des charbons actifs où les gaz dissous seront récupérés. L'ensemble d'extraction utilisé comprend 6 lignes de ce type. L'installation est étuvée (2 h à 160°C) et pompée pour atteindre une pression inférieure à 10^{-5} Torr. Au moment de l'extraction, chaque ligne est isolée du groupe de pompage. L'extrêmité inférieure du tube de cuivre est alors réouverte permettant ainsi à l'eau de s'écouler dans le ballon. Une cuve à ultrasons est ensuite placée sous le ballon qui servira à accélérer le dégazage de l'eau. A l'autre extrêmité de la ligne, l'ampoule est placée dans un bain d'azote liquide. L'hélium, ainsi que l'ensemble des gaz dissous s'y transfèrent : la majeure partie des gaz s'adsorbe sur les charbons à 77 K. L'hélium, qui n'est pratiquement pas piégé sur les charbons actifs, est transféré dans l'ampoule par le flux de vapeur d'eau qui s'établit entre l'échantillon d'eau et l'ampoule, où cette vapeur d'eau se congèle. Ce système fonctionne à la manière d'une pompe à diffusion. La présence d'un capillaire à l'entrée de l'ampoule empêche toute rétrodiffusion de l'hélium.

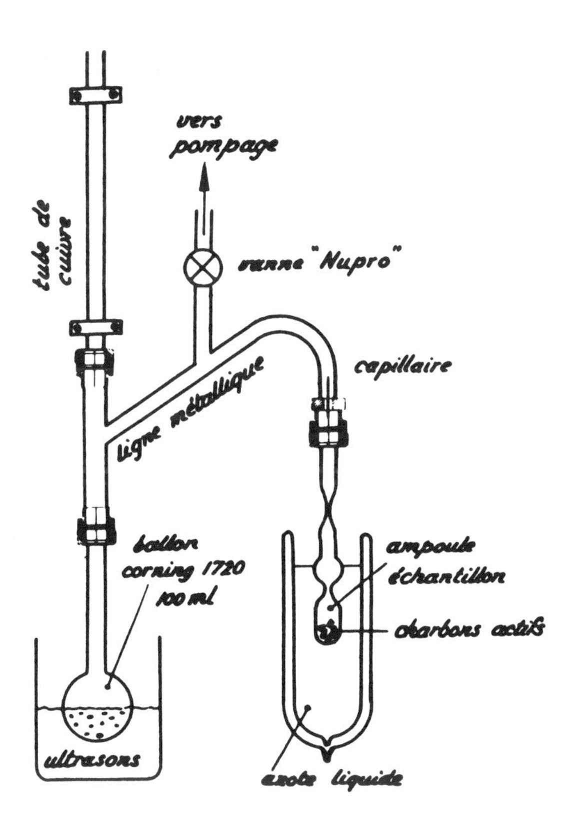

Fig. 2 : Ligne d'extraction des échantillons d'eau de mer
(*Extraction device for dissolved gases in seawater samples*)

Le taux de récupération de l'hélium a été évalué à partir de l'hélium résiduel dans l'échantillon d'eau et dans la ligne d'extraction. Il est supérieur à 99 %.

Après 15 mn d'extraction, l'ampoule est scellée au chalumeau et la ligne est réouverte sur le groupe de pompage. L'eau contenue dans le ballon subit alors un dégazage final pendant quelques minutes puis le ballon est à son tour scellé au chalumeau. L'échantillon d'eau, totalement dégazé, est stocké pendant la durée nécessaire à une accumulation suffisante d'hélium-3, compte-tenu de la limite de détection du spectromètre de masse soit en pratique une durée de l'ordre de 6 mois à 1 an.

PROCEDURE ANALYTIQUE DE LA MESURE DES QUANTITES D'HELIUM OCEANIQUE ET DU RAPPORT $^3He/^4He$:

L'appareil utilisé est un spectromètre de masse VG 3000 à secteur magnétique (90°) et double collection conçu pour la mesure simultanée des isotopes de l'hélium $^3He/^4He$.

Le spectromètre, muni d'une source à haute sensibilité (type Nier) fonctionne en mode statique.

Le courant ionique $^4He^+$, typiquement 10^9 ions/sec pour 40 cm^3 d'eau, est recueilli dans une cage de Faraday tandis que les ions $^3He^+$ (1000 ions/sec) sont collectés simultanément sur un multiplicateur d'électrons (ETP - AEM/1000) utilisé en mode impulsionnel.

La figure 3 décrit la ligne de préparation et d'introduction des échantillons : après la mise en place de l'ampoule échantillon sur la ligne, celle-ci est pompée jusqu'à une pression résiduelle inférieure à 10^{-5} Torr. La ligne est alors isolée des groupes de pompage (pompes à diffusion et ionique) et l'ampoule est cassée. L'échantillon gazeux est d'abord transféré sur un piège à vapeur d'eau (77 K) (1 mn) puis sur un piège en U à charbons actifs à 77 K (5 mn) avant l'introduction dans le spectromètre. Ainsi, des gaz dissous dans l'échantillon d'eau de mer initial, seuls sont admis dans le spectromètre l'hélium et une fraction du néon.

Les quantités d'hélium 3 et 4 des échantillons sont déterminées par comparaison avec des standards d'air atmosphérique de rapport isotopique connu ($R_A = 1.384.10^{-6}$). Le volume de l'aliquote d'air utilisé peut être ajusté afin de renfermer une quantité d'hélium approximativement égale à celle contenue dans les échantillons d'eau de mer (environ $1.6 \cdot 10^{-6}$ cm^3 TPN d'hélium-4 dans 40 g d'eau). Ceci permet de minimiser les corrections de non-linéarité dans la réponse du spectromètre en fonction des quantités de gaz analysées. Chaque analyse (standard ou échantillon) consiste en une série de 50 comptages de 5 secondes des courants collectés $^4He^+$ et $^3He^+$ et une série de 20 comptages sur la ligne de base. Le rapport isotopique est exprimé en δ % c'est-à-dire par la déviation en pourcentage au rapport isotopique de l'air

$$\delta \% = \left[\frac{(^3He\ /\ ^4He)_{ech}}{(^3He\ /\ ^4He)_{air}} - 1 \right] \times 100$$

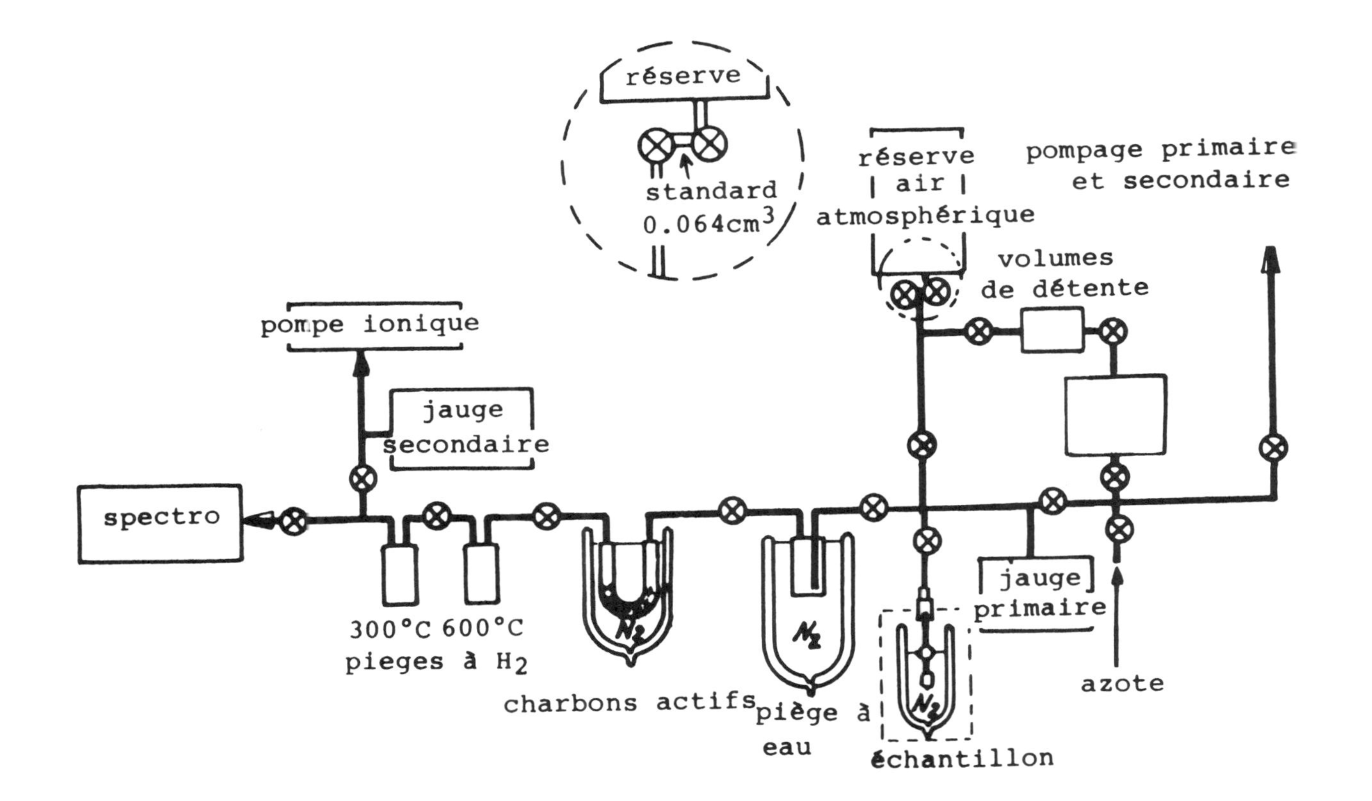

Fig. 3 : Ligne d'introduction des échantillons vers le spectromètre de masse
(*Introduction device for the gas samples into the spectrometer*)

avec $(^3He/^4He)_{air} = R_a = 1.384 \cdot 10^{-6}$.

La figure 4 montre un exemple d'enchaînement de standards et d'échantillons obtenus pour une journée. L'ensemble de la procédure de mesure -introduction de l'échantillon ou du standard dans la ligne de préparation, déclenchement des vannes pneumatiques, centrage des pics, acquisition des données- est automatisée : ce pilotage de l'installation est effectué par un micro-ordinateur HP 86. Ceci permet un gain de temps appréciable au niveau des analyses et une meilleure reproductibilité des mesures. En routine, l'erreur statistique sur la mesure de δ % est égale à ± 0.15 %. La reproductibilité des mesures, évaluée à partir de l'écart-type de la distribution du rapport isotopique $^3He/^4He$ des standards sur une journée de mesure, est de l'ordre de ± 0.2 % (Tableau 1).La reproductibilité globale sur la mesure de δ % testée sur 10 échantillons provenant de la même eau est de ± 0.4 %. Le Tableau 1 indique également les performances au niveau des quantités totales d'hélium.

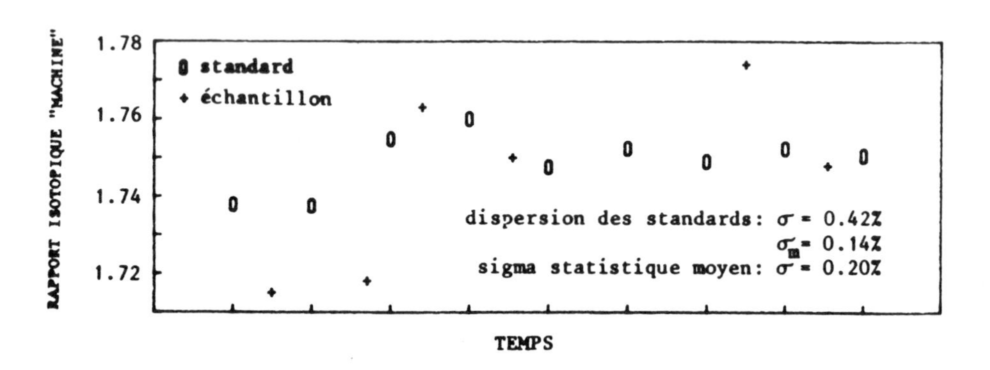

Fig. 4 : Exemple d'enchaînement des rapports isotopiques de standards et d'échantillons pour une journée. (*Plot of isotopic ratios of standard and samples over one day(an example)*).

Tableau 1 : Incertitude sur la mesure de δ % (en valeurs absolues)

erreur statistique	reproductibilité des standards	reproductibilité globale
± 0.15 %	± 0.2 %	± 0.4 %

Incertitude sur la mesure de l'hélium total

precision statistique	reproductibilité des standards
0.1 %	0.3 %

PROCEDURE ANALYTIQUE DE LA MESURE DU TRITIUM :

Pendant le stockage de l'eau de mer dégazée, le tritium se désintègre en donnant un atome d'hélium-3 suivant la réaction

${}^{3}_{1}H \longrightarrow {}^{3}_{2}He + {}^{0}_{-1}e^{-}$. Le temps de demi-vie du tritium est 12.427 ans.

Les teneurs en tritium sont couramment exprimées à partir du rapport isotopique T/H. Une unité tritium (UT) correspond à un rapport T/H = 10^{-18}, ce qui équivaut, dans le cas d'eau tritiée, à une concentration de 1.11 . 10^{-16} moles HTO/kg d'eau soit une activité de 0.12 Bq/kg.

Pour un échantillon stocké pendant une durée t, le nombre d'atomes d'hélium-3 générés, N_{He-3}, est lié au nombre de molécules HTO présentes au temps initial par la simple relation de décroissance radioactive :

$$N_{He-3} = N_{HTO}(1 - e^{-\lambda t})$$

A titre d'exemple, un échantillon de 40 cm^3 à 1 UT stocké un an produit 5.4 x $10^{-15} cm^3$TPN d'hélium-3.

L'analyse de l'hélium-3 de décroissance au spectromètre de masse s'effectue suivant une procédure semblable à celle exposée au paragraphe précédent : le ballon contenant l'échantillon est placé sur la ligne d'introduction du spectromètre comme précédemment. La séquence de passage dans les différentes parties de la ligne de préparation est identique. Dans ce cas, les quantités d'hélium-3 à mesurer étant nettement moindres (entre 100 et 1000 fois plus faibles), les standards utilisés sont obtenus à partir du même aliquote d'air que précédemment auquel on fait subir un certain nombre de détentes dans des volumes adaptés. Cette procédure permet d'obtenir différents types de standards dont la quantité d'hélium-3 couvre une gamme d'échantillons de teneur en tritium comprise entre 0.05 UT et 10 UT.

L'analyse consiste en 10 séries de 12 comptages de 5 secondes sur les pics alternées avec 10 comptages sur la ligne de base.

Idéalement, l'échantillon ne devrait contenir que de l'hélium-3. Dans la réalité, on observe également un pic résiduel d'hélium-4 qui correspond au "blanc" de la méthode. La nature de ce blanc a été étudiée précisément et son origine est détaillée sur la figure 5. Il apparait que la majeure partie provient du résidu de dégazage de l'eau de mer initiale : la valeur de cette contribution correspond à un résidu non dégazé de l'ordre de 0.25 % de la quantité totale d'hélium dissous. La contribution dûe à la diffusion de l'hélium à travers le verre du ballon est considérablement limitée par l'utilisation d'un verre spécial (CORNING 1724), particulièrement étanche à l'hélium. Des expériences de diffusion effectuées à partir de ballons vides scellés et recuits à différentes températures (cf Tableau 2) ont permis de vérifier le coefficient de perméation de l'hélium.

(1)	(2)	(3)	(4)	(5)

Composition du blanc Ra.^{4}He (unité : ccNTP d'hélium-3)
(1) blanc du spectro seul = $1.5.10^{-16}$
(2) contribution de la ligne d'introduction = 3.10^{-16}
(3) contribution de la ligne de dégazage = 3.10^{-16}
(4) résidu de dégazage (à t=0) = $1.5.10^{-15}$
(5) résidu de dégazage (après stockage) = 6.10^{-15}

Fig.5: Diagramme des différentes contributions en hélium au "blanc" de la méthode lors des analyses tritium. (*Different contributions to the helium blank during tritium analyses*).

Tableau 2 : Expériences de diffusion de l'hélium dans du verre type CORNING 1724 (ballon recuit sous atmosphère d'azote : épaisseur : 1.6 mm, diamètre = 60 mm).

Température (°C)	durée du recuit (heures)	quantité d'hélium mesurée (cm^3TPN d'He-4)	Coefficient de diffusion (cm^2/s)
617	17	1.02×10^{-7}	6.9×10^{-7}
617	49	1.03×10^{-7}	6.9×10^{-7}
440	8	6.40×10^{-8}	1.0×10^{-7}
330	31	5.39×10^{-8}	1.8×10^{-8}

A - 20°C (température de stockage des échantillons), cette contribution au "blanc", évaluée à partir des expériences ci-dessus est totalement négligeable (de l'ordre de 0.8×10^{-17} cm^3 TPN d'hélium-3).

Il a été vérifié d'autre part que le rapport isotopique du "blanc" était sensiblement égal au rapport atmosphérique Ra, à la précision des mesures près. Ceci permet, à partir de la mesure des quantités totales d'hélium-3, He-3, et d'hélium-4, He-4, dans le ballon, de déduire la quantité d'hélium-3 générée par la désintégration du tritium :

$$\text{He-3}_{\text{Tritium}} = \text{He-3} - \text{Ra} \times \text{He-4}$$

La précision des mesures dépend essentiellement de la statistique de comptage de l'hélium-3, cette dernière étant elle-même liée au bruit de fond du multiplicateur d'électrons (courant d'obscurité). Les performances actuelles de l'installation sont reportées sur le Tableau 3.

Tableau 3 : Incertitude sur les mesures de tritium (à partir de 40 cm^3 d'eau stockés pendant au moins 6 mois)

limite de détection ^3He	erreur statistique	reproductibilité des standards	reproductibilité globale
$1.10^{-16} cm^3$ d'He-3 TPN ou 0.04 UT	0.05 UT	0.1 UT	0.2 UT

La dernière colonne indique la reproductibilité globale de la méthode, testée sur 5 échantillons d'une eau identique, de teneur en tritium égale à 2 UT et stockés environ six mois.

Notre précision actuelle est donc comparable aux meilleures installations de comptage bas-niveau. Indépendamment des améliorations envisagées de la technique de mesure au spectromètre de masse, cette méthode renferme en elle-même la possibilité d'atteindre des limites de détection encore plus basses, en particulier par l'accroissement de la taille des échantillons.

APPLICATION DE LA METHODE A LA DETERMINATION DE L'AGE D'UNE MASSE D'EAU :

Par la nature même de leur filiation radioactive le couple des traceurs tritium/hélium-3, en l'absence d'hélium-3 hydrothermal, constitue un outil de choix pour accéder à l"âge" d'une parcelle d'eau, c'est-à-dire au temps écoulé entre la date où elle a quitté la surface de l'océan et la date du prélèvement

$$\tau = \lambda^{-1} \log \left[1 + \frac{(^3He)}{(^3H)} \right]$$

où (^3He) et (^3H) sont les quantités respectives d'hélium-3 (d'origine tritiumgénique) et de tritium dans la parcelle d'eau considérée.

Cette méthode permet d'abaisser sensiblement le seuil de détermination de l'âge d'un échantillon proche de la surface : des âges de l'ordre du mois peuvent être ainsi évalués alors que la mesure du tritium seul ne pourrait permettre d'accéder à un temps d'isolement inférieur à 4 mois compte-tenu de notre limite de détection actuelle (0.1 à 0.2 TU).

De façon générale, l'"âge" déterminé par la méthode $^3H/^3He$ ne doit pas être considéré comme l'âge réel d'une masse d'eau. Il est, en effet, très affecté par les processus de mélange. Cependant, les deux exemples qui suivent sont une illustration de la méthode décrite ci-dessus et montrent l'intérêt de l'utilisation de l'âge $^3H/^3He$ dans des conditions particulières :

Le couple tritium/hélium-3 est bien adapté à l'étude de la

circulation générale en Atlantique Nord. Celle-ci est essentiellement gouvernée par le tourbillon anticyclonique subtropical, les eaux étant alimentées en tritium lors des effleurements d'hiver essentiellement par des processus d'advection le long des isopycnes. C'est ainsi que nous avons mis en évidence une nette assymétrie Nord Est-Sud Ouest dans l'âge des eaux thermoclinales de part et d'autre des Açores [3, 4]. Cette assymétrie est le résultat d'un équilibre entre les processus de ventilation et la pénétration du tritium en profondeur (importants au Nord Est et faibles au Sud Ouest). Ce travail conduit à deux valeurs extrêmes caractérisant les échelles de temps de la circulation des eaux intermédiaires dans la zone étudiée : 35 ans pour le temps de ventilation des eaux du Labrador et 200 ans pour les eaux les plus vieilles situées au Sud-Est de la zone.

Enfin, les données hélium-3 et tritium des campagnes PHYCEMED en Méditerranée permettent de calculer pour les eaux profondes du Golfe du Lion un âge tritium-hélium de 12 ans. Ce résultat est en accord avec celui obtenu en ne considérant que les données tritium et un modèle vertical d'advection - diffusion qui évaluait à 11 ans le temps de renouvellement des eaux profondes [5].

BIBLIOGRAPHIE

[1] W.B. CLARKE, W.J. JENKINS and Z. TOP : Determination of tritium by mass spectrometric measurements of $^{3}He^{+}$. International Journal of Applied Radiation and Isotopes, (1976) 27, pp. 515-522.

[2] W.J. JENKINS : Mass spectrometric measurements of tritium and ^{3}He.Proc. consult. group - Meeting on low-level tritium measurements (IAEA-Vienna) (1981) pp. 179-189.

[3] C. ANDRIE and L. MERLIVAT : Use of tritium and helium-3 for oceanographic processes study. An example : The Northeastern Atlantic Ocean (ce volume).

[4] C. ANDRIE, P. JEAN BAPTISTE and L. MERLIVAT : Tritium and Helium-3 in the Northeastern Atlantic Ocean during the 1983 Topogulf cruise. J. of Geophysical Research, 1988 (à paraître).

[5] C. ANDRIE and L. MERLIVAT : Tritium in the Western Mediterranean Sea during 1981 Phycemed cruise. Deep-Sea Research, (1988), 35(2) ; 247-267.

LES ISOTOPES DU BERYLLIUM DANS LE MILIEU MARIN ET QUELQUES APPLICATIONS

F. Yiou, G.M. Raisbeck and D. Bourlès
Centre de Spectrométrie Nucléaire et de Spectrométrie de Masse
Bât. 108 - 91406 CAMPUS ORSAY - FRANCE

RESUME

In addition to the stable isotope, ^{9}Be, two radioactive isotopes ^{7}Be (half-life 53.1 days) and ^{10}Be (1.5 million years) are input to the ocean. ^{9}Be is transported to the ocean by erosion (rivers, wind). ^{7}Be and ^{10}Be are formed by nuclear interactions of cosmic rays with the earth's atmosphere, and transported to the ocean mainly in precipitation. The technique of accelerator mass spectrometry (AMS) now makes it feasible to measure ^{10}Be in $\sim$ 10 l of ocean water, and < 1g of ocean sediment. This opens up a number of applications of ^{10}Be, such as dating marine sediments, and as a proxy indicator of the past changes in cosmic ray intensity, which are being pursued by our group. Some measurements of ^{7}Be in ocean water have been made previously, using traditional radioactivity decay techniques, although the sample volumes required were substantial (> 10^{3} l). We report at this meeting a newly developed AMS technique for ^{7}Be, which has a detection limit $\sim$ 100 times lower than with decay counting. The greater sensitivities offered by AMS open up the possibility of using $^{10}Be/^{9}Be$ and $^{7}Be/^{10}Be$ as tracers of large scale circulation and rapid particule scavenging in the ocean respectively. Such applications however, require a substantially better knowledge of beryllium ocean geochemistry than presently exists. We present here a broad outline of the potential applications and problems, including our present understanding of beryllium behavior in the ocean.

INTRODUCTION

Un certain nombre de propriétés spécifiques au beryllium, confèrent à cet élément un intérêt tout particulier en géochimie marine.
Le beryllium est amphotère, insoluble pour des pH 6-9. Dans l'eau de mer, il va donc avoir tendance à se fixer sur les particules.
Le beryllium possède 3 isotopes utilisables en géochimie marine: ^{9}Be stable, ^{7}Be radioactif (période de 53,17 j) et ^{10}Be ($1,5x10^{6}$ ans).
Les isotopes radioactifs sont des isotopes cosmogéniques, c'est à dire fabriqués par le rayonnement cosmique. Leur apport dans l'océan a une origine extérieure à l'océan, et est relativement constant dans le temps et dans l'espace. Le ^{10}Be dans l'océan peut être considéré comme stable tandis que le ^{7}Be va servir d'horloge pour des temps < 1 an.
L'ensemble de ces propriétés étaient jusqu'à récemment encore très peu exploitées. Cela était dû principalement à la rareté de ces isotopes et aux difficultés de leur détection, en particulier celle de ^{10}Be. La situation a beaucoup évolué ces dernières années, grâce au développement d'une nouvelle technique, dite technique de SMA ou Spectrométrie de Masse par Accélérateur (1,2).

Cette présentation a pour but de donner une vue d'ensemble des travaux réalisés par notre groupe sur les isotopes du beryllium en général et dans le milieu marin en particulier.
Avant de décrire l'intérêt de ces 3 isotopes, quelques mots seront dits sur la façon de les mesurer.

MESURE DES ISOTOPES DU BERYLLIUM

Le ^{9}Be, stable, est mesuré par absorption atomique sans flamme dans un four de graphite (3), ou par chromatographie gazeuse (4).

Le ^{7}Be, décroît avec une période de 53,17 j; seulement 10.4% de sa radioactivité produit un γ caractéristique de 477 keV mesurable dans des détecteurs Ge (Li) et NaI. Il a été mesuré par exemple, dans l'eau de mer en utilisant des détecteurs NaI (5).

Le ^{10}Be, décroît par émission β avec une période de $1,5x10^{6}$ ans. La détection de cet isotope par sa seule radioactivité β nécessiterait des tonnes d'eau de mer. Nous avons mis au point depuis plusieurs années une technique qui permet de s'affranchir de ce problème, en comptant les atomes, un par un; c'est la Spectrométrie de Masse par Accélérateur mentionnée précédemment. L'originalité de cette technique consiste à utiliser un accélérateur de particules. Les techniques de physique nucléaire des ions lourds peuvent ainsi être utilisées pour identifier les noyaux de ^{10}Be et compter un par un les atomes de ^{10}Be. L'ensemble de ces techniques réussit la performance de séparer ces ^{10}Be d'un bruit de fond 10^8 à 10^{15} fois plus grand constitué de l'isotope stable ^9Be et de l'isobare stable ^{10}B. Ce développement auprès d'accélérateurs de particules, cyclotrons, Tandems, s'est concrétisé par l'achat par la France (CNRS, CEA, IN2P3) d'un Tandétron, appareil conçu pour faire la SMA. Ce Tandétron est installé sur le campus du CNRS à Gif sur Yvette. Il est depuis deux ans, tout à fait opérationnel pour détecter les isotopes de ^{26}Al, ^{10}Be et ^{14}C. Sa sensibilité est de 1 million d'atomes de ^{10}Be (6). Plus de 1200 échantillons de ^{10}Be ont été mesurés dans des filtres d'air, pluie, eau de mer, sédiments lacustres, marins, glace antarctique, arctique, poussières cosmiques, coraux, plancton, etc.

PRODUCTION DES ISOTOPES DU BERYLLIUM

^9Be, isotope stable est incorporé comme tous les éléments stables, dans le système solaire au moment de sa formation. C'est un élément extrêment rare puisque $^9Be/Si \sim 10^{-6}$. On le trouve dans les roches où il se trouve parfois concentré dans certains minéraux, comme le beryl. Libéré par l'érosion, il est transporté vers l'océan soit dans les rivières, soit dans les poussières d'origine éolienne.

^{10}Be, ^7Be. L'histoire est toute différente pour les isotopes radioactifs ^{10}Be et ^7Be qu'on observe maintenant dans notre environnement terrestre; ces isotopes sont en effet formés au cours de réactions nucléaires induites par le rayonnement cosmique sur l'azote et l'oxygène de l'atmosphère.

Quel est le sort des isotopes cosmogéniques ^{10}Be et ^7Be: 70% sont formés dans la stratosphère (7); étant très réactifs ils se fixent rapidement sur les aérosols et suivent ainsi le trajet des aérosols. Par précipitation ils vont donc se retrouver à la surface des océans; de là ils subissent une série de processus physico-chimiques d'adsorption sur les particules et de dissolution qui les conduisent vers les sédiments marins où ils se trouvent incorporés définitivement.

Où retrouve-t-on les isotopes du beryllium: partout pour ^{9}Be, un peu partout aussi pour ^{10}Be et ^{7}Be, dans la mesure bien sûr où le réservoir ne dépasse pas 15 millions d'années pour ^{10}Be, 2 années pour ^{7}Be; c'est ainsi qu'on les retrouve dans l'air, la pluie, l'océan, les rivières, la glace des pôles, les sédiments marins et lacustres.

INTERETS DES ISOTOPES COSMOGENIQUES ^{10}Be et ^{7}Be

Nous les avons abordés suivant 4 axes différents:

- formés par le rayonnement cosmique irradiant l'atmosphère, ces isotopes (essentiellement ^{10}Be vue sa période), sont un moyen d'étudier les variations dans le passé de ce rayonnement cosmique et les phénomènes qui en sont la cause: la variation du rayonnement cosmique galactique lui-même, celle de l'activité solaire qui modifie le rayonnement cosmique entrant dans la cavité solaire et enfin la variation du champ géomagnétique.
- par sa période, ^{10}Be est un outil potentiel en datation, jusqu'à environ 15 millions d'années.
- ^{10}Be (et ^{26}Al) permettent l'identification et la caractérisation de la matière extraterrestre.
- ^{10}Be et ^{7}Be sont des traceurs atmosphériques et océaniques.

Ces 4 points sont abordés depuis plusieurs années par notre groupe (8), et toutes ces études sont réalisées actuellement auprès du Tandétron (6). Si le 4e point est directement lié à cette conférence, les 3 autres points peuvent par contre sembler s'en éloigner. Ils ont été cités en fait car ils présentent des liens très étroits avec le 4e point de cette étude. En effet, l'un des réservoirs où sont réalisées ces études sont les sédiments marins. Ainsi pour mettre en évidence et étudier les variations mentionnées précédemment, dans les sédiments, il est fondamental de connaître la résolution en temps que l'on peut atteindre, et pour cela connaître le temps de résidence du ^{10}Be dans l'océan depuis son introduction à la surface jusqu'à son incorporation définitive dans les sédiments. Cela nécessite de connaître le mode de transport du ^{10}Be, et déterminer le rôle du transport par particules.

Pour dater les sédiments, il ne suffit pas de mesurer la concentration du ^{10}Be, sujette aux fluctuations de la sédimentation. Il est nécessaire de mesurer ^{10}Be par rapport à l'isotope stable ^{9}Be, tout comme dans la datation par ^{14}C on mesure le rapport $^{14}C/^{12}C$. Dans le cas du carbone, il n'y a pas de problème d'homogénéisation; en effet dès sa formation le carbone forme un gaz CO_2 et le ^{14}C est en équilibre avec ses isotopes stables. Le cas du beryllium n'est pas aussi favorable car ^{10}Be et ^{9}Be n'arrivent pas de la même façon dans les océans. Le ^{10}Be pénètre dans l'océan en phase soluble et une partie va s'adsorber sur les particules. Le ^{9}Be est transporté par l'érosion ou introduit dans l'océan par les sources hydrothermales. Une partie se trouvera sous forme soluble tandis qu' une autre partie se trouvera incorporée dans les particules détritiques. Il n'y a donc aucune raison, a priori, pour que ces deux isotopes, soient mélangés de façon homogène et constante dans les sédiments: là réside la principale difficulté pour l'utilisation du beryllium en géochronologie. Pour cette raison, nous avons été amenés à isoler, dans les sédiments, les phases solubles où le beryllium a des chances d'être homogénéisé et donc utilisable pour la datation, de la phase détritique où se retrouve le ^{9}Be provenant de l'érosion.

Telles sont les raisons qui nous ont amenés à étudier le beryllium en géochimie marine; nous avons ainsi abordé un domaine en plein développement: l'étude des traceurs radioactifs dans l'océan.

ETUDE DU BERYLLIUM EN GEOCHIMIE MARINE

Etude de ^{9}Be et ^{10}Be dans les sédiments marins de surface

Nous avons élaboré une chimie sélective, par "leaching", permettant d'isoler les phases dites authigéniques, (organiques, carbonates, Opale, hydroxydes de Fe, Mn), de la phase dite détritique.

Ces résultats montrent que ^{10}Be est effectivement enrichi dans les phases authigéniques, alors qu'une fraction importante du ^{9}Be se retrouve dans la phase détritique (9).

L'une des conséquences de ces mesures est qu'il est nécessaire, pour utiliser le ^{10}Be dans les sédiments marins soit comme témoin des variations du rayonnement cosmique, soit comme horloge géologique, de mesurer le rapport $^{10}Be/^{9}Be$ dans les phases authigéniques des sédiments.

Etude du transport de ^{10}Be dans l'océan

Comme mentionné au début, le beryllium, insoluble aux pH de l'eau de mer, va avoir tendance à se fixer par adsorption sur les particules, biogéniques ou détritiques. Le transport du beryllium jusqu'au fond des océans va donc suivre le sort de ces particules. Cependant il faut noter que lorsqu'il y a dissolution des particules biogéniques, le ^{10}Be sera remis en solution, puis s'adsorbera sur de nouvelles particules. Il est donc fort probable que le ^{10}Be subisse ainsi plusieurs fois des cycles de dissolution, adsorption.

Nous avons cherché à savoir le rôle des particules dans le transport du ^{10}Be de deux façons différentes:
a) en mesurant le ^{10}Be à la fois dans l'eau de mer filtrée et dans les particules recueillies sur ces filtres (9,10,11)
b) en mesurant le ^{10}Be directement dans les particules recueillies dans les pièges à sédiments(12).

Les résultats de ces expériences montrent que même si la fraction de ^{10}Be associée aux particules est faible ($<5\%$) ces mêmes particules sont essentielles pour le transport du ^{10}Be vers le fond des océans.

Pour terminer, nous évoquerons l'intérêt du 3e isotope de la trilogie des isotopes du beryllium, le ^{7}Be, et en particulier le rôle du couple ^{7}Be - ^{10}Be, en géochimie marine.

Le couple ^{7}Be - ^{10}Be en géochimie marine

Le ^{7}Be, avec sa période de 53,17 jours, est en principe le traceur idéal pour étudier le trajet rapide des particules dans l'océan. Mais là encore, on se trouve confronté au même problème que celui mentionné pour le ^{10}Be comme outil de datation. La détermination d'un "âge", ici le temps de transport des particules, nécessite une mesure relative par rapport à un isotope stable. A ce point, la nature nous est particulièrement favorable; en effet ^{7}Be et ^{10}Be, sont des isotopes cosmogéniques. Leur origine est donc identique, extérieure à l'océan et leur apport relatif dans l'océan $^{7}Be/^{10}Be$ est connu. Le ^{10}Be, vue sa longue période, peut être considéré comm stable, pour les échelles de temps envisagées ici. Le rapport $^{7}Be/^{10}Be$ est donc un outil de datation potentiel en géochimie marine, en particulier pour l'étude du transport par particule.

Ainsi que nous l'avons mentionné précédemment, la technique de SMA permet la mesure de $\sim 10^{6}$ atomes de ^{10}Be. Par contre, la détection de ^{7}Be par comptage radioactif nécessite $\sim 10^{7}$ atomes. D'après les mesures de ^{7}Be (5) et ^{10}Be (13) dans l'eau de surface de l'océan, on peut estimer le rapport $^{7}Be/^{10}Be < 0,1$. Dans ces conditions il est évident que la détermination de ce rapport est limitée par les mesures de ^{7}Be.

Très récemment, nous avons réussi à mesurer ^{7}Be par SMA auprès du Tandétron de Gif. Ceci est une première mondiale, communiquée pour la première fois à ce congrès. La limite de détection de ^{7}Be est de presque un facteur 100 inférieure à la technique de mesure par radioactivité, avec un bruit de fond tout à fait négligeable (14). Une telle sensibilité permet d'envisager la mesure du rapport $^{7}Be/^{10}Be$ associé aux particules collectées par filtrage ou dans les pièges à sédiments. Cela va certainement donner un coup de fouet à l'exploitation des isotopes du beryllium comme traceurs radioactifs de l'océan.

REFERENCES

1. Raisbeck, G.M., Yiou, F., Fruneau, M. and Loiseaux, J.M., Beryllium 10 - Mass Spectrometry with a Cyclotron, Science, 1978, 202, 215-217

2. Raisbeck, G.M., Yiou, F., Bourlès, D., Lestringuez, J. and Deboffle D., Measurement of ^{10}Be with a Tandetron accelerator operating at 2MV, Nucl. Instr. and Meth., 1984, B5, 175-178

3. Raisbeck, G.M., Yiou, F. and Bourlès D., Evidence for an increase in cosmogenic ^{10}Be during a geomagnetic reserval, Nature, 1985, 315, 315-317

4. Measures, C.I. and Edmond, J.M., Determination of beryllium in natural waters in real time using electron capture detection gas chromatography, Anal. Chem., 1986, 2065-2069

5. Young, J.A. and Silker, N.B., Aerosol deposition velocities on the Pacific and Atlantic Oceans calculated from ^{7}Be measurements, <u>Earth Planet Sci. Lett.</u>, 1980, <u>50</u>, 92-104

6. Raisbeck, G.M., Yiou, F., Bourlès, D., Lestringuez, J. and Deboffle, D., Measurements of ^{10}Be and ^{26}Al with a Tandetron AMS Facility <u>Nucl. Instr. and Meth.</u> 1987, <u>B29</u>, 22-26

7. Lal, D. and Peters, B., Cosmic Ray produced radioactivity on the Earth, Handbuch der Physik, ed. K. Sitte, Springer, Berlin, 1967, pp 551-612

8. Raisbeck, G.M. and Yiou, F., Production of long lived cosmogenic nuclei and their applications, Nucl. Instr. and Meth., 1984, <u>B5</u>, 91-99

9. Bourlès, D., Etude de la géochimie de l'isotope cosmogénique ^{10}Be et de son isotope stable ^{9}Be en milieu océanique. Application à la datation des sédiments marins. Thèse - janvier 1988

10. Bourlès, D., Raisbeck, G.M., Yiou, F., Loiseaux, J.M., Lieuvin, M., Klein, J. and Middleton, R., Investigation of the possible association of ^{10}Be and ^{26}Al with biogenic matter in the marine environment, <u>Nucl. Instr. and Meth.</u>, 1984, <u>B5</u>, 365-370

11. Yiou, F., Raisbeck, G.M., Bourlès, D. and Bacon, M., Particle associated ^{10}Be in the ocean, présentation par Poster à cette Conférence et en préparation

12. Raisbeck, G.M., Yiou, F., Honjo, S., Klein, J. and Middleton, R., ^{10}Be fluxes from sediment traps, présentation orale à cette Conférence et en préparation

13. Raisbeck, G.M., Yiou, F., Fruneau, M., Loiseaux, J.M. and Lieuvin, M., ^{10}Be concentration and residence time in the Ocean surface layer, Earth. Planet. Sci. Lett., 1979, 43, 237-240

14. Raisbeck, G.M. and Yiou, F., Measurement of ^{7}Be by accelerator mass spectrometry, Eart. Planet. Sci. Lett. in press

DETERMINATION OF CARBON-14 BY ACCELERATOR MASS-SPECTROMETRY: OCEANOGRAPHIC APPLICATIONS.

J. C. Duplessy, E. Bard, M. Arnold and P. Maurice
Centre des Faibles Radioactivités
Laboratoire mixte CNRS-CEA
91198 Gif-sur-Yvette Cedex
FRANCE

ABSTRACT

Accelerator mass-spectrometry (AMS) enables the determination of $^{14}C/^{12}C$ ratios with only one mg of carbon, i.e. with samples 1000 times smaller than those required for conventional radiocarbon ß-counting. For sediment dating, radiocarbon ages can thus be measured on monospecific samples of 1000-2000 foraminifera picked by hand. This enables an age to be assigned to the same sample which is to be used in the oxygen isotope measurement or to the dominant species which drive sea surface paleotemperature estimates. For studying the oceanic circulation within the main thermocline and the carbon cycle, ^{14}C measurements can be obtained with water samples of only 100 ml, without major chemical treatment on board.

INTRODUCTION

Among cosmonuclides, ^{14}C with its half-life of 5,730 years has been widely-used for sediment dating and to explore the global ocean circulation in relation with the oceanic part of the carbon cycle. Until recent years, the $^{14}C/^{12}C$ ratio was only measured by the conventional ß-counting method, which provides an accurracy up to 4 per mil but requires samples as large as 1–5 g of carbon. This relatively large size of the samples was the cause of several difficulties in either sample collection or in interpretation of the geochemical significance of the analytical result.

The development of AMS in recent years permits now to characterise ^{14}C atoms by their mass and to count the $^{14}C/^{12}C$ ratio in samples as small as 1 mg of carbon with an accuracy of 1% for modern activities. The accuracy of this measurement should be improved in the near future. This dramatic reduction in the size of the samples required to make one analysis results in a burst of new applications. We shall describe in this paper some major advances in oceanography connected with the determination of ^{14}C in sediment and water samples using the Tandetron of Gif sur Yvette, a small tandem accelerator devoted to the measurement of cosmonuclides for geochemical studies in Earth's sciences. The routine procedure for AMS measurements has been described by Arnold et al. (1).

DEGLACIAL WARMING OF THE NORTH ATLANTIC

Over the last million years, the Earth's climate experienced major changes, with alternation of glacial and interglacial periods. These climatic variations are recorded in the variations of the fossil planktonic foraminiferal shells, which accumulate on the sea floor : On the one hand, as the composition of the fauna in ocean surface water depends mainly on the temperature, the composition of the fossil fauna in a deep sea core may be used to estimate the past sea surface temperature (2). On the other hand, continental ice-volume changes may be reconstructed by oxygen isotopic analysis of the same foraminiferal shells (3).

The timing of the deglaciation has important consequences relating to mechanisms of climate change. However, details of the climatic evolution of the ocean and of the continental ice-volume during the last glacial to interglacial transition are not recorded in most sediment cores because of the short duration of this event. Althrough the sediment material is deposited at a rate on the order of a few centimeters per thousand years, a major problem encountered when attempting to record such a rapid climatic change in deep sea records is created by the natural process of bioturbation: The activity of benthic organisms disturbs the original stratification of the deposited shells and mixes the upper few centimeters of sediment. As a result, the variations of the paleontological and isotopic compositions in the sedimentological record are both smoothed and disturbed.

Radiocarbon ages can be measured by A. M. S. on monospecific samples of 1,000-2,000 foraminifera picked by hand. This technique offers several advantages: First, the foraminiferal samples can be much more pure than those used for the classical method of ß-counting, which required samples so large that usually the carbonate fraction larger than 60 μ was analyzed. The presence within the North Atlantic sediment of ice-rafted carbonate, or of wind-, river- and current-transported carbonates (derived from old continental rocks), biased classical ^{14}C ages towards older values. Second, as stable isotopes are also measured on monospecific samples, this enables an age to be assigned to the same sample which is to be used in the $\partial^{18}O$ measurement (4). A simple deconvolution model can then be used to date precisely the internal stages of the deglaciation and to measure the volocity of the climatic changes in the North Atlantic Ocean (5).

Fig. 1 displays the oxygen isotope record measured in the foraminifer *Globigerina bulloïdes*, the ^{14}C ages measured in the same species and the sea surface temperature (S.S.T.) estimates in core SU 81-18 (37°46' N, 10°11' W). The sedimentation rate during the deglaciation is so high in this core (35 cm/kyr), that the bioturbation smoothing is negligible. The deglaciation (defined on the isotopic record) began about 14,500 years ago. The melting of continental ice first resulted in a S.S.T. drop of 6°C for both summer and winter S.S.T. The end of the first phase of the deglaciation, after 12,500 ± 150 B.P., was marked by a sharp temperature rise with a mean rate of roughly 4°C per century. This warm phase was followed by a dramatic cooling, well-known as Younger Dryas cold event (from 11,000 ± 170 B.P. to 10,400 ± 130 B.P.). The readvance of the cold surface water corresponds to a temperature drop of 0,5 to 1°C per century. The final S.S.T. warming, which led to modern conditions, was less abrupt than the first one with a mean rate of temperature rise close to 1°C per century until 9,360 ± 130 B.P..

Fig 2A displays the oxygen isotope records of both *G. bulloides* and *N. pachyderma* (left coiling) in core CH 73-139 C (54°38' N, 16°21' W). The interpretation of the isotopic date is more difficult than in core SU 81-18, because the bioturbational disturbance is not negligible. For example, the $\partial^{18}O$ values of *G.bulloides* exhibit

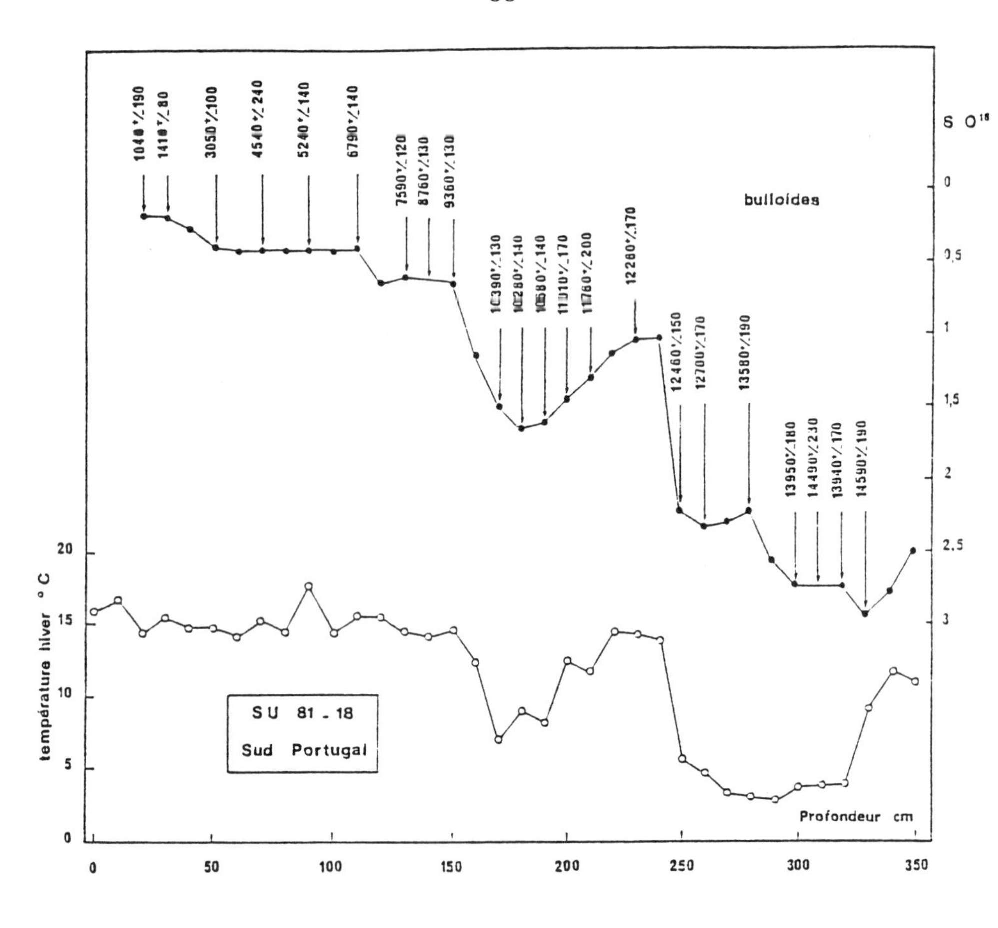

Fig. 1. - Oxygen isotope record of G. bulloides, ^{14}C ages of G. bulloides, and sea surface temperature estimates in core SU 81-18 off Portugal.

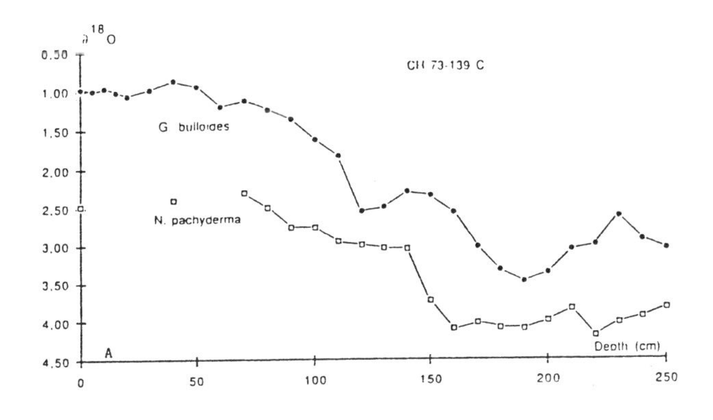

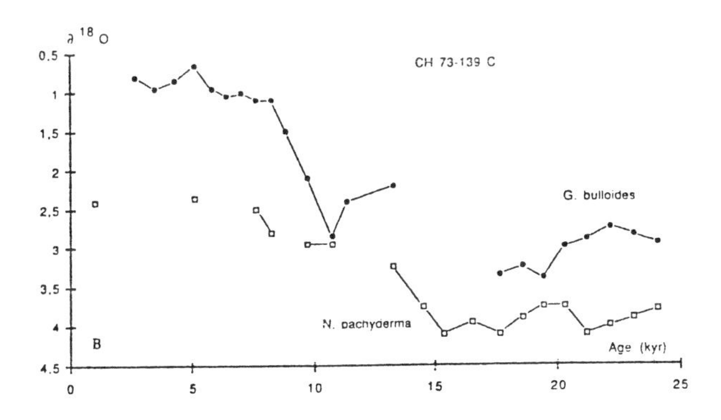

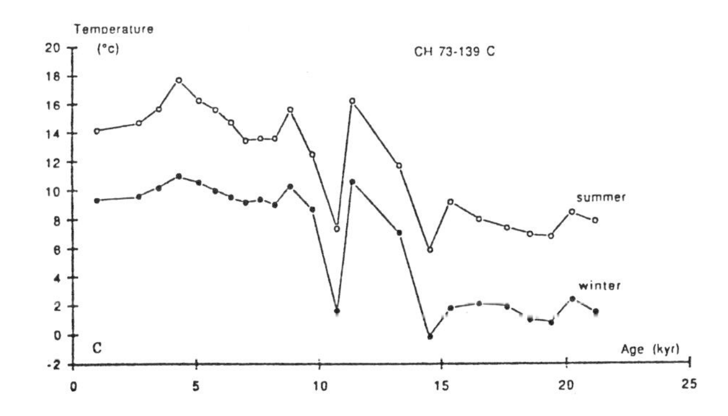

Fig.2 - A : oxygen isotope records of G. bulloides and N. pachyderma (left coiling) plotted vs depth in core CH 73-139 C off Ireland.

B : same record deconvolved according to the method of Bard et al (5) and plotted vs age.

C : sea surface temperature estimates deduced from faunal analysis in core CH 73-139 C plotted vs age after deconvolution of bioturbation mixing.

their first decrease at a depth 30 cm below that of *N. pachyderma*.. A.M.S. ^{14}C dates demonstrate that this 30 cm lead of the *G. bulloïdes* record over the *N. pachyderma* record is an artifact of bioturbation. The inconsistancy between the two isotopic records roughly disappears when both records are deconvolved and plotted with respect to the ^{14}C ages measured on the same foraminiferal species (Fig. 2B). This time scale places the beginning of the deglaciation between 15,000 yr B.P. and 14,500 yr B.P. and basically fits that of core SU 81-18.

The abundances of fourteen foraminiferal species have been deconvolved with the same bioturbation parameters which successfully explain the discrepancies between the isotopic records. The transfer function has been applied to these restored foraminiferal data in order to generate an unmixed S.S.T. record (Fig. 2C), which exhibits the same trends as the raw S.S.T. estimates but emphasizes the large amplitude of the temperature changes betwen 15,000 yr B.P. and 10,000 yr B.P..

The comparison of the paleoclimatic records of core CH 73-139 C and SU 81-18 may be used to estimate the rate of climatic change and the kinematics of the retreat of the cold water mass in the high latitude North Atlantic. The polar front is defined as the steep thermal gradient which separate the warm Gulf Stream waters from the cold polar water mass. The CLIMAP reconstruction (6) shows that core SU 81-18 is located in the southern part of the polar front during the last glacial maximum. Assuming that the passage of the polar front across the core will constitute the dominant thermal event within the general deglacial warming, A.M.S. ^{14}C ages of *G. bulloïdes* indicate that its retreat from the latitude of Portugal is dated at about 12,500 yr B.P. (7).

A.M.S. ^{14}C ages measured on G. *bulloides* between the levels 130 and 160 cm in core CH 73-139 C demonstrate that those levels correspond in fact to the dating of a single "pulse-like" abundance maximum for this species, of which shells have been bioturbated between 130 and 160 cm. The four radiocarbon ages are not statistically different and correspond to a mean value of 11,540 ± 170 yr B.P., which dates the invasion of warmer waters in the North Atlantic at 55°N. We therefore assume that this event marks the passage of the polar front at the location of core CH-73-139 C. We thus conclude that the North Atlantic polar front began its retreat at about 12,500 yr B.P. at the latitude of Portugal and reached the latitude of Ireland almost 1,000 years later, retreating with a mean velocity of about 2 km/year. Recently however, Broecker (pers. comm.) measured by A.M.S. an age of about 12,500 yr B.P. for the first warming phase in a core close to core CH 73-139 C. This discrepancy with our estimate may be due either to an artifact of bioturbation in Broecker's core or to a small sedimentation hiatus in core CH 73-139 C. The retreat velocity of the polar front that we determined is therefore a minimal value.

The cooling associated with the Younger Dryas cold event and the warming which followed have been synchroneous (± 400 years) between 37°N and 55°N, since we do not observe any significant age difference for the temperature variations recorded in both cores. This favours theories explaining the Younger Dryas by instabilities within the climatic system itself, e.g. a catastrophic influx of polar ice into the North Atlantic. The high velocity of the advance and the retreat of cold waters suggests that these temperature changes had affected only the most surficial waters and not the whole hydrologic structure of the North Atlantic ocean.

PENETRATION OF BOMB RADIOCARBON IN THE OCEAN

Under natural conditions, the distribution of ^{14}C in the deep ocean is influenced by many processes. Bottom water formation in the Norwegian Sea and the Weddell

Sea provides a direct input of surface-water ^{14}C . Additional input of ^{14}C to the deep sea occurs by transport along isopycnal surfaces, by lateral and vertical mixing in the main thermocline of the ocean, by dissolution of carbonate skeletons, and by oxidation of organic material from sinking particles. As compared to other oceanographic markers, ^{14}C adds a measure of time because it decays at the rate of 1% every 80 years. As a consequence, the distribution of ^{14}C in the abyssal waters of the world ocean has been used to estimate the replacement times for Pacific, Indian and Atlantic ocean deep waters (8).

The steady state of the carbon cycle was broken in the late fifties and in the sixties with the large scale testing of nuclear weapons. The ^{14}C activity of the atmosphere in the northern hemisphere increased by about 100% and has been decreasing since the end of the nuclear tests in the atmosphere. This atmospheric ^{14}C activity decrease is due to the homogeneization of the troposphere and to the penetration of CO_2 into other carbon reservoirs, mainly the ocean and the terrestrial biosphere. The bomb produced ^{14}C is therefore a good tracer to quantify the oceanic invasion by the atmospheric CO_2 excess resulting from human activities.

The conventional radiocarbon ß-counting requires the sampling of about 250 litres of sea water and chemical extraction of CO_2 on board the oceanographic vessel. The same ^{14}C measurements can be done with 100 ml water samples poisoned with 1 ml of saturated $HgCl_2$ solution (9). This operation is designed to supress isotope fractionation due to respiration and photosynthesis of microorganisms, which might affect the ^{13}C and ^{14}C content of the water sample. In the laboratory, the total dissolved CO_2 is extracted from the sea water in a vacuum line in which the 100 ml aliquot of sea water is acidified and flushed with pure helium gas. The evolving CO_2 is trapped in liquid nitrogen and the target is prepared by catalytic reduction of this CO_2 on iron powder (10).

As an example of the potential of measuring oceanic ^{14}C by A.M.S., we present here three profiles performed in the Indian Ocean during the INDIGO-2 cruise on board of the M/S Marion Dufresne. The radiocarbon activities (Fig. 3) are expressed by means of the $\Delta^{14}C$ scale (11). Stations 32 and 45 have the same locations as stations 424 and 420 of the GEOSECS expedition (12) and therefore will enable a comparison of the two sets of $\Delta^{14}C$ data.

The three profiles exhibit the presence of bomb ^{14}C in the upper part of the water column, until a depth of about 1000 m where $\Delta^{14}C$ values are equal to the GEOSECS values and represent the so-called pre-bomb level. Following Broecker et al. (13), we have estimated the water column inventories of bomb produced ^{14}C by integrating the area between the observed $\Delta^{14}C$ curve and the reconstructed $\Delta^{14}C$ curve versus depth for pre-nuclear time. At station 28, the INDIGO integrated amount of ^{14}C ($\Sigma^{14}C$) is 15.9 10^9 atoms/cm^2, in agreement with GEOSECS data obtained at the same latitude but not at the same location (Fig. 4). At station 32, the INDIGO $\Sigma^{14}C$ is 7.8 10^9 atoms/cm^2, identical with that measured at GEOSECS station 424 ($\Sigma^{14}C$ = 7.7 10^9 atoms/cm^2) although the two profiles exhibit significant differences. By contrast, at equatorial station 45, the INDIGO $\Sigma^{14}C$ is 10.3 10^9 atoms/cm^2, twice the value measured at the same location during the GEOSECS expedition ($\Sigma^{14}C$ = 4.8 10^9 atoms/cm^2.

The column inventories for bomb radiocarbon at the stations performed during the GEOSECS expedition showed pronounced minimal values in the low latitudes (Fig. 4). Oceanic models have showed that a sizable fraction of the bomb-^{14}C that entered the tropical ocean has been transported to the adjacent temperate zones and that the low-latitude deficit enables an equatorial upwelling component to be

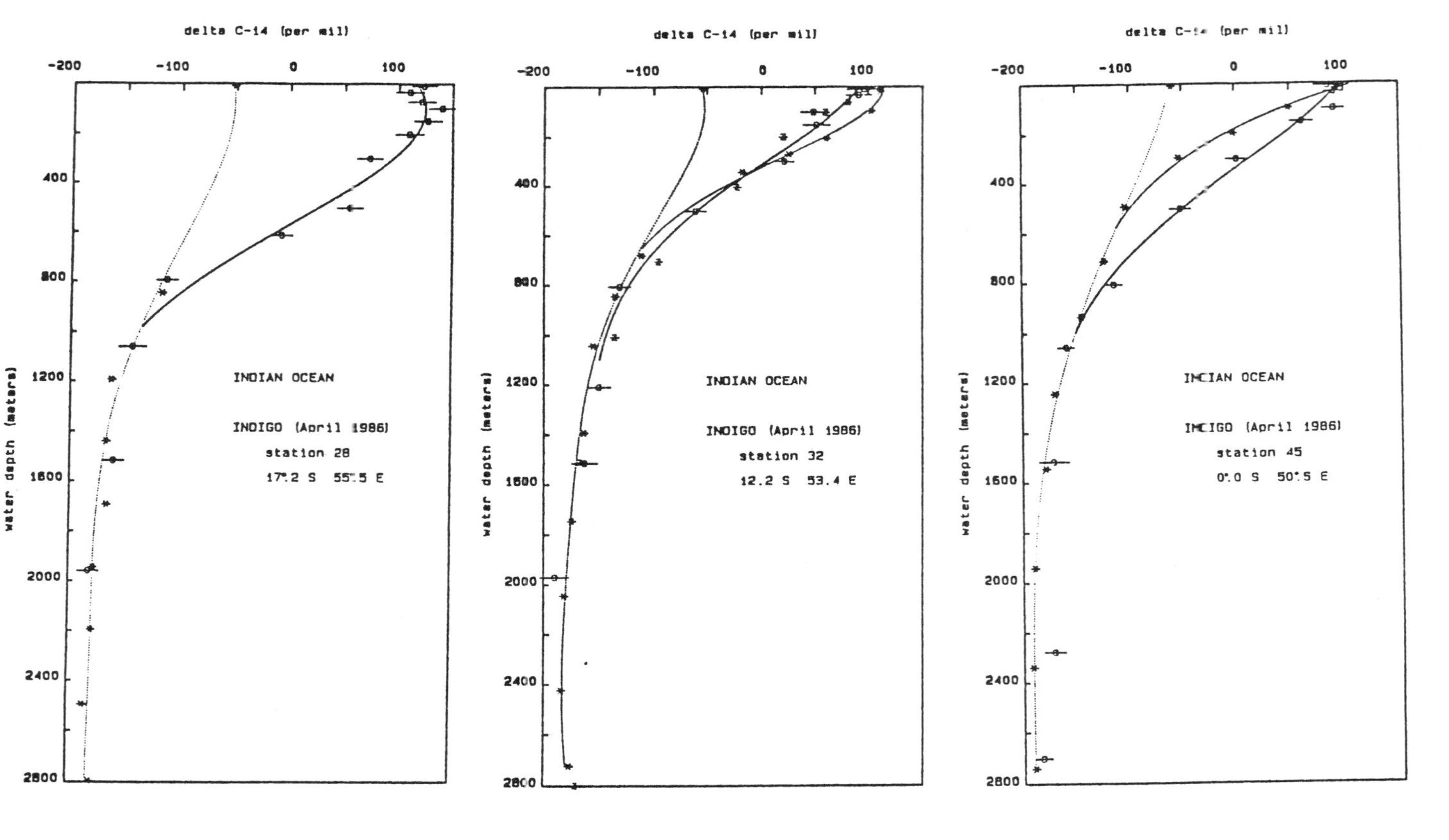

Fig. 3 - Profiles of Radiocarbon activities of sea water collected at 3 stations of tropical-equatorial Indian Ocean
- dotted lines indicate the estimated prebomb level
- stars with error bars indicate GEOSECS samples collected in 1978
- circles with error bars indicate INDIGO samples collected in April 1986 and measured by AMS with the Gif-sur-Yvette Tandetron.

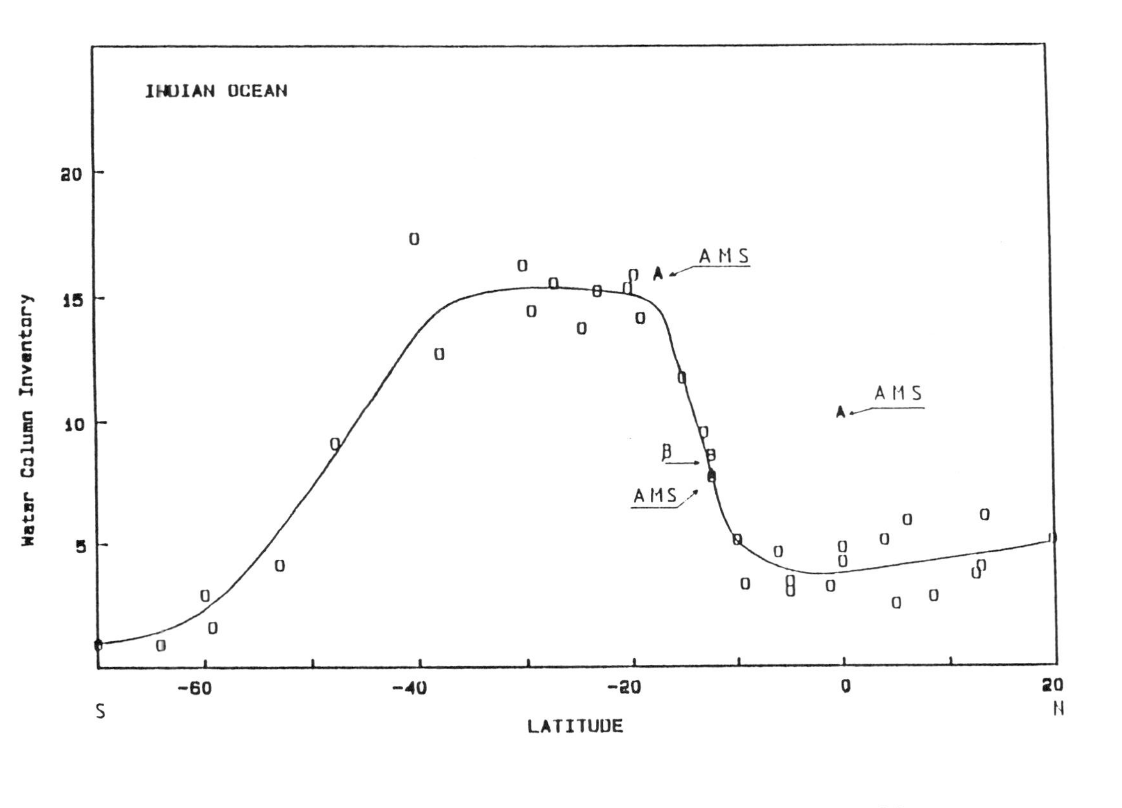

Fig. 4 – Variations with latitude of the integrated amount of ^{14}C in excess over the prebomb level in the Indian Ocean. Circles refer to GEOSECS samples collected in 1978 (ref. 12). A indicate samples collected in April 1986 and measured by AMS with the Gif-sur-Yvette Tandetron. At 12°5 S, β and A refer to the same station measured by β-counting by G.Ostlund (pers. comm.) and by AMS.

calculated (13,14). The three INDIGO stations selected in the narrow belt between 0°S and 20°S showed a steep $\Sigma^{14}C$ gradient at the time of the GEOSECS survey (1978). At the time of the INDIGO cruise (1986), the $\Sigma^{14}C$ value at the equatorial station 15 has roughly doubled during the last 8 years. Salinity and dissolved oxygen profiles demonstrate that the most likely explanation for the reduction of the equatorial deficit in the Indian Ocean is the advection of thermocline water called Banda Sea Water flowing from the east. This low-salinity water, which is rich in dissolved oxygen and in Tritium (15), is due to a throughflow from the North Pacific into the Indian Ocean, which results in its relatively high ^{14}C content.

We thus conclude that A.M.S. can be used successfully to trace the penetration of bomb produced ^{14}C into the ocean. This new technique should leach to the introduction at ^{14}C measurements into the WOCE program devoted to the understanding of the dynamics of the global ocean.

REFERENCES

1. Arnold M., Bard, E., Maurice, P., and Duplessy, J. C., ^{14}C dating with the Gif sur Yvette Tandetron accelerator : Status report. Nuclear Instrument and Methods (1987), P. 120-123.

2. Imbrie, J. and Kipp, N. G., A new micropaleontological method for quantitative paleoclimatology : Application to a late Pleistocene Carribean core. In the Late Cenozoic Glacial Ages, ed. K.K. Turrebian (1971), p. 71-181.

3. Shackleton, N. J. and Opdyke, N. D., Oxygen isotope and paleomagnetic stratigraphy of equatorial pacific core V 28-238 : oxygen isotope temperatures and ice volume on a 10^5 and 10^6 year scale. Quaternary Res., 1973, 3, 39-55.

4. Duplessy, J. C., Arnold, M., Maurice, P., Bard, E., Duprat, J. and Moyes, J., Direct dating of the oxygen isotope record of the last deglaciation by ^{14}C Accelerator mass spectrometry. Nature, 1986, 320, 350-352.

5. Bard, E., Arnold, M., Duprat, J., Moyes, J., and Duplessy, J. C., Reconstruction of the last deglaciation: deconvolved records of $\partial^{18}O$ profiles, micropaleontological variations and accelerator mass spectrometric ^{14}C dating. Climate Dynamics, 1987, 1, 101-112.

6. CLIMAP Project Members, Seasonal reconstructions of the Earth's surface at the last glacial maximum. Geol. Soc. Am. Map Series MC-36 (1981).

7. Duplessy, J. C., Bard, E., Arnold, M., and Maurice,P., A.M.S. ^{14}C- chronology of the deglacial warming of the North Atlantic Ocean. Nuclear Sciences and Methods, (1987), p. 223-227.

8. Stuiver, M., Quay, P. D., and Ostlund, H. G., Abyssal water carbon-14 distribution and the age of the world oceans. Science, 1983, 219, 849-851.

9. Bard, E., Arnold, M., Maurice, P. and Duplessy, J.C., Measurements of bomb radiocarbon in the ocean by means of accelerator mass-spectrometry : Technical aspects. Nuclear Instrument and Methods (1987), 297-301.

10. Vogel, J. S., Southon, J. R., Nelson, D. E. and Brown, T. A., Performance of catalytically condensend carbon for use in accelerator mass spectrometry. Nuclear Instrument and Methods, 1984, 233, 289-293.

11. Stuiver, M. and Pollach H. A., Discussion : Reporting of ^{14}C data. Radiocarbon 1977, 19, 355-363.

12. Stuiver, M. and Ostlund H. G., GEOSECS Indian Ocean and Mediterranean Radiocarbon. Radiocarbon, 1983, 25, 1-19.

13. Broecker, W. S., Peng, T. H., Ostlund, H. G. and Stuiver M., The distribution of bomb radiocarbon in the ocean. J. Geophys. Res., 1985, 90 (C4), 6953-6970.

14 Wunsch, C., An estimate of the upwelling rate in the Equatorial Atlantic based on the distribution of bomb radiocarbon and quasi-geostrophic dynamics. J. Geophys. Res., 1984, 89, 7971-7978.

15. Fine, R. A., Direct evidence using tritium date for throughflow from the Pacific into the Indian Ocean. Nature, 1985, 315, 478-480.

C.F.R. Contribution N° 867

Ra-226 AND Ba IN THE NORTH EAST ATLANTIC DEEP AND BOTTOM WATER LIMITATIONS OF Ra-226 AS A TIME TRACER

Monika Rhein and Wolfgang Roether

Institut für Umweltphysik der Universität Heidelberg
Im Neuenheimer Feld 366, 6900 Heidelberg, Fed.Rep.of Germany

ABSTRACT

Ra-226 and Ba data are presented for the NE Atlantic below 2000m depth from a section of stations 8°S to 46°N obtained by F.S. "METEOR" in 1981. The precision of the Ra-226 data (±1%) allows a resolution of structures in the Ra-226 distribution which is comparable to the resolution for Ba and the nutrients considered. Ra-226/Ba correlations as well as correlations with silica and alkalinity are presented. The Ra-226 excess over a linear Ra-226/Ba correlation in the deepest 800 to 1500m is attributed to the Ra-226 primary source by Th-230 decay in the sediment. From the Ra-226 deficit in intermediate depths, a deep water renewal time of 70 years is derived. From the Ra-226 correlations with silica and alkalinity, and from studying the property-property distributions across the Romanche fracture zone, it is concluded that $CaCO_3$ dissolution must be an important process of Ra-226 and Ba regeneration.

Box model calculations allow one to determine the Ra-226 and Ba sources in the deep East Atlantic, and for the first time an estimation of the Ra-226 regeneration source is given. Moreover, it appears that the dissolution of calcareous and siliceous skeletons contributes about 38-64% to the Ra-226 regeneration source and about 19-36% to the Ba input. The remaining part is attributed to dissolving barite.

INTRODUCTION

In 1958, Koczy proposed the use of Ra-226 (half-life 1600 a) as a time tracer for deep sea ventilation rates. But besides physical processes, the Ra-226 distribution is also governed by its involvement in the biogeochemical cycle. As a stable Ra isotope is lacking, Ba has been proposed in order to eliminate the influence of the biological cycle on the Ra-226 distribution. And indeed, rather uniform, linear Ra-226/Ba correlations have been found in many regions of the oceans, although input into the sea of these two earth alkaline metals are different : Ra-226 is produced by Th-230 decay in the sediment and thus mainly added to the deep sea, whereas Ba is brought partly into the surface waters via river input (70%) and partly into the deep sea (30%) by hydrothermal activities.

GEOSECS results show that Ra-226 and Ba have more in common with the hard part component of particulate matter (opal and calcareous skeletons) than with the tissue parts (nitrate and phosphate), (Chan et al., 1977 ; Chan et al., 1976). Works of Dehairs et al. (1980) and Collier and Edmond (1984) point to the fact that not only is the incorporation of Ra-226 and Ba in $CaCO_3$ and SiO_2, including dissolution at greater depths, important, but that the biogenically controlled production of barite ($BaSO_4$) also plays a major role for the downward flux of Ba, presumably similar to that of Ra.

Here, Ra-226 and Ba data from the North East Atlantic below 2000m are presented. They have been obtained on the RV "METEOR" 56/5 cruise in 1981. Tab.1 shows positions and depths of the stations.

Tab.1 Ra and Ba stations of cruise "METEOR" 56, leg 5.

Station No.	Position		Bottom depth (m)
497	5°03'S	24°03'W	5300
499	0°00'N	19°48'W	5200
501	2°52'N	19°49'W	5020
505	9°48'N	26°00'W	5300
509	16°47'N	27°01'W	4565
513	24°76'N	26°98'W	5450
515	28°74'N	25°16'W	5260
517	32°79'N	23°26'W	5330
521	40°30'N	18°56'W	5400
525	46°03'N	11°03'W	4775

This region has been chosen because of its simple hydrographic situation : the East Atlantic receives its deep water only through the Romanche Fracture Zone (Rfz, sill depth : 4000m ; 0°,18°W) near the equator from where it propagates polewards. The deep water inflow is a mixture of 70% Antarctic Bottom Water (AABW) and 30% North Atlantic Deep Water (NADW) ; it is slightly modified by vertical mixing along its poleward path producing minor, along basin, gradients in salinity and potential temperature. The circulation within the deep East Atlantic appears to be sufficiently sluggish, leading one to expect non-conservative behaviour of the tracers (Broecker et al, 1980). Moreover, the circulation scheme and the rate of input are approximately known from parallel C-14 inverse modeling (Schlitzer, 1987).

The improved measurement precision of the Ra-226 data of ±1% (Schlosser et al., 1985 ; Rhein et al., 1987) now allows a resolution of structures comparable to those of Ba and the nutrients considered here.

SAMPLE COLLECTION AND MEASUREMENT PROCEDURE

Ra-226 samples (20 l) were collected with Gerard-Ewing bottles (250 l) whereas Ba samples have been obtained from Niskin sampler casts. Silica and alkalinity were measured in every sample collected. The sampling depths of Niskin and Gerard-Ewing samples were always interspersed.

Measurement procedure for Ra-226 is described in Schlosser et al. (1985) and Rhein (1986). Ba concentrations were obtained using isotope dilution mass spectrometry (Chan et al., 1977) at a measurement uncertainty of ±0,7% and silica and alkalinity have been measured on board ship using standard procedures (Almgren et al., 1983) ; Koroleff, 1983).

As Ra-226 and Ba are generally derived from alternate depths, an interpolation procedure has been used to enable us to study the Ra/Ba correlations : Ra and Ba normally occurred linearly related to silica (see below), this fact has been used for interpolation. If linear correlations do not hold, interpolation has been done by depth (Rhein et al., 1987). The resulting error amounts to 1.5-3%.

THE RA-226 DISTRIBUTION

The Ra-226 isolines (in dpm/100kg, Fig.1) clearly show the bending of the isolines >17dpm/100kg between Stas. 499 and 501 at the inflow in the East Atlantic. In the deep NE basins higher Ra-226 and Ba as well as nutrient concentrations compared to the inflow concentrations are observed. But the increase in concentrations is not generally directed northwards indicating locally variable source terms.

Fig.1

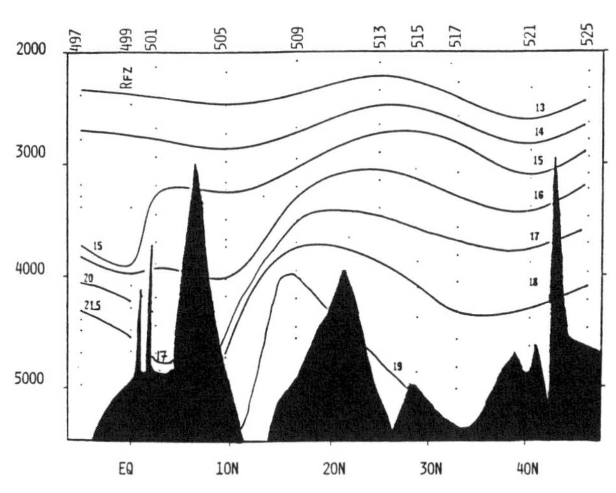

RA-226/BA CORRELATIONS

Ra/Ba correlations are found to be linear and very uniform : the mean slope amounts to $3.88\pm0.13*10^2$ nmolBa/dpmRa and the average intercept to 12.4 ± 2.6 nmolBa/kg (error : apparent standard deviation). But generally

Fig.2

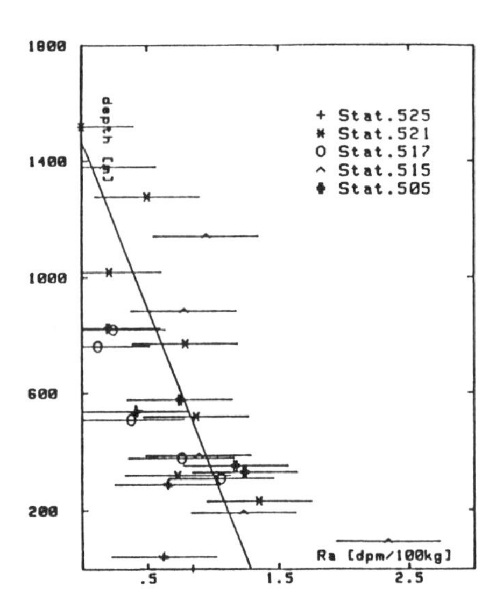

in the lowest 800 to 1500m above the sea floor, Ra-226 excess relative to a linear Ra/Ba relationship is present. Fig.2 presents this Ra excess which can be interpreted as a linear gradient dc/dz of $1*10^{-2}$dpm/m^2 ($\pm$20%). Neglecting decay and vertical advection the gradient can be converted into a mean Ra-226 flux Q out of the sediment with Q=K*dc/dz (K : vertical eddy diffusion coefficient : 2.5-5*$10^{-4}m^2/s$, Schlitzer, 1984). A Ra-226 primary source of 3.8$\pm$2*10^{-6}dpm/m^2s (=8*10^{-21}mol/m^2s) is obtained. This is comparable to the results of Cochran for the Atlantic (1-8.3*10^{-21}mol/m^2s) which he evaluated from the Ra-226 deficit compared to the abundance of Th-230 in the sediment (Cochran, 1980 ; Broecker and Peng, 1982).

Another deviation from a linear Ra/Ba correlation occurs at Stas. 521 and 525 in the Iberian basin in depths between 3600 and 4000m (521) and 3400 and 3900m (525) respectively (Fig.3). The observed Ra-226 deficit of 1.5-3% is in the same order of magnitude as the interpolation error, but nevertheless apparent. The deficit is attributed to radioactive decay of Ra-226 and can thus be converted into a deep water renewal time of about 70 years. This is an accordance with C-14 results (Schlitzer et al., 1985).

Fig.3

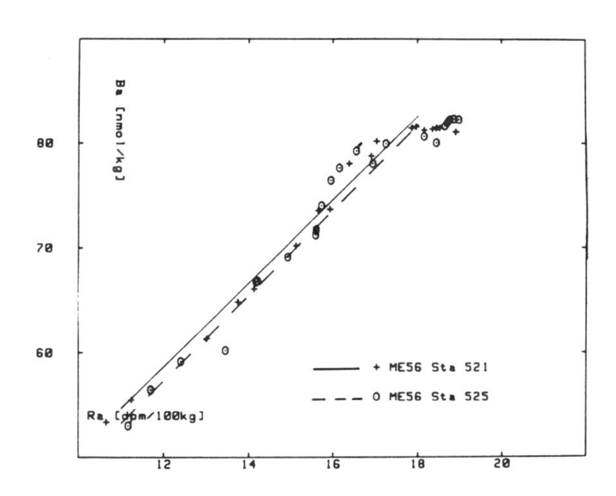

CORRELATIONS OF Ra AND Ba WITH SILICA AND ALKALINITY

The two earth alkaline metals are also linearly related to silica in the deep NE Atlantic, but these correlations are not uniform : between the equatorial West Atlantic (Stas. 497, 499) and NE Atlantic (Stas. 501, 505) an increase in the Ra/Si (and Ba/Si) slope of about a factor of two is observed (from 10 to 20*10^{-4}dpmRa/µmolSi) and, proceeding further to the north, the Ra/Si (Ba/Si) slope increases again by about 20%. The increasing slopes are accompanied by decreasing intercepts (from 10dpmRa/100kg to 6.5dpm/100kg).

In contrary to this feature, the linear correlations of Ra-226 and Ba with alkalinity appear to be rather uniform within the error of 10-15% ; this high uncertainty (compared to the correlations with silica : Ra/Si and Ba/Si : 5%) originates in the low dynamic range of alkalinity (variations in the deep NE Atlantic : 50µeq/kg, error :

4±eq/kg). Nevertheless, an increase of 100% as observed in the Ra/Si correlation between West and East Atlantic would be detectable and is indeed obvious in the linear Alk/Si correlations (Rhein et al., 1987).

Because of the greater similarity of the Ra and Ba distribution with alkalinity one can conclude that dissolution of calcareous skeletons plays an important role for the Ra and Ba distribution in the deep NE Atlantic. This finding is further supported by the similarity of the correlations of alkalinity, Ra-226 and Ba with Potential Temperature (Fig.4) for the stations near the Rfz (Stas. 497, 499, 501). On the other hand, the Si/Tpot correlations show a conservative behaviour of silica (Fig.5) during transition between West and East Atlantic.

These plots can also be used to determine the tracer concentrations of the inflowing deep water into the East Atlantic (Schlitzer al., 1985, Rhein et al., 1987)

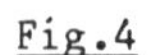

Fig.4

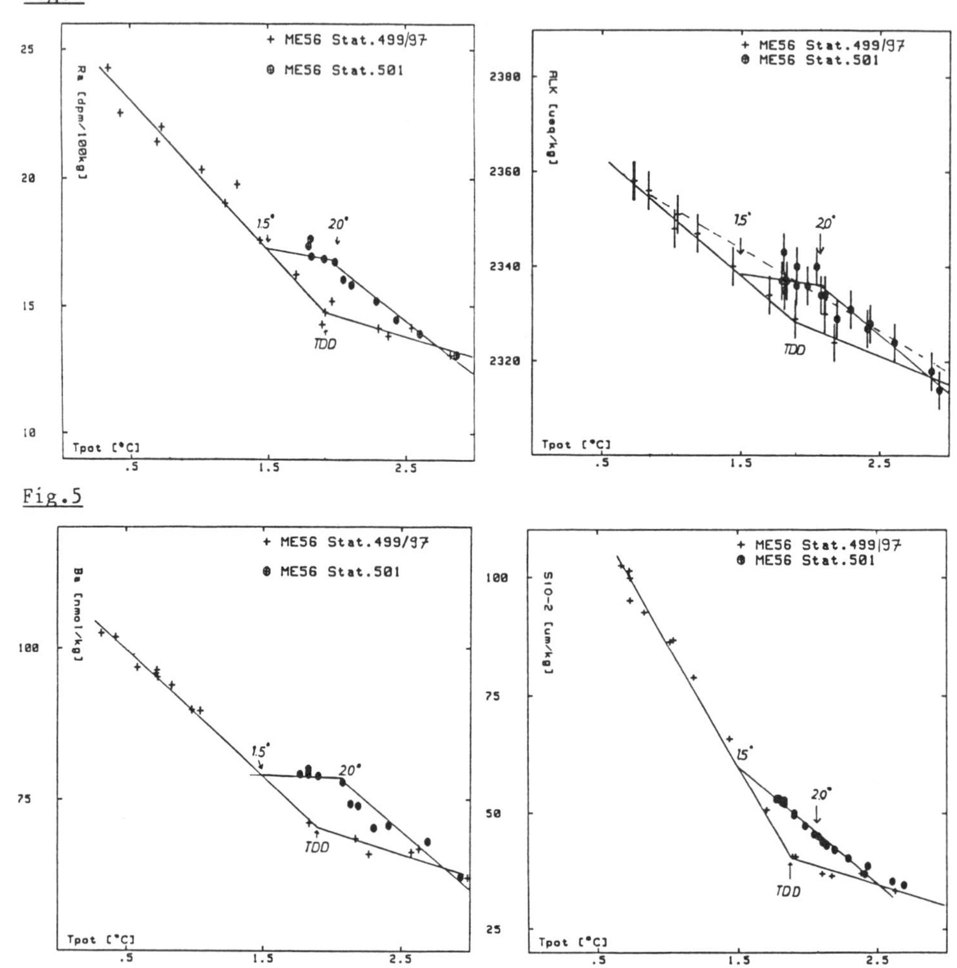

Ra-226 AND Ba REGENERATION SOURCES IN THE DEEP EAST ATLANTIC

As mentioned before, the rate of inflow and the circulation scheme of the deep East Atlantic are approximately known by inverse modeling of C-14 data (Schlitzer, 1987). Here, the same geometry as of the inverse box model is applied (Fig.6) and the C-14 derived water flows and eddy diffusion coefficients are used as input data for the calculations of the Ra-226 and Ba sources. The measured Ra-226 and Ba concentrations of the model region, as well as the concentrations of the boundary conditions, are shown in Fig.7. For the South East Atlantic, GEOSECS data from Stas. 107 and 111 have been used (Broecker et al., 1976 ; Chan et al., 1977).

Fig.6

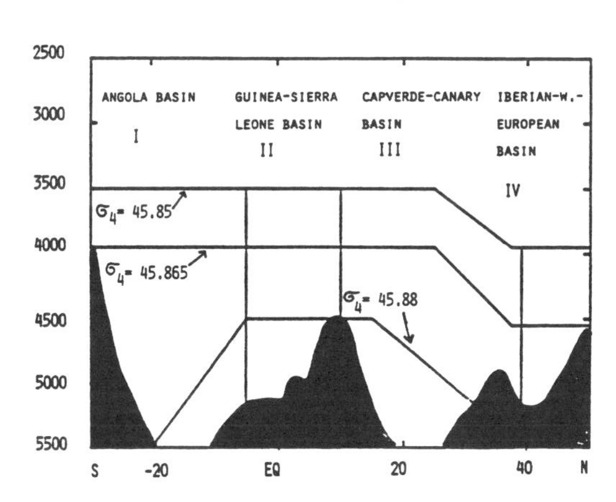

Under stationary conditions (dc/dt=0) the equation for each box is :

$$0 = J_{ADV} + J_{DIFF} + Q - S$$

with : J_{ADV} : advective fluxes, J_{DIFF} : diffusive fluxes

Q : sources and S : sinks of the tracers considered

These box equations are solved simultaneously, and by variation of the source terms an attempt has been made to reproduce the measured Ra and Ba distributions. Accordance between measured and calculated concentrations is tested with a chi-square test at a confidence level of 95%.

Fig.7

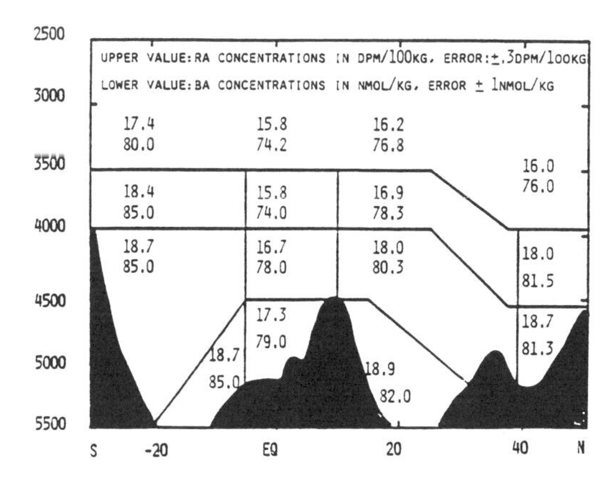

The deep water inflow through the Rfz ranges from 2.6 to $5.1*10^9$kg/s (Schlitzer, 1987). Using the higher value increases the Ra-226 source by about 25 % compared to the smaller inflow. A range of $35-53*10^{-21}$ mol/m^2s for the Ra-226 input and a Ba input of $32-60*10^{-13}$ mol/m^2s is received. Subtracting the Ra-226 primary source, a regeneration source of 27-45* 10^{-21} mol/m^2s is obtained for Ra-226.

These Ra-226 and Ba sources are about 10 times higher than those previously reported for the Pacific (Tab. 2). This could be caused by two reasons : first, the primary production rate in the Pacific may be lower compared to that of the East Atlantic, and second, the applied advection-diffusion models apparently point to lower source terms (Ku et al., 1980). For example, these models estimate a Ra-226 primary source in the North East Pacific of $3.2*10^{-21}$ mol/m^2s (Ku et al., 1980), whereas measurements of Cochran (1980) show a source of $135*10^{-21}$ mol/m^2s. Moreover the NE Pacific shows the highest oceanic Ra concentrations and is therefore believed to be a major Ra-source region with an input higher than the mean global average (Chung, 1980, Tab. 2).

Tab. 2 Estimations of Ra-226 and Ba sources in the deep sea.

Authors	Region	Method	Primary flux	Regeneration sources	
			Ra-226[1]	Ra-226[2]	Ba[3]
Ku et al, 80	E-Pacific	adv-diff model	3.2	0.65	1.4
Chan et al, 77	Pacific	adv-diff model	-	2.5	3.7
Chung, 80a	E-Pacific	adv-diff model	-	5.3	-
Chung, 80b	World ocean	balance consid.	20	-	-
Cochran, 80	NE-Pacific	Ra-deficit in	135	-	-
	Atlantic	sediment	1-8	-	-
Brewer et al 80	W-Atlantic	sediment traps	-	-	7.4
Dehairs et al 80	W-Atlantic	barite measurem. balance consid.	-	-	16.5
Rhein et al 87	NE-Atlantic	Ra-excess over Ba	8	-	-
Rhein and Schlitzer, 87	E-Atlantic	box model	-	23-37	26-49

1) Ra-226 primary flux $*10^{-21}$ mol/m^2s
2) Ra-226 regeneration source $*10^{-27}$ mol/kgs
3) Barium regeneration source $*10^{-19}$ mol/kgs

The Ba/Ra ratio of the source terms ($0.4-1.0*10^8$gBa/gRa) is different from that of the dissolved concentrations in the model region and of the boundary conditions ($1.3-1.4*10^8$gBa/gRa). However the sources cannot determine the ratios in the deep East Atlantic because of the relative short deep water renewal time of about 70 years compared to the source strength : it would take about 600 years to replace all Ba in the model region by the source and about 250 years to replace all Ra-226.

CONTRIBUTION TO THE Ra-226 AND Ba INPUT BY REDISSOLUTION OF SILICEOUS AND CALCAREOUS SKELETONS

Another point of interest is the contribution of dissoluting calcareous and siliceous matter to the Ra-226 and Ba sources. Schlitzer (1987) determined the magnitude of the East Atlantic silica and carbon source to be at maximum $1*10^4$ mol/s silica and $1.2*10^5$ mol/s carbon. On the other hand, Ra and Ba concentrations in biogenic produced hard parts vary considerably by about two orders of magnitude : $0.1\text{-}16*10^{-12}$gRa/g dry matter and 6-300 ppm Ba have been found (Szabo, 1967 ; Shannon and Cherry, 1971 ; Martin and Knauer, 1977 ; Krishnaswami et al., 1976 ; Dehairs et al., 1980 ; Knauss and Ku, 1983 ; Collier and Edmond, 1984). These variations are partly related to different regions and partly to different species measured. In Tab. 3 the highest Ra-226 and Ba concentrations reported for open ocean Atlantic samples have been used to calculate the contributions of silica and carbonate regeneration in the deep sea. It turns out that these two sources, together with the Ra primary flux, could only be responsible for 32-49 % of the total Ra input and to 19-36 % of the Ba source. The remainig part has to be attributed to dissolving barite. A lack of barite data makes it impossible to prove this assumption directly. However, this finding is in accordance with the results of Collier and Edmond (1984) who stated that dissolution of siliceous and calcareous matter could only explain one third of the difference in Ba concentration between surface and deep water, and it is also comparable to flux estimations made by Dehairs et al. (1980).

Tab. 3 Contribution from the redissolution of siliceous and calcareous matter to the Ra-226 and Ba source.
assumed Ra-226[1] concentration in silica : $50*10^{-12}$g/gSiO_2
assumed Ra-226[1] concentration in $CaCO_3$: $2.4*10^{-12}$g/g$CaCO_3$
assumed Barium[2] concentration in silica : 260 ppm
assumed Barium[2] concentration in $CaCO_3$: 200 ppm

	Ra-226 source $*10^{-21}$ mol/m^{2}s	Ba-source $*10^{-13}$ mol/m^{2}s
from SiO_2 dissolution	8.9	0.6
from $CaCO_3$ dissolution	8.5	11
total	17.4	11.6

1 : from Krishnaswami et al., 1976, 2 : from Dehairs et al., 1980.

CONCLUSIONS

- Between 2000 and 3400 to 4700 m depth depending on position, Ra-226 and Ba are linearly correlated.

- The Ra-226 deficit of 1.5 to 3 % relative to a linear correlationship with Ba is in the same order of magnitude as the uncertainty of the data, in spite of improved Ra-226 measurement precision but is nevertheless apparent. It yields a deep water renewal time in the NE Atlantic of about 70 years.

- The Ra-excess over Ba 800 to 1500 m above the sea floor can be converted into a Ra-226 flux out of the sediment, which is comparable to previous estimates. The evaluated Ra-226 primary source covers about 14-20 % of the total Ra input. It is high enough to compensate for the Ra decay in the deep NE Atlantic, but lower than the necessary global average.

- The Ra-226 regeneration source is in the range between $27\text{-}45 \cdot 10^{-21}$ mol/m^2s, the Ba input amounts $32\text{-}60 \cdot 10^{-13}$ mol/m^2s. The increase of deep water inflow into the deep East Atlantic by a factor of two increases the total Ra-226 source by 25 %.

- The Ba/Ra ratios of the source terms are significantly lower than the ratio of the dissolved concentrations.

- The redissolution of calcareous and siliceous skeletons contributes 38-64 % to the Ra-226 regeneration source and 19-36 % to the total Ba input. The remaining part is attributed to dissolving barite.

REFERENCES

Almgren, T., D. Dryssen, S. Fonselius (1983) Determination of alkalinity and total carbonate in : Methods of seawater Analysis, 2nd edition, M. Grasshoff, M. Ehrhardt, K. Kremling, editors, Verlag Chemie, Weinheim, 99-122

Brewer, P.G., Y. Nozaki, D.W. Spencer and A.P. Fleer (1980) Sediment trap experiments in the deep North Atlantic. Isotopic and elemental fluxes. Journal of Marine Research, 38, 703-728

Broecker, W.S., J. Goddard and J.L. Sarmiento (1976) The distribution of Ra-226 in the Atlantic Ocean. Earth and Planetary Science Letters, 32, 220-235

Broecker, W.S., T. Takahashi and M. Stuiver (1980) Hydrography of the Central Atlantic II, Waters beneath the two degree discontinuity. Deep Sea Research, 27A, 397-419

Broecker, W.S. and T.H. Peng (1982) Tracers in the Sea, Lamont Doherty Geological Observatory, Columbia University, Palisades, N.Y.

Chan, L.H., J.M. Edmond, R.F. Stallard, W.S. Broecker, Y.C. Chung, R.F. Weiss and T.L. Ku (1976) Radium and Barium at GEOSECS stations in the Atlantic and Pacific. Earth and Planetary Science Letters, 32, 258-267

Chan, L.H., D.Drummond, J.M. Edmond and B. Grant (1977) On the Barium data from the Atlantic GEOSECS expedition. Deep Sea Research 24, 613-649

Chung, Y. (1980a) Ra-Ba-Si correlations ans a two dimensional radium model for the world ocean. Earth and Planetary Science Letters, 49, 309-318

Chung, Y. (1980b) A Ra section across the East Pacific Rise. Earth and Planetary Science Letters, 49, 319-328

Cochran, K. (1980) The flux of Ra-226 from deep sea sediments. Earth and Planetary Science Letters, 49, 381-392

Collier, R.W. and J.M. Edmond (1984) Plankton deposition and trace element fluxes from the surface ocean. In : Trace Metals in Seawater, E. Goldberg, editor, Plenum Press, New York, pp 789-809

Dehairs, F., R. Chesselet, J. Jedwab (1980) Discrete suspended particles of barite and the barium cycle in the open ocean. Earth and Planetary Science Letters, 49, 528-550

Knauss, K., T.L. Ku (1983) The elemental composition and decay series radionuclide content of plankton from the East Pacific. Chemical Geology 39, 125-145

Koczy, F.F. (1958) Natural radium as a tracer in the ocean. Proceedings Second United Nations International Conference for Peaceful Uses of Atomic Energy (Geneva) 18, 351

Koroleff, F., (1983) Determination of silicon, in : Methods of seawater analysis, 2nd edition, M. Grasshoff, M. Ehrhardt, K. Kremling editors, Verlag Chemie, Weinheim, 174-185

Krishnaswami, S., D. Lal and B.L.K. Somayajulu (1976) Investigations of gram quantities of Atlantic and Pacific surface particulates. Earth and Planetary Science Letters, 32, 403-419

Ku, T.L., C.A. Huh and P.S. Chen (1980) Meridional distribution of Ra-226 in the Eastern Pacific along GEOSECS cruise tracks. Earth and Planetary Science Letters, 42, 239-308

Martin, J.H. and G.A. Knauer (1973) The elemental composition of plankton. Geochimica et Cosmochimica Acta, 37, 1639-1653

Rhein, M. (1986) Ra-226 im tiefen Nordostatlantik. Doctoral Dissertation, Universität Heidelberg, Heidelberg, FRG

Rhein, M., L.H. Chan, W. Roether and P. Schlosser (1987) Ra-226 and Ba in Northeast Atlantic Deep Water. Deep Sea Research, in press

Rhein, M and R. Schlitzer (1987) Ra-226 and Ba sources in the deep East Atlantic, in preparation

Schlitzer, R. (1984) Bestimmung der Tiefenwasserzirkulation des Ostatlantiks mittels einer Boxmodell-Auswertung von C-14 und anderen Tracerdaten. Doctoral Dissertation, Universität Heidelberg,Heidelberg, FRG

Schlitzer, R. (1987) The renewal of East Atlantic Deep Water by inversion of C-14 data. Journal of Geophysical Research 92 (C3), 2953-2969

Schlitzer, R., W. Roether, U. Weidmann, P. Kalt and H. Loosli (1985) A meridional C-14 and Ar-39 section in North East Atlantic Deep Water. Journal of Geophysical Research, 90, 6945-6952

Schlosser, P., M. Rhein, W. Roether and B. Kromer (1984) High precision measurement of oceanic Ra-226. Marine Chemistry, 15, 203-216

Shannon, L.V. and R.D. Cherry (1971) Ra-226 in marine phytoplankton. Earth and Planetary Science Letters, 11, 339-343

Szabo, B.J. (1967) Radium content of plankton and seawater in the Bahamas. Geochimica and Cosmochimica Acta, 31, 1321.

DETERMINATION OF THORIUM IN SEAWATER BY NEUTRON ACTIVATION ANALYSIS AND MASS SPECTROMETRY

Chih-An Huh
College of Oceanography
Oregon State University
Corvallis, OR 97331
USA

ABSTRACT

The recent development of neutron activation analysis and mass spectrometric methods for the determination of ^{232}Th in seawater has made possible rapid sampling and analysis of this long-lived, non-radiogenic thorium isotope on small-volume samples. The marine geochemical utility of ^{232}Th, whose concentration in seawater is extremely low, warrants the development of these sensitive techniques. The analytical methods and some results are presented and discussed in this article.

INTRODUCTION

Applications of thorium isotopes to various studies in marine geochemistry and geochronology have long received wide attention. For instance, ^{230}Th (half-life = 75,200 yrs) has been extensively used to determine accumulation rates of deep-sea sediments and ferromanganese oxide concretions (1-2, and references therein). Deficiencies of ^{234}Th (half-life = 24 days) and ^{228}Th (half-life = 1.9 yrs) relative to their soluble parents (^{238}U and ^{228}Ra, respectively) have been applied to index the removal rates of particles and reactive elements in surface ocean and coastal waters (3-8). Excess ^{228}Th (over ^{228}Ra) and excess ^{234}Th (over ^{238}U) have been used to evaluate accumulation rates and mixing rates of near-shore sediments (9-11). More recently, the distributions of these isotopes in water columns and between dissolved and particulate forms have been investigated (12-18), leading to important understandings of the rate and mechanism of deep-sea scavenging, a process that is believed to be important in regulating the oceanic concentrations and distributions of a large variety of trace metals and anthropogenic pollutants.

The utilities of the aforementioned thorium isotopes (^{234}Th, ^{230}Th, and ^{228}Th) are dictated by the fact that they are reactive daughter products of non-reactive progenitors (^{238}U, ^{234}U, and ^{228}Ra). Therefore, once produced, these daughter products can be effectively removed from seawater to cause radioactive disequilibrium within these parent-daughter pairs. The extent of disequilibrium can be taken as a measure of the general scavenging rate. Another important advantage of utilizing these parent-daughter pairs is that the production rates of these radiogenic Th isotopes, which depend on the concentration and distribution of their parents, are either nearly constant everywhere (viz., for ^{234}Th and ^{230}Th) or determined by circulation and mixing of water masses (viz., for ^{228}Th). Further benefits can be gained in certain applications if more than one Th isotopes are measured, which often leads to a better constraint of the time scales or rate constants concerned.

However, when using radiogenic Th isotopes as proxies to study the general scavenging of trace metals, it is circumstantial to make meaningful direct comparisons between them, because their pathways of input to the oceans are very different. In order to provide a link, one would like to be able to determine the water-column distribution of the nonradiogenic isotope ^{232}Th, which, like most of the trace metals, is delivered to the ocean by the normal fluvial and aeolian routes.

Up until now, the measurement of ^{232}Th in seawater has long been hampered by its exceedingly low concentration. Conventionally, ^{232}Th concentration in seawater is determined by α-spectrometry, which requires large volumes ($\sim 10^3 \ell$) of samples, but also often results in very high and uncertain blanks. To circumvent these shortcomings, two sensitive methods have been developed in the past three years. Huh and Bacon developed a neutron activation analysis (NAA) technique, which enables one to measure ^{232}Th in 10-ℓ samples (19). Chen and co-workers developed a mass spectrometric method, which is even more sensitive; not only ^{232}Th, but also ^{230}Th can be measured on sub-liter samples using this method (20-21). The purpose of this paper is to briefly introduce these two new analytical tools, presenting some results obtained thereby for discussion, and prospecting future studies along this line of research.

METHODOLOGY

Neutron Activation Analysis

Principle

^{232}Th, with a half-life of 14 billion years, can be literally regarded as a "stable" nuclide. It is awkward to detect such an element via its natural decay, especially when its concentration is low in the sample of interest, as in seawater. Upon irradiation with thermal neutrons, ^{232}Th can be transformed into short-lived daughter products via the reaction:

$$^{232}\text{Th}\ (n, \gamma)\ ^{233}\text{Th} \xrightarrow{\beta} {}^{233}\text{Pa}$$

The induced activities of the daughter products can be easily managed to reach over five (for ^{233}Pa) to seven (for ^{233}Th) orders of magnitude higher than the target ^{232}Th activity. Accurate and precise determination of ^{232}Th can then be attained via its daughter products.

In principle, ^{232}Th content in the target sample can be indexed by either ^{233}Th (half-life = 22.2 min) or ^{233}Pa (half-life = 27 days). Although the induced activity of ^{233}Th is normally two orders higher than that of ^{233}Pa, ^{233}Pa assay is more desirable due to its longer half-life and hence greater convenience in working time. The generated activity of

^{233}Pa in the course of irradiation can be described by the equation:

$$\lambda_3 N_3 = \phi \sigma_1 N_1 \left[\frac{\lambda_3}{\lambda_2 - \lambda_3} \exp(-\lambda_2 t) + \frac{\lambda_2}{\lambda_3 - \lambda_2} \exp(-\lambda_3 t) + 1 \right]$$

where N indicates the number of atoms; λ is the decay constant; ϕ is the thermal neutron flux; σ is the thermal neutron capture cross section; t is the duration of irradiation; and the subscripts 1, 2, and 3 represent ^{232}Th, ^{233}Th, and ^{233}Pa, respectively. In practice, however, ^{232}Th standards were prepared and irradiated in parrallel with samples, and the results were used to establish a calibration curve to eliminate the necessity of absolute determination and calibration of various parameters (e.g., flux density, slow neutron capture cross section, counting efficiency, delay in counting, etc.).

Chemical procedures

^{233}Pa is to be identified by its 312 KeV γ-radiation. If a seawater sample is not radiochemically processed, the induced activity of interfering nuclides, emitting γ-radiations in the energy window of 300-320 KeV, will likely be over five orders of magnitude higher than that of ^{233}Pa. Consequently, the ^{233}Pa signal could be completely swamped. Therefore, radiochemistry is absolutely necessary. In this sense, this method should be referred to as "radiochemical neutron activation analysis", as opposed to "instrumental neutron activation analysis".

Our method involves both preirradiation and postirradiation radiochemical separations. Th isotopes were separated from the samples and purified during the preirradiation chemistry. ^{229}Th or ^{230}Th were added as the yield determinant. Sea salts and virtually the entire periodic table of elements were quantitatively removed during the preirradiation procedure by coprecipitation, ion exchange, and solvent extraction. The purified Th source were prepared by mounting TTA/benzene onto high-purity (Puratronic, 99.9995% purity) Al foils for α-spectrometry to determine the recovery. After that, they were packed in polyethylene vials for irradiation.

After irradiation and subsequent cooling off, the samples were subjected to postirradiation chemistry, which entails separation and purification of Pa by ion-exchange and solvent extraction. ^{231}Pa was added as the yield determinant. The final sources were plated on stainless steel planchets and counted.

The preirradiation chemistry has to be executed with great care. All labware were acid-washed and the reagents were of the purest grade. Still, the blanks and potential contamination procedures had to be controlled carefully. After irradiation, ^{233}Pa became the nuclide of interest, which was not prone to contamination, and thus the chemistry can be performed in a regular manner using reagent-grade chemicals.

A complete description of the neutron activation analysis method has been given in Huh and Bacon (19). Interested readers are referred to that paper for more details.

Mass Spectrometric Analysis

Our procedure for the mass spectrometric analysis were basically similar to that for NAA, with some minor modifications. Briefly, the sample solution was made to 8N in HNO_3 (ULTREX grade) and passed through an anion column (AG1x8, 100-200 mesh). After adequate elution with fresh 8N HNO_3, Th was eluted off the column with 12N HCl (ULTREX grade) and then evaporated to dryness. The above procedure was repeated two more times, with the third time performed using a micro-column (4 mm i.d., 1 cm length). Th was finally eluted off the micro-column with ∿200 μl of 12N HCl, dripped directly onto a teflon pad (3/4 inch dia.), and evaporated to dryness. One drop (∿50 μl) of 8N HNO_3 was added to the teflon pad to dissolve the final refinate, which was then evaporated to ∿1 μl in the presence of a single 150-μl anion exchange resin bead. The bead was subsequently recovered using a rhenium wire, loaded onto a rhenium filament and placed into the mass spectrometer. Following carbonization of the resin bead at low temperatures, thorium was volatilized at high temperatures with the different isotopes being counted in their respective mass regions of the spectrum.

REVIEW OF RESULTS

As mentioned earlier, to measure the thorium isotope composition of seawater by the conventional α-particle counting technique, a very large sample must be processed with minimal blank contribution. Recent work (e.g., 12-14, 22) has used MnO_2-coated media to concentrate Th isotopes from seawater to yield activities equivalent to those in volumes of the order of 10^3 liters. In addition to the sampling of large volumes, blank corrections have been taken more seriously. Frequently, it has been found that, compared with "sample" ^{232}Th activities, blank levels are high and variable (for example, 0.026-0.073 dpm per sample reported by Knauss and co-workers [7]). In many samples virtually no ^{232}Th could be detected above even the low end of blank levels. Those results seen to be above blank levels are often associated with fairly high uncertainties. Nevertheless, it has been more or less agreed that the ^{232}Th content of surface ocean water is $\leq 2\times10^{-5}$ dpm/l as set by Kaufman (23). Moore has concluded that the ^{232}Th content of both surface and deep waters in the Pacific Ocean may be $(1-2)\times10^{-5}$ dpm/l but has yet to be measured unequivocally (22).

Besides knowing the actual content of ^{232}Th in seawater, another equally important question is: What is its distribution between dissolved and particulate forms? Figure 1 summarizes previous data determined by α-spectrometry; it shows that near-shore waters usually had much higher, but variable, ^{232}Th content than open-ocean waters. In view of the general observation that the concentration of suspended particulate matter increases when the land is approached, I tend to suspect that the reported ^{232}Th contents could merely reflect the particulate concentration of the sampled waters. It has to be noted that the data presented in Fig. 1 were obtained using a variety of sampling techniques. The large fluctuation in the ^{232}Th concentration near the continents might be the result of different degree of removal of particulates. An ability to determine ^{232}Th in small volumes of seawater would allow the use of fine membrane filters to separate the particulate material more cleanly and could make feasible other experiments, such as observation of the dialysis behavior of Th in seawater.

Our laboratory at OSU has been engaged in the determination of Th concentration in seawater by both the NAA and the MS methods since 1984. The Caltech group under Drs. Wasserburg and Chen has also been undertaking the MS analysis of Th in seawater in recent years. To date, there are only two profiles published in the literature by these two groups. One is the profile of dissolved ^{232}Th in the Eastern Caribbean Sea, measured by Huh and Bacon using the NAA method (19); the other is that of total ^{232}Th

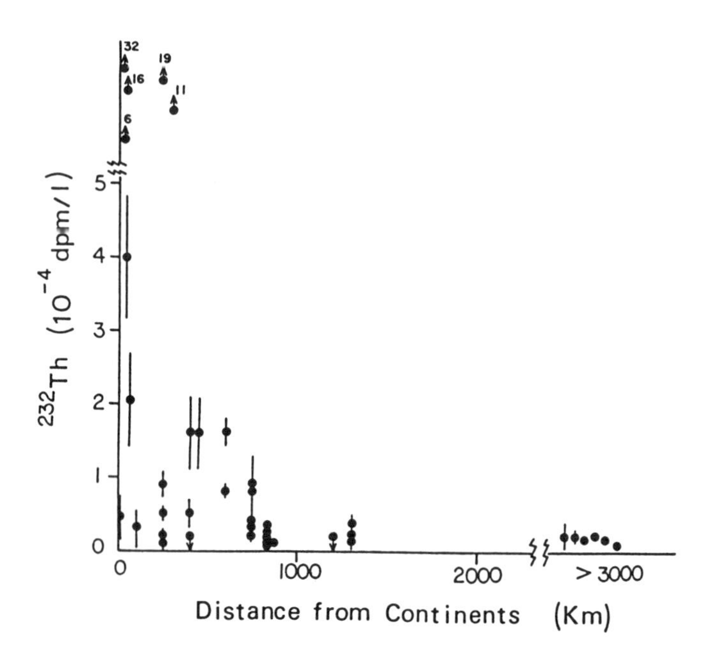

Figure 1. Reported seawater ^{232}Th concentration measured by alpha-spectrometry

in the open Atlantic Ocean reported by Chen and co-workers based on the MS method (20). The Caribbean Sea profile (Figure 2) showed relatively high concentrations in the upper 200 m with a maximum of 8.7×10^{-5} dpm/l near the base of the mixed layer. Average concentrations in deep water (2.7×10^{-6} dpm/l) were uniform and atypically low, perhaps because of unusually rapid scavenging in the restricted basin. The profile in the open Atlantic Ocean did not show any systematic trends, with an average concentration of 2.9×10^{-5} dpm/l. In addition to the above, there have been some discrete data points published. Greenberg and Kingston (24) developed a multi-element neutron activation analysis method and measured Th, along with fourteen other metals, in seawater samples from the Chesapeake Bay. The ^{232}Th concentration they measured (4×10^{-5} dpm/l) was the lowest ever reported for coastal waters. The Caltech group reported, for the first time, ^{232}Th content in hydrothermal fluids from the East Pacific Rise at 21°N and Guaymas Basin (21). Based on the concentrations they measured (2.4×10^{-5} to 1.0×10^{-3} dpm/l), the authors concluded that hydrothermal input contributes little to the oceanic Th flux.

Using the MS method, our laboratory has obtained profiles of ^{230}Th and ^{232}Th, in both dissolved and particulate forms, in the water column of Santa Monica Basin. A surface enrichment was observed for both dissolved ^{230}Th and dissolved ^{232}Th, strongly suggesting a terrestrial source for both long-lived Th isotopes. The concentration of dissolved ^{232}Th dereased by a factor of ∿4 from the surface (5.1×10^{-5} dpm/l) to 5 m off the bottom (1.4×10^{-5} dpm/l). The concentration levels, averaging 2.8×10^{-5} dpm/l, were comparable to most published data; but I tend to consider that the dissolved ^{232}Th concentration in the SM basin, which is only ∿30 km

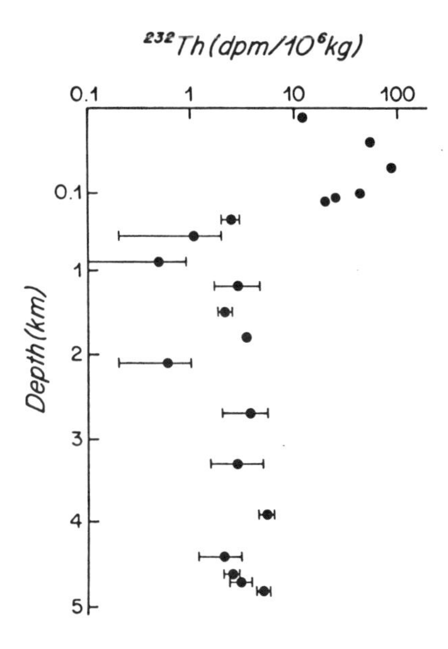

Figure 2. Dissolved ^{232}Th profile in the Eastern Caribbean Sea

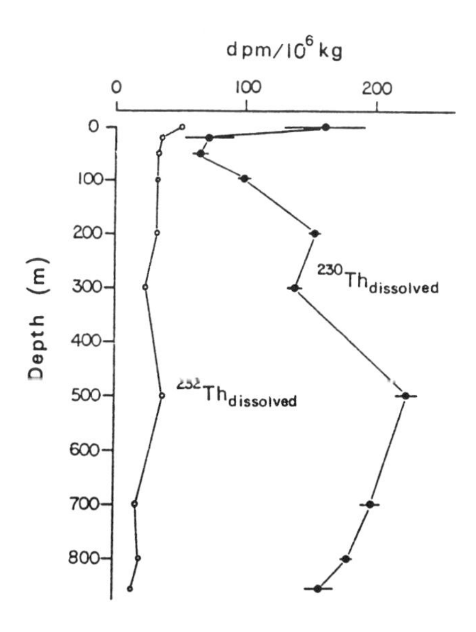

Figure 3. Profiles of dissolved ^{230}Th and ^{232}Th in the Santa Monica Basin

offshore, could very well be much higher than in open ocean waters. This conjecture, however, will have to be tested by future work.

In conclusion, the late technical advancement, outlined above, has made possible rapid sampling and analysis of samples with reduced potential contamination problems. Oceanographically-consistent ^{232}Th data are starting to emerge. We are currently planning to measure the distribution of ^{232}Th (as well as ^{230}Th) in both major oceans, at different distance away from the continents, and across redox boundaries. By comparing the oceanic distribution of Th with those of other biogeochemically active and inactive chemical species, we intend to gain a first order appreciation of the biogeochemical behavior of ^{232}Th in the ocean.

ACKNOWLEDGEMENTS

This research was supported by grants from the U.S. National Science Foundation (OCE83-14706 and OCE85-15631) and Department of Energy (DE-FG-05-85ER60340). My association with Drs. Bacon and Beasley inspired this work.

REFERENCES

1. Ku, T.L., The uranium-series methods of age determination. Ann. Review of Earth Planet. Sci., 1976, 4. 347-79.

2. Ivanovoch, M. and Harmon, R.S., Uranium Series Disequilibrium: Applications to Environmental Problems, Claredon Press, Oxford, 1982, 571 pp.

3. Bhat, S.G., Krishnaswami, S., Lal, D., Rama and Moore, W.S., $^{234}Th/^{238}U$ ratios in the ocean. Earth Planet. Sci. Lett., 1969, 5. 483-91.

4. Moore, W.S., Measurement of ^{228}Ra and ^{228}Th in sea water. J. Geophys. Res., 1969, 74. 694-704.

5. Broecker, W.S., Kaufman, A. and Trier, R.M., The residence time of thorium in surface water and its implications regarding rate of removal of pollutants. Earth Planet. Sci. Lett., 1973, 20. 35-44.

6. Matsumoto, E., ^{234}Th-^{238}U radioactive disequilibrium in the surface layer of the ocean. Geochim. Cosmochim. Acta, 1975, 39. 205-12.

7. Knauss, K.G., Ku, T.L. and Moore, W.S., Radium and thorium isotopes in the surface waters of the East Pacific and coastal Southern California. Earth Planet. Sci. Lett., 1978, 39. 235-49.

8. Li, Y.H., Feely, H.W. and Santschi, P.H., ^{228}Th-^{228}Ra radioactive disequilibrium in the New York Bight and its implications for coastal pollution. Earth Planet. Sci. Lett., 1979, 42. 13-26.

9. Bruland, K.W., Franks, R.P., Landing, W.M. and Soutar, A., Southern California inner basin sediment trap calibration. Earth Planet. Sci. Lett., 1981, 53. 400-8.

10. Koide, M., Bruland, K.W. and Goldberg, E.D., $^{228}Th/^{232}Th$ and ^{210}Pb

geochronologies in marine and lake sediments. Geochim. Cosmochim. Acta, 1973, 37. 1171-87.

11. Huh, C.A., Zahnle, D.L., Small, L.F. and Noshkin, V.E., Budgets and behaviors of uranium and thorium series isotopes in Santa Monica Basin sediments. Geochim. Cosmochim. Acta, 1987, 51. in press.

12. Bacon, M.P. and Anderson, R.F., Distribution of thorium isotopes between dissolved and particulate forms in the deep sea. J. Geophys. Res., 1982, 87. 2045-56.

13. Anderson, R.F., Bacon, M.P. and Brewer, P.G., Removal of ^{230}Th and ^{231}Pa from the open ocean. Earth Planet. Sci. Lett., 1983, 62. 7-23.

14. Anderson, R.F., Bacon, M.P. and Brewer, P.G., Removal of ^{230}Th and ^{231}Pa at ocean margins. Earth Planet. Sci. Lett., 1983, 62. 73-90.

15. Nozaki, Y., Horibe, Y. and Tsubota, H., The water column distributions of thorium isotopes in the western North Pacific. Earth Planet. Sci. Lett., 1981, 54. 203-16.

16. Nozaki, Y., Yang, H.S. and Yamada, M., Scavenging of thorium in the ocean. J. Geophys. Res., 1987, 92. 772-8.

17. Nozaki, Y. and Horibe, Y., Alpha-emitting thorium isotopes in northwest Pacific deep waters. Earth Planet. Sci. Lett., 1983, 65. 39-50.

18. Coale, K.H. and Bruland, K.W., ^{234}Th:^{238}U disequilibrium within the California Current. Limnol. Oceanogr., 1985, 30. 22-33.

19. Huh, C.A. and Bacon, M.P., Thorium-232 in the Eastern Caribbean Sea. Nature, 1985, 316. 718-20.

20. Chen, J.H., Edwards, R.L. and Wasserburg, G.J., ^{238}U, ^{234}U and ^{232}Th in seawater. Earth Planet. Sci. Lett., 1986, 80. 241-51.

21. Chen, J.H., Wasserburg, G.J., von Damm, K.L. and Edmond, J.M., The U-Th-Pb systematics in hot springs on the East Pacific Rise at 21°N and Guaymas Basin. Geochim. Cosmochim. Acta, 1986, 50. 2467-79.

22. Moore, W.S., The thorium isotope content of ocean water. Earth Planet. Sci. Lett., 1981, 53. 419-26.

23. Kaufman, A., The Th-232 concentration of surface ocean water. Geochim. Cosmochim. Acta, 1969, 33. 717-24.

24. Greenberg, R.R. and Kingston, H.M., Trace element analysis of natural water samples by neutron activation analysis with chelating resin. Anal. Chem., 1983, 55. 1160-65.

SORBENTS FOR THE EXTRACTION OF RADIONUCLIDES FROM NATURAL WATERS

R.M. Gershey and D.R. Green
Seakem Oceanography Ltd.
Argo Bldg., Bedford Institute of Oceanography
P.O. Box 696
Dartmouth, Nova Scotia B2Y 3Y9
CANADA

ABSTRACT

Two solid phase sorbents for the extraction of radionuclides from natural waters have been adapted for use with a microprocessor-controlled in-situ sampler. Consideration was given to the following performance criteria: efficiency of sorption, flow characteristics, mechanical stability, and cost.

The sorbents developed were: 1) potassium cobalt ferrocyanide on silica gel ($KCoFC/SiO_2$) for the extraction of cesium and 2) manganese dioxide on XAD-7 macroreticular acrylic resin (MnO_2/XAD-7) for the extraction of radium, thorium and actinium. MnO_2-coated acrylic beads can be prepared easily and reproducibly, and have packing and flow characteristics that are superior to those of coated acrylic fiber currently in use for radium extraction. We have adapted an existing method for the extraction of cesium from water with $KCoFC/SiO_2$ for use with the in-situ sampling pump.

The concentration profiles in our MnO_2/XAD-7 extraction columns indicate quantitative extraction of radium from both tap water and seawater samples spiked with Ra-226. The removal of cesium from seawater and freshwater samples spiked with Cs-137 also is essentially quantitative under our test conditions.

INTRODUCTION

The extraction columns described here were developed specifically for use with the Seastar in-situ water sampler. They may be used, however, in a variety of sampling applications when due consideration is given to flow rates and volumes of water sampled.

We studied two columns for the extraction of radionuclides from seawater:

I. Potassium cobalt ferrocyanide on silica gel for the extraction of cesium

II. Manganese dioxide on XAD-7 macroreticular acrylic resin for the extraction of radium, thorium and actinium

In both cases, the technology used has been adapted from laboratory methods that previously have found acceptance in the scientific community. The following pages briefly describe the physical specifications and extraction efficiency of these columns.

EXTRACTION COLUMNS FOR RADIUM

Early in the last decade, the common procedure for extracting Ra-226 from natural waters was co-precipitation with $BaSO_4$ or by stripping 20-litre water samples of Rn-222, the immediate daughter of Ra-226 (1,2). Interest in the measurement of Ra-228 (which is 1 to 2 orders of magnitude lower in concentration than Ra-226) called for a procedure which would be able to process samples that were thousands of liters in volume (2). Thus the idea of using a solid phase sorbent held by a support of some kind was conceived. The first such system used iron hydroxide supported on acrylic fibre (3). In attempts to find more efficient sorbents, barium sulfate, various zirconium salts (OH^{1-}, PO_4^{3-}, and Mo_4^{2-}) and manganese dioxide were tested. The most effective of these proved to be MnO_2-coated acrylic fibre (4). Following optimization of the process for making the impregnated fibre (5), the method has found wide use in sampling seawater (6) and fresh water (7).

The sorbent we have selected for the extraction of radium from water is manganese dioxide (MnO_2) supported on Amberlite[1] XAD-7 macroreticular acrylic resin rather than acrylic fibre. The MnO_2-coated acrylic beads can be prepared easily and reproducibly, and have packing and flow characteristics that are superior to those of coated acrylic fibre. A brief discussion of the development of the sorbent follows.

The high efficiency of MnO_2 in the extraction of radium is well accepted but little attention has been paid to the support on which it is carried. Presently two methods are in general use: 1) pumping or otherwise exposing coated acrylic fibres to the sample or 2) freshly precipitating MnO_2 in the sample and filtering it through a cartridge filter. The second method is unsuitable for use with **in-situ** sampling techniques and was not considered further.

Our initial efforts were directed to the task of adapting the coated acrylic fibre technology for use with the Seastar **in-situ** sampler or other **in-situ** pumping systems. In evaluating this material we considered the following criteria:

- efficiency of sorption
- flow characteristics
- mechanical stability
- ease and reproducibility of preparation
- cost

---

[1]Rohm & Haas Co., Philadelphia, PA.

MnO_2-coated acrylic fibre has proven to extract radium with efficiencies ranging from 85% to effectively quantitative recovery by various workers. There are, however, problems with maintaining a uniform flow of sample through columns packed with this material (7). In addition, there is a significant and variable amount of "washout" of MnO_2 particles that are not firmly bonded to the fibre. To coat the acrylic fibre with MnO_2, it is immersed in a solution of $KMnO_4$ at a given temperature for a certain amount of time. The fibre is bulky when dry and hard to handle when wet, making the process a difficult one to perform reproducibly and efficiently. After drying, the product must be "fluffed" to restore its former volume; a process that is neither convenient nor reproducible. All of the above problems were experienced in our lab when producing and testing MnO_2-coated acrylic fibre.

Seeking a more practical alternative to the use of acrylic fibre as a support, we investigated the use of Amberlite XAD resins. This choice was made on the basis of our previous experience with XAD-2 resin as a sorbent for trace organics in water. This material satisfies the mechanical and flow requirements for use in columns designed for **in-situ** sampling.

Considering that the permanganate oxidation process should be able to coat almost any organic substrate with MnO_2 (5), we attempted to use XAD-2 polystyrene resin as a support. We were not able to successfully coat this material with MnO_2 and next tried XAD-7 acrylic ester resin. XAD-7 is rapidly coated with a uniform and mechanically stable coating of MnO_2 upon treatment with potassium permanganate solution at room temperature. The treated resin contains 5% by weight MnO_2, resulting in an almost identical amount per unit volume as contains the treated acrylic fibre. The acrylic beads have uniform packing and flow characteristics giving this support a distinct advantage over the acrylic fibre. XAD-7 resin is thus a more suitable support than the fibre based on at least three of the five criteria listed above. The following sections of this report describe experiments that were done to determine the sorption efficiency of MnO_2-coated XAD-7 resin, the most important criterion.

The physical characteristics of the columns are given below:

Column dimensions:	37 cm x 1.9 cm dia. column = 85 cm^3
Packing material:	Amberlite XAD-7 macroreticular acrylic resin coated with manganese oxides
Mesh size:	20-50
Packing density:	0.34 g/cm^3

For use with the Seastar sampler, sorption columns must accommodate flow rates ranging from 50 to 150 ml/min. The pressure drops across a full column of coated XAD-7 were tested at 50, 100 and 150 ml/min while pumping degassed, distilled water at room temperature (20°C). In the Seastar sampler, the sample is "pulled" through the column by a pump downstream. This is done in order to avoid contamination from the pump or flowmeter. The resultant pressure drop was measured with a closed-end manometer connected to the low pressure end (i.e. the outlet) of the column. Figure 1 shows the results of these tests along with the results obtained when testing coated acrylic fiber.

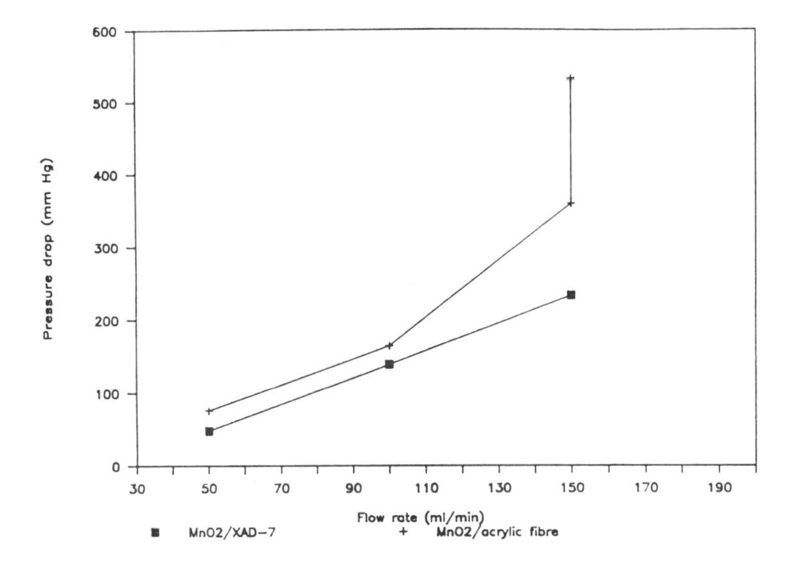

Figure 1. Pressure drop as a function of flow rate for columns loaded with 95 ml MnO_2-coated XAD-7 resin. Column dimensions as noted above. Acrylic fibre collapsed at 150 ml/min.

The resin beads have a clear advantage over the fiber, which compressed at the higher flow rates resulting in an unacceptably high and inconstant pressure drop. Based on earlier results for XAD-2 resin, these pressure drop data for the XAD-7 resin can be extrapolated linearly to flow rates of at least 400 ml/min.

Radium Extraction Efficiency

The extraction efficiencies of the MnO_2 columns were determined by analyzing the concentration profiles obtained by extruding the sorbent and dividing it into sections. The concentration expected to be found in each of these sections can be calculated with a function of the following form:

$$[Ra]_i = C_0 E (1-E)^{Xi}$$

where $[Ra]_i$ is the radium concentration per gram of sorbent in each section, C_0 is the initial radium concentration in the water sample, **E** is the extraction efficiency (percent radium removed from the water sample per unit weight of sorbent), and **X** is the cumulative median weight of each section. **E** is a useful parameter in that it describes the overall efficiency of the sorption columns, encompassing the combined effects of flowrate, column dimensions and sorbent efficiency. The value of **E** was determined by plotting the logarithm of $[Ra]_i$ against X_i and applying linear least squares analysis.

Figure 2 and table 1 show the results of experiments that were done to test the extraction efficiency of the MnO_2/XAD-7 columns using 3-litre water samples spiked with 4780 Bq Ra-226. The sample was pumped throught the extraction column at a rate of 150 ml/min.

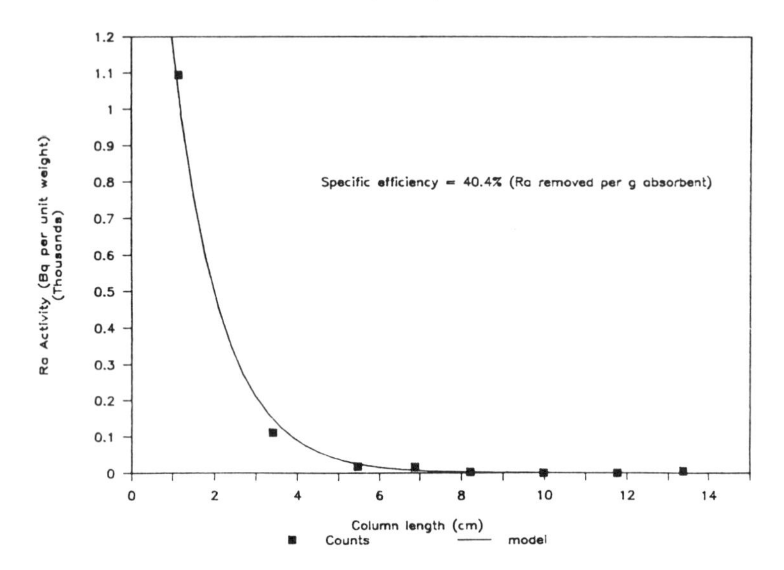

Figure 2 Concentration profile of radium recovered from sea water spiked with 4780 Bq Ra-226. Sorbent: MnO_2-coated XAD-7 resin. Inlet of column is to the left of the figure. Each gram of sorbent is equivalent to 1.02 cm of column length.

Table 1 Recovery of radium from spiked sea water sample. Details as given in Figure 2.

COLUMN PROFILE:			**COLUMN EFFICIENCY:**	
SECTION	ACTIVITY (Bq)	WT (g)	LENGTH (cm)	% RECOVERY
1	4181	3.82	3.90	87.5
2	416	3.73	7.70	96.2
3	56	3.13	10.89	97.3
4	26	1.57	12.49	97.9
5	8	2.87	15.42	98.1
6	1	3.03	18.51	98.1
7	ND	2.93	21.50	98.1
8	13	2.40	23.95	98.4

Radium was completely recovered from spiked tap water samples within the first half (15 cm) of the sorption columns with an average specific efficiency of 51 %/gram sorbent. The recoveries were slightly lower (41 %/g) in seawater but 100% recovery was still achieved in the first half of the column. When tap water samples spiked with 0.58 mg Ca^{++}/l as a carrier were used, 60 %/g recovery was attained.

Tests were run to determine how efficiently the sorbed radium could be removed from the resin. When the sorption column was rinsed with 2N nitric acid, 91% of the radium was removed in the first 95 ml. A further 200 ml acid yielded an additional 7% giving a total recovery of 98%.

The recovery was unexpectedly lower when the resin was removed from the column and treated with 8N nitric acid in a beaker. This procedure entirely removes the MnO_2 coating from the resin but only 73% of the radium was in the acid wash. The resin was found to contain a residual amount of radium that could not be removed by the acid, suggesting that this harsh treatment may result in a modification of the resin matrix that sequesters the radium.

EXTRACTION COLUMNS FOR CESIUM

Several methods for the determination of radiocesium, strontium and other fission products have been used to study the distribution of these nuclides in fresh water and seawater. Ammonium molybdophosphate (AMP) has been used to treat discrete water samples (8) while a variety of ferrocyanide ion-exchangers have been used to sorb cesium **in-situ** (9). Granular forms of copper, cobalt, zirconium, molybdenum, vanadium, nickel, zinc, iron, tin and tungsten ferrocyanides have been tested for their efficacy in sorbing cesium directly when they are exposed to water in a variety of ways (10). In these systems, the water contacts the ferrocyanide passively or is pumped over the sorbent.

Ion exchange resins or silica gel supports treated with various ferrocyanides have also been used for **in-situ** collection of cesium (11). These materials offer several advantages over the pure compounds for the concentration of cesium from large volumes of natural waters, including improved efficiency and selectivity.

The sorbent we have selected for the extraction of cesium from water is potassium cobalt ferrocyanide on a silica-gel support ($KCoFC/SiO_2$). In earlier work, the granular form of the compound was used directly without the use of a support. The superior flow characteristics of columns packed with a ferrocyanide-coated support, however, recommends that this approach be used. We have adapted an existing method (11) for the extraction of cesium from water by $KCoFC/SiO_2$ for use with the Seastar sampler. The physical characteristics are given below:

Column dimensions: 15 cm x 1.9 cm dia. column = 25 cm^3
Packing material: Chromatographic grade silica gel treated with potassium ferrocyanide and cobalt chloride
Mesh size: 60-80
Packing density: 0.58 g/cm^3

Figure 3 shows the pressure drop characteristics of 15 cm-long columns packed with KCoFC/SiO_2 along with pressure drop data for longer (37 cm) sorption columns packed with KCoFC on an Amberlite IRA-400 resin support. The resin has superior flow characteristics (lower pressure drop) than the silica gel support, but the sorption efficiency of this material for cesium was found to be inferior to that of KCoFC/SiO_2. While silica gel is a finer mesh size material, shortening the length of the column reduced the pressure drops to reasonable values. Based on earlier experience, these pressure drop data for can be extrapolated linearly to flow rates of at least 400 ml/min.

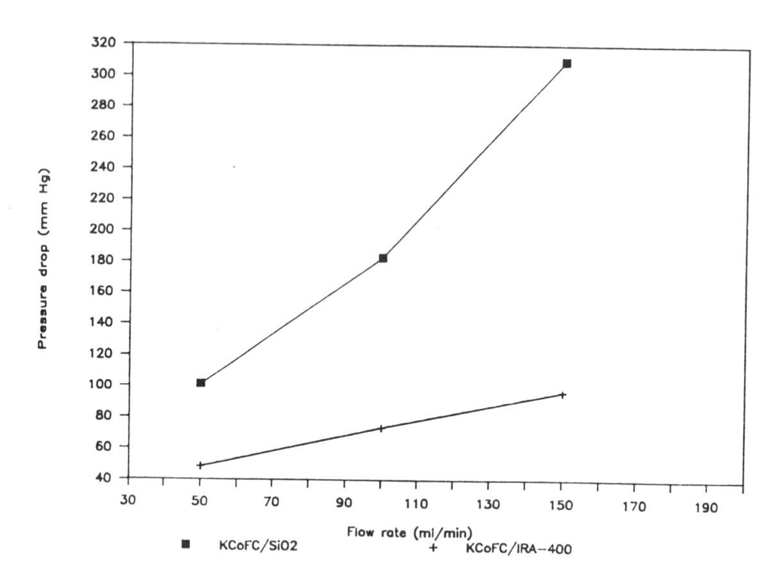

Figure 3. Pressure drop as a function of flow rate for columns loaded with Amberlite IRA-400 resin (20-40 mesh) and silica gel (60-80) mesh. Column dimensions as noted above.

Cesium Extraction Efficiency

The extraction efficiencies of the KCoFC/SiO_2 columns for cesium were determined in the same manner as for the MnO_2 columns. Three-litre water samples were spiked with approximately 7.5 kBq Cs-137 and pumped through the sorption columns at 150 ml/min. The results are shown in figure 4.

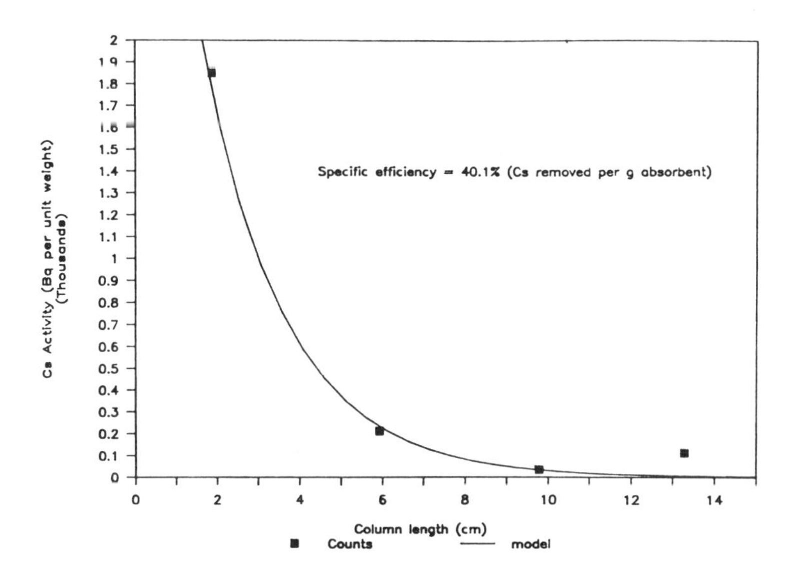

Figure 4 Concentration profile of cesium recovered from sea water spiked with 8498 Bq Cs-137. Sorbent: KCoFC on silica gel. Inlet of column is to the left of the figure. Each gram of sorbent is equivalent to 0.60 cm of column length.

Table 2 Recovery of cesium from spiked sea water sample. Details as given in Figure 4.

COLUMN PROFILE:			COLUMN EFFICIENCY:	
SECTION	ACTIVITY (Bq)	WT (g)	LENGTH (cm)	% RECOVERY
1	6791	3.67	2.20	79.9
2	903	4.28	4.77	90.5
3	115	3.27	6.73	91.9
4	404	3.61	8.90	96.7

Cesium was recovered with an efficiency of about 96% from spiked sea water samples within the first 8 cm of the sorption columns containing KCoFC on silica gel, with a specific efficiency of 40 %/gram sorbent. The recoveries were slightly higher (45-50 %/g) from spiked tap water samples (with or without stable cesium carrier) resulting in slightly more than 96% recovery in the first 8 cm of the column.

REFERENCES

1. Broecker, W.S., An application of natural radon to problems in ocean circulation. In Oceans and Fresh Waters, ed. T. Ichiye, Lamont-Doherty Geological Observatory, Palisades, N.Y., 1965, pp. 116-145.

2. Moore, W.S., Measurement of Ra-228 and Th-228 in sea water. J. Geophys. Res., 1969, 74. 694-704.

3. Krishnaswami, S., Lab, D., Somayajulu, B.L.K., Dixon, F.S., Stonecipher, S.A., and Craig, H., Silicon, radium, thorium, and lead in sea-water: In-situ extraction by synthetic fiber. Earth Planet. Sci. Lett., 1972, 16. 84-90.

4. Moore, W.S. and Reid, D.F., Extraction of radium from natural waters using manganese-impregnated acrylic fibers. J. Geophys. Res., 1973, 78. 8880-86.

5. Moore, W.S., Sampling of radium-226 in the deep ocean. Deep-Sea Res., 1976, 23. 647-51.

6. Reid, D.F., Key, R.M. and Schink, D.R., Radium, thorium and actinium extraction from seawater using an improved manganese-oxide-coated fiber. Earth and Planet. Sci. Lett., 1979, 43. 223-6.

7. Moore. W.S. and Cook, L.M., Radium removal from drinking water. Nature, 1974, 253. 262-3.

8. Folsom, T.R., Hansen, N., Tatum, T.J., and Hodge, V.F., Recent improvements in methods for concentrating and analyzing radiocesium in sea water. J. Radiat. Res., 1975, 16. 19-27.

9. Folsom, T.R. and Sreekumaran, C., Some reference methods for determining radioactive and natrual caesium for marine studies. In Reference methods for marine radioactivity studies. Tech. Rep. 118, IAEA, Vienna, 1970, pp. 129-177.

10. Bhargava, S.S. and Venkateswarlu. K.S., Ion adsorption properties of molybdenum ferrocyanide, tungsten ferrocyanide and multiloaded molybdenum ferrocyanide-Dowex exchangers. Ind. J. Chem., 1976, 14A. 800-3.

11. Terada, K., Hayakawa, H., Sawada, K., and Kiba, T., Silica gel as a support for inorganic ion-exchangers for the determination of caesium-137 in natural waters. Talanta, 1970, 17. 955-63.

MARINE SCAVENGING OF Th AND Pa - EVIDENCE FROM MID-OCEAN RIDGE, CONTINENTAL MARGIN AND OPEN OCEAN SEDIMENTS

G.B. Shimmield and N.B. Price

Grant Institute of Geology
University of Edinburgh
West Mains Road
Edinburgh, EH9 3JW
Scotland, UK

ABSTRACT

Thorium, protactinium and uranium isotopes have been measured in sediments from two important, but contrasting, areas of the Pacific identified as being preferential sinks for Pa-231 relative to Th-230. Fifteen box-cores were collected from transects at the East Pacific Rise (10° & 20° S) and the Baja California continental margin. U, Fe and Mn contents increase towards the ridge crest confirming the hydrothermal nature of the sediment, whereas these same elements reflect increasing diagenesis shorewards at the margin transect. The sediment record of hydrothermal activity at 20° S is one of pulsed episodes of approximately 1000 years duration. Under both oceanic regimes high measured to predicted fluxes increase from typical open ocean values of 0.69 for Th-230 and 0.49 for Pa-231, to 5.02 and 10.82 respectively at the ridge crest, and 2.56 and 3.59 respectively at the ocean margin. These results indicate enhanced scavenging by Fe and Mn oxyhydroxides of hydrothermal and diagenetic origin which remove Th-230 and Pa-231 without fractionation from seawater. Local hydrographic conditions are important for lateral supply of dissolved Th and Pa.

INTRODUCTION

Recent evidence suggests that two radionuclides, Th-230 and Pa-231, commonly used to study the removal of reactive elements in the ocean, in fact follow surprisingly different removal pathways (1-4). Both nuclides are the daughter products of U decay and are removed from the ocean on a time scale many times less than their half-lives. As a result, radioactive decay in the water column is insignificant and therefore the theoretical production Th-230/Pa-231 ratio is 10.8. Water column studies have shown the highly reactive behaviour of Th-230 towards open ocean particulate matter (1,5) with a residence time of 30 years, whereas Pa-231 has less affinity, and undergoes lateral transport to other environments where intense scavenging may occur (2,3). Consequently, its residence time of 150-200 years is longer than that of Th-230 (6). Within surficial sediment such removal processes are reflected in Th-230/Pa-231 ratios deviating from 10.8. Higher values may result from preferential Th-230 scavenging, bioturbation, sediment erosion and core top loss (4), but lower values uniquely indicate preferential Pa-231 removal. A recent compilation of data (4) for the Pacific suggests that 30-40% of the sea floor is an effective sink for Pa-231. Apart from the Antarctic, the continental margins, and possibly the mid-ocean ridge (MOR) (7) areas, account for this sink. Unfortunately, quantitative study of actinide removal to the sediments in these areas is lacking, and therefore only broad statements concerning the relative importance of (i) high aluminosilicate flux, (ii) Mn and Fe redox recycling, (iii) high biogenic fallout, and (iv) resuspension effects have been made. By studying the distribution and behaviour of Th-230 and Pa-231 at both an ocean margin, where factors 1 and 2 dominate, and a hydrothermally-active MOR where Fe and Mn inputs to the water column are intense, we will demonstrate that enhanced Pa-231 removal is indeed the result of chemical scavenging by both Fe and Mn particulates. In addition, the importance of active MORs on the flux of Th-230 and Pa-231 to the sea floor is discussed.

Table 1. Sample locations, water depths and measured radionuclide activities.

Station	Location	Depth (m)	Sample (cm)	U (ppm)	U-234 (dpm/g)	Th(ex)-230 (dpm/g)	Pa(ex)-231 (dpm/g)	Th(ex)-230 / Pa(ex)-231
East Pacific Rise - 10°S, TGT-154								
154-4	12°30'S 134°50'W	4606	6-7	1.68±0.06	1.22±0.05	76.7±1.7	2.84±0.12	27.0
154-5	12°20'S 125°57'W	3885	1-2	0.27±0.03	0.24±0.02	27.0±0.8	1.88±0.10	14.4
154-6	12°06'S 119°48'W	3711	2-3	0.35±0.02	0.31±0.02	15.2±0.4	1.18±0.08	12.9
154-8	10°49'S 113°53'W	3395	0-1	0.26±0.04	0.27±0.04	12.1±0.4	1.33±0.09	9.1
154-10	10°17'S 111°19'W	3216	0-1	0.71+0.03	0.59+0.03	13.5+0.3	2.17+0.11	6.2
East Pacific Rise - 20°S, TGT-154								
154-20	19°25'S 117°45'W	3470	0-1	0.39±0.03	0.36±0.03	27.6±0.9	1.56±0.07	17.7
154-19	19°50'S 116°38'W	3334	0-1	0.50±0.05	0.37±0.03	29.3±1.3	2.23±0.13	13.1
154-18	20°05'S 114°05'W	3129	0-2	4.16±0.16	3.72±0.13	16.1±0.5	3.21±0.17	5.0
Baja California - TGT-163								
163-9	24°16'N 115°04'W	3685	0-1	2.66±0.10	1.86±0.07	38.2±1.3	4.1±0.20	9.3
163-10	24°42'N 114°04'W	3402	0 10^{+}	2.68±0.13	2.09±0.08	23.8±0.7	3.0±0.15	7.9
163-14	24°49'N 113°45'W	3168	0-11^{+}	2.82±0.08	2.17±0.06	21.3±1.1	3.0±0.16	7.1
163-7	24°55'N 113°25'W	3647	0-13^{+}	3.00±0.11	2.11±0.08	19.7±0.8	3.0±0.18	6.6

Quoted uncertainties are based on 1-sigma counting statistics.
All activities are corrected for dilution by sea salt.
$^{+}$mixed layer depth and corrected mixed layer mean activities (3).

SAMPLING LOCATIONS AND METHODS

In order to examine the distribution and behaviour of Th and Pa in hydrothermal and ocean margin sediments two well-documented Pacific ocean localities were chosen for study. The R/V Thomas G. Thompson collected cores from the East Pacific Rise (EPR) along two transects at 10° and 20°S in 1980 (TGT-154), and from the Baja California margin (BCM) in 1981 (TGT-163). Core locations and water depths are given in Table 1. On each cruise box-cores were used to preserve the sediment-water interface intact. The western end of each transect represents the most pelagic, or open ocean-type, sediments of the cores studied. Sub-samples of the box-cores were sectioned, dried, and ground prior to analysis. Al, Ca, Fe and Mn were determined by X-ray fluorescence spectrometry. U, Th and Pa isotopes were measured by alpha particle spectrometry following complete digestion in HF, HNO_3 and $HClO_4$ acids, ion exchange chromatography and electrodeposition onto stainless steel planchettes. Pa-231 was determined indirectly via Th-227 which is assumed to be in secular equilibrium (8). $CaCO_3$ was calculated from excess Ca over the Ca/Al ratio in shale. Water contents, salt and dry sediment bulk densities of the TGT-154 samples were calculated from their $CaCO_3$ contents and the empirical relationship determined by Lyle and Dymond (9); similar TGT-163 data was determined by weight loss on drying the wet sediment and assuming a particle density of 2.6 g/cm^3.

RESULTS AND INTERPRETATION

Table 1 presents the U contents and surficial activities of the isotopes measured. Unsupported activities (Th(ex)-230 and Pa(ex)-231) were calculated by subtracting parent U activity equivalents (U-234 and U-235, respectively). At the EPR, U contents increase towards the ridge crest from an average value of 0.35 ppm, reaching a maximum of 4.16 ppm in 154-18, except at 154-4 where phillipsite is abundant. High U contents of MOR sediments are well known (7,10-12), and therefore we believe this indicates the extreme hydrothermal nature of the sediment in core 154-18 (see below). Table 1 also indicates that the activity ratio (Th(ex)-230/Pa(ex)-231) decreases towards the ridge crest from 27.0 and 17.7 in 154-4 and 154-20 on the 10°and 20°S transects respectively, to 6.2 and 5.0 in 154-10 and 154-18. Close to the ridge crest this activity ratio is less than the theoretical production ratio of 10.8. At the BCM surficial U contents increase slightly landwards from 2.66 ppm in 163-9 to 3.00 ppm in 163-7, perhaps due to increasing organic carbon and associated diagenesis in the more proximal cores (3). The activity ratio also decreases landwards from 9.3 to 6.6, always being less than the production ratio. These low activity ratios suggest preferential Pa-231 removal which we wish to investigate by studying the geochemical nature and the behaviour of the sediment during deposition in these two oceanographic regimes.

Off Baja California the sediment geochemistry records a pattern of increasing diagenesis landwards. Bioturbation causes sediment mixing down to a depth of 15-20 cm. The majority of the sediment input is terrigenous in origin with biogenic calcite increasing above 5 wt.% only in the last glacial episode (3). Low accumulation rates allow Mn recycling from the upper slope oxygen minimum window (14) and within the hemipelagic sediments (13) to dominate the diagenetic pattern of the sediment. Very little Fe remobilisation occurs, most being associated with detrital smectite (13), and no hydrothermal input is detectable. Depth profiles of Th(ex)-230 and Pa(ex)-231 have been published elsewhere (3) and display biomixing with log-linear decreases below. Core 163-9, taken to approximate to pelagic, open ocean sedimentation, is unmixed. Accumulation rates have been derived from the calcite stratigraphy of the cores and several C-14 ages (3).

In contrast, the EPR sediments display high calcite levels (Fig.1) of approximately 90 wt.% where the water depth is less than the CCD. In 154-18 the cyclical variation of calcite may be attributed to dilution by hydrothermal inputs (see below), whilst 154-4 indicates calcite

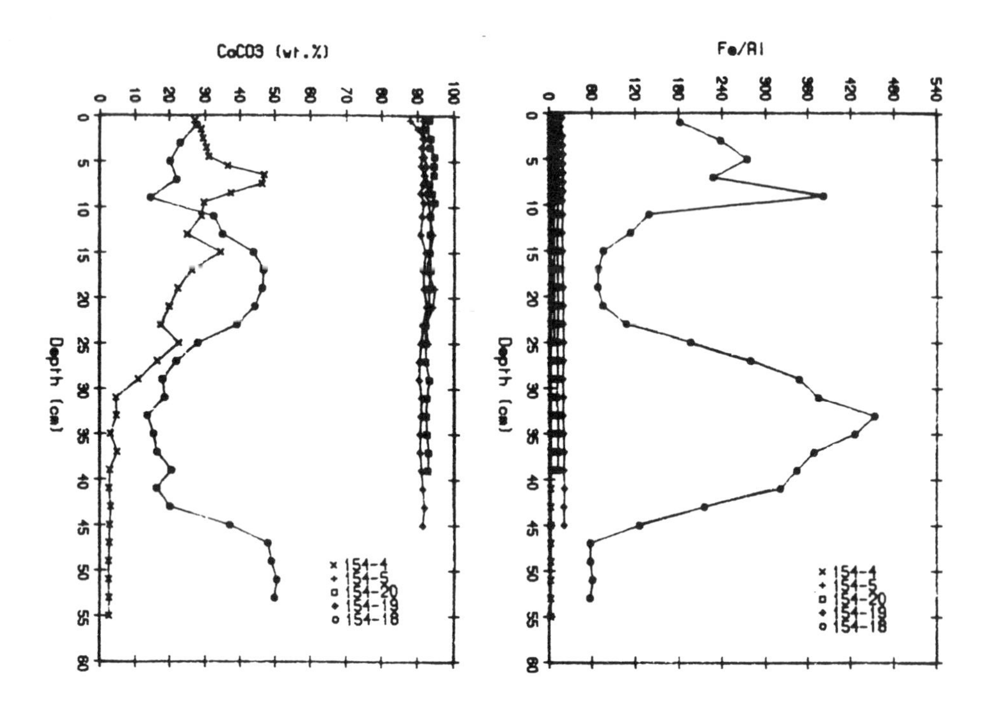

Fig.1 Calcite trends with depth in selected EPR cores.

Fig.2 Fe/Al trends with depth in selected EPR cores.

dissolution is occurring. The hydrothermal nature of these sediments is well illustrated in Figure 2 where the Fe/Al ratio increases above the deep-sea clay value (0.77) to achieve a mean maximum of 200 in 154-18. We interpret this data to indicate the input of hydrothermal Fe to the EPR sediments which decreases in a westwards direction consistent with water mass movement, He-3 (15) and hydrothermal plumes (16). An assymetric distribution of Fe- and Mn-rich hydrothermal sediments about the EPR at 10-20° S has been demonstrated (17,18). In addition, the antithetic covariation of Fe/Al and calcite in 154-18 suggests that pulsed input may be occurring with high Fe/Al during any particular hydrothermal episode (at 3-10 cm and 25-44 cm depth), returning to preferential calcite deposition during periods of less intense activity. Figure 3 presents the measured excess activities for Th-230 and Pa-231 in cores 154-4, 154-19, 154-20 and 154-18. In all but the latter log-linear decreases with depth are seen, with no evidence of biomixing in the upper sediments. Absence of bioturbation is to be expected due to the central gyre location of the EPR transects and hence low productivity and organic matter within the sediments. In a similar way, we believe the cyclical variation of Fe and Mn in 154-18 is not a result of diagenesis during organic matter degradation. Th(ex)-230 and Pa(ex)-231 also display cyclical variation in activity with peaks appearing coincident with the calcite profile (Fig.2).

In order to interpret the measurement of radionuclides in terms of flux, both at each station location and with depth (time) in 154-18, it is necessary to establish sediment accumulation rates for each core studied. Log-linear decreases in Th(ex)-230 and Pa(ex)-231 displayed in Figure 3

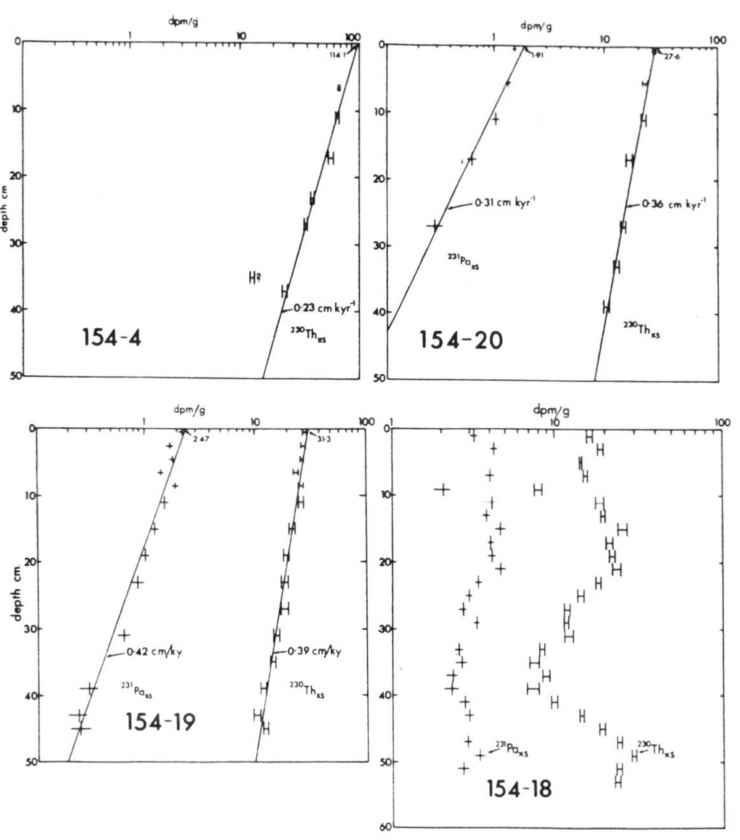

Fig.3 Th(ex)-230 and Pa(ex)-231 distributions with depth, and calculated accumulation rates.

may be fitted to an exponential function by least-squares regression in order to establish sediment accumulation rates, given the half-lives of Th-230 and Pa-231. However, the variation with depth in 154-18 in Figure 3 requires a different approach to estimate apparent accumulation rates at each sample interval down the core. This will allow us to define the rate of hydrothermal input and Th(ex)-230 and Pa(ex)-231 flux to the sediments. Over the geographical area of the EPR transects, atmospheric Al input is likely to be constant due to the lack of any continentally-derived detritus in the sediments. Hence on a carbonate-free basis (to avoid local calcite dissolution effects), the product of Al concentration [C] (Fig.4) and sediment density [p] should be inversely proportional to the sediment accumulation rate (19). This assumption rests on the flux of Al being constant in space and time, that Al behaves conservatively within the sediment, and that sediment redistribution by bottom currents is negligible. Flux calculations from 154-19 (F=0.0043 g Al/cm^2-ky) allows us to calculate the apparent accumulation rates in the other cores. Hence:

$$S = \frac{F}{C \cdot p} \qquad 1$$

These estimates of apparent accumulation rate are compared to rates derived from radionuclide depth decrease, where measured (Table 2).

The two methods of estimating sediment accumulation rates agree extremely well, allowing us to use the constant Al flux model to calculate accumulation rate changes down core in 154-18. These apparent rates and the age of each sample horizon are shown in Figure 5. From Table 2 and Figure 4 it is apparent that the accumulation rates gradually increase towards the ridge crest, but in 154-18 achieve maximum rates of 20 cm/ky. Comparison with the other figures reveal that high accumulation rates result in lower calcite and Th(ex)-230 and Pa(ex)-231 contents, but higher Fe/Al (and Mn/Al) ratios. Therefore the hydrothermal pulses interpreted from Figure 2 have a duration of approximately 1000 years (Fig. 4) during the Holocene.

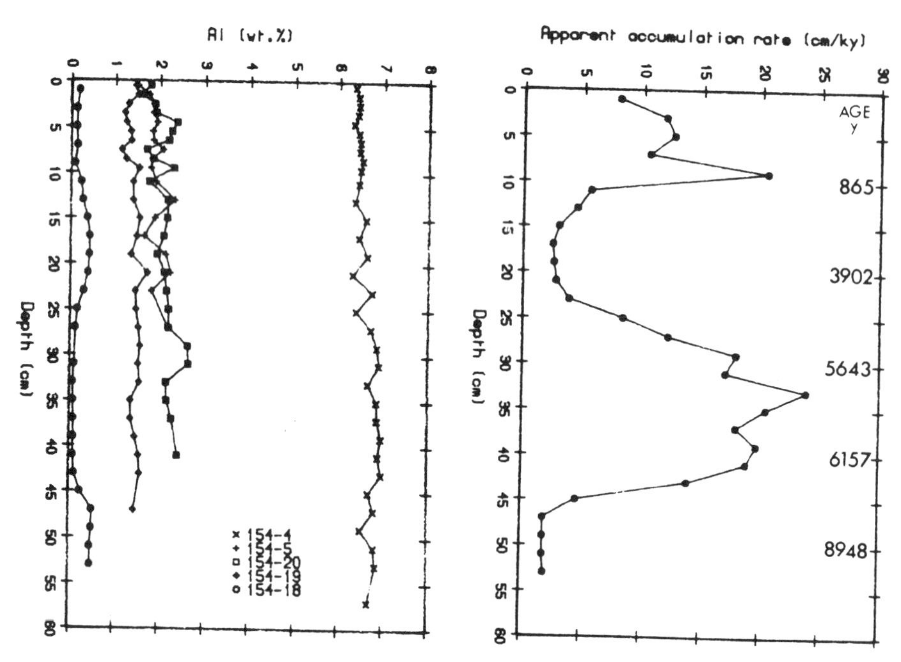

Fig.4 Al (wt.%) contents in EPR cores (calcite-free).

Fig. 5 Apparent sediment accumulation rates in 154-18.

Table 2. A comparison of constant Al flux and radionuclide determinations of sediment accumulation rates.

	154-4	154-5	154-20	154-19
Sp (cm/ky)	0.23	0.30	0.31	
Sm (cm/ky)	0.23	-	0.33	0.40

Sp - Sediment accumulation rate calculated from equation 3.
Sm - Mean sediment accumulation rate from depth decreases in Th(ex)-230 and Pa(ex)-231.

DISCUSSION

In this discussion we will present evidence for the enhanced sedimentary sink of both Th-230 and Pa-231 at the Baja California continental margin and the East Pacific Rise. By the calculation of input flux of these radionuclides it is possible to quantify this removal, and to compare the behaviour of the two actinides. Specific attention is drawn to scavenging by hydrothermal processes.

Table 3 presents the calculated radionuclide fluxes and the values predicted assuming vertical scavenging from the overlying water column. For both the EPR and the BCM transects the ratio of measured to predicted flux increases towards the ridge crest and landwards, respectively. Similarly, the measured flux of Pa(ex)-231 increases relative to Th(ex)-230 with increased efficiency of removal. In "open ocean" sediments (154-4,154-5,154-20) the average measured to predicted flux of Th(ex)-230 is 0.67, and 0.49 for Pa(ex)-231. This compares to Kadko's (20) average for Th(ex)-230 for 41 pelagic cores of 0.77$\pm$0.39. Anderson et al. (2) estimate open ocean scavenging of Th(ex)-230 to be 0.77 of that predicted, and 0.28 for Pa(ex)-231. Approaching the ocean margin, the removal of Th(ex)-230 is x2.56 greater than predicted (average of 163-10, 163-14 and 163-7) and x3.59 for Pa(ex)-231. From the Panama and Guatemala Basins Anderson et al. (2) estimate Th(ex)-230 and Pa(ex)-231 removal to be x1.85 and x3.70, respectively. At the EPR ridge crest (154-18) removal is the highest measured with Th(ex)-230 at x5.02 and Pa(ex)-231 at x10.82 that predicted. This data confirms that enhanced scavenging is occurring off Baja California and at 20° S on the EPR, relative to more open ocean environments.

We believe the preferential removal of Pa-231 is due to Mn recycling and precipitation via the oxygen minimum window off Baja California, and by hydrothermal plumes of Fe and Mn oxyhydroxides emanating from vents on the EPR. Further evidence for oxyhydroxide scavenging at the EPR (154-18) is seen in Figure 6. Age-corrected Th(ex)-230/Pa(ex)-231 ratios generally fall in the range 3-6 (except near the base of the core where the hydrothermal input is weak, Fig.2). MnO_2-coated fibres immersed in sea water (1) and Mn nodules (21) remove Th-230 and Pa-231 from sea water in this ratio, which is also approximately the dissolved ratio in open ocean water (1,2,6). Therefore we believe that the removal process operating at 20°S results in no fractionation of Th-230 and Pa-231, in contrast to the majority of the Pacific Ocean (4) where aluminosilicate debris is the dominant flux.

If hydrothermal Fe and Mn is such an efficient scavenger of Th-230 and Pa-231, the question arises as to why the depth profile of Th(ex)-230 and Pa(ex)-231 specific activities in Figure 3 do not co-vary with the hydrothermal Fe/Al signal (Fig.2). In Figure 6 the apparent flux of Th(ex)-230 has been calculated and indicates that although the sediment specific activity of Th(ex)-230 is less during hydrothermal episodes, the flux is greater. However, we believe the lower apparent activities result from the hydrothermal plume never rising above 2600 m (16), and therefore having a limited vertical water column depth from which to scavenge Th-230 and Pa-231, in a similar way to that observed for rare-earth scavenging in the same area (22). Nevertheless, the high fluxes are maintained by sea floor absorption of the two actinides, and additional advection of water

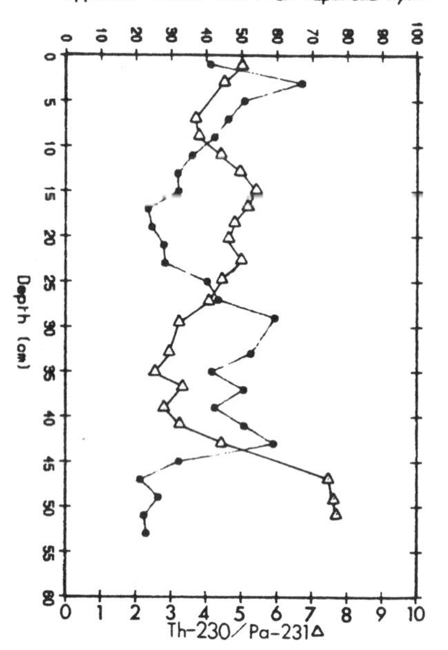

Fig.6 The apparent flux of Th(ex)-230 and Th(ex)-230/Pa(ex)-231 ratios with depth in 154-18.

Table 3. A comparison of measured and predicted fluxes of Th(ex)-230 and Pa(ex)-231

Station	Fm(Th)	Fm(Pa)	Fp(Th)	Fp(Pa)	Fm/Fp(Th)	Fm/Fp(Pa)
	(dpm/cm -ky)		(dpm/cm -ky)			
East Pacific Rise						
154-4	8.4	n.a.	12.1	-	0.69	-
154-5	5.8	0.41	10.2	0.95	0.57	0.43
154-20	6.8	0.47	9.1	0.85	0.74	0.55
154-19	9.2	0.74	8.7	0.81	1.06	0.91
154-18	41.2	8.22	8.2	0.76	5.02	10.82
Baja California						
163-9	9.6	1.0	9.6	0.9	1.01	1.16
163-10+	24.8	3.5	9.1	0.8	2.73	4.35
163-14+	17.7	2.7	8.2	0.8	2.16	3.39
163-7+	26.4	2.7	9.5	0.9	2.78	3.03

Fm = A_o.S.p where A_o is the surface or regression intercept activity, S is the accumulation rate, and p is the dry sediment bulk density

Fp(Th) = 2.63D, Fp(Pa) = 0.244D (D = depth in km.)

+ $A = A_o \left(1 + \frac{\lambda . L}{S}\right)$ to correct for biomixing of sediment (3)

mass containing the dissolved nuclides from the east (16). At the ocean margin off Baja California the horizontal transport of Th-230 and Pa-231 along isopycnals (2,6) supplies the excess Th-230 and Pa-231 which is also removed with little fractionation by Mn cycling and by the high aluminosilicate flux.

ACKNOWLEDGEMENTS

We would like to thank the Captain and crew of the R/V Thomas G. Thomson for their efforts, and Dr J.W. Murray for allowing us to participate on his cruises. Analytical expertise and stamina was provided by Frances Lindsay and Mike Saunders. GBS acknowledges NERC support on grant GR3/6175.

REFERENCES

1. Anderson, R.F., Bacon, M.P. and Brewer, P.G., Removal of Th-230 and Pa-231 from the open ocean. Earth Planet. Sci. Lett., 1983a, 62. 7-23.

2. Anderson, R.F., Bacon, M.P. and Brewer, P.G., Removal of Th-230 and Pa-231 at ocean margins. Earth Planet. Sci. Lett., 1983b, 66. 73-90.

3. Shimmield, G.B., Murray, J.W., Thomson, J., Bacon, M.P., Anderson, R.F. and Price, N.B., The distribution and behaviour of Th-230 and Pa-231 at an ocean margin, Baja California, Mexico. Geochim. Cosmochim. Acta, 1986, 50. 2499-2507.

4. Yang, H-S., Nozaki, Y., Sakai, H. and Masuda, A., The distribution of Th-230 and Pa-231 in the deep-sea surface sediments of the Pacific Ocean. Geochim. Cosmochim. Acta, 1986, 50. 81-89.

5. Nozaki, Y., Horibe, Y. and Tsubota, H., The water column distributions of thorium isotopes in the western North Pacific. Earth Planet. Sci. Lett., 1981, 54. 203-216.

6. Nozaki, Y. and Nakanishi, T., Pa-231 and Th-230 profiles in the open ocean water column. Deep-Sea Res., 1985, 32. 1209-1220.

7. Kadko, D., A detailed study of some uranium series nuclides at an abyssal hill area near the East Pacific Rise at 8° 45'N. Earth Planet. Sci. Lett., 1980, 51. 115-131.

8. Mangini, A. and Sonntag, C., Pa-231 dating of deep-sea cores via Th-227 counting. Earth Planet. Sci. Lett., 1977, 37. 251-256.

9. Lyle, M.W. and Dymond J., Metal accumulation rates in the southeast Pacific - errors introduced from assumed bulk densities. Earth Planet. Sci. Lett., 1976, 30. 164-168.

10. Veeh, H., Depostion of uranium from the ocean. Earth Planet. Sci. Lett., 1967, 3. 145-150.

11. Rydell, H., Kraemur, T., Bostrom, K. and Joensuu, O., Post-depositional injections of uranium-rich solutions into East Pacific Rise sediments. Mar. Geol., 1974, 17. 151-164.

12. Lalou, C. and Brichet, E., Anomalously high uranium contents in the sediment under Galapagos hydrothermal mounds. Nature, 1980, 284. 251-253.

13. Shimmield, G.B. and Price, N.B., The behaviour of molybdenum and manganese during early sediment diagenesis - offshore Baja California, Mexico. Mar. Chem., 1986, 19. 261-280.

14. Sawlan, J.J. and Murray, J.W., Trace metal remobilisation in the interstitial waters of red clay and hemipelagic sediments. Earth Planet. Sci. Lett., 1983, 64. 213-230.

15. Lupton, J.E. and Craig, H., A major helium-3 source at 15°S on the East Pacific Rise. Science, 1981, 214. 13-18.

16. Klinkhammer, G. and Hudson, A., Dispersal patterns for hydrothermal plumes in the South Pacific using manganese as a tracer. Earth Planet. Sci. Lett., 1986, 79. 241-249.

17. Bostrom, K. and Peterson, M.N.A., The origin of aluminium-poor ferromanganoan sediments in areas of high heat flow on the East Pacific Rise. Mar. Geol., 1969, 1. 427-447.

18. Dymond, J., Geochemistry of Nazca plate surface sediments: An evaluation of hydrothermal, biogenic, detrital and hydrogenous sources. Geol. Soc. Amer. Memoir, 1981, 154. 133-173.

19. Krishnaswami, S., Authigenic transition elements in Pacific pelagic clays. Geochim. Cosmochim. Acta, 1976, 40. 425-434.

20. Kadko, D., Th-230, Ra-226 and Rn-222 in abyssal sediments. Earth Planet. Sci. Lett., 1980, 49. 360-380.

21. Moore, W.S., Ku, T.L., Macdougall, J.D., Burns, V.M., Burns, R., Dymond, J., Lyle, M.W. and Piper, D.Z., Fluxes of metals to a manganese nodule: radiochemical, chemical, structural, and mineralogical studies. Earth Planet. Sci. Lett., 1981, 32. 151-171.

22. Ruhlin, D.E. and Owen, R.M., The rare earth element geochemistry of hydrothermal sediments from the East Pacific Rise: Examination of a seawater scavenging mechanism. Geochim. Cosmochim. Acta, 1986, 50. 393-400.

EFFECT OF NATURAL COLLOIDAL MATTER ON THE EQUILIBRIUM ADSORPTION OF THORIUM IN SEAWATER

Sherry E. H. Niven and Robert M. Moore

Department of Oceanography
Dalhousie University
Halifax, NS
Canada B3H 4J1

ABSTRACT

The equilibrium partitioning of Th among dissolved, colloidal, and particulate size fractions was studied under conditions similar to those of coastal marine waters in laboratory experiments using ^{234}Th tracer and ultrafiltration techniques. In colloid-free alumina suspensions (*i.e.* prepared in ultrafiltered seawater), distribution coefficients (K_d's) were constant over the range of particle concentrations studied (0.1 – 5.0 mg l^{-1}). Adsorption of Th by the alumina involved hydrolysis species that are retained by a 1,000 nominal molecular weight ultrafilter. Th adsorption was decreased by the presence of natural dissolved organic matter (*Phaeodactylum tricornutum* exudates) and by pre-existing colloidal matter.

INTRODUCTION

Background

The concentration, distribution, and transport of both naturally occurring and anthropogenic chemicals in the oceans are controlled by biogeochemical processes that partition solutes between the solid and aqueous phases. The dominant mechanism of this solid-solution partitioning is considered to be adsorption by suspended and settling particles; a process usually modelled as the complexation of dissolved metals by homogeneous –OH ligands on particle surfaces (1,2). Experimental partitioning data generally agree with this simple model; however, discrepancies between experimental observation and model prediction have been reported (3).

Solid-solution partitioning studies frequently use the equilibrium distribution coefficient (K_d : the ratio of particulate to dissolved metal), as an empirical measure of adsorption (4). According to the surface complexation model, K_d values should be independent of particle concentration as long as the metal concentration is low compared to the concentration of free surface sites. However, laboratory experiments with natural particles repeatedly show that K_d values decrease with an increase in particle concentration (4).

Recent work has suggested that colloidal-size particles play an important role in the partitioning of trace metals between solution and particulates, and that they may account for the discrepancies noted between experimental observation and theoretical prediction (3,5). The surface complexation models used to explain solid-solution partitioning define solids and solutes by thermodynamic considerations. However, the standard separation techniques used in geochemistry consider only large or filterable (*e.g.* > 0.45 μm) particles in the solid phase. Submicron or colloidal-size particles are included with truly dissolved species in the solution phase. For many geochemical processes, colloidal matter can be considered to be effectively dissolved since it does not settle from solution and it moves by advective and diffusive processes. However, in terms of understanding the solid-solution partitioning of trace metals, colloidal particles must be considered as part of the solid phase.

Morel and Gschwend (3) have considered the effect of inadvertently including colloidal-size particles with the solution phase in K_d measurements, and they concluded that experimental artifacts of phase separation could explain the thermodynamically inconsistent partitioning results observed in experiments. Little direct evidence of the importance of the colloidal fraction to the solid-solution partitioning of trace metals exists for natural waters; however, a study of PCB adsorption onto lake sediments showed a decrease in the observed K_d when colloidal particles were present in the samples (6).

Objectives of Paper

This paper presents experimental results that demonstrate the importance of colloidal-size particles in the equilibrium partitioning of trace metals between solid and aqueous phases. The experiments were part of a laboratory investigation of the marine geochemistry of Th in which solution complexation and particle sorption were studied under natural marine conditions with ^{234}Th tracer (7). Particulate, colloidal, and dissolved fractions of the samples were separated by sequential filtration through a 0.2 μm Nuclepore filter and Amicon ultrafilters, and all three fractions were analyzed for ^{234}Th by liquid scintillation counting. Separation of the dissolved and the colloidal Th species provided important information about the solid-solution partitioning of Th in seawater.

EXPERIMENTAL TECHNIQUES

Methods

Th isotopes are frequently used as analogues for other particle-reactive elements in studying their removal from the oceanic water column (8,9). Laboratory investigations of the marine geochemistry of Th can, therefore, improve the understanding of trace metal removal processes by providing a controlled and systematic study of the variables influencing Th partitioning between aqueous and solid phases. Laboratory studies of trace metal geochemistry are often complicated by the use of higher than natural concentrations of metal. However, the high specific activity of ^{234}Th (half-life of 24.1 days) allows liquid scintillation counting of its β-decay at low molar concentrations (detection limit of 10^{-17}M). Consequently, in these experiments, the partitioning of ^{234}Th among the dissolved, colloidal, and particulate fractions could be measured in reasonable sample volumes (100 ml aliquots from 2 l bottles) without increasing the natural concentration of Th (10^{-15}M ^{234}Th spikes were added to ultrafiltered coastal seawater with a ^{232}Th concentration of approximately 10^{-13}M).

Experiments were carried out under constant natural conditions of pH, salinity, and temperature; and variable concentrations of dissolved organic matter (DOM), colloidal, and particulate matter. Two liter polyethylene bottles were acid cleaned and soaked with ultrafiltered, photo-oxidized seawater for one week prior to use. Replicate bottles were then filled with fresh ultrafiltered seawater (with known DOM concentration) and colloidal and/or particulate matter was added from stock suspensions. Suspensions were equilibrated for 24 hours at 4°C before being spiked with ^{234}Th tracer. Acid stabilized ^{234}Th stock solution (100 μl) was added to give initial sample activities of $2-4\times10^4$dpm l^{-1}. The original pH of the seawater (7.9) was not significantly affected by the tracer spike. Bottles were stored upright in the dark at 4°C and gently mixed on a rotating table (25 rpm). Aliquots (100 ml) were withdrawn from the bottles at specific time intervals until equilibrium was reached. Equilibrium samples were taken after 25 days.

Particulate, colloidal, and dissolved fractions of ^{234}Th were separated and operationally defined in this study by filtration. Particulate Th (Th_p) is defined as that portion of Th removed from solution by a 0.2 μm Nuclepore filter. Colloidal Th is the size fraction of Th in the filtered sample that is retained by ultrafiltration (with an Amicon model TCF10 Thin-Channel ultrafiltration unit). Three different colloidal fractions were separated in the experiments by filtration through Amicon YM10 ultrafilters (nominal molecular weight (NMW) of 10,000) and Amicon YM2 ultrafilters (1,000 NMW). The colloidal fractions are defined as Th_c (1,000 NMW–0.2 μm), Th_{c1} (1,000 – 10,000 NMW), and Th_{c2} (10,000 NMW–0.2 μm). The dissolved fraction is defined as that portion of Th that passes the YM2 ultrafilter (< 1,000 NMW).

Loss of ^{234}Th was monitored by mass balance checks throughout the experiments. Filter blanks, determined by sequential filtration of samples through two 0.2 μm Nuclepore filters, were consistently low and insignificant (< 1%). Adsorptive loss during ultrafiltration was sometimes significant; however, any loss was small (typically 6%) and known since the activities of the original sample, ultrafiltrate, and retentate were all measured. Loss to the ultrafilters was minimized by rinsing and soaking with 0.05 M HCl, Super-Q water, 1% NaCl, Super-Q water, and ultrafiltered seawater. Before ultrafiltration of a sample, the ultrafiltration unit and filter were rinsed with Super-Q water, ultrafiltered seawater, and a 50 ml aliquot of the sample.

Loss of ^{234}Th to the walls of the sample bottles was monitored by measuring the total ^{234}Th activity of each sample immediately after spiking and at every sampling interval. Wall adsorption was < 10% for all the samples for the first 10 days. Loss then increased to 15 – 25% at 25 days with the greatest losses occurring at low particle concentrations.

The partitioning results are reported as percentages of the total ^{234}Th in the sample at equilibrium and as distribution coefficients, K_d's. The K_d's were calculated as Th_p per gram of added particles/(Th_d+Th_c) per gram of solution. Error bars in the figures represent one standard deviation around the average of replicate samples (3-4).

Uncertainties in the results due to ultrafiltration and counting errors are 3% for the particulate fraction and 6% for the dissolved and colloidal fractions.

Materials

Coastal seawater from Dalhousie University's Aquatron system was photo-oxidized and ultrafiltered (nominal molecular weight cut-off of 1,000) to prepare batch seawater free of organic material and particles (both colloids and > 0.2 μm). This seawater was used as a control and to prepare samples with specific concentrations of dissolved organic matter (DOM), colloids, and particles.

^{234}Th was prepared by extraction from uranyl nitrate using a procedure based on that described by Cospito and Rigali (10) and modified to ensure a good separation of ^{234}Th from ^{238}U (11). Further purification of the ^{234}Th was accomplished by ion exchange (12). Oxidation of the ^{234}Th stock solution with hot nitric acid, final dissolution in 0.1 M HCl, and filtration through a 0.2 μm Nuclepore filter ensured that ^{234}Th tracer was added to experiments in ionic form (13). Depending on the activity of the sample, ^{234}Th colloidal and dissolved fractions were either measured directly in 2 ml aliqouts or were concentrated by co-precipitation with $Fe(OH)_3$. Particulate ^{234}Th and filter blanks were measured by counting the Nuclepore filters (soaked for 1 hour in 0.02 M HCl before adding cocktail).

The β-decay (E_{max} = 0.191 MeV) of ^{234}Th was counted with a Beckman LS1801 liquid scintillation counter using *H#* as a quench monitor. The activity of the ^{234}Th quench standard was determined to 1.6% accuracy by counting its 63 and 93 KeV γ-rays on a Ge(Li)-detector calibrated with Amersham Mixed Standard 1000. Background for the liquid scintillation counter was 38 cpm.

Exudates from batch cultures of the diatom *Phaeodactylum tricornutum* were used as dissolved (< 1,000 NMW) organic matter in the experiments. *Phaeodactylum tricornutum* exudates have been shown to be chemically similar to marine humics by ^{1}H-NMR and ^{13}C-NMR spectroscopy (14) and have proven to be a good analogue for natural marine DOM in complexation of Th (7). ^{14}C-labelled exudates were prepared by growing batch cultures of *Phaeodactylum tricornutum* in ultrafiltered, photo-oxidized seawater (*i.e.* seawater free of pre-existing organic matter, colloids and particles) with $NaH^{14}CO_3$. At senescence, the cultures were centrifuged and filtered through a 1 μm filter. The filtrate was then acidified to pH 2 with HCl and bubbled with air to remove inorganic ^{14}C. After adjusting the pH to 7.9 with NH_4OH, sodium azide (2 g l^{-1}) was added to inhibit bacterial activity and the solution was filtered through a 0.2 μm Nuclepore filter. This filtrate was then ultrafiltered with a YM10 (10,000 NMW cut-off) ultrafilter to concentrate colloidal organic matter (COM). Dissolved organic matter (DOM) was separated by further ultrafiltration through a YM2 ultrafilter (1,000 NMW cut-off). Ultrafiltered photo-oxidized seawater and ultrafiltered *Phaeodactylum* exudates were mixed to prepare seawater with a DOM (< 1,000 NMW) concentration of 1.0 mg C l^{-1} . Different concentrations of COM were produced by diluting the stock colloid suspension. Organic carbon (DOM and COM) concentrations were determined at the beginning and end of each experiment by a UV-photooxidation autoanalyzer (15). The ^{14}C activity of the dissolved, colloidal, and particulate fractions was measured at all sampling intervals by dual liquid scintillation counting with ^{234}Th.

The particulate matter used in the experiments was a commercial product, Alon (supplied courtesy of Cabot Corp., Boston, MA), which consists predominantly of γ-Al_2O_3 (16). Alon has previously been used as a model oxide surface in studies of adsorption of organic matter and some trace metals onto particle surfaces (16,17). Alon particles are reported to be nonporous and approximately spherical with a specific surface area of 120 m^2 g^{-1} (17). Scanning electron microscopy showed that the Alon particles provided for this study consisted of particles with a range of diameters from 0.04 – 0.5 μm. Consequently, to ensure addition of only > 0.2 μm particles in the experiments, a cleaned Alon suspension (17) was fractionated by a series of filtrations. Alon was added to experiments to give particle concentrations typical of near-shore surface waters (0.1 – 5.0 mg l^{-1}).

Particle concentration was monitored by light scattering measurements using a fluorometer. After 25 days, no significant change in scattering was measured in the photo-oxidized seawater samples. A decrease in scattering of $\leq 16\%$ was observed with time in the 1.0 mg C l^{-1} samples however. This decrease may be due to dissolution of the alumina particles or to the attachment of particles to the walls of the bottles.

RESULTS

To determine the effect of colloidal matter on the equilibrium partitioning of Th, ^{234}Th spikes were first added to Alon suspensions that had been prepared in colloid-free seawater (ultrafiltered through a 1,000 NMW filter). Figure 1 shows the partitioning results for Alon suspensions (0 – 5.0 mg l^{-1}) in colloid-free seawater with dissolved organic matter (DOM) concentrations of 0.1 mg C l^{-1} (Figure 1a) and 1 mg C l^{-1} (Figure 1b). Separation of the samples into dissolved (< 1,000 NMW), colloidal (1,000 NMW–0.2 μm), and particulate (> 0.2 μm) fractions showed that with increasing Alon concentration, colloidal Th (Th_c) decreased concurrently with an increase in particulate Th (Th_p). Alon concentration had little effect on the percentage of Th in the dissolved fraction (Th_d).

The colloidal fraction of ^{234}Th was found, by sequential filtrations, to pass ultrafilters with nominal molecular weight cut-offs of $\geq 5,000$ and to be retained by the YM2 ultrafilter (1,000 NMW). Under the conditions of Figure 1 (*i.e.* colloid-free seawater), hydrolysis complexes and organic complexes are the only species of Th present in solution (13). Therefore, the colloidal ^{234}Th measured in the photo-oxidized seawater samples (Figure 1a) was assumed to consist of Th hydroxide complexes. The colloidal Th in the 1.0 mg C l^{-1} DOM samples (Figure 1b) was assumed to be predominantly bound in hydroxides as well since < 5% of the total sample ^{14}C was detected in the colloidal fraction. This concentration of *Phaeodactylum tricornutum* exudates (< 0.05 mg C l^{-1}) would complex < 3% of the total Th in the sample (7). Amicon YM2 ultrafilters have been reported to retain species with molecular weights much lower than their nominal molecular weight cut-off of 1,000 (down to 200 MW) and to have a 50% retention at 380 MW (18). Therefore, these hydroxide complexes may be $Th(OH)_4$, the dominant species of the Th in seawater (13), or polynuclear hydroxide complexes.

The dissolved Th species in Figure 1 are probably organic complexes or hydrolysis complexes not retained by the YM2 ultrafilter. *Phaeodactylum tricornutum* exudates have been found to form strong complexes with Th; at 1.0 mg C l^{-1}, 47% of total Th is bound in dissolved organic complexes (7).

From a study of Th adsorption by metal oxides in seawater, Hunter *et al.* (19)

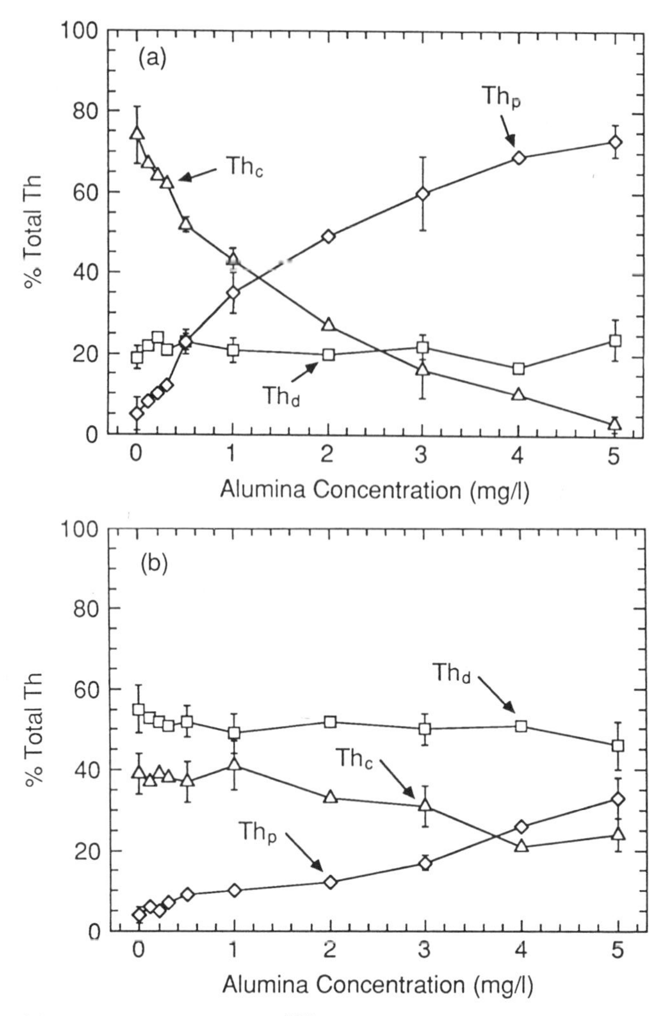

Figure 1. Equilibrium partitioning of ^{234}Th as a function of Alon concentration in ultrafiltered (colloid-free) seawater with DOC concentrations of 0.1 mg C l^{-1} (a) and 1.0 mg C l^{-1} (b). Error bars represent one standard deviation around the average of replicate samples; absence of error bars implies unreplicated samples.

concluded that $Th(OH)_4$ is the form of Th adsorbed by particles at the pH of seawater (7.9 – 8.2). Model organic ligands (EDTA and CDTA) were found to decrease Th adsorption by complexing Th and reducing the amount of $Th(OH)_4$ available for adsorption; organic complexes were not sorbed by the particles. The results in Figure 1 agree well with this work. Th hydroxides retained by YM2 ultrafilters (Th_c) appear to be the form of Th adsorbed by the Alon particles and adsorption is seen to be decreased by *Phaeodactylum tricornutum* DOM. Organic Th complexes, monitored by counting of ^{234}Th associated with the particulate fraction. Consequently, the K_d values calculated for the samples (Table 2) show an inverse correlation with colloid concentration. Aggregation of the colloidal matter into filterable (> 0.2 μm) particles, as indicated by ^{14}C measurements, was $\leq 7\%$.

Table 2. Effect of colloid concentration on K_d values calculated for 5.0 mg l^{-1} Alon suspension.

Colloid Conc. (mg C l^{-1})	0	0.65	1.0	2.5
$K_d(\times 10^5)$	4.3 ± 0.8	2.5 ± 0.5	1.8 ± 0.3	1.2 ± 0.4

SUMMARY AND DISCUSSION

The experimental results presented in this paper demonstrate the need to consider reactions competing with particle adsorption when studying the solid-solution partitioning of trace metals in natural waters. Th adsorption by alumina particles in seawater was found to be primarily of small molecular weight hydrolysis colloids (*i.e.* hydrolysis complexes retained by a 1,000 NMW ultrafilter). Complexation of Th by natural DOM decreased the amount of Th present as hydrolysis complexes and, since the organic Th complexes were not adsorbed by the alumina, resulted in decreased adsorption (Figure 1) and a lower distribution coefficient or K_d (see Table 1). Natural colloidal matter also decreased the amount of Th adsorbed by the alumina particles by competing for the available Th hydrolysis complexes. In a suspension of constant Alon concentration (Figure 2), an increase in colloid concentration caused an increase in colloidal adsorbed Th (Th_{c2}) and a decrease in the fraction of Th adsorbed by the alumina particles (Th_p). The K_d values consequently decreased with increasing colloid concentration (Table 2).

The significance of colloid adsorption must be addressed both in terms of its direct effect on the speciation and transport of trace metals in the ocean as well as its effect as an "artifact" in relating results from partitioning experiments to theoretical prediction. Studies of trace metal adsorption should be carefully designed to avoid adding particles smaller than those operationally defined as "particulate" in the experiments (6). In addition, adsorption by any colloidal matter already present in suspension (*e.g.* particles not removed by filtration of the experimental solution) should be considered.

K_d values determined from batch sorption experiments with suspended sediments are commonly used to calculate metal transport in sediment pore waters. Colloidal metal species are usually included together with truly dissolved species in these experiments; however, the mobility of colloidal metal may be considerably different from that of the metal ion or dissolved complexes of the metal (5). Knowledge of the colloidal fraction of metals is particularly important in determining the mobility of transuranic elements which may be largely transported in colloidal form in sediment pore waters (20).

The importance of colloidal matter to the geochemistry of trace metals in natural waters is speculative at present due to a lack of data. If significant adsorption by colloidal matter does occur, models used to interpret trace metal field data and to estimate residence times will have to be revised to include metal sorption by several size classes of particles (from colloidal particles to large aggregates). Tsunogai and Minagawa (21) have proposed such a scheme to explain the removal of particle reactive

^{14}C, were found not to be adsorbed by the Alon.

The K_d values calculated for the Figure 1a samples ([DOM]=0.1 mg C l^{-1}) were constant for Alon concentrations from 0.1 to 5.0 mg l^{-1} (see Table 1). K_d values for unreplicated samples are not included in Table 1 because of the unknown error in the measurements. The variability in the K_d's of the 1.0 mg C l^{-1} samples (Figure 1b) may be due to error associated with using the concentration of particles added to the samples rather than the actual equilibrium particle concentration in the K_d calculations. Light scattering measurements had suggested some loss of Alon during the experiment in the *Phaeodactylum* DOM samples while no loss was observed in the photo-oxidized seawater samples.

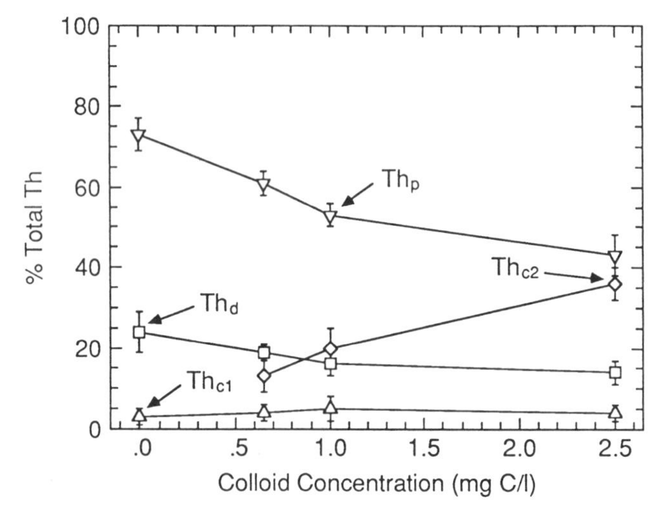

Figure 2. Equilibrium partitioning of ^{234}Th in 5 mg l^{-1} Alon suspensions as a function of colloid concentration: Th_d(< 1,000 NMW), Th_{c1}(1,000 – 10,000 NMW), Th_{c2}(10,000 NMW–0.2 μm), and Th_p(> 0.2 μm).

Table 1. K_d values for colloid-free Alon suspensions.

Alon Conc.	(mg l^{-1})	0.5	1.0	3.0	5.0
$K_d(\times 10^5)$	Fig.1a	4.4 ± 0.6	4.3 ± 1.0	4.3 ± 1.6	4.3 ± 0.8
	Fig.1b	1.1 ± 0.2	0.64 ± 0.10	0.50 ± 0.09	0.91 ± 0.18

The effect of pre-existing colloidal matter on the adsorption of Th by Alon is seen in Figure 2. Alon suspensions (5 mg l^{-1}) were prepared in ultrafiltered, photo-oxidized (DOM=0.1 mg C l^{-1}) seawater containing different concentrations of colloidal organic matter (10,000 NMW–0.2 μm fraction concentrated from ^{14}C-labelled *Phaeodactylum tricornutum* cultures). The samples were spiked with ^{234}Th and its partitioning among the dissolved, colloidal (Th_{c1} and Th_{c2}), and particulate fractions was measured at equilibrium. Figure 2 shows an increase in the Th_{c2} (10,000 NMW–0.2 μm) fraction of ^{234}Th with increasing colloid concentration and a concurrent decrease in the percentage metals from seawater; however, information on the sorption capacity of natural colloidal matter and on the aggregation/disaggregation processes between the particle classes is still needed.

CONCLUSIONS

Th adsorption by alumina particles in seawater was found to be predominantly of Th hydrolysis complexes that are retained by a 1,000 nominal molecular weight ultrafilter. Competing solution complexation of Th by *Phaeodactylum tricornutum* exudates and sorption by colloidal matter decreased Th adsorption by the alumina.

The sorption of metals by colloidal matter (pre-existing in the experimental solution or added with particles) must be considered in laboratory determinations of K_d values.

Field data on the concentration, composition, and aggregation reactions of naturally occurring colloidal matter are required to determine the effect of colloid adsorption on the marine geochemistry of trace metals.

REFERENCES

1. Balistrieri, L., Brewer, P.G. and Murray, J.W., Scavenging residence times of trace metals and surface chemistry of sinking particles in the deep ocean. Deep-Sea Res., 1981, 28A. 101-121.

2. Whitfield, M. and Turner, D.R., The role of particles in regulating the composition of seawater. In Aquatic Surface Chemistry: Chemical Processes at the Particle-Water Interface, ed. W. Stumm, Wiley- Interscience, New York, 1987, pp. 457-493.

3. Morel, F.M.M. and Gschwend, P.M., The role of colloids in the partitioning of solutes in natural waters. In Aquatic Surface Chemistry: Chemical Processes at the Particle-Water Interface, ed. W. Stumm, Wiley-Interscience, New York, 1987, pp. 405-422.

4. Li, Y.-H., Burkhardt, L., Buchholtz, M., O'Hara, P. and Santschi, P.H., Partition of radiotracers between suspended particles and seawater. Geochim. Cosmochim. Acta, 1984, 48. 2011-2019.

5. Higgo, J.J.W. and Rees, L.V.C., Adsorption of actinides by marine sediment: Effect of the sediment/seawater ratio on the measured distribution ratio. Environ. Sci. Technol., 1986, 20. 483-490.

6. Gschwend, P.M. and Wu, S.-c., On the constancy of sediment-water partition coefficients of hydrophobic organic pollutants. Environ. Sci. Technol., 1985, 19. 90-96.

7. Niven, S.E.H., The partitioning of thorium among dissolved, colloidal, and particulate fractions in seawater. Ph.D. dissertation, Dalhousie University, 1988.

8. Broecker, W.S., Kaufman, A. and Trier, R.M., The residence time of thorium in surface sea water and its implications regarding the fate of reactive pollutants. Earth Planet. Sci. Lett., 1973, 20. 35-44.

9. Santschi, P.H., Adler, D., Amdurer, M., Li, Y.-H. and Bell, J.J., Thorium isotopes as analogues for "particle-reactive" pollutants in coastal marine environments. Earth Planet. Sci. Lett., 1980, 47. 327-335.

10. Cospito, M. and Rigali, L., Determination of thorium in natural waters after extraction with Aliquat-336. Anal. Chim. Acta, 1979, 106. 385-388.

11. Moore, R.M. and Hunter, K.A., Thorium adsorption in the ocean: reversibility and distribution amongst particle sizes. Geochim. Cosmochim. Acta, 1985, 49. 2253-2257.

12. Anderson, R.F., The marine geochemistry of thorium and protactinium. Ph.D. dissertation, M.I.T./W.H.O.I., 1981, 40-43.

13. Langmuir, D. and Herman, J.S., The mobility of thorium in natural waters at low temperatures. Geochim. Cosmochim. Acta, 1980, 44. 1753-1766.

14. Wilson, M.A., Gillam, A.H. and Collin, P.J., Analysis of the structure of dissolved marine humic substances and their phytoplanktonic precursors by ^{1}H and ^{13}C nuclear magnetic resonance. Chem. Geol., 1983, 40. 187-201.

15. Gershey, R.M., MacKinnon, M.D., Williams, P.J. le B and Moore, R.M., Comparison of three oxidation methods used for the analysis of the dissolved organic carbon in seawater. Mar. Chem., 1979, 7. 289-306.

16. Davis, J.A., Complexation of trace metals by adsorbed natural organic matter. Geochim. Cosmochim. Acta, 1984, 48. 679-691.

17. Davis, J.A., Adsorption of natural dissolved organic matter at the oxide/water interface. Geochim. Cosmochim. Acta, 1982, 46. 2381-2393.

18. Wilander, A., A study on the fractionation of organic matter in natural waters by ultrafiltration techniques. Hydrologie, 1971, 34. 190-200.

19. Hunter, K.A., Hawke, D.J., and Choo, L.K., Equilibrium adsorption of thorium by metal oxides in marine electrolytes. Geochim. Cosmochim. Acta, 1988, 52. 627-636.

20. Kim, J.I., Chemical behaviour of transuranic elements in natural aquatic systems. In Handbook on the Physics and Chemistry of the Actinides. Vol. 4, eds. A. J. Freeman and C. Keller, Elsevier Science Publishers B. V., Amsterdam, 1986, pp. 413-456.

21. Tsunogai, S. and Minagawa, M., Settling model for the removal of insoluble chemical elements in seawater. Geochem. J., 1978, 12. 47-56.

TEMPORAL CHANGES OF Th-234 CONCENTRATIONS AND FLUXES IN THE NORTHWESTERN MEDITERRANEAN

P. Buat-Ménard*, H.V. Nguyen*, J.L. Reyss*, S. Schmidt*, Y. Yokoyama*
J. La Rosa**, S. Heussner**, S. W. Fowler**.
*Centre des Faibles Radioactivités, Laboratoire mixte CNRS-CEA
BP 1, F 91190 Gif sur Yvette, France.
** International Laboratory of Marine Radioactivity, IAEA,
C/o Musée Océanographique, MC 98000, Principauté de Monaco

ABSTRACT

The half life of Th-234 (t 1/2 = 24.1 d) is optimal for studying biogeochemical processes occuring in the euphotic zone on time scales of one to hundred days. Such time scales allow the assessment of temporal variability for primary productivity, zooplankton grazing activity and the resultant production of fecal pellets. As part of the DYFAMED programme, profiles of dissolved and particulate Th-234 between 0 and 200 m, as well as measurements of particulate Th-234 fluxes were made in 1986 and 1987 at two sites, one off the coast of northwestern Corsica and the other off the French Riviera, near Nice, in a 2000 m water column. Suspended matter and sediment trap samples were analyzed for a suite of radionuclides by non-destructive gamma ray spectrometry. Filtered 30 l water samples were acidified on board ship and later analyzed for dissolved Th-234. Results indicate that such gamma counting techniques permit detection of vertical changes in dissolved and particulate Th-234 concentrations with an accuracy of ± 10%, as well as temporal changes in Th-234 particulate fluxes on a time scale of 6 days or less. The Th-234 specific activity in sediment trap samples ranged from 800 to 3000 dpm/g. Th-234 particulate fluxes at 200 m depth were closely correlated with total particulate mass fluxes. Our results are in agreement with studies in other oceanic areas which demonstrates the usefulness of Th-234 as a tracer of scavenging and particulate transport processes in marine surface waters.

INTRODUCTION

The formation, dissolution and sinking of biogenic particulate matter are fundamental processes for regulating the composition of seawater. In surface waters biogenic particles which dominate the detrital phases, can remove dissolved organic and inorganic species by both metabolic uptake and passive adsorption mechanisms. Depending on the relative amount of recycling within the upper ocean a variable fraction of the material scavenged onto particles will be transported to greater depths, primarily associated with rapidly sinking particles, e.g. fecal pellets, amorphous aggregates (1). As a result of the biological cycling, the residence times of many dissolved trace elements in surface waters are much shorter than that of the water itself and can be expected to exhibit a strong spatial and temporal variability. Over the last twenty years, our understanding of these surface water scavenging rates has come mostly from the use of natural U and Th series radionuclides such as Th-234, Th-228, Pb-210 and Po-210 (2-11). Recently, it has been demonstrated that Th-234, with a half-life of 24.1 days, is an ideal particle reactive tracer for studying the scavenging of thorium from surface waters (12-13). Th-234 is continuously produced from the alpha decay of U-238 in seawater. Dissolved uranium is relatively unreactive to particulate adsorption whereas dissolved thorium exists as the hydrolysis product Th $(OH)_4^0$ (14) which is a particle reactive species that can be removed from the dissolved state by adsorption onto particles. It follows that the study of the Th-234/U-238 disequilibrium in dissolved and particulate fractions can be used to quantify the rates of Th-234 transformations from dissolved to particulate forms as well as the rates of removal of particulate Th-234 from the surface euphotic zone. Coale and Bruland (12) and Bruland and Coale (13) have applied a simple surface water scavenging model, assuming essentially irreversible scavenging of dissolved Th-234 onto fine suspended particles, followed by particle packaging (essentially zooplankton grazing) and subsequent large particle removal. These authors have shown that for a wide variety of Pacific surface waters, the dissolved Th-234 residence time varied from 6 to 220 days and that the first order scavenging rate of dissolved thorium is proportional to the rate of primary production. The residence time of particulate Th-234, of the order of weeks, appeared to be governed by the rate of zooplankton grazing and the type of zooplankton present.

The purpose of the present investigation was to utilize Th-234/U-238 disequilibria to investigate the dependance of Th-234 scavenging rates and particulate thorium fluxes from the surface waters of the northwestern Mediterranean sea. Though these waters are known to be oligotrophic, a significant temporal variability in primary productivity has been shown to occur, essentially during short but intense spring blooms (15). Upper water column sampling as well as sediment trap deployments started in spring 1986, as part of the DYFAMED programme. Sampling and analytical protocols used in this work are slightly different from those previously used (12-13) and more adapted to work at sea using small-size coastal research vessels. Although preliminary, the results obtained to date provide a first insight into the temporal variability of Th-234 scavenging and particulate transport processes in the open Mediterranean sea.

SAMPLING AND ANALYSIS

The samples were collected between April 1986 and May 1987 during several cruises (1-3 days duration) at two sites: approximately 15 nautical miles off the coast of Calvi, Corsica (42°44'N, 08°31'E) in a 2,200 m water column and approximately 8 nautical miles off the French Riviera, near Nice (43°36'N, 07°29'E) in a 2,000 m water column.

At both sites, an automated time-series cylindrical sediment trap (0.125 m^2 opening and height-diameter aspect ratio of 2.5) manufactured by TECHNICAP (16) was moored at a depth of 200 m. Each of the six trap collector cups (volume 250 ml) contained a buffered formaline solution to avoid decomposition losses (17), and was set to sample sinking particles for consecutive periods ranging from 6 to 10 days. After retrieval of the traps, the particulate material was immediately processed upon return to shore. A small aliquot (1/8 to 1/16 of the total sample) was subsampled for direct optical observations of the collected material. The remaining fraction was freezed-dried, weighed and then analysed for radioactivity by non- destructive γ-spectrometry within 3 months following retrieval. The total dry weight per trap sample ranged between 50 and 200 mg.

Seawater samples were collected with 30 litre Niskin or Go-Flo samplers (General Oceanics). After collection, seawater was passed through a polypropylene filter holder supporting an acid-washed 15 cm glass fiber filter (Whatman GF/C). The use of such filters allowed the filtration to be performed within 10 to 30 minutes. Their collection efficiency was found to be nearly equal to that obtained with Millipore membrane filters (0.4 μm pore size). The use of the latter filters required excessive filtration times especially in the upper 20 m due to the elevated suspended particulate loads (> 1mg/l). The rapid filtration technique was therefore used when considering the other sampling requirements (trace metals, etc) which had to be performed within a few hours at the sites. As for the trap samples, the filter samples were analysed for radioactivity by non-destructive γ spectrometry within a few weeks following the sampling.

Dissolved samples (generally 20 litres) were collected in cubitainers and immediately acidified with 50 ml of concentrated hydrochloric acid. On shore a few hours after sampling, they were transferred to 25 litres buckets and 0.8 dpm of Th-229 yield tracer in a solution of Fe as $FeCl_3$ were added to the samples which were then shaked vigorously during 1-2 hours for spike equilibration. The pH was then adjusted to 8 by adding concentrated ammonia. After shaking the $Fe(OH)_3$ precipitate formed was allowed to settle overnight and then separated from the supernatant. The $Fe(OH)_3$ precipitates were returned to the laboratory 3-4 days after sampling and immediately processed to separate thorium from uranium. This is required because uranium precipitates together with thorium and as a result produces Th-234 through "ingrowth" which adds to the Th-234 initially present . Using this protocol, the "ingrowth" correction was found to be less than 10%.

The chemical separation was made as follows: the precipitate was filtered, dissolved in 100-150 ml of concentrated HNO_3 and passed through an anion exchange column (Dowex 1×8 100-200 mesh). The column was washed with 8N HNO_3 which allowed retaining Th while removing U and Fe from the column. Th was then eluted with 20 ml of 8N HCl. After evaporation and dissolution in 0.1 HNO_3, Th was extracted into TTA (18). This organic phase was plated by evaporation onto a thin aluminium foil draped around a brass-ring. The aluminium foils were measured for their alpha activities essentially to determine Th-229 and thus the yield of the Th-234 extraction and eventually to detect other thorium isotopes (Th-228, Th-230). Alpha counting involved the use of silicon solid state detectors coupled to a multi-channel pulse analyser.

As for the trap samples and the suspended particulate samples, the dissolved Th-234 activity was determined by γ-spectrometry. We used different kinds of germanium detectors (well-type germanium detector for the aluminium foils (19); high purity coaxial and planar germanium detectors for the particulate samples(20)) coupled with multichannel analysers. Th-234 was measured from its 63 and 92 keV gamma-rays. Most samples were counted for 1 to 4 days depending of their activity. The detectors were calibrated with several standards in the same geometry and weight as the samples.

Errors reported in this work include i) the errors derived from the counting statistics (1σ); ii) the uncertainty on the detection yields of the germanium detectors, and iii) for dissolved Th-234 the uncertainties in the calibration of the tracer solutions and in the dissolved Th extraction efficiency. The average total error is 3-5% for Th-234 in sediment trap samples, 5-10% for dissolved Th-234 and 10-20% for particulate Th-234.

For dissolved and particulate Th-234, the errors reported by Bruland and Coale (12) based on β activity measurements are significantly lower than ours, especially for particulate thorium. This is due principally to the smaller sample size used here for particulate Th-234 (20-30 l compared to 55-110 l). Because of sample size limitations inherent in our cruises on small ships, we plan to lower our errors in the future through longer counting times, better counting geometries and the use of detectors with higher detection yields. Such improvements are necessary since Bruland and Coale (12) have convincingly demonstrated that accurate determinations of residence times of both dissolved and particulate Th-234 required precision of about 2-3% on the Th-234 determinations.

RESULTS AND DISCUSSION

Vertical profiles of dissolved and particulate Th-234 (A^d(Th) and A^p(Th) respectively) for both sites are presented in Fig. 1. These results can be compared to the activity of U-238 which has been determined on several cruises and found to be constant with A (U-238) = 2.70 ±0.05 dpm/l (dashed line, Fig. 1). This value is higher than in the open waters of other oceans due to the higher salinity in the Mediterranean (∿38°/∘∘). As expected there is a marked deficiency of dissolved Th-234. With the exception of the October 8, 1986 profile off Calvi, the observed deficiency is fairly uniform with depth between the surface and 200 m. The particulate thorium activity is always lower than that of dissolved thorium. The observed values of the $A^p(Th)/A^d(Th)$ ratio vary between 0.13 and 0.75. This is in agreement with what has been found in other oceanic areas (13).

In Table 1 the degree of radioactive disequilibrium between Th-234 and U-238 is given by the calculated value of the deficiency of total Th-234 (dissolved + particulate) A^{Σ}(Th). Because of the errors in our measurements, only three surface samples appear to show a deficiency in thorium. For the other samples, Th-234 appears close to equilibrium, which is typical of central ocean oligotrophic surface waters (13). The deficiency observed for the October 8, 1986 surface sample corresponds to the existence of a well-developed mixed-layer of 20 m thickness. This mixed layer was not observed for the other profiles. Therefore, the deficiencies observed on March 30, 1987 and May 14, 1987 may rather result from enhanced primary productivity conditions during the spring, as reflected by the chlorophyll data obtained at that time (Nival and Andersen, pers. comm.). This interpretation is supported by results from the bay of Calvi in January and February 1987, where significant deficiencies in total Th-234 have been observed 1 mile offshore.

The remaining data suggest that below a depth of 20 m, little or no net scavenging of Th-234 occurs. This situation corresponds to what is generally found in deep ocean waters. However the activity ratio $A^p(Th)/A^d(Th)$ is 3-4 times higher than in deep marine waters, indicating enhanced adsorption of Th onto suspended matter. This is not surprising since the suspended matter loads which we generally observed between 20 and 200 m are of the order of 50-200 μg/litre, significantly higher than what is generally found in deep ocean waters.

Although limited, the present data set suggests that appreciable removal of Th-234 from the surface waters of the northwestern Mediterranean Sea occurs essentially in the top 20 m. As suggested by previous works (12,13) this removal flux should be associated with large sinking particles, e.g. fecal pellets and aggregates. Initial adsorption

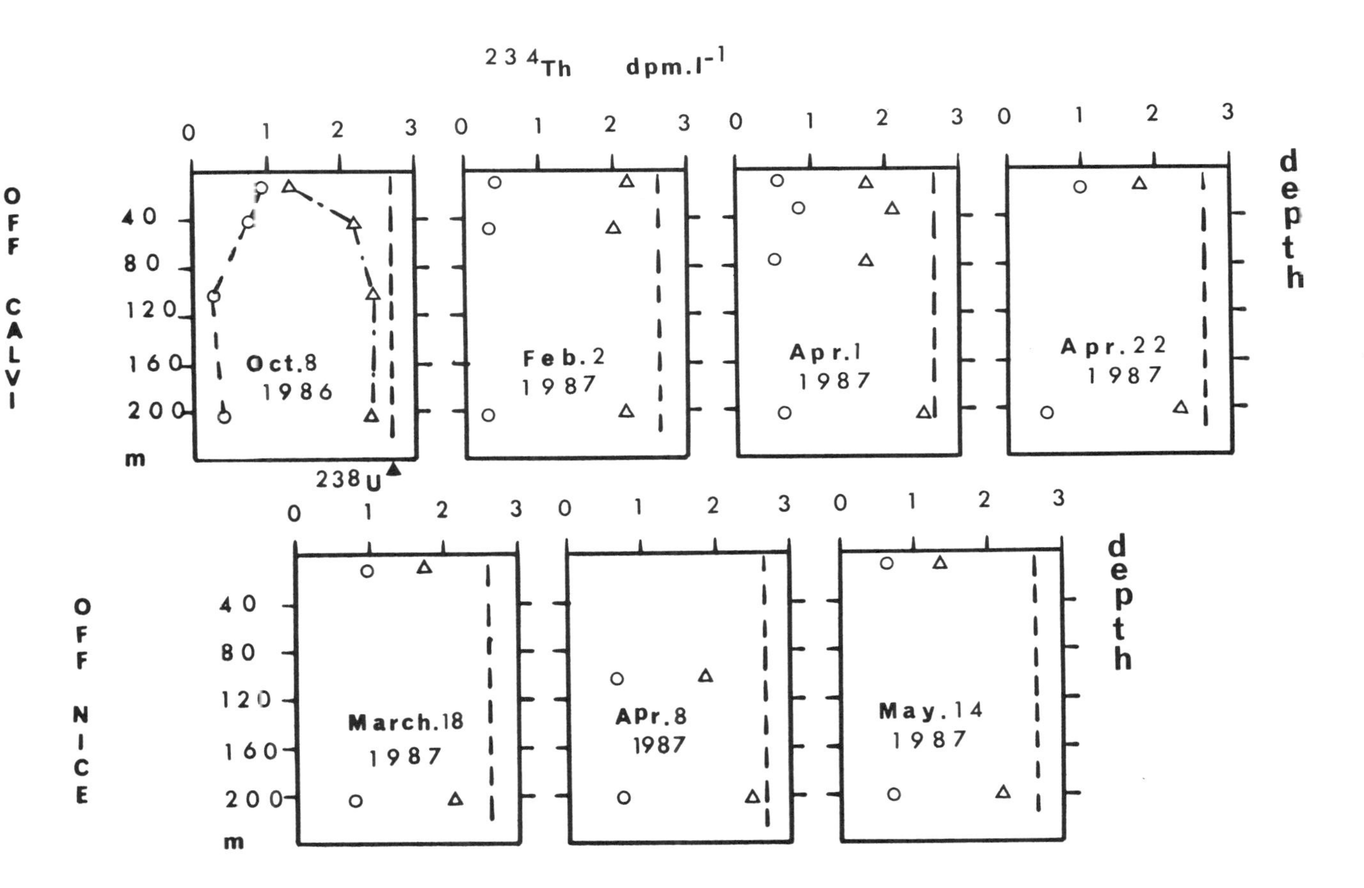

Figure 1: Vertical profiles of dissolved (Δ) and particulate (O) Th-234 in the upper water column of the Northwestern Mediterranean. Dashed line is activity of U-238

of Th onto particulate matter occurs principally on small particles, which have the highest surface adsorption properties. For sinking and removal to occur, these small particles must be packaged into the large particles probably as the result of zooplankton grazing activity. If this the case, the vertical flux of particulate Th-234 in these waters should exhibit temporal variabilities similar to that of the total mass flux.

Date	Depth	$A^d/A(U)$	$A^p/A(U)$	$A^{\Sigma}/A(U)$
Off Calvi				
10/08/86	10 m.	0.50±0.04	0.36±0.05	0.86±0.09
	40 m.	0.80±0.05	0.27±0.03	1.07±0.08
	100 m.	0.94±0.06	0.12±0.03	1.06±0.09
	200 m.	0.99±0.11	0.16±0.04	1.05±0.15
02/03/87	10 m.	0.83±0.06	0.15±0.06	0.98±0.12
	50 m.	0.73±0.09	0.13±0.06	0.86±0.15
	200 m.	0.80±0.05	0.10±0.04	0.90±0.09
03/30/87	10 m.	0.64±0.04	0.21±0.04	0.85±0.08
	30 m.	0.78±0.06	0.30±0.06	1.08±0.12
	75 m.	0.66±0.08	0.18±0.06	0.84±0.14
	200 m.	0.96±0.07	0.25±0.08	1.21±0.15
04/22/87	10 m.	0.65±0.03	0.37±0.05	1.02±0.08
	100 m.	0.93±0.03	0.19±0.04	1.12±0.07
Off Nice				
03/18/87	10 m.	0.68±0.07	0.39±0.07	1.07±0.14
	200 m.	0.80±0.09	0.32±0.03	1.12±0.12
04/08/87	100 m	0.70±0.05	0.25±0.07	0.95±0.12
	200 m.	0.94±0.06	0.28±0.06	1.22±0.12
05/14/87	10 m.	0.52±0.05	0.23±0.06	0.75±0.11
	200 m.	0.80±0.03	0.28±0.08	1.08±0.11

Table 1: Observed activity ratios, A(Th)/A(U) for dissolved, particulate and total Th-234

This interpretation is strongly supported by the sediment trap data presented in Fig. 2. The sampling depth of 200 m below the surface was choosen since it has been suggested that vertical fluxes of organic matter passing that depth should be equivalent to new production (21,22). Temporal variations of about one order of magnitude are observed for both the total mass flux and the particulate Th-234 flux. The range of mass fluxes which we observe, 50-350 mg/m²/d is within the range of results obtained for open ocean waters (22). It can be seen that the pattern of the temporal variation of the Th-234 flux is similar to that of the total mass flux. In other words, the specific activity of Th-234 in the trap samples is fairly constant within approximately a factor of 2. Moreover, the mean specific activity, 1500 dpm/g of dry material is close to that found in other open ocean areas (12,13) and to that predicted in biogenic debris (23) due to passive uptake by phytoplancton cells.

The highest mass fluxes and particulate Th-234 fluxes are observed during late winter and early spring (Fig. 2), when primary productivity generally increases in the northwestern Mediterranean (15). Indeed, during the period of trap deployment off Nice, chlorophyll measurements indicated the development of a plankton bloom starting March 16 and almost disappearing by April 8 (Nival and Andersen, pers. comm.). We therefore

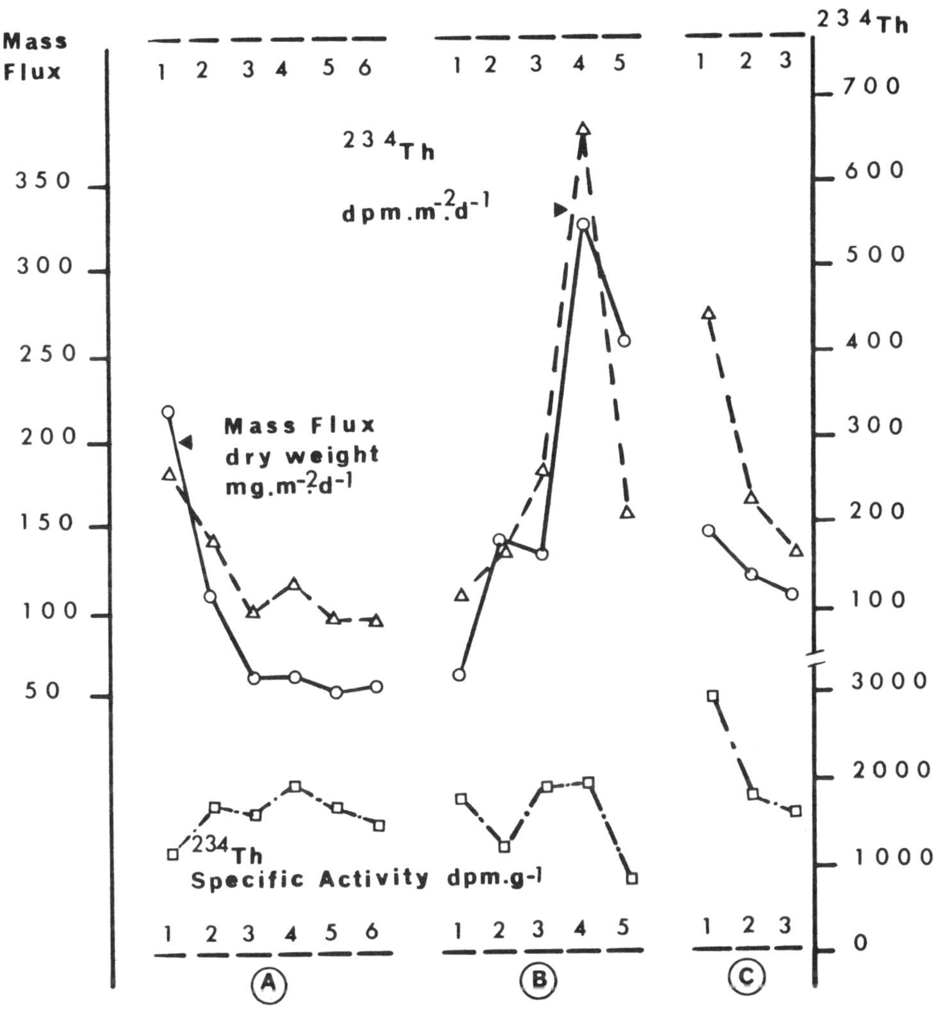

Figure 2: Temporal variations of total mass flux (○), particulate Th-234 flux (△) and Th-234 specific activity (□) at 200 m depth. A: off Calvi, from April 13 to May 20, 1986; 6 samples; 6.25 days per sample. B: off Calvi, from January 18 to March 3, 1987; 5 samples; 10 days per sample. C: off Nice, from March 26 to April 13, 1987; 3 samples; 6 days per sample.

suggest that the observed increases in particulate fluxes reflected enhanced zooplankton grazing activity.

On March 27 and April 30, 1987, zooplankton was sampled from the surface layers at the trap site and living individuals (mostly copepods) were immediately placed in special devices to collect fecal pellets (24). The Th-234 activity in the fecal pellets (300-600 dpm/g dry weight) was about 50 times higher than that of the zooplankton producing them. Microscopic examination of the trap samples (25) indicated that they were generally rich in fecal material which made up between 10 and 70% of the total dry weight in these samples. Therefore, when considering the specific activity of Th-234 in small suspended particles (600-4000 dpm/g dry weight), it is likely that the ingestion of small particles by zooplankton and their subsequent packaging into fecal pellets constitutes a significant fraction of the particulate Th-234 flux at 200 m depth. The remaining fraction is probably associated with large, amorphous aggregates (marine snow) which consisted of inorganic and detrital particles including fecal matter (25).

The above considerations strongly suggest that there should be a close similarity between the temporal variability of the zooplankton biomass and that of the particulate fluxes measured in our trap samples. One question which then arises is what is the time lag between the production of large sinking particles in the surface layers and their arrival at a depth of 200 m. According to our results on the downward movement of Chernobyl radioactivity in spring 1986 at the trap site off Calvi (25), it can be inferred that the transit time for particle reactive radionuclides between the surface and 200 m is of the order of a week. From this we conclude that the temporal variations of particulate Th-234 fluxes at 200 m should mimic closely those of the zooplankton biomass in the overlying water column. Work presently in progress with frequent collections of zooplankton material at the trap sites will allow testing this hypothesis.

Finally, the scavenging model of Coale and Bruland (12) can be used to compare model-derived Th-234 fluxes from the euphotic zone to those determined from the trap samples. However, this model can only be applied to oceanographic situations with a well developed mixed layer. Unfortunately, with our present data set, this situation was only met on October 8, 1986, off Calvi (see above). If we assume steady state in a 20 m thickness of the mixed layer, one obtains from the data given in Table 1 a residence time of 34 days for dissolved thorium, 85 days for particulate thorium, a removal flux (J_{Th}) from the dissolved to the particulate form of 790±70 dpm/m^2/d and a particulate Th-234 flux out of the mixed layer of 230±140 dpm/m^2/d. This value is within the range of measured fluxes at 200 m depth (90-660 dpm/m^2/d). This comparison may support the above mentioned hypothesis that most of the net scavenging of Th-234 in the Northwestern Mediterranean occurs in the top 20 m. Field experiments currently underway should allow testing our arguments.

ACKNOWLEDGEMENTS

This research was supported in part by the Centre National de la Recherche Scientifique, the Commissariat à l'Energie Atomique and grants from the French Institut des Sciences de l'Univers, Flux de Matière dans l'Océan (DYFAMED) and U.S. National Foundation VERTEX Programmes. We thank P. Bourdon, A. Vigot and the crew and officers of R.V. Korotneff, R.V. Catherine Laurence and R.V. Recteur Dubuisson for logistic support. We also thank for support the Musée Océanographique at Monaco. The International Laboratory of Marine Radioactivity operates under an agreement between the International Atomic Energy Agency and the Government of the Principauté de Monaco. This is CFR contribution n°873.

REFERENCES

1. Fowler, S.W. and Knauer, G.A., Role of large particles in the transport of elements and organic compounds through the oceanic water column. Prog. Oceanogr., 1986, 16. 147-194.

2. Bhat, S.G., Krishnaswami, S., Lal, D., Rama and Moore, W.S., Th-234/U-238 ratios in the ocean. Earth Planet. Sci. Lett., 1969, 5. 483-491.

3. Broecker, W.S., Kaufman, A. and Trier, R.M., The residence time of thorium in surface seawater and its implications regarding the fate of reactive pollutants. Earth Planet. Sci. Lett., 1973, 20. 35-44.

4. Matsumoto, E., Th-234-U-238 radioactive disequilibrium in the surface layers of the ocean. Geochim. Cosmochim. Acta, 1975, 39. 205-212.

5. Knauss, K.G., Ku, T.L. and Moore, W.S., Radium and thorium isotopes in the surface waters of the East Pacific and coastal Southern California. Earth Planet. Sci. Lett., 1978, 39. 235-249.

6. Kaufman, A., Li, Y.H and Turekian, K.K., The removal rates of Th-234 and Th-238 from waters of the New York Bight. Earth Planet. Sci. Lett., 1981, 54. 385-392.

7. Li, Y.H., Feely, W.H. and Santschi, P.H., Th-228/Ra-228 radioactive disequilibrium in the New York Bight and its implications for coastal pollution. Earth Planet. Sci. Lett., 1979, 42. 13-26.

8. Feely, H.W., Kipphut, G.W., Trier, R.M. and Kent, C., Ra-228 and Th-228 in coastal waters. Estuarine Coastal Mar. Sci., 1980, 11. 179-205.

9. Nozaki, Y., Horibe, Y. and Tsubota, H., The water column distributions of thorium isotopes in the Western North Pacific. Earth Planet. Sci. Lett., 1981, 54. 203-216.

10. Bacon, M.P. and Anderson, R.F., Distribution of thorium isotopes between dissolved and particulate forms in the deep sea. J. Geophys. Res., 1982, 87, 2045-2056.

11. Chesselet, R. and Lalou, C., The use of natural radionuclides in oceanography: an overview. This volume.

12. Coale, K.H. and Bruland, K.W., Th-234:U-238 disequilibria within the California current. Limnol. Oceanogr., 1985, 30, 22-33.

13. Bruland, K.W. and Coale, K.H., Surface water Th-234/U-238 disequilibria: spatial and temporal variations of scavenging rates within the Pacific Ocean. In: Dynamic Processes in the chemistry of the upper ocean, eds. J.D. Burton, P.G. Brewer and R. Chesselet, Plenum Press, New York, 1986. 159-172.

14. Turner, D.R., Whitfield, M. and Dickson, A.G., The equilibrium

speciation of dissolved components in freshwater and seawater at 25°C and 1 atm pressure. Geochim. Cosmochim. Acta., 1981, 45. 855-881.

15. Jacques, G. and Treguer, P., Ecosystèmes pélagiques marins, Masson, Paris, 1986, 243 pp.

16. Heussner, S., Monaco, A., Fowler, S.W., Buscail, R., Millot, C., and Bojanowski, R., Flux of particulate matter in the Northwestern Mediterranean (Gulf of Lion). Oceanol. Acta, 1987, in the press.

17. Knauer, G.A., Karl, D.M., Martin, J.H. and Hunter, C.N., In situ effects of selected preservatives on total carbon, nitrogen and metals collected in sediment traps. J. Mar. Res., 1984, 42. 445-462.

18. Turekian, K.K., Cochran, J.K., Kharkar, D.P., Cerrato, R.M., Rimas Vaisnys, J., Sanders, H.L., Grassle, J.F. and Allen, J.A., Slow growth rates of deep-sea clam determined by Ra-228 chronology. Proc. Nat. Acad. Sci. USA., 1975, 72. 2829-2832.

19. Brichet, E., Le Foll, D., Reyss, J.L., and Lalou, C., Age determination of spider-crabs by Ra-228 chronology: possible extension to other crustaceans. In preparation.

20. Yokoyama, Y. and Nguyen, H.V., Direct and non-destructive dating of marine sediments, manganese nodules, and corals by high resolution gamma-ray spectrometry. In: Isotope Marine Chemistry, eds. E.D. Goldberg, Y. Horibe and K. Saruhashi, Rokakvo, Tokyo, 1980. 259-289.

21. Eppley, R.W. and Peterson, B.J., Particulate organic matter flux and planktonic new production in the deep ocean. Nature, 1979, 282. 677-680.

22. Hargrave, B.T., Particle sedimentation in the ocean. Ecol. Model., 1986, 30, 229-246.

23. Fisher, N.S., Cochran, J.K., Krishnaswami, S. and Livingston, H.D., Vertical flux of radionuclides in marine systems mediated by biogenic debris, This volume.

24. Small, L.F., Fowler, S.W. and Unlu, S.W., Sinking rates of natural copepod fecal pellets. Mar. Biol., 1979, 51. 233-241.

25. Fowler, S.W., Buat-Ménard, P., Yokoyama, Y., Ballestra, S., Holm, E. and Nguyen, H.V., Rapid removal of Chernobyl fall-out from Mediterranean surface waters by biological activity. Nature, 1987, 329, 56-58.

SCAVENGING AND BIOTURBATION IN THE IRISH SEA FROM MEASUREMENTS OF $^{234}Th/^{238}U$ AND $^{210}Pb/^{226}Ra$ DISEQUILIBRIA

P. J. Kershaw, P. A. Gurbutt,
A. K. Young and D. J. Allington

Ministry of Agriculture, Fisheries and Food
Directorate of Fisheries Research
Fisheries Laboratory
Lowestoft, Suffolk NR33 OHT
England

ABSTRACT

The disequilibria which arise in the distribution of parent-daughter radionuclides, because of differences in their chemical behaviour, can be used to advantage to study physical and chemical processes in coastal waters. The degree of element scavenging and the extent and rate of biological mixing (bioturbation) within the sea bed have been investigated in the eastern Irish Sea from the determination of $^{234}Th/^{238}U$ and $^{210}Pb/^{226}Ra$ disequilibria in sea water and/or sediments.

Estimates have been made of the scavenging coefficient and residence time of dissolved ^{234}Th, the rate of removal of ^{234}Th to the sea bed, and the rate at which sea-bed sediment is resuspended. Systematic changes in these rates are observed as the coastline is approached and the suspended load increases. There appears to be a net horizontal flux of ^{234}Th, and perhaps ^{234}Th bearing sediment, into or within the coastal zone.

The extent and rates of sediment turnover by bioturbation in the upper layers of the sea bed have been assessed from vertical profiles of unsupported ^{234}Th, assuming bioturbation to be analogous to a simple diffusion process. Measurements of $^{210}Pb/^{226}Ra$ in cores have provided additional estimates of bioturbation rates, integrating the mixing process over a longer time scale and to a greater depth. The vertical distribution of unsupported ^{234}Th and ^{210}Pb, and in particular the occurrence of sub-surface concentration maxima, is discussed with regard to the behaviour of those benthic organisms thought chiefly responsible for most of the sediment movement, and an alternative method of modelling their effects is presented.

INTRODUCTION

Measurements of the disequilibria which arise as a result of the differing chemical behaviours exhibited by the naturally-occurring isotope of uranium, ^{238}U ($t_{1/2}$ = 4.5 x 10^9 years), and its daughter radionuclide ^{234}Th ($t_{1/2}$ = 24 days) provide an elegant and useful method of assessing the residence time of particle-reactive elements in turbid coastal waters, the rate of sediment resuspension, and the rate of sediment mixing in the upper layers of the sea bed (1,2). Similarly, measurements of the disequilibria between ^{226}Ra ($t_{1/2}$ = 1600 years) and ^{210}Pb ($t_{1/2}$ = 22.2 years) can provide information on sediment mixing and accumulation, integrating the effects of these processes over a longer period and greater depth, commensurate with the longer half-life of ^{210}Pb (3,4).

A study has been made of the disequilibria of $^{234}Th/^{238}U$ in sea water, suspended particulate, and sea-bed sediment, and of $^{210}Pb/^{226}Ra$ in sea-bed sediment, in the Irish Sea. The aim was to assess the degree of scavenging and bioturbation of particle-reactive nuclides as part of a comprehensive investigation into the fate of artificial radionuclides released from the British Nuclear Fuels plc (BNFL) reprocessing plant at Sellafield, Cumbria, UK.

METHODS

Eleven Reineck cores were collected from the eastern and western Irish Sea in April 1984. Water samples and cores were collected from a further 5 sites in the eastern Irish Sea in November 1985 along a transect normal to the Cumbrian coast (Figure 1). The 1984 cores were sub-sampled at 1 or 2 cm intervals to provide detailed radiochemical profiles. The 1985 transect cores were sub-sampled in triplicate to provide 6 cm-deep samples for ^{234}Th inventory determinations. Water samples were taken from 2 m below the sea surface and ~ 2 m above the sea bed using rosette-mounted 30 l Niskin bottles and filtered (0.22 μm). ^{234}Th was determined by β-counting its daughter ^{234m}Pa (β^- - 2.29 MeV) and ^{238}U by α-spectrometry on a silicon-surface barrier detector. Full details of the analytical procedures adopted have been reported elsewhere (5).

RESULTS

Scavenging in seawater

The concentrations of both dissolved and particulate ^{234}Th varied systematically with distance from the coast by over a factor of ten (Table 1). The distribution coefficient (K_D) was calculated to assess the degree of partitioning of ^{234}Th between suspended particles and sea water, defined in this instance as:

$$K_D = \frac{\text{concentration on particles (Bq kg}^{-1}\text{ dry sediment)}}{\text{concentration in sea water (Bq l}^{-1}\text{)}}. \quad (1)$$

K_D values were in the range 1.2 to 4.5 x 10^6 (Table 1). The relative constancy of the ^{234}Th K_D (mean 2.04 ± 1.0), despite large variations in concentration, implies that the adsorption of ^{234}Th onto particle surfaces is controlled by a reversible adsorption/desorption mechanism. But we are not able, from this limited data set, to define more precisely a range of K_D values which would be compatible with this hypothesis.

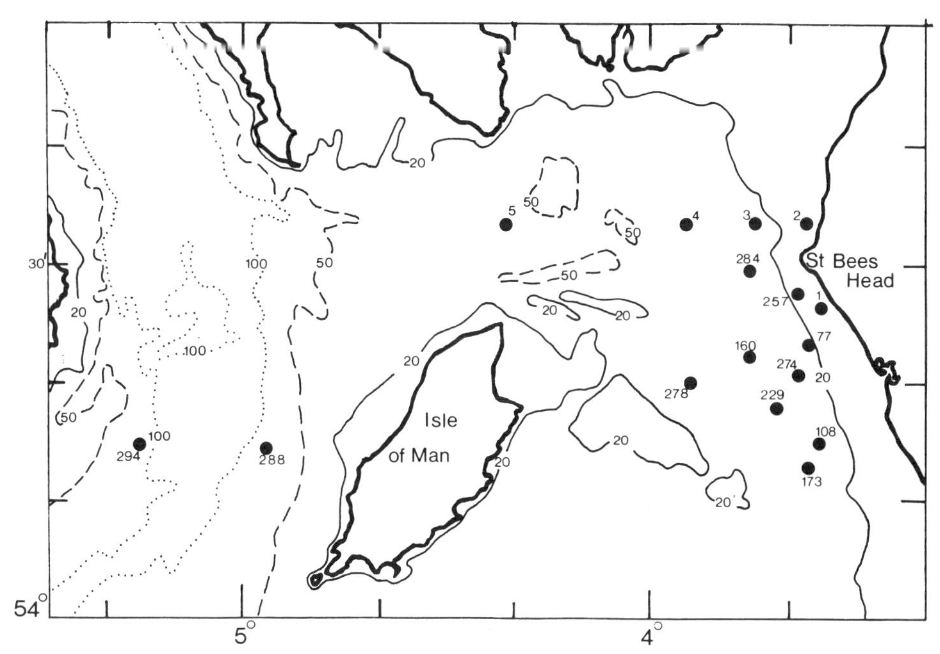

Figure 1. Sample locations and bathymetry (water depth, metres): sites 77 to 294, sediment cores (April, 1984); sites 1 to 5, water samples and sediment cores (November, 1985).

Table 1 Concentrations of ^{238}U and ^{234}Th in sea water. Errors quoted for radiochemical measurements ± 1 σ propagated combined sampling and counting error and for non-radiochemical measurements ± 1 σ sampling error.

Site / Water depth (m)	Sampling depth (m)	Sea water ^{238}U (mBq l^{-1})	Sea water ^{234}Th (mBq l^{-1})	Suspended particulate ^{238}U (mBq l^{-1})	Suspended particulate ^{234}Th (mBq l^{-1})	Suspended load (mg l^{-1})	^{238}U k_D (l kg^{-1})	^{234}Th K_D (l kg^{-1})
TH01	2	40.0±1.3	0.8±0.1	0.33±0.02	8.96±0.39	6.80±0.21	(1.2±0.05)E3	(1.7±0.05)E6
21	19	38.6±1.4	1.1±0.1	0.25±0.02	7.99±0.37	6.51±0.25	(1.0±0.05)E3	(1.2±0.03)E6
TH02	2	39.8±0.1	0.6±0.1	0.44±0.02	8.25±0.28	6.78±0.20	(1.6±0.06)E3	(1.9±0.03)E6
21	19	42.0±1.6	0.6±0.1	0.88±0.04	10.74±0.49	6.90±0.25	(3.0±0.08)E3	(2.8±0.08)E6
TH03	2	38.1±0.5	1.1±0.1	0.11±0.01	6.22±0.36	3.58±0.20	(0.8±0.04)E3	(1.6±0.02)E6
26	24	39.8±1.5	1.5±0.1	0.07±0.01	5.46±0.29	2.85±0.14	(0.6±0.05)E3	(1.3±0.02)E6
TH04	2	39.2±1.4	4.1±0.2	0.03±0.00	6.37±0.74	0.99±0.11	(0.8±0.07)E3	(1.6±0.02)E6
36	34	40.9±1.9	4.0±0.2	0.06±0.01	10.28±1.16	1.78±0.20	(0.8±0.07)E3	(1.4±0.02)E6
TH05	2	40.9±1.5	5.4±0.2	0.07±0.02	15.54±3.52	0.64±0.15	(2.8±0.24)E3	(4.5±0.06)E6
61	54	37.7±1.3	6.8±0.3	0.03±0.01	5.95±1.99	0.36±0.12	(2.3±0.19)E3	(2.4±0.04)E6

From the eastern Irish Sea data it was possible to calculate a scavenging coefficient (K, a^{-1}), for dissolved ^{234}Th onto particles, by assuming that the supply of dissolved ^{234}Th was balanced by decay and particle scavenging (1). Dealing with numbers of atoms:

$$\lambda_U N_U^D = \lambda_{Th} N_{Th}^D + KN_{Th}^D \tag{2}$$

where λ_U, λ_{Th} are the decay constants (a^{-1}) and N_U^D, N_{Th}^D the number of atoms (l^{-1}) of dissolved ^{238}U and ^{234}Th. Converting to concentrations:

$$K = (C_U^D - C_{Th}^D) \frac{\lambda_{Th}}{C_{Th}^D}, \tag{3}$$

where C_U^D, C_{Th}^D are the dissolved ^{238}U and ^{234}Th concentrations (Bq l^{-1}).

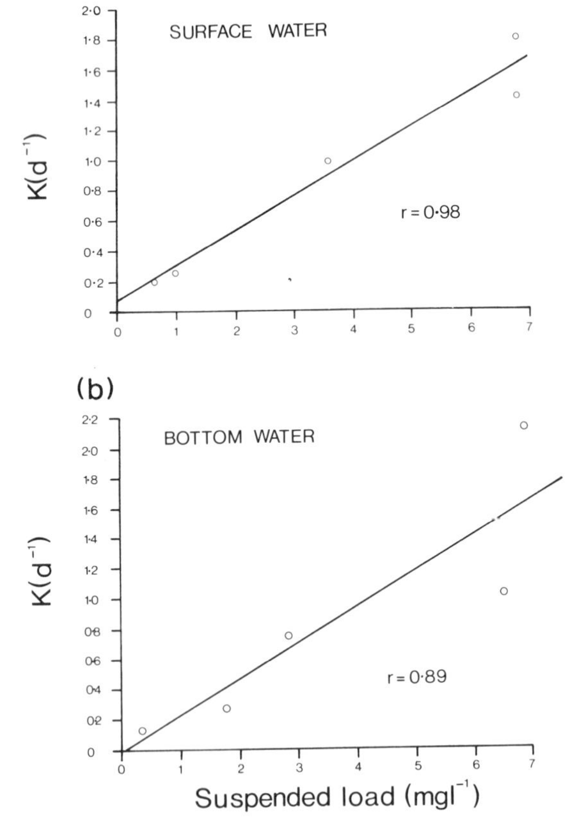

Figure 2. Variation of the scavenging coefficient (K, d^{-1}) with suspended load (mg l^{-1}): (a) for surface waters; (b) at ~ 2 m above the sea bed.

The scavenging coefficient was proportional to the number of particles in suspension (Figure 2). The mean residence time of dissolved ^{234}Th (t_R), defined as the reciprocal of the scavenging coefficient K, varied from 0.5 to 7.7 days, increasing with distance from the coast as the suspended load decreased (5). The establishment of an equilibrium K_D value may be limited by the rate of adsorption if t_R is significantly longer than the mean particle residence time (t_p) (6). Following the work of Santschi et al. (2), t_p is defined as:

$$t_p = \frac{z.h}{S}, \tag{4}$$

where z is the suspended load (kg m^{-3}), h is the height of the water column (m) and S the net quantity of material (kg m^{-2} a^{-1}) resuspended from, and returned to, the sea bed. S can be estimated by considering the balance of ^{234}Th and ^{238}U between the water column and the sea bed (2). The number of atoms of particulate ^{234}Th (N^P_{Th}, l^{-1} sea water) is determined by the decay of particulate ^{238}U ($\lambda_U N^P_U$, l^{-1} sea water) and the adsorption of dissolved ^{234}Th (K N^D_{Th}, l^{-1} sea water). There will be an additional contribution from resuspended sea-bed sediment which will depend on (S) and the number of atoms of sediment-bound ^{234}Th (N^S_{Th}, kg^{-1} dry). These contributions are assumed to be balanced by the radioactive decay of particulate ^{234}Th (λ_{Th} N^P_{Th}, l^{-1} sea water) and the sedimentation of ^{234}Th-bearing particles back to the sea bed (S $N^{P'}_{Th}$, kg^{-1} dry):

$$\lambda_U N^P_U + K N^D_{Th} + \frac{S N^S_{Th}}{h} = \lambda_{Th} N^P_{Th} + \frac{S N^{P'}_{Th}}{h}. \tag{5}$$

Converting to concentrations (Bq l^{-1}, Bq kg^{-1}) and substituting for K (from equation 3):

$$S = \frac{\lambda_{Th}.h \left((C^P_U + C^D_U) - (C^P_{Th} + C^D_{Th})\right)}{(C^{P'}_{Th} - C^S_{Th})}. \tag{6}$$

Values of S in the eastern Irish Sea were proportional to the observed suspended load and fell in the range 0.5 to 6 kg m^{-2} a^{-1}, the rate of resuspension increasing as the coast line was approached (5). Substituting for S in equation 4 gave t_p values of 6 to 27 days, significantly longer than those of t_R, indicating that the removal of ^{234}Th was not limited by the rate of adsorption and that there was sufficient time to establish an equilibrium K_D, hence the relatively constant K_D values which were measured.

Sea-bed inventories

Inventories (Bq m^{-2}) were determined from the unsupported ^{234}Th concentrations of both the 1985 transect samples and the 1984 core profiles. These were compared with the estimated production of ^{234}Th (Bq m^{-2}) in the overlying water column, using the measured ^{238}U sea-water concentrations for the 1985 inventories and an assumed concentration of 38 Bq m^{-3} for the 1984 inventories when no sea-water measurements were made.

^{234}Th sea-bed inventories far exceeded the estimated water column production near the Cumbrian coast (Figure 3). Along the transect, inventories increased systematically in proportion to several interrelated factors: a decrease in water depth and increases in the suspended load, the proportion of fine-grained sediment, and the rate of resuspension. These all, collectively, increase the probability of ^{234}Th being scavenged out of the water column. The inventories - as a percentage of water column production - were relatively constant for water depths below about

30 to 35 m (Figure 4). In shallower waters, inventories increased inversely with depth, perhaps reflecting the increase in wave energy and consequent increase in the rate of resuspension. The inventories of the muddiest sediments sampled, west of the Isle of Man, were significantly lower than the overlying water production (60% and 78%) reflecting the relatively deep water, low tidal streams (7) and consequent low probability of particle resuspension.

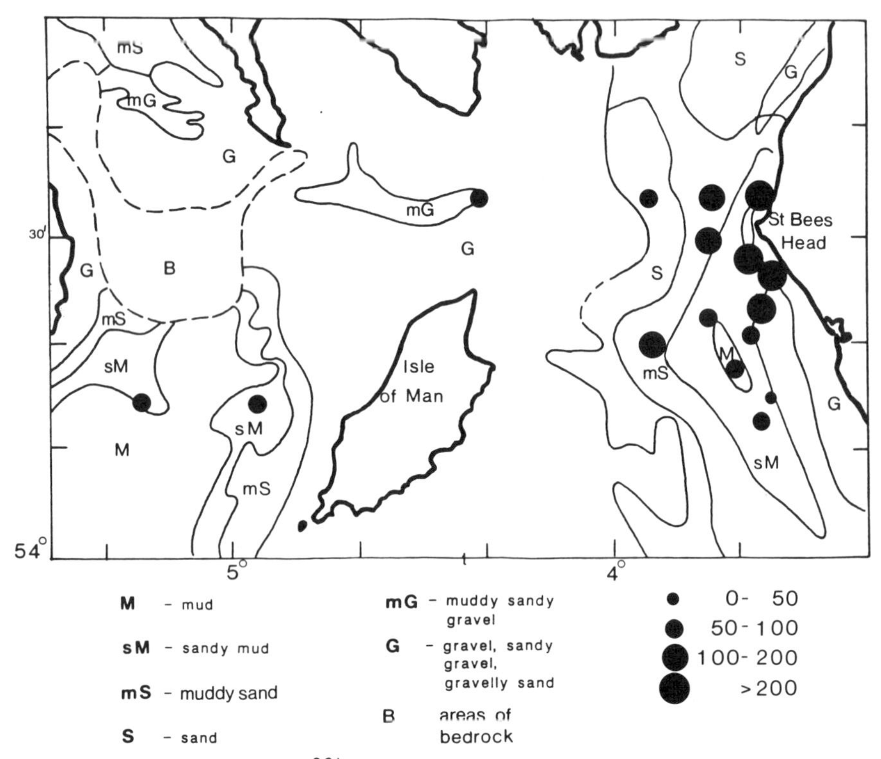

Figure 3. Unsupported ^{234}Th inventories expressed as a percentage of the overlying water column production, with the distribution of bottom sediments adapted from maps published by the British Geological Survey, Keyworth, Nottingham, UK.

On the basis of the unsupported ^{234}Th inventories, the inshore sites represent a net sink for ^{234}Th. This may result from the horizontal transport of ^{234}Th-enriched surface sediment within the coastal zone, or from enhanced scavenging by fine-grained resuspended sediment from water passing through the region which initially has a low suspended load and high dissolved ^{234}Th concentration.

Mixing within the sea bed

The unsupported radionuclide profile can be used to estimate the rate of biological mixing in terms of a solid particle biodiffusion coefficient, D_B (cm^2 a^{-1}) (8,9):

$$\frac{\partial C}{\partial t} = \frac{\partial}{\partial x}\left(D_B \frac{\partial C}{\partial x}\right) - \omega \frac{\partial C}{\partial x} + P - \lambda C, \tag{7}$$

where C is the nuclide concentration (Bq kg^{-1}), P the production rate from the parent nuclide (Bq kg^{-1}), λ the decay constant (a^{-1}), x the depth (cm) below the sediment-water interface (x = 0), and ω the sediment accumulation rate (cm a^{-1}). At steady state $\frac{\partial C}{\partial t} = 0$.

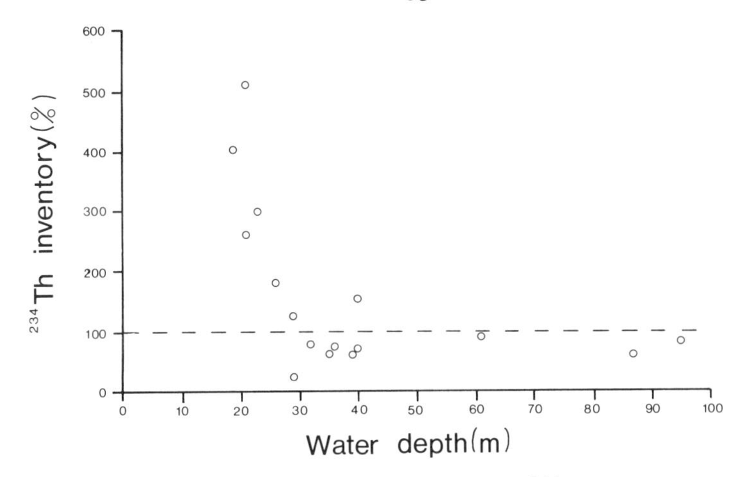

Figure 4. Variation of the unsupported ^{234}Th inventory, expressed as a percentage of the overlying water column production, with water depth.

There is no published, unequivocal evidence that net sediment accumulation is taking place extensively in the eastern Irish Sea at present. Radiocarbon age profiles of the sediment carbonate tend to have constant values down to 1.5 m (10) but there is evidence from age profiles of Turritella communis shells of slow rates of accretion of 0.02 to 0.07 cm a^{-1} over the past 3000 years (P. J. Kershaw, unpublished data). In view of the slow accumulation rates observed, and the uncertainty as to their general applicability, sedimentation rates were assumed to be zero for the purpose of estimating D_B values.

Unsupported ^{234}Th profiles were quasi-exponential (Figure 5) with D_B values (Table 2) varying within a factor of 4 (42 ± 16 cm^2 a^{-1}). An exception was core 108 (257 cm^2 a^{-1}) which was a sandy sediment(~ 80% > 63 µm) with a very low ^{234}Th inventory.

Unsupported ^{210}Pb profiles were more variable with several exhibiting prominent sub-surface concentration maxima (Figure 6). Application of the diffusion equation produced D_B values which varied by more than 2 orders of magnitude (Table 2).

At each of 7 of the 1984 sites a series of 5 Reineck cores was collected for benthic macrofaunal (> 2 mm) analysis. This provided an opportunity to compare bioturbation rates with the total numbers of animals (m^{-2}), the numbers of particular species (m^{-2}) and the grain-size composition. Values of ^{234}Th and ^{210}Pb D_Bs were unrelated to the total number of animals, which varied by a factor of 10, and there were no apparent systematic variations of D_B with the numbers of any of the main species (11). ^{234}Th D_Bs were unrelated to the percentage of sand (r = -0.3 excluding station 108) but ^{210}Pb D_Bs, excluding station 77, varied inversely with the sand content (r = 0.6).

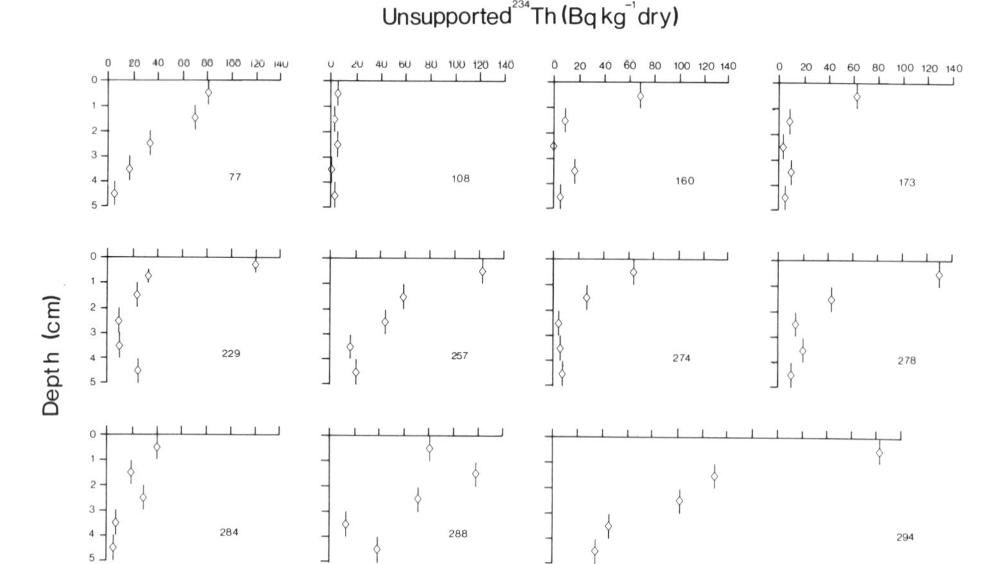

Figure 5. Profiles of unsupported ^{234}Th concentrations (Bq kg^{-1}, dry).

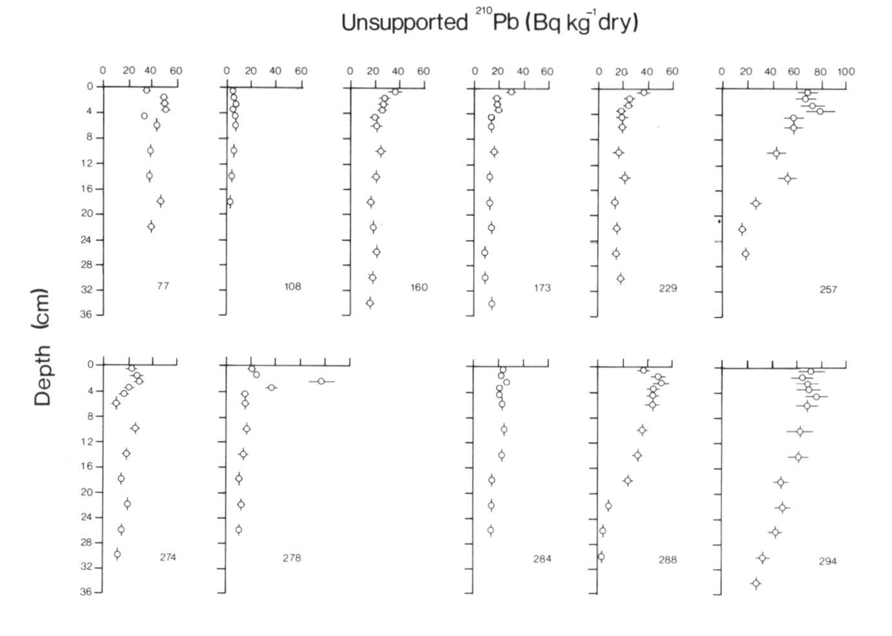

Figure 6. Profiles of unsupported ^{210}Pb concentrations (Bq kg^{-1}, dry).

The reasonable approximation of ^{234}Th concentrations to an exponential form in many core profiles suggests that bioturbation, in the upper 5 cm of the sea bed, proceeds as a series of random mixing events and that the penetration of unsupported ^{234}Th from the sediment surface into

Table 2. Biodiffusion coefficients (D_B) calculated from profiles of ^{234}Th and ^{210}Pb assuming that no advective mixing takes place

Site	^{234}Th		^{210}Pb	
	depth range (cm)	D_B ($cm^2\ a^{-1}$)	depth range (cm)	D_B ($cm^2\ a^{-1}$)
77	0-5	23	0-23	2299
108	0-5	257	0-19	22
160	0-5	52	0-35	144
173	0-5	40	0-35	86
229	0-5	75	0-31	90
257	0-5	45	0-27	9
274	0-5	31	0-31	109
278	0-5	32	0-27	15
284	0-5	40	0-27	73
288	0-5	78	0-31	4
294	0-5	40	0-35	45

the sea bed can be represented as a biodiffusion process. The relative constancy of D_B values obtained implies that such random mixing is not restricted by the grain size composition of the sediment and is not limited by the occurrence of particular species, or associations of species, of macrofauna.

Below 5 cm mixing is dominated by 2 species, the Echiuran worm _Maxmüllleria lankesteri_ and the Thalassinid shrimp _Callianassa subterranea_ (11,12). The former is frequently found at depths of 30 to 40 cm and the latter has been observed down to 150 cm. _M. lankesteri_ feeds by extending its probosis onto the sediment surface and ingesting sediment particles. A proportion of this material is defaecated within the burrow, becoming incorporated in the burrow lining, leading to enhanced concentrations of $^{239,240}Pu$ and ^{241}Am in the burrow lining relative to adjacent sediment at the same depth (12). _C. subterranea_ excavates extensive burrow systems and transports sediment to the surface, effectively burying freshly deposited particles. Both animals bring about the net transfer of surface sediment by an advective or 'conveyor-belt' mechanism, leading to difficulties if applying a diffusive model to describe the effect of bioturbation on a 'tracer' delivered at the sediment surface. Such profiles will be dependent on the frequency and rate of burrow formation and can be expected to show considerable variation given the relatively low densities of the two principal, deep bioturbating organisms (_M. lankesteri_ 0 to 11 m^{-2}, _C. subterranea_ 0 to 83 m^{-2} (11)).

The considerable variability of the ^{210}Pb profiles implies that an equilibrium has not been reached.

The term P in equation 7, described earlier as the rate of _in-situ_ production (Bq kg^{-1} a^{-1}), may include removal processes such as the transfer, by burrowing animals, of sediment-bound radionuclides from the sediment surface to a deeper layer in the sediment column. As such, P may be a function of the time varying concentration field. Again the advection term is ignored because the low sediment accumulation rates are unlikely to have a great effect on the shape of the ^{210}Pb profiles.

Equation 7 was solved using a subroutine (DO 3PAF) from the NAG library (13), with a grid spacing of 1 mm giving adequate resolution (14) and the model values being averaged over the appropriate slice thickness for comparison with measured core profiles.

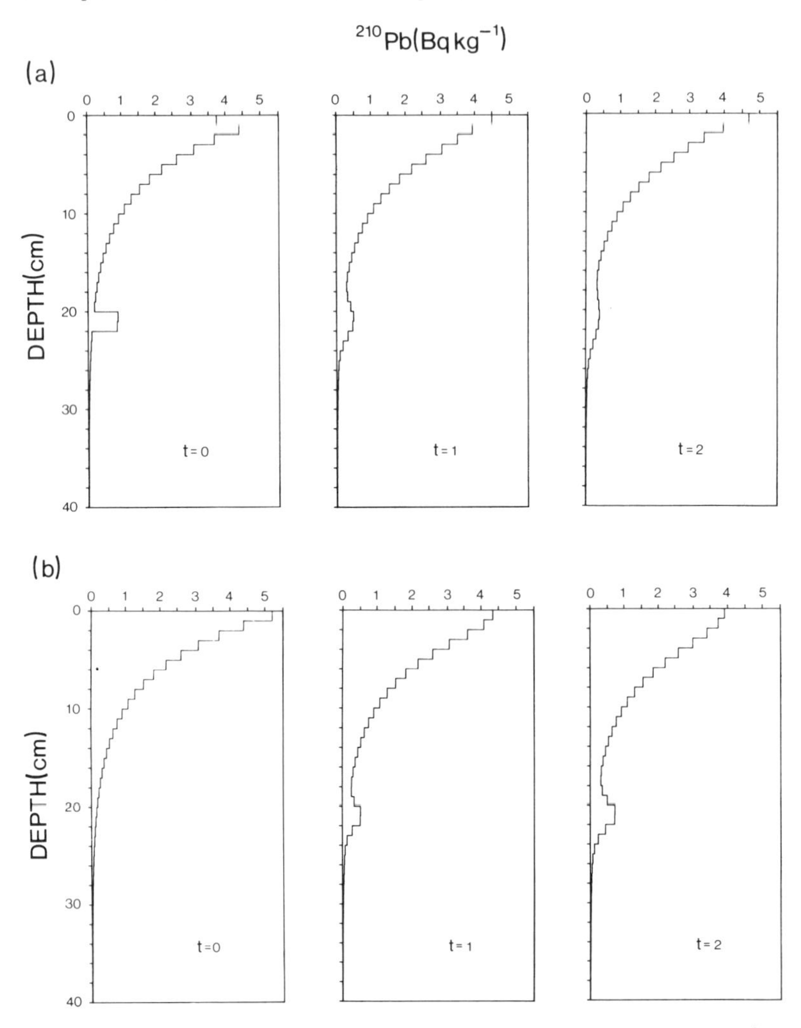

Figure 7. Model prediction of the effect of 'conveyor-belt' mixing on an equilibrium profile of ^{210}Pb with D_B = 1 cm^2 a^{-1} and at times (t) of 0, 1 and 2 years after initiation of injection(s): (a) a single injection; (b) continuous injection.

Smith et al. (15) used various forms of P to match ^{210}Pb profiles in NE Atlantic cores. A similar approach has been adopted in the present study. The model used simulates two simultaneous processes: continual mixing, analagous to diffusion processes (biodiffusion), plus the injection of material from the sediment surface to a specified depth within the sediment. It has been used to interpret the existence of sub-surface concentration maxima. Such maxima either result from a single, recent injection of surface material or from a continuous injection to the same depth. In the former case, the peak concentration is rapidly dissipated by biodiffusion (Figure 7(a)), with a D_B of only 1 $cm^2 a^{-1}$. Preservation of such a peak requires the repeated injection of surface material (Figure 7(b)). Observations in the eastern Irish Sea do not indicate any obvious pattern in the measured ^{210}Pb profiles suggestive of a population of animals redistributing sediments in this way. Instead, they appear to represent the effects of many single injection events with biodiffusion causing individual peaks to be 'smeared-out', resulting in the net downward transport of ^{210}Pb and a great variety of individual profiles, each of which is a 'snap-shot' of a continuing process. Estimated D_B values will be unrealistically high if it is assumed that the profiles are the result solely of biodiffusion.

CONCLUSION

Measurements of the $^{234}Th/^{238}U$ disequilibria in sea water and suspended and sea-bed sediment in the eastern Irish Sea have provided an insight into scavenging processes in coastal waters. The rate of scavenging of ^{234}Th was proportional to the number of particles in suspension, with a consequent increase in the residence time of dissolved ^{234}Th (0.5 to 7.7 days) with increasing distance from the Cumbrian coast. The removal of ^{234}Th was not limited by the rate of adsorption and the measured K_D values represented equilibrium conditions.

Unsupported ^{234}Th sea-bed inventories, as a percentage of the overlying water column production, were relatively constant for water depths below about 30 to 35 m. In shallower waters inventories increased inversely with water depth, reflecting an increase in the probability of sediment resuspension, and near the coast were greatly in excess of the estimated production. This suggests that there was a net horizontal transfer of ^{234}Th into the coastal region.

The incorporation of ^{234}Th into the sea bed by bioturbation may be represented as a biodiffusion process, with D_B values varying within a factor of 4 (42 ± 16 $cm^2 a^{-1}$). Unsupported ^{210}Pb profiles penetrated to greater depths, integrating mixing events over a longer period, and were more variable, with D_B values varying by more than 2 orders of magnitude. The redistribution of ^{210}Pb into the sea bed was better described by a model incorporating advective, or 'conveyor-belt', mixing which simulated the known behaviour of one of the principal bioturbating organisms, *M. lankesteri*.

REFERENCES

1. Aller, R. C. and Cochran, J. K., $^{234}Th/^{238}U$ disequilibrium in near-shore sediment: particle reworking and diagenetic time scales. *Earth Planet. Sci. Lett.*, 1976, **29**, 37-50.

2. Santschi, P. H., Li, Y.-H. and Bell, J., Natural radionuclides in the water of Narragansett Bay. *Earth Planet. Sci. Lett.*, 1979, **45**, 201-213.

3. Benninger, L. K., Aller, R. C., Cochran, J. K. and Turekian, K. K., Effects of biological sediment mixing on the ^{210}Pb chronology and trace metal distribution in a Long Island Sound sediment core. Earth Planet. Sci. Lett., 1979, 43, 242-259.

4. Carpenter, R., Peterson, M. L. and Bennett, J. T., ^{210}Pb-derived sediment accumulation and mixing rates for the Washington continental slope. Mar. Geol., 1982, 48, 1155-1172.

5. Kershaw, P. J. and Young, A. K., Scavenging of ^{234}Th in the eastern Irish Sea. J. Environ. Radioact., 1988, 6, 1-23.

6. Nyffeler, U. P., Santschi, P. H. and Li, Y.-H., The relevance of scavenging kinetics to modelling of sediment-water interactions in natural waters. Limnol. Oceanogr., 1986, 31, 277-292.

7. Howarth, M. J., Currents in the eastern Irish Sea. Oceanogr. Mar. Biol., Ann. Rev., 1985, 22, 11-53.

8. Schink, D. R. and Guinasso, N. L., Redistribution of dissolved and adsorbed materials in abyssal marine sediments undergoing biological stirring. Am. J. Sci., 1978, 278, 687-702.

9. Officer, C. B., Mixing, sedimentation rates and age dating for sediment cores. Mar. Geol., 1982, 42, 261-276.

10. Kershaw, P. J., Radiocarbon dating of Irish Sea sediments. Estuar. Cstl. Shelf Sci., 1986, 23, 295-303.

11. Swift, D. J. and Kershaw, P. J., Bioturbation of contaminated sediments in the north-east Irish Sea. ICES C.M. 1986/E:18, 12 pp. (mimeo).

12. Kershaw, P. J., Swift, D. J., Pentreath, R. J. and Lovett, M. B., The incorporation of plutonium, americium and curium into the Irish Sea sea bed by biological activity. Sci. Total Environ., 1984, 40, 61-81.

13. NAG, Numerical Algorithms Group Fortran Library Manual, Mark II, Vol. 2, NAG, Oxford, 1984, D02K-E02.

14. Gurbutt, P. A., Kershaw, P. J. and Durance, J. A., Modelling the distribution of soluble and particle-adsorbed radionuclides in the Irish Sea. In Radioactivity and Oceanography, Elsevier Applied Science Publishers, London, 1987 (this volume).

15. Smith, J. N., Boudreau, B. P. and Noshkin, V., Plutonium and ^{210}Pb distributions in northeast Atlantic sediments: subsurface anomalies caused by non-local mixing. Earth Planet. Sci. Lett., 1986, 81, 15-28.

$^{234}Th/^{238}U$ DISEQUILIBRIUM TO ESTIMATE THE MEAN RESIDENCE TIME OF THORIUM ISOTOPES IN A COASTAL MARINE ENVIRONMENT

A. Battaglia
CISE-Tecnologie Innovative SpA
Segrate (Milano)
I
and
W. Martinotti, G. Queirazza
ENEL - Italian Electricity Board, Center for Thermal
and Nuclear Research
Via Rubattino, 54 - 20134 Milano
I

ABSTRACT

The biogeochemical behaviour in aquatic environments of radionuclides with high "reactivity" for suspended matter can be considered to be similar to that of the thorium isotopes. In this paper are presented the results of some field measurements of the mean residence time of thorium using the $^{234}Th/^{238}U$ disequilibrium method in the Tyrrhenian Sea near the Montalto di Castro nuclear power plant.

Sea water was filtered with porous membrane filters (0.3 μm) in order to remove completely the suspended matter. Dissolved thorium isotopes were recovered by sequential $BaSO_4$-activated alumina beds. The efficiency of recovery was determined with field and laboratory tracer tests. Filters and alumina beds were analyzed by gamma spectrometry using the 92.6 keV line of ^{234}Th to evaluate separately the adsorption (τ_S) and sedimentation (τ_P) mean residence times. The experimental values were determined at different seasons and ranged between 1 and 70 days. These low values indicate that the $^{234}Th/^{238}U$ ratio is more suitable than the $^{228}Th/^{228}Ra$ ratio for evaluating the mean residence time of thorium isotopes in the coastal marine environment under consideration.

The sedimentation rate values calculated from the residence time values are compared with those obtained by ^{210}Pb excess technique in sediment cores.

INTRODUCTION

The natural series radionuclides can be used to trace the processes and mechanisms which determine the distribution of radionuclides released into coastal marine environment. Bender et al.[1] described the chemical, biological and physical transfer processed involved. The natural radionuclides to be considered are:

1) dissolved in sea water: ^{238}U, ^{235}U, ^{234}U, ^{228}Ra, ^{226}Ra and ^{222}Rn
2) associated with the suspended matter: ^{234}Th, ^{228}Th, ^{231}Pa, ^{210}Pb, and ^{210}Po.

Thorium-234 has proven to be useful in examining rates of removal processes in coastal marine environments[2,3]. ^{238}U dissolved in sea water produces ^{234}Th which is adsorbed by particles settling through the water column to the sea bed. This mechanism produces a radioactive disequilibrium which offers the possibility of estimating the mean residence of the radionuclide in the water column.

The purpose of this work is to determine the mean residence time of ^{234}Th in order to predict the behaviour of artificial radionuclides which exhibit a similar reactivity with respect to the suspended matter.

The present study was carried out in the Tyrrhenian Sea near Montalto di Castro, where a BWR nuclear power plant is under construction. In this paper we show and discuss the ^{234}Th-^{238}U disequilibrium recorded on five cruises during the period 1984-1986. In addition the sampling and analytical methods for ^{234}Th are briefly described. Furthermore some of our sedimentation rate values, estimated by sediment age measurements using ^{210}Pb were compared with these obtained for thorium containing particles.

THEORY

The residence time of thorium in sea water can be determined by means of ^{234}Th-^{238}U and ^{228}Th-^{228}Ra disequilibrium: the choice between these two approaches should be made on the basis of the expected residence time values. Broecker[4] was the first to use ^{228}Th-^{228}Ra disequilibrium in coastal waters and the open sea to evaluate thorium isotope residence times. His formulation is:

$$\tau = \frac{A_{Th}/A_{Ra}}{1,5-(A_{Th}/A_{Ra})} \cdot \tau_{\,^{228}Th} \qquad (1)$$

where:

τ is the Thorium mean residence time in the water column connected with the removal process; $\tau_{^{228}Th}$ is radioactive mean life of ^{228}Th $(1/\lambda_{^{228}Th})$; A_{Th} and A_{Ra} are the activity of the respective radionuclides and 1.5 is their ratio at the transient equilibrium existing in sea water. The same expression can be applied to the nuclide pair ^{234}Th-^{238}U and the equilibrium factor is, in this case, one.

Both methods are quite applicable to sea water studies: both can be used in order to obtain the best accuracy for the residence time evaluations. In a coastal environment, where τ values approach 10-100 days,(12) ^{234}Th-^{238}U disequilibrium should be used while in the open sea, where τ values reach up to 1 year, ^{228}Th-^{228}Ra disequilibrium should be preferred.

If the activities are measured in filtered water the τ values relate to the sorption process of thorium by suspended particles (τ_S); in the case of unfiltered water the τ values include the removal of particles by sedimentation (τ_T). The mean residence time of ^{234}Th adsorbed on the suspended particles, τ_p, is related to τ_T as follows: $\tau_T = \tau_p/f_p$, where f_p is the fraction of the radionuclide in the particulate. Santschi et al.(5) utilized τ_T values to calculate the thorium flux from water column to the bottom sediments by settling particles using the equation:

$$J = \lambda_T \cdot h \cdot \Sigma C \qquad (2)$$

where:

J = thorium flux (Bq/m^2.day);

λ_T = first order removal rate constant of Th in the water column $= \frac{1}{\tau_T}$ (day^{-1});

h = water column height (m);

ΣC = total concentration of Th in the water column (Bq/m^3).

The equation (2) neglects other removal mechanisms such as adsorption at the sediment water interface and horizontal exchange processes. According to this formulation it is possible to calculate the vertical flux of settling particles F (g/m^2 . day) = J/As where As = concentration of thorium isotope in suspended matter (Bq/g) and J is the above defined thorium flux.

The sedimentation rate (SR) (mm/year) can be related to the particle flux (F) by means the apparent density of the bottom sediments (ρ): $SR = F/\rho$ where ρ(g/cm^3) depends on the porosity according to the relation $\rho = 2.65(1 - \Phi)$, and Φ is the fractional porosity of the bottom sediments.

These sedimentation rate values are not necessarily a measure of the true sediment accumulation rate because the process of resuspension is included. As the sediments of the marine environment under investigation show fractional porosity values of about 0.7, ρ equals 0.8 g/cm^3.

SAMPLING AND ANALYTICAL METHODS

Figure 1 shows the principal physico-chemical characteristics of the sea water of the study area in relation to depth and season. In summer the thermocline appears at about 20 m depth; the water quality is good, as indicated by the high levels of dissolved oxygen and positive redox potentials.

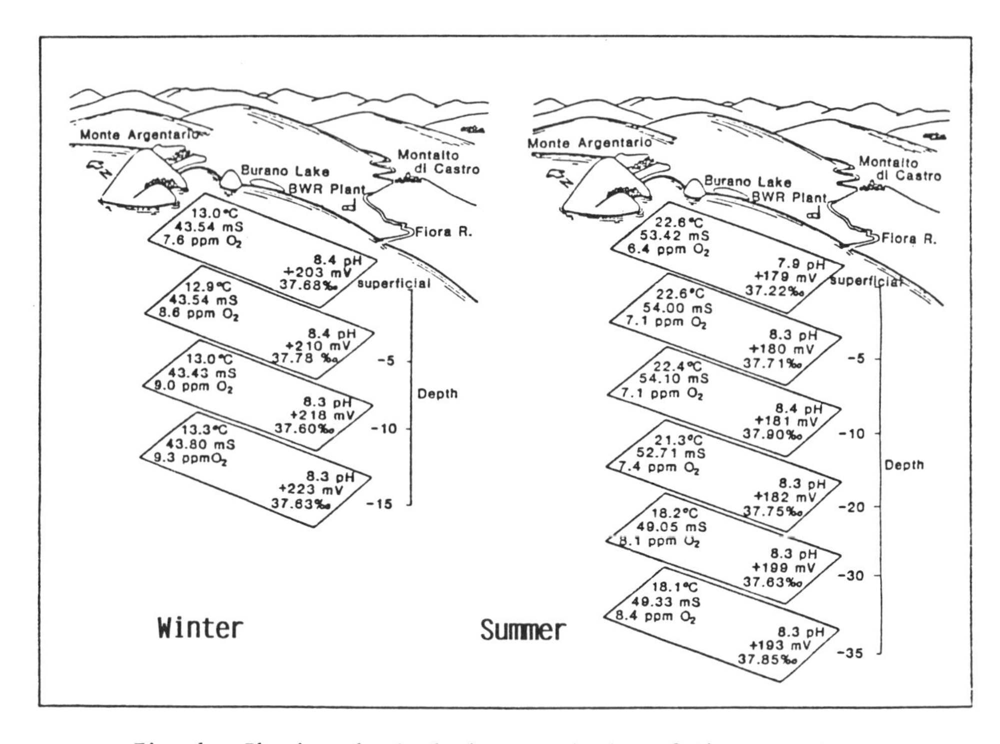

Fig. 1 - Physico-chemical characteristics of the sea water

In the study area, five cruises were made during 1984-1986. Water samples were collected by means of a large volume water sampler (up to 6 m^3). The radionuclides were extracted by pumping sea water through a specially designed filtration-sorption apparatus. A general outline of this device is shown in Fig. 2: water was sequentially filtered through 1.9 and 0.3 micrometer pore-size membrane cartridges, and was pumped through four $BaSO_4$ impregnated Al_2O_3 sorption beds for the quantitative recovery of thorium and radium isotopes[6,7,8]. The filtered water was also passed through two MnO_2 impregnated fiber cartridges to quantitatively retain radium isotopes and other natural and artificial radionuclides present in sea water(9).

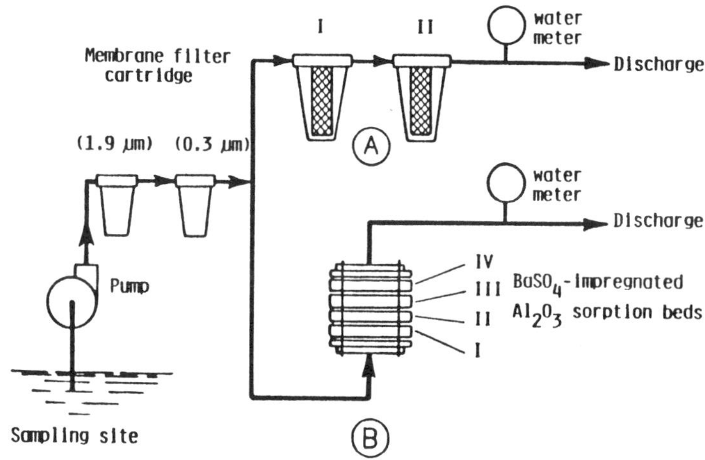

Fig. 2 - Large Volume Water sampler Layout

The equipment can operate at a flowrate of 20 litres per minute for both sorption systems. Under laboratory conditions the recovery yield for barium sulfate impregnated alumina beds was:

^{88}Y	^{210}Pb	^{234}Th	^{210}Pb	^{226}Ra
0.66	0.83	0.76	0.82	0.75

while under field conditions the recovery values were determined by applying the formula: $R = (A_n - A_{n+1})/A_n$, where A_n and A_{n+1} are the activities on two consecutive beds. The total activity was calculated by means of the equation(7):

$$A = A_1 + \dots A_n + A_{n+1}/R \qquad (3)$$

In this case the recovery was slightly reduced in comparison with laboratory results.

The MnO_2 cartridges showed high efficiency for radium isotopes but low efficiency for thorium isotopes: their use is at present in an experimental stage.

The determination of gamma emitting radionuclides in the sorption beds was performed by gamma spectrometry within two to three days from sampling to eliminte large corrections for the ^{234}Th ingrowth from the decay of the ^{238}U partially recovered in the alumina beds. The ^{234}Th amounts were determined by the 92.5 keV energy line, neglecting the contribution of the 93.3 keV ^{228}Ac line given the relatively low concentration of ^{228}Ra in sea water. The ^{238}U concentration in the study area was not determined but it was derived from published data: 3.2 ppb (39.5 Bq/m^3)[10].

RESULTS AND DISCUSSION

The results of a test performed on June 13, 1986 are reported in Table 1. Detectable concentrations of ^7Be, ^{228}Th, ^{228}Ra, ^{234}Th as well as ^{134}Cs, ^{137}Cs and ^{140}Ba coming from the Chernobyl accident were found. Concentration values are affected by small statistical fluctuations and particularly the ^{234}Th residence time is estimated with a relative standard deviation of 15% and the sedimentation rate with a standard deviation of about 20%.

Table 1. Concentration values and standard deviations of radionuclides at 30 m depth (June 13th, 1986). Suspended matter 0.5 mg/l

Nuclide	Water concentration (Bq/m^3)		(3) Particulate concentration	(4)(*) Kd
	(1) particulate	(2) solution	(Bq/g)	(cm^3/g)
^7Be	0.22 ± 0.03	3.17 ± 0.16	0.44 ± 0.06	1.4×10^5
^{134}Cs	0.030 ± 0.003	0.91 ± 0.05	0.060 ± 0.006	6.6×10^4
^{137}Cs	0.058 ± 0.007	2.79 ± 0.13	0.116 ± 0.014	4.2×10^4
^{140}Ba	0.063 ± 0.015	0.69 ± 0.05	0.13 ± 0.03	1.8×10^5
^{228}Th	0.048 ± 0.005	N.D.	0.10 ± 0.01	N.D.
^{228}Ra	0.023 ± 0.007	0.58 ± 0.04	0.05 ± 0.01	7.9×10^4
^{226}Ra	0.040 ± 0.004	1.76 ± 0.04	0.08 ± 0.01	4.5×10^4
^{234}Th	13.1 ± 1.4	2.0 ± 0.4	26 ± 3	1.3×10^7

N.D. = Not Determined; (*) $Kd = \frac{(3)}{(2)} \cdot 10^6$

ESTIMATED 234-THORIUM PARAMETERS

τ_p = 18.6 ± 2.6 days　　J = 21.1 ± 3.9 Bq/m^2 · day
τ_s = 1.8 ± 0.5 days　　F = 0.81 ± 0.18 g/m^2 · day
τ_t = 21.5 ± 3.4 days　　SR = 1.0 ± 0.2 mm/year

Table 2 reports the results of contemporaneous samplings performed at four different depths: 2 m, 10 m, 18 m and 26 m at a location with a total water depth of 30 m. The samples collected at 2 m and at the maximum depth were filtered and treated with $BaSO_4$ impregnated alumina beds; only the suspended matter was sampled at the intermediated depths. The concentrations of dissolved ^{228}Ra and ^{226}Ra are about the same; although ^{7}Be and ^{234}Th show a slightly higher concentration at the 2 m sampling point, the dissolved radionuclides seem to be homogeneously distributed throughout the water column. The particulate ^{234}Th concentration in sea water shows a significant increase at the 26 m depth. Nevertheless a lower concentration of particulate radionuclides at the 18 m depth sampling point was measured compared to the concentration recorded at 10 m depth.

Table 2. Concentration (Bq/m^3) of radionuclides at different depth

Depth	Water phase	^{7}Be	^{228}Th	^{228}Ra	^{226}Ra	^{234}Th
2 m	P	0.12±0.02	0.033±0.004	0.026±0.009	0.031±0.003	14.2±1.5
2 m	S	5.20±0.21	N.D.	0.76±0.07	2.07±0.05	3.2±1.3
10 m	P	0.45±0.06	0.079±0.010	0.073±0.019	0.106±0.009	18.5±1.1
18 m	P	0.40±0.04	0.045±0.007	0.040±0.007	0.054±0.005	8.4±0.4
26 m	P	0.90±0.11	0.260±0.020	0.180±0.040	0.240±0.020	39.5±2.3
26 m	S	2.76±0.16	N.D.	0.83±0.06	2.10±0.06	2.1±1.0
Mean Value	P	0.47±0.32	0.10±0.10	0.08±0.07	0.11±0.09	20±13
	S	4.0	N.D.	0.80	2.08	2.6

P = Particulate   S = Solution   N.D. = Not Determined

The ^{234}Th concentrations in particulate and dissolved forms recorded for the five cruises carried out during 1984-1986 are summarized in Table 3. The reported data are the calculated mean values obtained for each cruise with the respective standard deviation. It can be seen that the mean residence time of dissolved ^{234}Th , τ_s, ranges between 1 and 7 days. These values imply rapid scavenging and, moreover, seem to be independent of season. The ^{234}Th in particulate form shows a different behaviour: the τ_p values in winter are about 5 days as opposed to 15-71 days during the summer cruises. It can be observed that the ^{234}Th flux values (J) to the bottom sediments appear slightly higher in winter than in summer; this phenomenon could be explained in terms of particulate loading. The sedimentation rate values (SR) show a difference between winter (4.1 and 1.2 mm/year) and summer (0.38, 0.11 and 0.19 mm/year) and it can be inferred that the suspended particles are mostly deposited in the bottom sediments during winter.

Seasonal determinations of chlorophyll-a concentration, an indication of living phytoplankton activity, showed very high values in summer (200 μg/m^3) and low values in winter (0.5 μg/m^3)(11). This fact and the low sedimentation rates recorded in summer indicate that non-living particles are mainly responsible for the removal of the ^{234}Th dissolved in the sea water in the study area.

Table 3. ^{234}Th residence times calculated from 1984-1986 cruises

Cruise	Water phase	Concentration ^{234}Th (Bq/m^3)	Mean residence time τ(days)	Flux J(Bq/m^2.day)	Particulate load (mg/l)	Sedimentation Rate SR (mm/year)
	P	5.1 ± 1.1	5.3 ± 1.5			
Feb.'84	S	1.4 ± 0.1	1.3 ± 0.1		1.1	
	T	6.5 ± 1.1	6.8 ± 1.5	33 ± 9		4.1 ± 2.1
	P	11 ± 4	15 ± 7			
Jun.'84	S	1.6 ± 0.8	1.5 ± 1.2		0.3	
	T	12 ± 4	16 ± 8	27 ± 16		0.38 ± 0.36
	P	4.8 ± 1.1	6.1 ± 1.6			
Mar.'85	S	6.5 ± 3.6	6.8 ± 5.4		0.5	
	T	11 ± 4	14 ± 8	28 ± 18		1.2 ± 0.8
	P	25 ± 17	71 ± 56			
Jun.'86	S	2.2 ± 0.4	2.0 ± 0.5		0.5	
	T	27 ± 17	77 ± 66	12 ± 13		0.11 ± 0.18
	P	20 ± 11	41 ± 28			
Sept.'86	S	2.4 ± 0.5	2.2 ± 0.5		0.5	
	T	22 ± 11	45 ± 35	17 ± 16		0.19 ± 0.21

P = Particulate S = Solution T = Total (Solution + Particulate)

Analysis of depth profiles of excess ^{210}Pb in cores collected in the study area provided sedimentation rate values reported in Table 4. These values seem to be in accordance with sedimentation rate values derived from ^{234}Th data for the winter period.

Table 4. Sedimentation Rates (mm/year) obtained from the analysis of depth profiles of excess ^{210}Pb in cores collected in the study area.

Cruise	Depth (m)	S.R. (mm/year)	Cruise	Depth (m)	S.R. (mm/year)
Sept. 84	30	2.55±0.48	Febr. 84	30	3.29±0.40
Mar. 85	30	3.48±0.36	Jun. 84	30	6.9±2.9
Mar. 85	30	5.60±0.50	Mar. 85	30	2.65±0.50
Jun. 84	40	2.29±0.21	Jun. 84	40	4.2±1.0

REFERENCES

1. M. Bender; T. Church, D. Edington, M.G. Gross, D. Haidvogel, J. McCarthy, G. Needler, A. Robinson, E. Schneider and J. Steel (1979), Coastal and open ocean science, Proceed. of a Workshop on Assimilative Capacity of U.S. Coastal Waters for Pollutants, Crystal Mountain, Washington July 29-August 4.

2. B.A. McKee, D.J. De Master and C.A. Nittrouer (1984), The use of $^{234}Th/^{238}U$ disequilibrium to examine the fate of particle-reactive species on the Yangtze continental shelf, Earth Planet. Sci. Lett. 68, 431-442.

3. Santschi, P.H., Li, Y.H., Bell, J. (1979), Natural radionuclides in the water of Narragansett Bay, Earth Planet. Sci. Lett. 45, 201-213.

4. Broecker, W.S., Kaufman, A., Trier, R.M. (1973), The residence time of Thorium in surface sea water and its implications regarding the rate of reactive pollutants, Earth Planet. Sci. Lett. 20, 35-44.

5. Santschi, P.H., Li, Y.H. (1981), Removal pathways of Th and Pu isotopes in coastal marine environments, in: Natural Radiation Environment, K.G. Vohra et al. eds., Wiles Eastern Ltd., pp.643-650.

6. Silker, W.B., Perkins, R.W., Rieck, H.G. (1971), A sampler concentrating radionuclides from natural waters, Ocean. Engineering 2, 49-55.

7. Nevissi, A., Schell, W.R. (1975), Efficiency of large volume water sampler for some radionuclides dissolved in salt and fresh water, Radioecology and Energy Resource, Fourth Nat. Symp. on Radioecology, May 14-14, Oregon State Univ. Corvallis, Oregon; C.E. Cushing Ed., pp.277-282.

8. Perkins, R.W. (1968), Radium and radiobarium measurement in sea water and fresh water by sorption and direct multidimensional gamma-ray spectrometry, BNWL 1951, pt. 2, pp.21-27.

9. Reid, D.F., Key, R.M., Schink, D.R. (1979), Radium, Thorium and Actinium extraction from sea water using an improved manganese-oxide-coated fiber, Earth Planet. Sci. Letters 43, 223-226.

10. Marangoni, G., Paulucci, G. (1980), Determinazione del contenuto di uranio nelle acque costiere italiane. Atti del Simposio sulle metodiche radiometriche e radiochimiche nella radioprotezione, Pavia 28-29 aprile.

11. ENEL (1985), Progetto REM, Radioecologia Marina. Stato di avazamento delle ricerche al febbraio 1985.

12. Santschi, P.H., Nyffeler, U.P., Li, Y.H., O'Hara, P. (1984), Radionuclide Cycling in natural Waters: Relevance of Sorption Kinetics, in Third Intern. Symp. "Interactions between Sediments and Water", Geneva, August 1984.

SOURCES AND SINKS OF LEAD-210 IN PUGET SOUND

A.E. Nevissi
Laboratory of Radiation Ecology, WH-10
University of Washington
Seattle, Washington 98195
U.S.A.

ABSTRACT

The Pb-210 budget to Puget Sound sediments is calculated using atmospheric and advective fluxes and comparing them with the input of this radionuclide to sediments. The mean atmospheric flux of Pb-210 measured on the northwest coast of the United States is 0.44 ± 0.24 $dpm \cdot cm^{-2} \cdot yr^{-1}$. The Pb-210 flux due to advection, calculated from the literature values, is 0.94 ± 0.1 $dpm \cdot cm^{-2} \cdot yr^{-1}$. The average flux of Pb-210 to Puget Sound sedmients, obtained from Pb-210 dated sediments, is 2.02 ± 1.0 $dpm \cdot cm^{-2} \cdot yr^{-1}$. The combined atmospheric and advective fluxes are 68% of the Pb-210 input to sediment, and the remaining 32% is attributed to sediment focusing in Puget Sound. Depending on the type of coring devices used, large variations in sedimentation rates and Pb-210 flux to sediments are encountered.

INTRODUCTION

Lead-210 is a member of the U-238 natural radioactive series, and is supplied to sea water from atmospheric input, rivers and runoff, and the in situ decay of Ra-226 precursor in the water column. The source of atmospheric Pb-210 for the west coast of the United States is the Eurasian continent. Radon and its decay products, including Pb-210, are supplied to the atmosphere from this source and are then transported by the prevailing winds eastward over the North Pacific Ocean.

EXPERIMENTAL

Atmospheric Pb-210 Flux. Measurements of the monthly atmospheric flux of Pb-210 in Seattle, Washington, were conducted over a period of seven years. The mean annual flux of Pb-210 for this period was 0.44 ± 0.24 $dpm \cdot cm^{-2} \cdot yr^{-1}$ (1). Direct correlation existed between monthly rainfall and Pb-210 deposition. However, for a given amount of rainfall the

Pb-210 flux varied from year to year (Fig. 1).

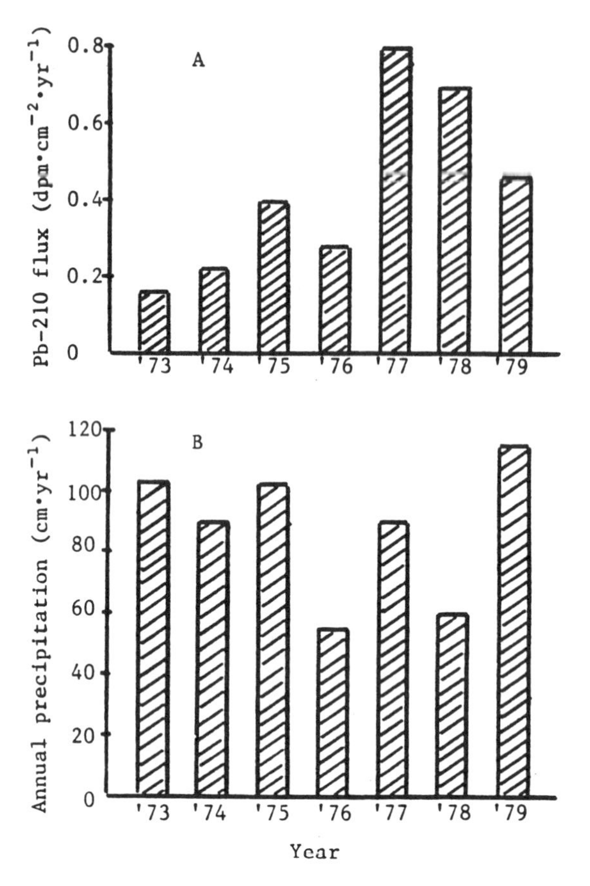

Figure 1. Variation of annual deposition of (A) Pb-210 and (B) annual precipitation in Seattle, WA. Annual cycle is computed from 1 September through 31 August.

Lead-210 Dating of Sediments. Measurement of sedimentation rate by the Pb-210 dating technique is widely used for dating both marine and freshwater sediments. The basic assumptions for the Pb-210 dating technique are that there is a constant influx of Pb-210 from water column to the sediment, a constant sedimentation rate, and immobilization of Pb-210 within the sediment (2).

$$A(z) = \frac{P}{\rho S(t)} e^{-\lambda t} \tag{1}$$

where A(z) is the activity of excess Pb-210 per unit weight of dry sediment ($dpm \cdot g^{-1}$) at depth z, P is the rate of deposition of unsupported Pb-210 per unit area and unit time ($dpm \cdot cm^{-2} \cdot yr^{-1}$), S(t) is the in situ sedimentation rate ($cm \cdot yr^{-1}$), ρ is the in situ density of sediment, dry weight per unit wet volume ($g \cdot cm^{-3}$), λ is the radioactive decay constant of Pb-210 (yr^{-1}), and t is the time (yr). For uniform sedimentation rate the time of sediment accumulation in equation (1) can be replaced by Z/S(t) where Z (cm) is the depth of sediment column accumulated at the time t.

When the log excess Pb-210 activity ($dpm \cdot g^{-1}$) of dry sediment is plotted as a function of depth of wet sediment accumulated (cm), the slope of this line represents the sedimentation rate ($cm \cdot yr^{-1}$). The linear or in situ sedimentation rate, S ($cm \cdot yr^{-1}$), is related to the mass deposition rate, W ($g \cdot cm^{-2} \cdot yr^{-1}$), using the following equation:

$$W = S \cdot \rho \tag{2}$$

where ρ is the in situ density of sediment.

The specific activity of Pb-210, G ($dpm \cdot g^{-1}$), at the sediment-water interface is obtained from the intercept of the fitted line and the Pb-210 axis (Fig. 2). The flux of unsupported Pb-210, P ($dpm \cdot cm^{-2} \cdot yr^{-1}$), is calculated from the product of specific activity, G, and the mass deposition rate W.

$$P = G \times W \tag{3}$$

Experimental details of Pb-210 dating technique are presented elsewhere (3).

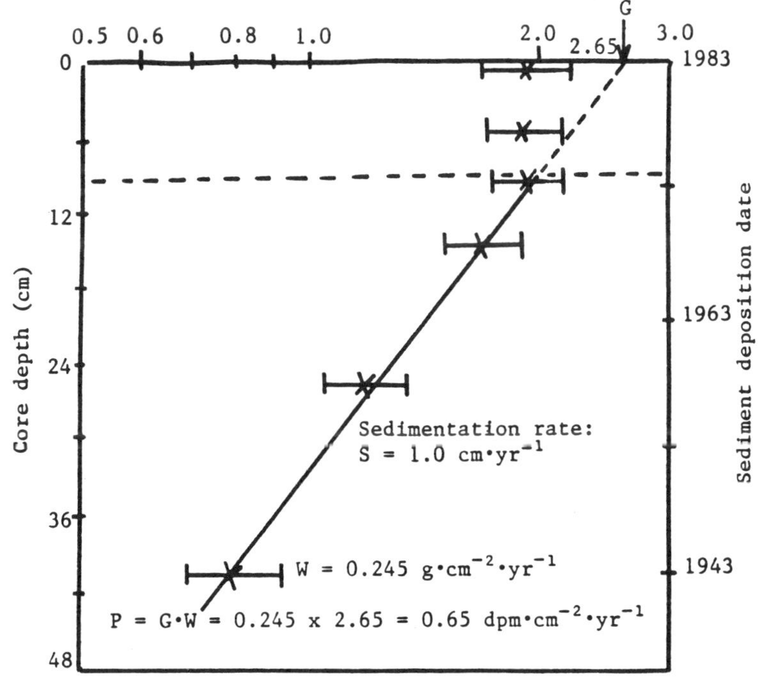

Figure 2. Example of Pb-210 dating of sediment core and calculation of Pb-210 flux ($dpm \cdot g^{-1}$) at the sediment-water interface.

RESULTS AND DISCUSSION

The annual atmospheric flux of Pb-210 measured during the seven-year period fluctuated 4.4-fold. For example, in 1974 the precipitation was 91 $cm \cdot yr^{-1}$ and the Pb-210 deposition was 0.232 $dpm \cdot cm^{-2}\ yr^{-1}$, whereas in 1977, for the same amount of precipitation the Pb-210 flux was 0.81 dpm $\cdot cm^{-2}\ yr^{-1}$ (Fig. 1). Clearly, the 3.5-fold increase in Pb-210 deposition is due to factors other than the quantity of rainfall. The explanation of the fluctuation in Pb-210 deposition is outside the scope of this paper; however, this fluctuation should be taken into account when this radioisotope is used as a tracer of natural processes. For the purpose of budget calculation, the mean annual Pb-210 flux of 0.44 ± 0.24 $dpm \cdot cm^{-2} \cdot yr^{-1}$ was used in this work.

The sinks of Pb-210 in Puget Sound are sediments. After incorporation into biota or adsorption on Mn and Fe hydroxides, Pb-210 is finally precipitated into the sediments. The flux of this radionuclide to bottom sediments can be measured by the Pb-210 dating of sediments.

Using equations (1), (2), and (3), the mean deposition rate of unsupported Pb-210 to Puget Sound sediments, together with related parameters, was computed from the results of some 35 gravity cores that were dated by Pb-210 dating technique (Table 1). The mean flux of Pb-210 from the water column to Puget Sound sediments, 2.02 ± 1.0 $dpm \cdot cm^{-2}\ yr^{-1}$, is 4.6 ± 3.4 times higher than the atmospheric flux of Pb-210 to Puget Sound surface waters, which is 0.44 ± 0.24 $dpm \cdot cm^{-2}\ yr^{-1}$. This is due to additional input of Pb-210 from the overlying water column and sediment focusing.

The sedimentation rates measured during the course of this study ranged from 1.0 to 20 $mm \cdot yr^{-1}$ with a mean value of 5.2 ± 4 $mm \cdot yr^{-1}$. Compared with the coastal areas of the United States, the Puget Sound values may seem to be very high. For example, sedimentation rates for three sites off the west coast of the United States near Farallon Islands at depths of 914-1829 m were reported to be 1-3 $mm \cdot yr^{-1}$ (4). On the east coast of the United States, the Pb-210 sedimentation rate measured in Hudson Canyon at 4000 m deep ranged from 0.1 to 2.0 $mm \cdot yr^{-1}$ (5). On the other hand, the average sediment mass deposition rate for the Washington

Table 1. The mean annual flux of Pb-210 to Puget Sound sediments together with other related parameters obtained from 35 sediment cores which were dated by Pb-210 dating technique (P = flux of unsupported Pb-210 to sediments, S = in situ sedimentation rate, W = mass deposition rate, G = the specific activity of Pb-210, H = coring depth).

P $(dpm \cdot cm^{-2} \cdot yr^{-1})$	S $(mm \cdot yr^{-1})$	W $(g \cdot cm^{-2} \cdot yr^{-1})$	G $(dpm \cdot g^{-1})$	H (m)
2.02 ± 1.0	5.2 ± 4	0.145 ± 0.1	13.9 ± 9.8	145 ± 80

State continental slope was reported to be 220 ± 230 $mg \cdot cm^{-2} \cdot yr^{-1}$ for the canyons and 53 ± 25 $mg \cdot cm^{-2} \cdot yr^{-1}$ for the open slope areas (6). The mean sediment mass deposition rate for Puget Sound, 145 ± 100 $mg \cdot cm^{-2} \cdot yr^{-1}$ (Table 1) compares favorably with the canyon values.

Sedimentation rate studies using the Pb-210 dating technique were also conducted by other investigators along the axis of the main basin of Puget Sound, about 200 m deep, using a Kasten corer (7). The mean values of the sedimentary parameters for 25 cores were reported to be as follows: Sedimentation rate, 1.62 ± 0.74 $cm \cdot yr^{-1}$; mass deposition rate, 0.88 ± 0.39 $g \cdot cm^{-2} \cdot yr^{-1}$; and specific activity of Pb-210, 10.5 ± 2.37 $dpm \cdot g^{-1}$ (7). Both the reported sedimentation rate and the mass deposition rate were much higher than the values in Table 1. These large differences between the results of the two studies may be due to the use of different coring devices and collection depths.

The in situ production of Pb-210 in the water column is due to decay of Ra-226. However, relative to the activity of Ra-226, the Pb-210 concentration in the surface waters is augmented by the atmospheric input, and near the bottom, the Pb-210 concentration is depleted as a result of absorption of this radionuclide to suspended particulate matter and deposition of the latter on the bottom sediments. Lead-210 profiles in the Pacific, Atlantic, and Indian Oceans have been measured as part of the GEOSECS program (8,9,10). For the stations closest to the west coast of the United States (stations 202 and 212), concentrations ranged from 180 to 330 $dpm \cdot m^{-3}$, and at station 201, values ranged from 130 to 200 $dpm \cdot m^{-3}$ (8,9). These concentrations are much higher than the values reported (see below) for Puget Sound (11). The low concentration may be the result of upwelling of Pb-210-depleted ocean water that enters Puget Sound.

The circulation of water in Puget Sound is surface outflow, approximately 0-50 m depth, and deep inflow of oceanic water, below 50 m depth, through Admiralty Inlet and the Strait of Juan de Fuca (Fig. 3). The

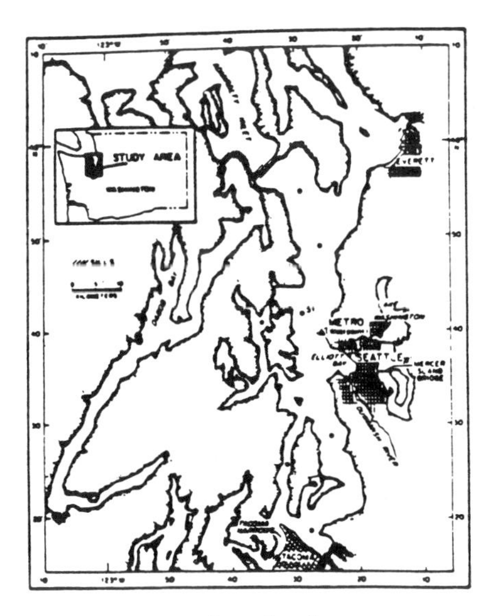

Figure 3. Study area.

Pb-210 input to sediments by advection can be calculated from the difference between the concentration of this radionuclide in deep oceanic water entering the Puget Sound basin and the concentration in the surface water leaving the Sound. The Pb-210 concentrations of surface and deep waters at three stations in the Strait of Juan de Fuca have been measured over a period of three years (11). The data showed the mean concentration of Pb-210 in 12 samples collected at 55-155 m depth was 55.8 ± 29 $dpm \cdot m^{-3}$, and for 16 samples collected at 2-45 m depth was 23.4 ± 11.4 $dpm \cdot m^{-3}$. Assuming the mean values represent the concentration of this radionuclide in the inflowing and outflowing waters, the difference, 32.4 ± 31.0 $dpm \cdot m^{-3}$, or 0.0324 ± 0.031 $dpm \cdot L^{-1}$, is the amount of Pb-210 which is deposited to sediment by advection.

At mean depth of core collection, 145 ± 80 m, the volume of overlying water over one square centimeter is 14.5 ± 8.0 L; hence, the amount of Pb-210 precipitated from this volume to sediment is 14.5 x 0.0324 = 0.47 ± 0.05 $dpm \cdot cm^{-2}$. If on the average Puget Sound water is replaced with oceanic water twice a year, then the advective Pb-210 input will be twice the above value, or 0.94 ± 0.1 $dpm \cdot cm^{-2} \cdot yr^{-1}$. The sum of the advective and the atmospheric Pb-210 input to sediment is 0.94 ± 0.44 = 1.38 ± 0.29 $dpm \cdot cm^{-2}$ yr^{-1}. This combined value accounts for 68% of the Pb-210 flux, 1.38/2.02 = 0.68, to Puget Sound sediment.

Because of a combination of sediment grain size, currents, and bottom topography, the sediments are not settled all over the area, but are focused in a much smaller area, which may be one half to two thirds of the total bottom area of Puget Sound.

The main basin of Puget Sound is a fjord-like basin, over 200 m deep, with relatively steep sides. The sources of sediments for deposition in the main basin are bluff erosion along the shoreline, submarine erosion of sidewalls, and transport of sediments by rivers and turbidity currents. Contributions from atmospheric deposition, and biological and wastewater particulate inputs seem to be small. The relative contributions by different sources to sediment accumulation in the main basin are not known. But it should be noted that bluff and submarine erosion do not contribute to the Pb-210 budget of sediment, since Pb-210 is in equilibrium with Ra-226 in soil and rocks.

Horizontal transport of sedimentary material and subsequent redeposition by turbidity currents are well known sedimentary processes in the deep ocean. For example, Hudson Canyon is both a sink and a transport conduit for sediments introduced to the head of the canyon (12). Sand, silt, and aggregates of clay and organic matter are largely retained in the upper canyon (<1,000 m), whereas settling particulate matter is transported by down-canyon currents. Such horizontal transport and deposition of fine particles containing excess Th-234, with subsequent reworking by benthic organisms, has been observed also in the shallow waters of Long Island Sound (13). Similar processes may be responsible for transport of fine particles from the steep sides of Puget Sound and estuaries to the deep areas of the main basin, which results in focusing of sediments to the deepest parts of the area.

Another source contributing to the Pb-210 inventory in Puget Sound is the riverine input. Part of the atmospheric Pb-210 deposited in the watershed may be carried by rivers and runoffs to Puget Sound. Attempts have been made to estimate the input from this source (14,15). However, the data were presented as total Pb-210 per unit weight of riverine suspended particulate matter, with no correction given for supported activity by radium. In addition, because of the large amount of suspended particulate matter present in the river water (up to a few hundred milligrams per liter) as compared with Puget Sound water (around 1 $mg \cdot L^{-1}$), riverine Pb-210 is probably deposited locally near the mouth of the rivers and does not contribute significantly to the overall flux of this radionuclide to sediment. Other factors such as decay of Pb-210 in the water column and ingrowth of Pb-210 from the radioactive decay of radium are negligible because of the short residence time of water in Puget Sound.

It should be noted that, although the rivers are the major contributor to the sedimentary budget of Puget Sound (via estuarine deposition and turbidity currents), riverine input of Pb-210 to the overall budget of this radionuclide in the sediment is not clear at this time.

A rather simplistic model is used for the above calculations. However, it should be noted that the largest error is due to calculation of Pb-210 flux (P) at the sediment-water interface from the Pb-210 dated sediment cores. Devices used to obtain sediment cores (gravity corer, box corer, piston corer, etc.) disturb the retrieved sediment samples and affect the sedimentation rate. Generally, a coring device may induce two different types of disturbances in the sediments: (a) Vertical mixing, which refers to mixing of the top layers of sediment with lower ones, and (b) shortening of sediment cores, which means the collected sediment core is shorter than the distance the corer penetrates into the sediment. In addition, natural in situ mixing of top sediment layers occurs as a result of bioturbations, turbidity currents, gas release from sediment, and storms. It is often difficult to distinguish between natural mixing of sediments and mixing as an artifact of the sampling processes. Discussions of coring problems and comparison of Pb-210 dating of sediment cores using different coring devices are presented elsewhere (16).

Shortening of sediment cores is an artifact of the coring process and occurs to some extent in almost any type of coring device. The core shortening affects the sediment rate measurements and is independent of any dating model used. Comparison of two types of coring devices, a 7.6-cm diameter corer and a 15 x 15 cm^2 cross-section corer, showed that the sedimentation rates obtained from the small diameter corer were 2-3 times smaller than the rates obtained from the large cross-section corer (16).

CONCLUSIONS

The average flux of Pb-210 to Puget Sound sediments, 2.02 ± 1.0 $dpm \cdot cm^{-2} \cdot yr^{-1}$, was computed from the Pb-210 dated sediment cores. The mean annual flux of atmospheric Pb-210 was measured at 0.44 ± 0.24 $dpm \cdot cm^{-2} \cdot yr^{-1}$. Using the literature values, Pb-210 deposition from the water column to sediments was calculated to be 0.94 ± 0.1 $dpm \cdot cm^{-2} \cdot yr^{-1}$. The sum of advection and atmospheric depositions accounts for 68% of the Pb-210 budget and the remaining 32% is attributed to sediment focusing in Puget Sound. The budget calculations depend on the type of coring devices used. A cross-calibration of coring techniques to correlate the sedimentation rates and other sedimentary parameters obtained using different coring devices seems to be necessary.

REFERENCES

1. Nevissi, A.E., Measurement of Pb-210 atmospheric flux in the Pacific Northwest. Health Phys., 1985, 48. 169-174.

2. Krishnaswami, S., Lal, D., Martin, J.M., and Meybeck, M., Geochronology of lake sediments. Earth Planet. Sci. Lett., 1971, 2. 407-414.

3. Schell, W.R., and Nevissi, A., Sedimentation in lakes and reservoirs. In Guidebook on Nuclear Techniques in Hydrology. IAEA Tech. Report 91, Austria, 1983, pp. 163-176.

4. Schell, W.R., and Sugai, S., Radionuclides at the U.S. radioactive waste disposal site near the Farallon Islands. Health Phys., 1980, 39. 475-496.

5. Schell, W.R., and Nevissi, A.E., Radionuclides at the Hudson Canyon disposal site. In Wastes in the Oceans, eds., P.K. Park, et al., Wiley Interscience, New York, 1983, pp. 183-214.

6. Carpenter, R., Peterson, M.L., and Bennett, J.T., ^{210}Pb-driven sediment accumulation and mixing rates for the Washington continental slope. Marine Geol. 48. 135-164.

7. Lavelle, J.W., Massoth, G.J., and Crecelius, E.A., Sedimentation rates in Puget Sound from Pb-210 measurements. National Oceanic and Atmospheric Administration, NOAA Technical Memorandum ERL PMEL-61, Seattle, Washington, January 1985. 43 pp.

8. Nozaki, Y., Turekian, K.K., and Von Damm, K., ^{210}Pb in Geosecs water profiles from the North Pacific. Earth Planet. Sci. Lett., 49. 393-400, 1980.

9. Chung, Y., and Craig, H., ^{210}Pb in the Pacific: Geosecs measurements of particulate and dissolved concentrations. Earth Planet. Sci. Lett., 65. 406-432. 1983.

10. Chung, Y., Pb-210 in western Indian Ocean: Distribution, disequilibrium, and partitioning between dissolved and particulate phases. Earth Planet. Sci. Lett., 85, 28-40. 1987.

11. Schell, W.R., Concentrations, physico-chemical states and mean residence times of Pb-210 and Po-210 in marine and estuarine waters. Geochim. Cosmochim. Acta, 1977, 41. 1019-1031.

12. Drake, D.E., Hatcher, P., and Keller, G., Suspended particulate matter and mud deposition in upper Hudson submarine canyon. In Sedimentation in Submarine Canyons, Fans, and Trenches, eds., Stanley, D.S., and Kelling, G., Dowden, Hutchinson and Ross, Stroudsburg, Pennsylvania, USA, 1978. pp. 33-41.

13. Aller, A.C., Benninger, L.K., and Cochran, J.K., Tracking particle associated processes in nearshore environments by use of Th-234/U-238 disequilibrium. Earth Planet. Sci. Lett., 47, 1980. 161-175.

14. Meyers, L.J., A biogeochemical study of lead-210 in Puget Sound. M.S. Thesis, University of Washington, Seattle, USA, 1982. 243 pp.

15. Carpenter, R., Peterson, M.L., Bennett, J.T., and Somayajulu, B.L.K., Mixing and cycling of uranium, thorium, and Pb-210 in Puget Sound sediments. Geochim. Cosmochim. Acta, 1984, 48. 1949-1963.

16. Nevissi, A.E., Shott, G.J., and Crecelius, E.A., Comparison of two gravity coring devices for sedimentation rate measurement by Pb-210 dating technique. Hydrobiologia, 1988. (In press)

THE GEOCHEMISTRY OF URANIUM IN MARINE SEDIMENTS

Christina Barnes and J. Kirk Cochran
Marine Sciences Research Center
State University of New York
Stony Brook, New York 11794-5000 USA

ABSTRACT

A fission track technique has been used to determine pore water uranium concentrations on 20 μl samples collected from sediments in the NW Atlantic Ocean, San Clemente Basin and Cariaco Trench. Pore water U profiles in box cores in which suboxic to anoxic diagenesis is occurring show depletions (relative to the overlying water value) and enrichments at depth in the core. The profiles show little consistent relationship to pore water Fe, Mn or NO_3 profiles. In contrast, pore water U profiles taken by in situ samples (probes, harpoons) show depletions relative to the overlying water value. U is removed from the pore water in sediments in which NO_3^- reduction is nearly complete, and in the anoxic, sulfide-rich sediments of the Cariaco Trench, reaches concentrations as low as 0.08 dpm/l. A comparison of box core and harpoon data for the same station in the San Clemente Basin shows higher pore water U values at depth in the box core than in the in situ samples. We conclude that oxidation of reduced U in core samples during pore water separation can cause high pore water U concentrations and that U can be reduced at pe's comparable to those encountered during Fe and Mn reduction. Based on in situ pore water U profiles, the uptake rate of U in deep-sea sediments is coupled to the redox environment. U fluxes are lowest (<10 $\mu g/cm^2/ky$) in oxic sediments or suboxic sediments in which organic matter diagenesis proceeds by NO_3^- reduction. Rates are 10-20 $\mu g/cm^2/ky$ in suboxic sediments in which Mn and Fe reduction is occurring, and highest values (~70 $\mu g/cm^2/ky$) are observed in anoxic, sulfide-rich sediments such as those of the Cariaco Trench.

INTRODUCTION

The redox chemistry of uranium in marine sediments has been the focus of considerable research in recent years (1-4). In the marine environment U can exist in two oxidation states (VI and IV) with strikingly different geochemical behaviors. Hexavalent uranium is present in seawater as the soluble uranyl tricarbonate, $UO_2(CO_3)_3^{4-}$. U(IV), on the other hand, should have a chemistry similar to that of thorium and be removed from solution on to particle surfaces.

In marine sediments, bacterial oxidation of organic matter uses electron acceptors other than O_2 in the sequence: NO_3, MnO_2, $Fe(OH)_3$ and $SO_4^=$. This sequence, predictable from the free energy yields of the redox reactions involved, has been documented through pore water profiles of O_2, NO_3^-, Mn^{+2}, Fe^{+2}, and $SO_4^=$. (5) Reduction of U can occur at pe's comparable to that for Fe Fe(III) → Fe(II) (6). Iron reduction commonly occurs in the top few centimeters or decimeters of marine sediments, and the similarity in redox potentials of U and Fe suggests that U reduction and removal from the pore water to the solid phase should occur within this zone as well. Yet previous studies have reported pore water U concentrations in excess of the overlying seawater value (2-4,7) even in sediments in which sulfate reduction is occurring. Taken at face value, these high pore water U activities imply that other geochemical processes are controlling uranium mobility despite thermodynamic predictions that U should be reduced and removed from solution.

A complicating factor in interpreting previous work comes from recent evidence that pore water U data may reflect sampling artifacts. Both pressure and oxidation effects resulting from the shipboard or laboratory sampling of box cores could alter in situ pore water U values. Toole et al. (8) reported that there was a pressure effect on U upon recovery of deep sea box cores such that pore water U values are lowered (possibly in association with alkalinity loss on decompression). Kinetic studies involving exposure of sediment samples to air, as might occur during box core handling, show that oxidation of solid phase U and release to the pore water occur rapidly (2,9,14).

We have analyzed pore water samples from a range of sedimentary environments in order to characterize the geochemistry of U with respect to changes in sediment redox chemistry. In this paper, we present a comparison of samples obtained both from box cores and in situ samplers such as harpoons (10) and use these data to address questions concerning sampling artifacts.

METHODS

Pore water uranium concentrations were determined using a fission track technique. The method we have used is an adaptation of the methods of Fleischer and Delaney (11) for seawater analysis, and more recently that of Toole et al. (8) for small volume pore water samples.

The fission track method requires no chemical pretreatment of the sample. A 20 μl aliquot of the pore water is evaporated on a piece of Lexan polycarbonate plastic. This plastic is a solid state nuclear track detector sensitive to the ^{235}U fission products produced during neutron irradiation. Following noutron irradiation, chemical etching enlarges the tracks so that they are easily counted using an optical microscope.

There is good agreement between the fission track technique and determination of U activities by radiochemical separation and alpha spectrometry. Fig 1A shows a plot of 238U activity (dpm/kg) vs. track density for a set of pore water samples collected in Buzzards Bay, Massachusetts (2). The U activity was determined by radiochemical analysis and the track density numbers are from replicate 20 μl samples. The two methods are well correlated over the range of U activities measured. Figure 1B shows the correlation between radiochemical and fission track determination of uranium standards prepared in artificial seawater. Again, the correlation is good over the 0.5-6 dpm/l range of activities considered. Replicate determinations yield an uncertainty in the value of a particular sample of $\pm$10-15%.

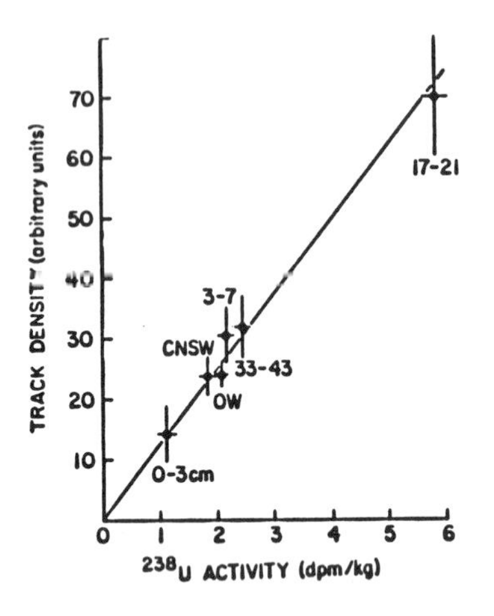

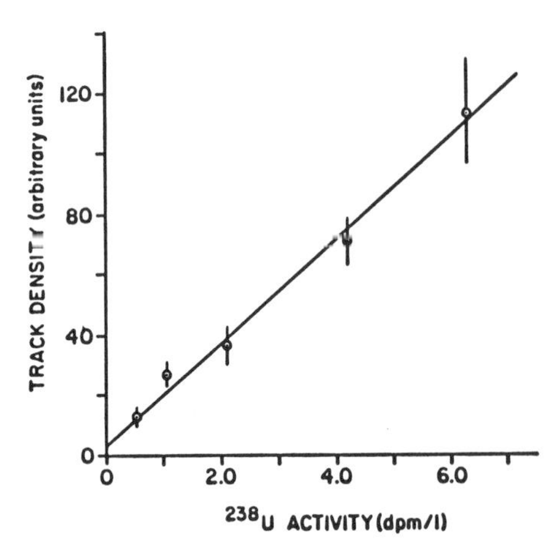

Figure 1: Fission track density vs dissolved ^{238}U activity. (A) Analyses of Buzzards Bay sea water and sediment pore water. (B) U standard prepared in artificial sea water.

RESULTS AND DISCUSSION

We have applied the fission track technique to small volume pore water samples obtained from a transect of box core extending from reducing muddy sediments of the NW Atlantic continental shelf to oxic sediments of the deep sea. These samples, part of the SEEP program, were provided to us by Dr. E. Sholkovitz, of the Woods Hole Oceanographic Institution. At sea, 50 x 50 cm single spade box cores were collected and stored whole at ambient temperature under a nitrogen or freon atmosphere for several days. Upon return to the laboratory the cores were sectioned in two cm intervals and the sediment transferred to a nitrogen flushed glove bag where it was homogenized and packed into bottles. Pore water was extracted by centrifugation.

Figure 2 shows pore water U activity as a function of depth in the sediment for four stations along the SEEP transect. Profiles of pore water Fe, Mn and NO_3 (12) illustrate changes in the redox environment. These profiles show that at the shallow water core (station A) Mn and Fe are reduced and come into solution within a few cm of the sediment-water interface. The depth in the sediment column at which Mn and Fe reduction occur becomes greater in cores from deeper water. In the abyssal plain core (station G) NO_3 is present in the pore water throughout the core, and the profile indicates that the core is characterized by suboxic diagenesis throughout. Except for this site, the pore water U profiles show decreasing concentrations from the overlying water value (2.4 dpm ^{238}U/l) to minima as low as 0.8 dpm ^{238}U/l in the upper 5-15 cm of the core. With increasing depth in the cores, U concentrations increase to as high as 7 dpm ^{238}U/l.

The pore water U data from the abyssal plain site Station G (Fig 2) show relatively little variation with depth and concentrations are comparable to the overlying water value. This pattern is consistent with little reduction of U in sediments in which organic matter is being oxidized principally by NO_3 reduction. The other profiles, however, show scatter in concentrations of adjacent samples and, at stations E and C, sharp U gradients across the sediment-water interface (Fig 2). The profiles do not display variations in U consistent with the core-to-core variations in pore water Fe and Mn.

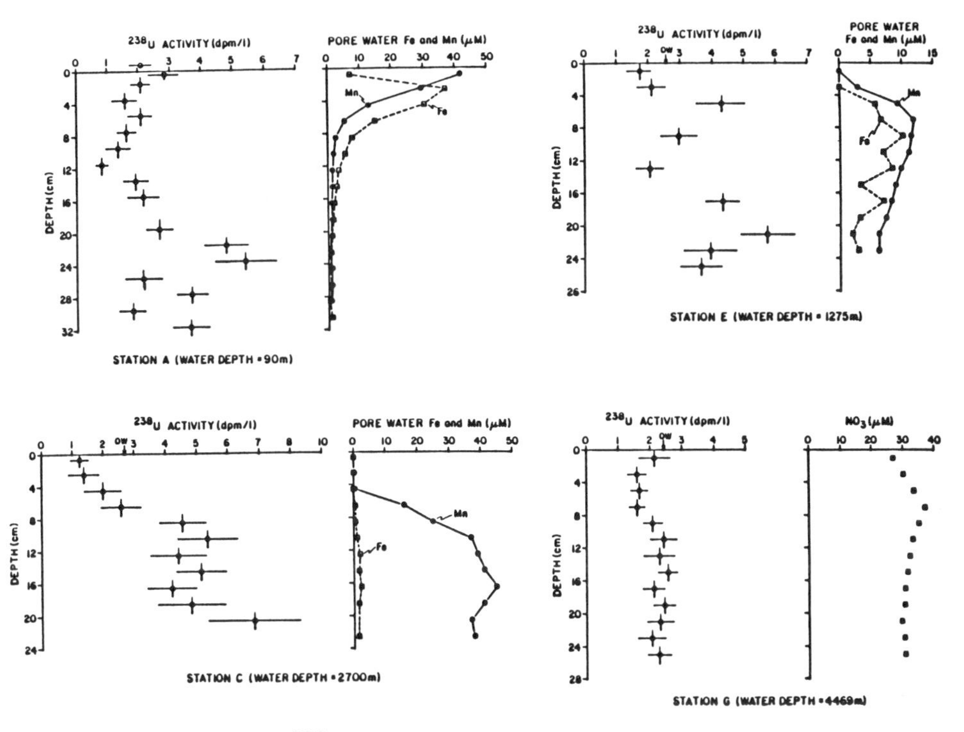

Figure 2. Pore water ^{238}U, Fe, Mn and NO_3 profiles in box cores from the NW Atlantic Ocean.

Oxidation of reduced U during processing of the sediment could lead to anomalously high pore water concentrations, as well as scatter in the data. To evaluate this possibility, we analyzed pore water samples taken in situ from the NW Atlantic shelf and slope. The stations were chosen close to the SEEP stations and samples were provided by Dr. F. Sayles of the Woods Hole Oceanographic Institution. The results (Fig 3) show a different pattern from the SEEP cores. As NO_3 becomes depleted in the pore water, and Mn and Fe reduction occur, U is removed from solution. This is demonstrated by station F (Oceanus 86, Fig 3) which shows pore water U activity decreasing from the overlying water value to 0.5 dpm/l at depth in the core. Cases in which pore water NO_3 persists throughout the core show relatively little change in U (Station H, Knorr 90, Fig 3), a pattern similar to SEEP station G (Fig 2). In general the pore water U profiles from the in situ samples suggest that U reduction occurs after NO_3 reduction. Such a pattern is consistent with the similarity in redox potential for Fe reduction and U reduction. It is also consistent with the mechanisms by which U migrates and is immobilized at redox fronts in pelagic turbidites (15,16).

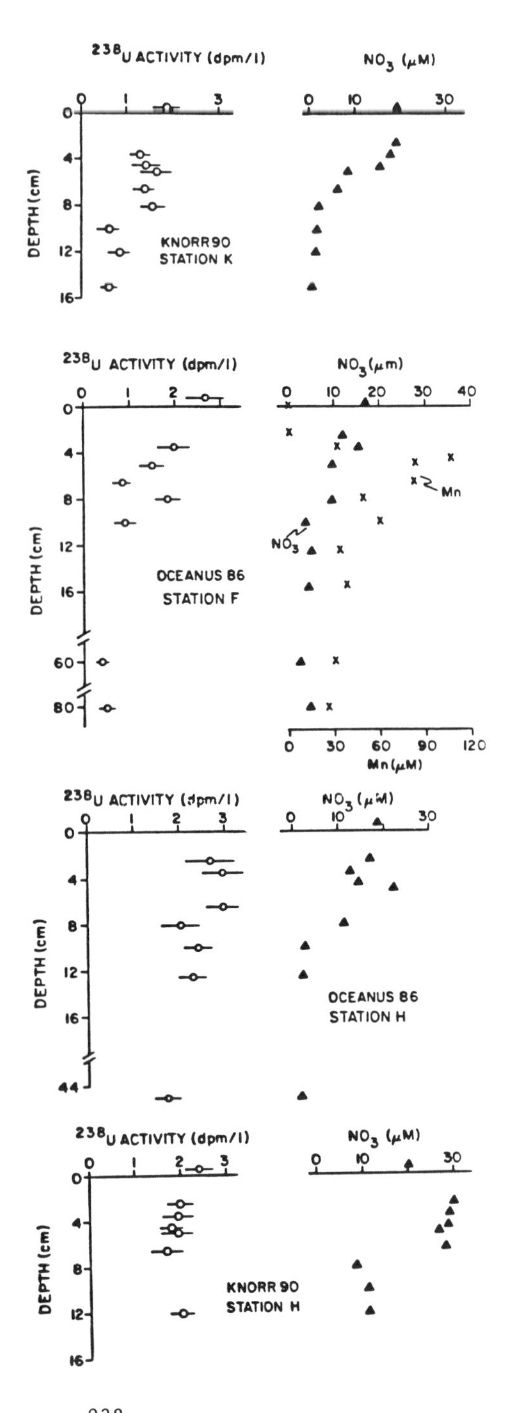

Figure 3. Pore water ^{238}U, NO_3 and Mn profiles in deep-sea sediments of the NW Atlantic Ocean. Samples were collected in situ.

In order to evaluate the behavior of U in sediments which are highly reducing, we have analyzed in situ samples taken by Dr. F. Sayles in the Cariaco Trench. The trench is characterized by anoxic bottom water. Sulfate reduction is extensive and pore water hydrogen sulfide concentrations are as great as 10 mmoles/l (Fig 4). Pore water U decreases from the overlying water value to values indistinguishable from zero by about 10 cm. It thus appears that U reduction is occurring and that U is removed from pore water to the solid phase in these anoxic sediments. This conclusion is consistent with solid phase U profiles of Cariaco Trench cores (14), which show increases in authigenic U with depth in the core.

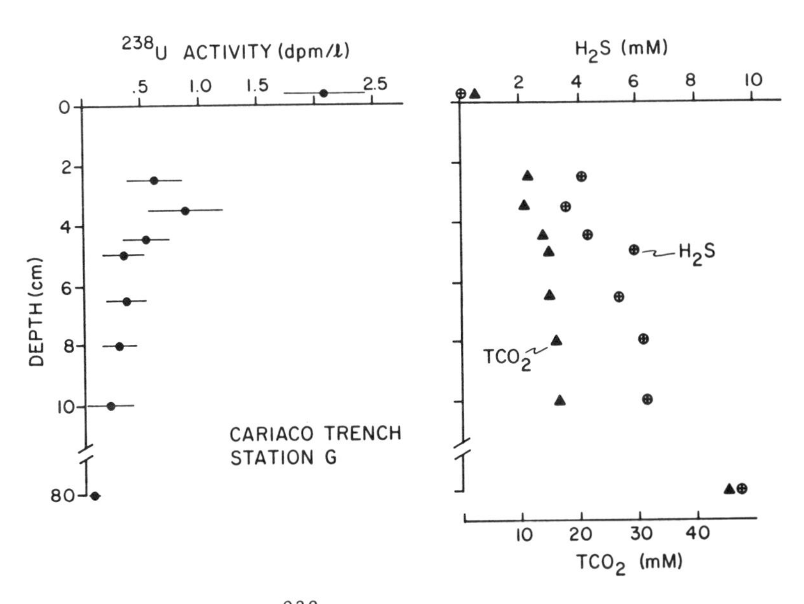

Figure 4. Pore water ^{238}U, H_2S and ΣCO_2 profiles in Cariaco Trench sediments. Samples were collected in situ.

The possibility that differences between in situ and box core U data are related to oxidation artifacts introduced during sample handling is perhaps most reliably evaluated by comparison of in situ and box core samples obtained at the same location and time. Fig 5 shows pore water U data from box core and in situ harpoon samples recovered in San Clemente Basin off California, USA. The samples were provided to us by Dr. S. Emerson of the University of Washington. The in situ samples show a pattern of decreasing pore water U to values less than 1 dpm/l by 4 cm. Activities remain low to the bottom of the core (35 cm). In contrast, the box core samples show scatter and anomalously high values at depth in the core. Significantly, the profiles diverge at depth in the core, suggesting that reduced U present in the solid phase is brought into solution by oxidation. The differences between the box core and in situ data from the San Clemente Basin are similar to the differences between the two sampling methods in the NW Atlantic, and we conclude that oxidation artifacts may complicate pore water U profiles obtained from cores.

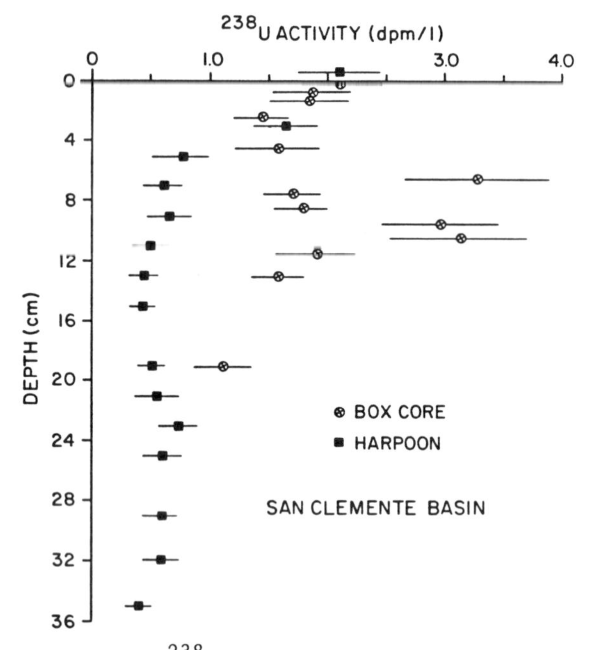

Figure 5. Pore water ^{238}U profiles in San Clemente Basin sediments. Open symbols: samples from box core. Closed squares: pore water samples recovered in situ.

If reduction of U and removal from solution is occurring, as the in situ profiles from the NW Atlantic sites, the Cariaco Trench, and the San Clemente Basin suggest, then removal rates of U in these environments can be estimated and placed in the context of the oceanic U balance. We have made flux calculations based on the concentration gradient from the overlying water to near-interface pore water samples. Table 1 shows that U fluxes into the sediment are lowest in the oxic to suboxic sediments of the NW Atlantic (0-20 $\mu g/cm^2/ky$) and are greatest in the anoxic sediments of the Cariaco Trench (~70 $\mu g/cm^2/ky$). The values for the San Clemente Basin and Cariaco Trench are somewhat lower than those given in the compilation of solid phase estimates from similar areas presented by Cochran (17). The difference may be due to uncertainties in solid phase authigenic U concentrations, sediment accumulation rate or pore water gradients. Moreover, in cores taken from areas in which the overlying water column is oxic, particle mixing and irrigation by the benthic fauna could enhance the uptake of U in the sediment.

The annual input of U to the oceans by rivers is 1-2x10^{10} g U/y (17). If this input is removed uniformly over the sea floor the required uptake is 2.8-5.5 μg U/cm^2/ky. A significant but uncertain fraction of the riverine input of U is probably removed by hydrothermal circulation of seawater through the mid-ocean ridges and by low temperature weathering of oceanic basalts. Removal of U in basins with an anoxic water column accounts for about 10-20% of the riverine input (17). Although our data suggest that U removal is not large in deep-sea sediments in which Fe reduction is not occurring, uptake rates of 10-20 $\mu g/cm^2/ky$ in suboxic slope sediments could be important. Further work is necessary to delineate the magnitude of U fluxes in such sediments.

Table 1. Uranium fluxes in marine sediments.

Core	Depth/Location	Redox Characteristics	U flux*($\mu g/cm^2/ky$)
Kn 90-H	4332 m, 40°56.2'N 62°46.2'W	Suboxic, NO_3-reduction	~0
Oc 86-H	1170 m, 39°5'N 72°14'W	Suboxic, NO_3-reduction	5
OC 86-F	2600 m, 38°50.5'S 71°48.8'W	Suboxic, NO_3 -Mn-reduction	23
Kn 90-K	2239 m, 42°7.4'N 63°52.0'W	Suboxic, NO_3-reduction	10
Cariaco Trench	1375 m, 10°34.6'N 64°47.7'W	Anoxic, $SO_4^=$ reduction	68
San Clemente Basin	33°N,118°W	Anoxic	32

* Values are U fluxes into the sediment calculated as flux = $D_s(\Delta U/\Delta x)$ where D_s is the molecular diffusion coefficient for U, corrected for tortuosity, and $\Delta U/\Delta x$ is the concentration gradient in U calculated from the upper few centimeters of the core. The value for D_s ($2.7 x 10^{-6} cm^2/s$) was calculated from that for uranyl (13), assuming a sediment porosity of 80%. Flux due to burial is neglected.

ACKNOWLEDGEMENTS

This research was supported by the U. S. National Science Foundation (grant OCE-8516393). We thank Drs. K. Buesseler, S. Emerson, F. Sayles and E. Sholkovitz for providing us with samples and for permitting us to use their supporting pore water data.

REFERENCES

1. Anderson R.F., Concentration, vertical flux and remineralization of particulate uranium in seawater. Geochim. Cosmochim. Acta, 1982, 46, 1293-1299.

2. Cochran J.K., Carey A.E., Sholkovitz E.R. and Surprenant L.D., The geochemistry of uranium and thorium in coastal marine sediments and sediment pore waters, Geochim. Cosmochim. Acta, 1986, 50, 663-680.

3. Kolodny Y. and Kaplan I.R., Deposition of uranium in the sediment and interstitial water of an anoxic fjord. In Proc. Symp. Hydrogeochem. Biogeochem. 1: Hydrogeochem., 1973, pp. 418-442. Clarke Co., Washington D.C.

4. Zhorov V.A., Boguslavskiy S.G., Babinets A.Y., Solov'yeva L.V., Kirchanova A.I., and Kir'yakov P.A., Geochemistry of uranium in the Black Sea. Geochim Internat., 1982, 19, 191-200.

5. Froelich P.N., Klinkhammer G.P., Bender M.L., Luedtke N.A., Heath G.R., Cullen D., Dauphin P., Hammond D., and Hartman B., Early oxidation of organic matter in pelagic sediments of the eastern equatorial Atlantic: Suboxic diagenesis, Geochim. Cosmochim. Acta, 1979, 43, 1075-1090.

6. Langmuir D., Uranium solution-mineral equilibria at low temperatures with applications to sedimentary ore deposits. Geochim. Cosmochim. Acta, 1978, 42, 547-569.

7. Boulad A.P. and Michard G., Etude de l'uranium, du thorium et de leurs isotopes dans quelques carottes du basin Angolais (Atlantique Sud-Est), Earth Planet. Sci. Lett., 1976, 32, 77-83.

8. Toole J.,Thomson J., Wilson T.R.S.,, Baxter M.S., A sampling artefact affecting the uranium content of deep-sea sampling porewaters obtained from cores, Nature, 1984, 308, 263-266.

9. Anderson R.F., personal communication.

10. Sayles F.L., Mangelsdorf P.C., Wilson T.R.S. and Hume D.N., A sampler for the in-situ collection of marine sedimentary pore waters, Deep-Sea Res., 1976, 23, 259-264.

11. Fleischer R.L., and Delaney A.C., Determination of suspended and dissolved uranium in water. Anal. Chem., 1976, 48, 642-645.

12. Buesseler K., Plutonium isotopes in the North Atlantic. Ph.D. Thesis (Woods Hole Oceanographic Institution), 1986, 220pp.

13. Li Y.H., and Gregory S., Diffusion of ions in sea water and in deep-sea sediments, Geochim. Cosmochim. Acta, 1974, 38, 703-714.

14. Anderson R.F., Redox behavior of uranium in an anoxic marine basin, In Concentration Mechanisms of Uranium in Geological Environments, 1985, Elsevier, pre-print, in press.

15. Wilson T. R. S., Thomson J., Colley S., Hydes D. J., Higgs N. C. and Sorensen J., Early organic diagenesis: the significance of progressive subsurface oxidation fronts in pelagic sediments, Geochim. Cosmochim. Acta, 1985, 49, 811-822.

16. Colley S. and Thomson J., Recurrent uranium relocations in distal turbidites emplaced in pelagic conditions, Geochim. Cosmochim. Acta, 1985, 49,. 2339-2348.

17. Cochran J.K., The oceanic chemistry of the U- and Th-series nuclides. In Uranium Series Disequilibrium: Applications to Environmental Problems. (eds. M. Ivanovich and R.S. Harmon), Clarendon Press, Oxford, 1982, pp. 384-430.

URANIUM IN THE MARINE ENVIRONMENT
A GEOCHEMICAL APPROACH TO ITS HYDROLOGIC AND SEDIMENTARY CYCLE
I. Theoretical considerations

Renata Djogić, Goran Kniewald and Marko Branica

Center for Marine Research Zagreb
"Rudjer Bošković" Institute
University of Zagreb
Zagreb
Yugoslavia

ABSTRACT

The fate of uranium in the marine environment is influenced both by the physico-chemical state of uranium species dissolved in seawater, as well as by the redox equilibria of the various oxidation states of this element. More than one oxidation state of uranium can react and coexist along the vertical profile of the seawater column and the underlying sediments. A uranium species distribution model was developed so as to enable a prediction on the speciation of dissolved uranium(VI) forms in seawater. The calculations show that a very significant percentage of uranyl species should be in the form of a mixed ligand complex, the uranyl carbonato-hydrogenperoxo complex, when coordination with the peroxide ligand of seawater concentration is taken into account. An assessment of the redox equilibria of uranium in the oxidation states VI, V and IV based on revised and critically reviewed thermodynamic data is also given. The enhanced stability of uranium(V), present in aqueous solutions as the hypouranyl ion UO_2^+, implies its possible role as a transitional oxidation state of uranium which may be encountered in reduced or anoxic sediments and incorporated into uranium minerals.

INTRODUCTION

Uranium is ubiquitously present in seawater at a uniform average concentration of 3.3 ug dm^{-3} (Goldberg 1965; Ku et al., 1977). This comparatively high level of dissolved uranium in the ocean is most probably due to its residence time of about 3 x 10^6 years (Goldberg et al. 1971), which in turn, may be explained in terms of the high solubility of uranium(VI) and the generally conservative behaviour of this element in the marine environment. Consequently, numerous investigations dealing with the possibilities of uranium recovery from seawater have been undertaken (Schwochau et al., 1976; Yamashita et al., 1980). Feasibility assessments of seabottom, deep trench or subseabed disposal of uranium--containing radioactive waste are currently being performed in a number of countries (Pritchard, 1961; Hollister et al., 1980; Templeton, 1980). It is here that the aqueous chemistry of uranium is of fundamental importance. The interaction of uranium with groundwaters and seawater, leading to its dissolution and remobilization, would be the first step in the event of an "accident" (i.e. containment breakdown or leakage) of a submarine geologic or seabed waste repository. Hence, all such endeavours must necessarily rely upon a sound understanding of the reaction pathways and the mass balance of uranium in the sea, both of which are under the direct influence of the physico-chemical state, redox equilibria and distribution patterns of uranium species in the watercolumn and in the underlying sediments.

Ever since the now classic paper by Starik and Kolyadin (1957), it has been widely accepted that the chemical form of dissolved uranium in seawater is the anionic uranyl tricarbonate complex $UO_2(CO_3)_3^{4-}$. As a direct consequence of the highly probable ion pairing reactions, the complex should more precisely be described as $Na_2|UO_2(CO_3)_3|_3^{2-}$ (Branica, unpublished results). Contrary to these well established hypotheses, a set of recently published calculations (Krylov et al., 1985) suggests uranyl hydroxo complexes as predominant uranium species in seawater of pH 8. These data are most probably based upon an uncritical or even erroneous application of hydroxide association constants. They do, however show that the problem of uranium species in seawater is of great importance, because the underlying mechanisms and equilibria are still not fully understood.

The work of Cooper and Zika (1983) on the photochemical formation of hydrogen peroxide in natural waters exposed to sunlight, indicates that H_2O_2 concentrations of 10^{-7} mol dm^{-3} are encountered in the photic layer of the ocean. The peroxide concentration can rise to 10^{-5} mol dm^{-3} in areas of particularly high organic production, but considerably lower H_2O_2 levels must be expected in the disphotic zone (below about 80 m, depending on the incident radiation and transparency of the water). On the other hand, the fact that uranium shows a particularly strong affinity for complex formation with peroxide (Martin-Frére, 1957, 1965; Komarov et al., 1959), prompted us to pursue the possibility of mixed ligand complex formation, involving both the carbonate and peroxide ligand. Although the peroxide ligand is thermodynamically the strongest complexing agent for the UO_2^{2+} ion, it has been shown (Shchelokov, 1975; Shchelokov et al., 1981) that an excess of carbonate leads to the formation of the mononuclear uranyl dicarbonato-peroxo complex $|UO_2(CO_3)_2(O_2)|^{4-}$. The mechanism probably involves the disruption of the bridging peroxo-group bond through the action of the carbonate ligand which coordinates the uranyl ion less stably than the chelated peroxo group.

In order to evaluate the formation and possible significance of uranyl peroxo complexes and mixed uranyl peroxo-carbonato complexes in seawater, a theoretical distribution model was developed. It enabled us to estimate the equilibria between the various chemical forms of uranium over a wide range of concentration and pH values.

Furthermore, all predictions and evaluations of uranium species distribution at or near the oxic/anoxic (suboxic) interface, must observe the redox equilibria between the three well defined oxidation states of uranium occurring in the hydrosphere. This is also true for the various oxygen-depleted environments such as anoxic marine or lacustrine sediments. Until recently, the oxidation states U(VI) and U(IV) have by the vast majority of authors been regarded as the only two stable redox forms of uranium in the marine environment (Krauskopf, 1967; Baturin, 1973). As expected, their geochemical behaviours are nonetheless strikingly different (Goldschmidt, 1958; Cochran and Barnes, 1987), particularly with respect to their complex formation and solubilities. The third well defined redox state, uranium(V) has been persistently neglected as a geochemical species of any significance. An obvious reason for such a view is its instability in aqueous solution at neutral pH values, the instability associated with disproportionation. Experimental evidence exists, however, that U(V) is formed from U(VI) and U(IV) in solution at high carbonate contents (Čaja et al., 1965). Moreover, several authors have pointed out that U(V) should be expected in some parts of the geochemical cycle of uranium, and that this oxidation state is stabilized in the solid phase of uranium compounds and minerals (Allen et al., 1978; Calas, 1979).

The possible significance of uranium(V) as a transitional oxidation state in the marine sedimentary environment has also been discussed (Kniewald and Branica, 1988). The principal cause for U(V) being looked upon as an unstable species is the difficulty of its experimental identification in the oxidation state V. In aqueous media of pH values within the range of natural water pH's the UO_2^+ ion (hypouranyl ion) quickly disproportionates to U(VI) and U(IV), rendering solution-based analytical techniques unsuitable for experimental identification, regardless of their sensitivity or degree of instrumental sophistication. Several disproportionation mechanisms have been suggested, but in seawater this probably involves proton transfer and may be described as originally proposed by Kern and Orlemann (1949):

$$UO_2^+ + H^+ \longrightarrow UO_2H^{2+}$$

$$UO_2^+ + UO_2H^{2+} \longrightarrow UO_2H^+ + UO_2^{2+}$$

$$UO_2H^+ \longrightarrow \text{stable uranium(IV) species}$$

Due to the single unpaired electron in the outer shell of the UO_2^+ ion, magnetic resonance methods have been used for the analytical identification of U(V), both in solution and in the solid phase (Selbin et al., 1968; Burton Lewis et al., 1973; Souliè et al., 1980). It has been shown that ESR spectroscopy can be applied for the determination of uranium(V) even in the presence of U(VI) and U(IV), at present only at a qualitative level (Kniewald and Branica, 1988). Care should be taken to avoid as far as possible interferences from paramagnetic impurities or constituent ions, particularly when natural samples are analysed.

Our aim here was to point out, based on theoretical considerations, one ionic uranyl peroxide complex and the uranium(V) redox species, as probably significant for the hydrogeochemical cycle of uranium.

THE THEORETICAL MODEL

Calculation of the theoretical distribution curves of uranyl species in seawater

The Pytkowicz model of seawater (Sipos et al., 1980), which is based on the assumption of major constituent interactions, has been used to derive the distributions of uranium species present. The constituent concentrations and respective stability constants were thus taken into account and included into the program data base.

In order to obtain a closer correspondence to conditions in nature (particularly those concerning carbon dioxide exchange and equilibration), the concept of a carbonate system for aqueous media open to atmospheric CO_2 - at a constant pCO_2 of 0.0003 atm - was incorporated into the model (Stumm and Morgan, 1981). This implies that the "used" or "bound" carbonate may be derived from the atmosphere. The equilibrium relationships between free carbonate, hydrogencarbonate and dissolved carbon dioxide are thereby satisfied for all pH values.

The stability constants of 19 uranyl complexes which did not include peroxo and peroxo-carbonato complexes (Djogić et al., 1986), were used for uranium distribution calculations. In additon, 5 new peroxo and mixed ligand complexes as well as the first and second dissociation constants of hydrogen peroxide were incorporated into the calculations. The values are given in Table 1. All equilibrium calculations were done using a modified Newton-Raphson iterative convergence procedure (Sipos et al., 1980).

Table 1. Stability constants of uranyl peroxo complexes, mixed ligand uranyl peroxo-carbonato complexes and hydrogen peroxide dissociation constants

Chemical species	K	Reference
$|UO_2(CO_3)_2(O_2)|^{4-}$	1.77×10^{21}	Souchay and Martin-Frère, 1965
$|UO_2(CO_3)_2(HO_2)|^{3-}$	2.5×10^{11}	Komarov et al. 1959
$UO_2(O_2)^{o}$	1.1×10^{32}	Moskvin, 1968
$UO_2(O_2)_2^{2-}$	1.4×10^{60}	Moskvin, 1968
$UO_2(O_2)_3^{4-}$	9.0×10^{72}	Gurevich and Susorova, 1968
$H_2O_2 \rightleftharpoons H^+ + HO_2^-$	$K_1 = 2.24 \times 10^{-12}$	Evans and Uri, 1949
$HO_2^- \rightleftharpoons H^+ + O_2^{2-}$	$K_2 = 2.24 \times 10^{-37}$	Evans and Uri, 1949

Evaluation of the stability range of uranium(V) in terms of pH and redox potential

A very convenient way of visualising and displaying equilibria and the corresponding stability fields of various redox and ionic (or complex) species of an element are Eh-pH (or pe-pH) diagrams. These are graphical extrapolations of thermodynamic data, using pH and redox potential as the two master variables. Such diagrams are extremely useful but obviously only as good as the data (based on experiments in the whole range of parameters) used in their construction. Nevertheless, data extrapolation is the main cause of misinterpretations with regard to the predominance of the chemical forms of an element. Thus even a brief look at Eh-pH representations prior to 1978, concerning the U-H_2O or the U-H_2O-CO_2 system, will suffice to notice that no stability field for U(V) or UO_2^+ is given at all. The reason for this is that no redox data, although existing, on the U(VI)/U(V) or U(V)/U(IV) were put into the diagrams (i.e. figure 5, Zoubov de, 1963). Due to its presumed instability, association constants and free energy data were just not being used in these diagrams.

A reevaluation of the standard redox potential for the reaction

$$U^{4+} + 2H_2O \rightleftharpoons UO_2^{2+} + 4H^+ + 2e^-$$

gives a value E^o = 0.273 V at 298 K. This data (Fuger and Oetting, 1976) is at variance with the more commonly known value of 0.33 V (i.e. Ahrland et al., 1973), and is a result of correction for hydroxide complexation and hydration of the uranyl ion.

Equilibrium constants of several uranium(V) species were computed from partial molar Gibbs free energies of formation, given in Table 2. Using data on free energy, enthalpy and entropy of another 35 soluble species and solid phases of uranium as revised and given by Langmuir (1978), the Eh-pH diagram given in Fig. 6 was constructed.

Table 2. Thermodynamic data for some uranium species at 298 K and 1 atm total pressure. Given are values for the free energy, enthalpy and entropy respectively

Species	ΔH^o kJ/mol	ΔG^o kJ/mol	S^o J/mol.deg
U^{4+}	- 591.1	- 530.9	- 174.9
UO_2^{2+}	- 1018.7	- 952.6	- 97.1
UO_2^+	- 1032.5	- 968.5	- 25.1
U_3O_8	- 3574.4	- 3369.4	282.6

RESULTS AND DISCUSSION

The results of the theoretical distribution calculations for U(VI) species in seawater are displayed in figures 1, 2 and 3. Figure 1 shows the accepted hypothesis on the pH dependent distribution of U(VI) forms in seawater when complexation with peroxide is not taken into account. The total uranium concentration is 10^{-8} mol dm^{-3}, and it can be observed that the uranyl tricarbonate complex is the predominant species while the trihydroxide and dicarbonate complexes are of subordinate importance at pH 8.

Figure 2 gives the distribution of uranyl species vs. pH when the peroxide complexation and formation of mixed peroxo-carbonato complexes are taken into consideration. The total uranium concentration is again 10^{-8} mol dm^{-3}, while the peroxide level is 10^{-7} mol dm^{-3}. Approximately

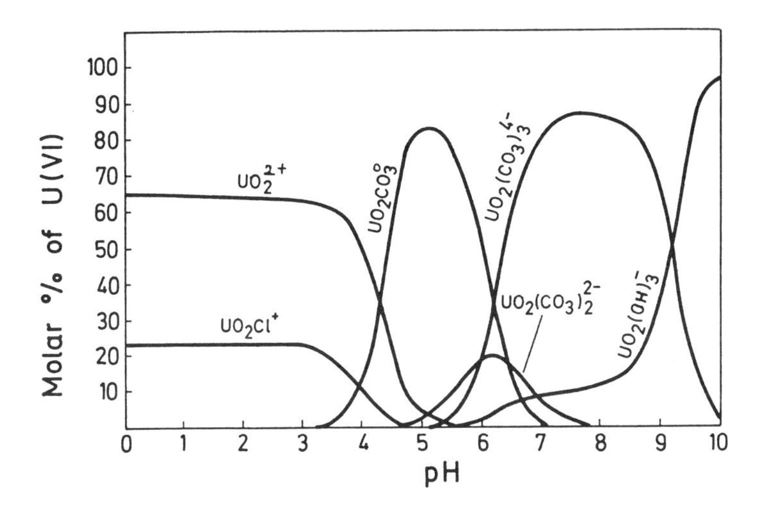

Figure 1. Theoretical distribution of uranium(VI) hydroxo and carbonato complexes in seawater. Hydrogen peroxide activity is not considered. The total uranium concentration is 10^{-8}mol dm^{-3}.

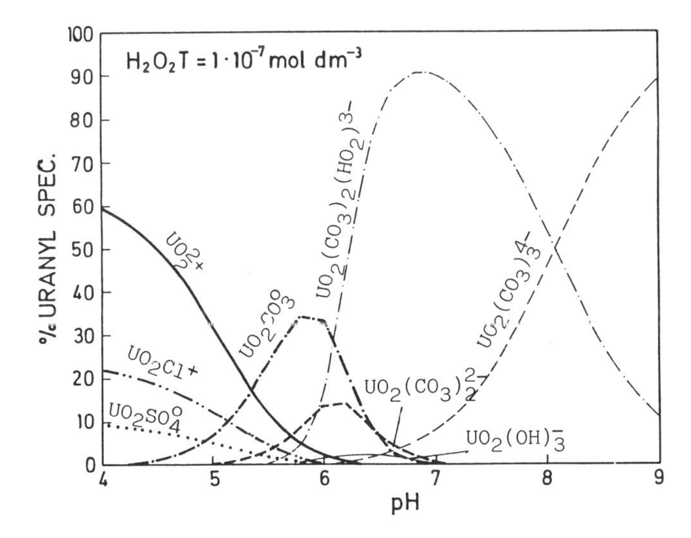

Figure 2. Theoretical distribution of uranium(VI) species in seawater. The total uranium concentration is 10^{-8} mol dm^{-3}.

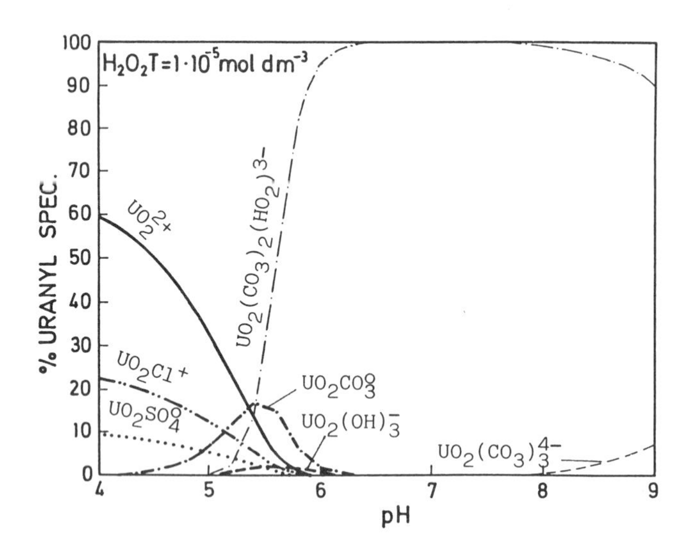

Figure 3. Theoretical distribution of uranium(VI) species in seawater. Total uranium concentration is 10^{-8} mol dm^{-3}.

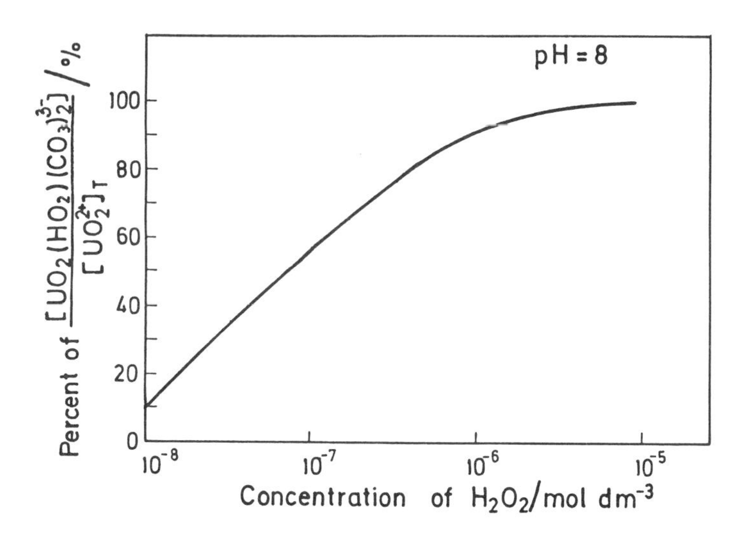

Figure 4. Percentage curve of the mixed ligand uranyl carbonato-hydrogenperoxo complex in seawater, depending on the hydrogen peroxide concentration.

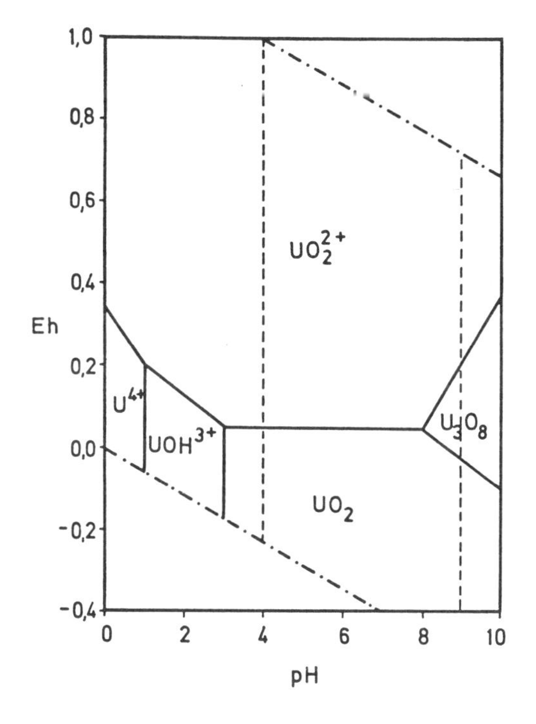

Figure 5. Eh-pH diagram based on thermodynamic data given by de Zoubov (1963). No stability field for U(V) is present.

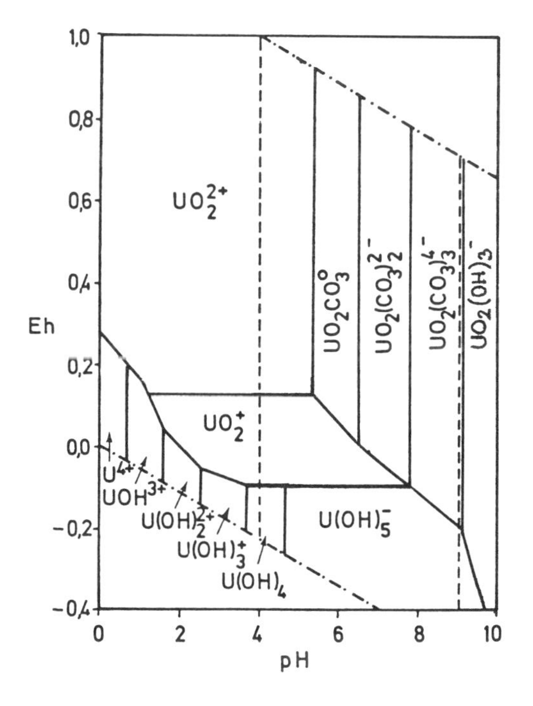

Figure 6. Eh-pH diagram for various complexed and redox species of uranium. The total uranium concentration is 10^{-8} mol dm^{-3}. pCO_2 = 0.0003 atm

50% of uranium would under these conditions and a pH of 8, be in the form of the mixed uranyl carbonato-hydrogenperoxo complex. When the hydrogen peroxide concentration rises to about 10^{-5} mol dm^{-3}, more than 99% of the uranium present should be in the form of the above mentioned mixed complex (figure 3), again at the pH value of seawater.

Figure 4 gives the percentage curve of total uranium(VI) in the form of the uranyl carbonato-hydrogenperoxo complex as a function of the hydrogen peroxide concentration at pH 8.

A comparison of figures 1 and 3 indicates significant differences in the distribution of uranyl species. At seawater conditions, the uranyl tricarbonate complex is displaced to a significant extent by the mixed ligand complex. Hydrolysis (i.e. the formation of hydroxo complexes) becomes negligible. A pronounced shift in equilibrium takes place when peroxide concentrations rise to approximately 10^{-5} mol dm^{-3} (areas with high DOM contents) so that almost all uranium is associated with the mixed uranyl carbonato-hydrogenperoxo complex (figures 3 and 4).

It can be concluded that our theoretical distribution model predicts a considerable influence of hydrogen peroxide present in seawater upon the speciation of the chemical forms of uranium in the ocean (photic zone). This hypothesis departs significantly from the accepted distribution models, so that experimental verification of the distribution calculations is warranted. We are at present in the course of providing experimental evidence to confirm our predictions.

Using thermodynamic data and derived stability constants of uranium redox species given in Table 2 as well as by Langmuir (1978), the Eh-pH diagram shown in figure 6 was constructed. The total uranium concentration is here also 10^{-8} mol dm^{-3}, pCO_2 = 0.0003 atm. Corrections for the variations of free carbonate concentration $(CO_3)^{2-}_F$ with pH were made. It can be seen from the diagram that uranium(V) has an appreciable stability range in low Eh and slightly acidic waters. Contrary to this, the Eh-pH diagram in figure 5 is based on data by de Zoubov (1963) where U(V) equilibria were disregarded, so that no stability range of UO_2^+ appears.

The large hypouranyl ion UO_2^+ forms with the same ligands probably considerably weaker complexes than either U^{4+} or UO_2^{2+}, so that the formation of soluble U(V) species is unfavourable for a wide range of natural conditions. However, the reduction of U(VI) to U(IV) in reducing environments could proceed via the oxidation state V. As U(V) seems to be stabilized in the solid phase, we argue that U(V) might prove to be a significant transitional redox state of uranium in the marine sedimentary cycle.

Further studies on the effects of early diagenesis on the oxidation state of uranium in the solid phase of sediments are being done, with the aim of rendering the experimental identification of uranium oxidation states in natural samples possible.

ACKNOWLEDGEMENTS

This study is part of the joint project "Environmental Research in Aquatic Systems" of the Institute of Applied Physical Chemistry, Nuclear Research Center (KFA) Jülich, Federal Republic of Germany, and the Center for Marine Research of the "Rudjer Bošković" Institute, Zagreb, Yugoslavia. Financial support by the International Bureau of the KFA Jülich, within the framework of the bilateral research agreement between FR Germany and SFR Yugoslavia is gratefully acknowledged. The study was in part supported by the US National Science Foundation and the Research Council of SR Croatia, through funds made available to the US-Yugoslav Joint Board on Scientific and Technical Cooperation under the project JFP(NSF-IRB)/679.

REFERENCES

Ahrland, S., Liljenzin, J.O. and Rydberg, J., Solution Chemistry of the Actinides. In: Comprehensive Inorganic Chemistry, Pergamon Press, Oxford, 1973, 465-541.

Allen, G.C., Griffiths, A.J. and Suckling C.W., Chem. Phys. Lett., 1978, 53, 309-312.

Baturin, G.N., Geokhymia, 1973, 9, 1362-1372.

Burton Lewis, W., Hecht, H.G. and Eastman, M.P., Inorg. Chem. 1973, 12, 1634-1639.

Calas, G., Geochim. Cosmochim. Acta, 1979, 43, 1521-1531.

Cochran, K.J. and Barnes, C., Proc. Intl. Symp. "Radioactivity and Oceanography", 1987, Cherbourg, France (this volume).

Cooper, W.J. and Zika, R.G., Science, 1983, 220, 711-712.

Čaja, J., Žutić, V.,Guidotti, G. and Branica, M., Proc. 2nd Yugosl. Symp. React. Mat., 1965, Herceg Novi, Yugoslavia.

Djogić, R., Sipos, L. and Branica, M., Limnol. Oceanogr. 1986, 31, 1122-1131.

Evans, M.G. and Uri, N., Trans. Farad. Soc., 1949, 45, 224-230.

Fuger, J. and Oetting, F.L., The Actinide Aqueous Ions, IAEA, Vienna, 1976.

Goldberg, E.D., Minor Elements in Seawater. In: Chemical Oceanography (Riley, J.P., Skirrow, G. Eds.) Academic Press, London, 1965.

Goldberg, E.D., Broecker, W.S., Gross, M.G. and Turekian, K.K., Radioactivity in the marine environment, Natl. Acad. Sci., Washington, 1971.

Goldschmidt, V.M., Geochemistry, Oxford University Press, London, 1958.

Gurevich, A.M. and Susorova, N.A., Radiokhim., 1968, 10, 211-221.

Hollister, C.D., Corliss, B.H. and Anderson, D.R., Submarine geological disposal of radioactive waste. IAEA, Vienna, 1980.

Kern, D.M.H. and Orlemann, E.F., J. Am. Chem. Soc., 1949, 71, 2102-2106.

Kniewald, G. and Branica, M., Mar. Chem., 1988, 24, 1-12.

Komarov, E.V., Preobrazhenskaya, L.D. and Gurevich, A.M., Zh. Neorg. Khim., 1959, 4, 1668-1673.

Krauskopf, K.B., Introduction to geochemistry, McGraw Hill, New York, 1967.

Krylov, O.T., Novikov, P.D. and Nesterova, M.P., Okeanologyia, 1985, 25/2, 242-244.

Ku, T.L., Knauss, K.G. and Mathieu, G.G., Deep Sea Res., 1977, 24, 1005-1017.

Langmuir, D., Geochim. Cosmochim. Acta, 1978, 42, 547-569.

Martin-Frère, H., Compt. Rend., 1957, 245, 848-856.

Martin-Frère, H., Bull. Soc. Chim. France, 1965, 2868-2874.

Moskvin, A.I., Radiokhim., 1968, 10, 13-21.

Pritchard, D.W., Health Phys., 1961, 6, 103-109.

Schwochau, K., Astheimer, L., Schenk, H.J. and Schmitz, J., Report No. 1415, Nucl. Res. Center Jülich, 1976, 71 pp.

Selbin, J., Ortego, J.D. and Gritzner, G., Inorg. Chem., 1968, 7, 976-982.

Shchelokov, R.N., Complex chemical behaviour of uranyl peroxo compounds, Akademia Nauk SSSR, Moscow, 1975, 153-158.

Shchelokov, R.N., Orlova, I.M. and Sergeyev, A.V., Koord. Khim., 1981, 7/6, 901-904.

Sipos, L., Raspor, B., Nürnberg, H.W. and Pytkowicz, R.M., Mar. Chem., 1980, 9, 37-47.

Souchay, P. and Martin-Frère, H., Bull. Soc. Chim. France, 1965, 10, 2874-2881.

Soulié, E., Folcher, G. and Kanellakopoulos, B., Can. J. Chem., 1980, 58, 2377-2379.

Starik, I.E. and Kolyadin, L.B., Geokhymia, 1957, 3, 245-247.

Stumm, W. and Morgan, J.J., Aquatic Chemistry, Wiley, New York, 1981.

Templeton, W.L., Proc. Symp. Radionucl. Marine Environ., IAEA, Vienna, 1980, 451-464.

Yamashita, H., Ozawa, Y., Nakajima, F. and Murata, T., Bull. Chem. Soc. Japan, 1980, 53, 1-5.

Zoubov de, N., In: Atlas d´equilibres électrochimiques (Pourbaix. M., Ed.) Gauthier-Villars, Paris, 1963, 198-212.

SOME ASPECTS OF THE MARINE GEOCHEMISTRY OF URANIUM

J Toole and M S Baxter
Scottish Universities Research and Reactor Centre, East Kilbride, Glasgow G75 0QU, UK

and

J Thomson
Institute of Oceanographic Sciences, Brook Road, Wormley, Godalming, Surrey GU8 5UB, UK

ABSTRACT

Conservative uranium behaviour during estuarine mixing is shown by linear uranium-salinity plots for three British systems - the Clyde, Isauld and Tamar, although slight coastal removal of uranium is occurring outwith the Clyde estuary. Significant localised uranium removal in another estuary, the Forth, is identified at low salinities, coincident with high particulate concentrations and phosphate removal.

In the underlying estuarine sediments, which are anoxic at shallow depth, uranium is removed from the pore waters and is released at depth as a result of bioturbational processes in the cores.

Uranium is observed to behave conservatively in an oxic, deep-sea core and it is shown that in-situ sampling is desired for reliable porewater uranium data due to a pressure-related artefact. In contrast to this, sediments which have anoxia give rise to porewater uranium levels which are elevated over the seawater value. One such core gives direct evidence of a sediment to seawater uranium flux, while others show high porewater uranium peaks, with little diffusion, due perhaps to the presence of high molecular weight organic complexes.

INTRODUCTION

In oxidising, carbonate-bearing waters, the natural uranium isotopes exist predominantly as the uranyl carbonate anions $UO_2(CO_3)_2^{2-}$ and $UO_2(CO_3)_3^{4-}$ (1). The long-term stability of these dissolved species is reflected in the constancy of the uranium concentration of open ocean seawater - 3.2 to 3.3 $\mu g l^{-1}$ (2, 3), a relatively low K_d of $\sim 10^3$ (4, 5) and a long residence time of ~2-4kyr (6). In river waters, however, the uranium content is highly variable, ranging from 0.02 to 6.6 $\mu g l^{-1}$ (7) and is believed to be controlled by the intensity of weathering of the source rocks (8). Thus, positive correlations with total dissolved solids (9, 10) and bicarbonate (11, 12) have been demonstrated. In the estuarine zone, where the above two water types mix, uranium is generally found to behave conservatively, its concentration varying linearly with salinity (12-18). Any observation of non-pullutant uranium removal has been restricted to the low salinity region (12, 17, 18) or to the source river itself (19, 20). When uranium-bearing waters encounter reducing conditions, however, U^{6+} may be reduced to insoluble U^{4+} and removed from solution (eg. 21). Such a process is particularly evident at roll-front deposits, where dissolved uranium is at its lowest down-gradient of the front where reducing conditions exist (22-24). Increases in solid-phase uranium concentration with depth in nearshore (25-27), hemipelagic (21, 28) and deep-sea sediments (29) have been interpreted as evidence for uranium removal from the aqueous phase. There are many other instances of uranium enrichment in marine sediments, either under an anoxic water column (30, 31) or within emplaced turbidites (32) or in sapropels (33, 34). Often a strong correlation exists between organic carbon and uranium contents in such sediments, with a tendency for the solid phase $^{234}U/^{238}U$ activity ratio to be characteristic of an authigenic source. This fixation of authigenic uranium may be due either to reduction of U^{6+} in the porewater or seawater to insoluble U^{4+} by the low Eh associated with high-organic sediments, or to the surface adsorption or complexation of U^{6+} by the organic matter present (31, 35), with some proportion of the uranium being reduced (36). Mobilisation of uranium in the porewaters can therefore be by release of some adsorbed or complexed U^{6+} as organic carbon combustion proceeds, or as soluble organo-uranium complexes. Since sediment porewaters are the most sensitive indicators of diagenetic reactions, it is the uranium concentration in this dissolved phase which best helps us to understand this elements diagenetic behaviour. Available results on porewater uranium in suboxic to reducing nearshore and deep sea sediments (28, 31, 36-42) reveal uranium enrichments (relative to the respective overlying water values) which would perhaps be unexpected from Eh considerations.

In this paper, we review some new data which pertain to the above aspects of uranium geochemistry. We present dissolved-phase uranium data from four UK river-estuarine systems and we compare and contrast the geochemical behaviour of this element in the estuarine mixing regime, in underlying reducing estuarine sediment and in both oxic and anoxic deep-sea sediments.

Sampling and Analysis

River/estuarine water samples (up to 10 litres) were collected, filtered (0.45 μm, 1 μm GF/B prefilters), acidified, spiked (^{232}U) and analysed α-spectrometrically following ion exchange purification and electrodeposition (18). Estuarine porewaters (St John's Lake, Tamar estuary) were taken at low water using 25 cm long perspex corers, squeezed within 90 minutes of collection in a nitrogen glove box, precautions being taken to guard against oxidation and temperature effects (43, 44). Deep sea porewaters were collected in the east Mediterranean and north-east Atlantic using (a) the I.O.S. Mark II in-situ porewater sampler - a modified version of that described in (45), (b) a 30 x 30 cm box corer for short cores with undisturbed interfaces (46) and (c) a 2 or 4 metre long 15 x 15 cm barrel Kastenlot corer, subsampling of (b) and (c) being by 4-inch square butyric coreliners and again, during collection and squeezing, precautions being taken to minimise temperature and oxidation effects. All porewater samples were filtered with 0.4 μm Millipore or Nuclepore filters and were analysed by the fission track technique (41, 47), which permits uranium analysis of low-volume (100 μl) porewater or seawater samples with a sensitivity better than 5 x 10^{-12} g U. Isotopic information however, is not achievable by this method.

Results

Estuarine Waters

Figure 1(a-d) shows the variation in uranium concentration with salinity change in the four UK river water-sea water mixing systems studied. In the Clyde estuary (Fig 1a), it can be seen that uranium concentration increases linearly (R^2 = 0.994) with salinity from a river end-member value of 0.16 $\pm$ 0.01 $\mu g l^{-1}$ to the local Firth of Clyde sea water end-member of 2.78 $\pm$ 0.01 $\mu g l^{-1}$ at 31.9°/oo this during a period of low river flow ($12.2 m^3 s^{-1}$). The accepted open ocean seawater uranium concentration is 3.2 - 3.3 $\mu g l^{-1}$ (2, 3) and thus there is evidence here of uranium removal outside the Firth of Clyde. There are, in fact, large areas of mud and muddy sand at ~ 100m depth in the Irish Sea and some sites of anoxic sediments due to sewage sludge dumping (up to 10^6 tonnes yr^{-1}) and these are available for uranium stripping of the water column. Such low U/S^o/oo ratios nearshore with respect to open ocean seawater are not uncommon (48–50). The $^{234}U/^{238}U$ activity ratios for the Clyde water samples also followed the theoretical mixing curve within the precision of the measurements, decreasing from 1.63 ± 0.06 at the river end-member to 1.13 ± 0.04 at the seawater end (normal seawater ratio 1.14 ± 0.02 (2)), indicating conservative behaviour also for ^{234}U, i.e. no additional isotopic fractionation during mixing. This was also found to be the case in the other systems studied (18).

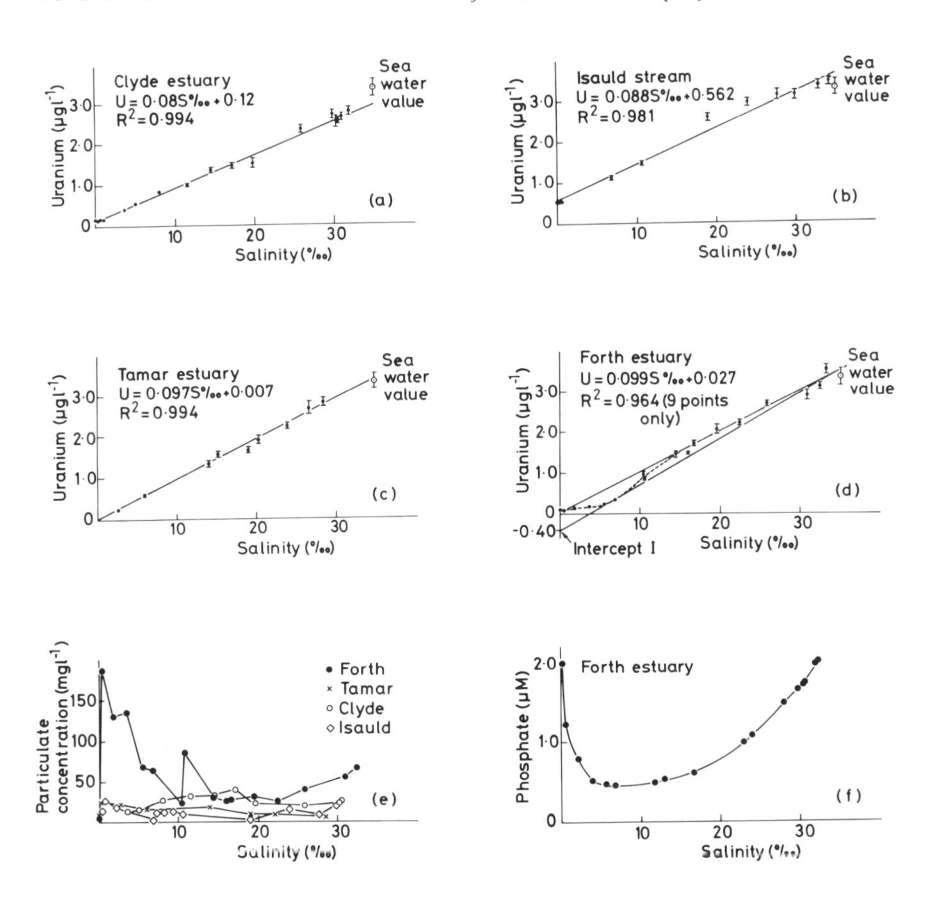

Figure 1 Uranium-salinity plots for (a) Clyde estuary, (b) Isauld system, (c) Tamar estuary and (d) Forth estuary. The particulate concentrations for each are shown in (e), with phosphate removal in the Forth estuary shown in (f).

Fig 1b shows the dissolved uranium data for the Isauld stream-seawater system, not a true estuary but one in which mixing of a stream draining a blanket peat catchment and thus high in dissolved organic matter occurs rapidly with seawater in a sandy bay. There does not appear to be any evidence of uranium removal at low salinity mixing ratios where flocculation of organic material might be expected to occur (51). The high uranium concentration of 0.54 $\pm$ 0.04 $\mu g l^{-1}$ in freshwater here is a combined result of high uranium content of the catchment rocks and enhanced bicarbonate content of the water (pH of the stream water ~ 7.8 to 8.0) the catchment basement rock consisting of cyclic outcrops of sandstones (calcite cement) dolostones and shales (high U) (52). Linear regression of the Isauld data points gives a corelation coefficient of 0.991, although there is slight evidence (at 1σ) of a uranium release from particulates at high salinity. The most saline data point had a uranium concentration of 3.45 $\pm$ 0.08 $\mu g l^{-1}$, within 1σ of the natural seawater value.

Figure 1c shows straightforward mixing of Tamar river water (0.04 $\pm$ 0.01 $\mu gU\ l^{-1}$) with coastal seawater (2.83 $\pm$ 0.10 $\mu gU\ l^{-1}$ at 28.6^{o}/oo). The best fit regression line (R^2 = 0.994) meets the open ocean end-member concentration of 3.35 $\pm$ 0.20 $\mu g l^{-1}$ (2). Slight removal of phosphate and silicate have been noted in continuous sampling surveys in the Tamar (53).

The last example of uranium behaviour across an estuary is from the Forth estuary, which as seen from Fig 1d does show a significant negative deviation from the theoretical mixing line which joins the river water value of 0.09 $\pm$ 0.01 $\mu g l^{-1}$ to the local coastal seawater value of 3.53 $\pm$ 0.11 $\mu g l^{-1}$. The maximum deviation occurs at a salinity of 6.8^{o}/oo. River flow data (10-12 $m^3 s^{-1}$) indicated low flow conditions at time of sampling and there were high particulate loadings, of up to 187 $\mu g l^{-1}$, at low salinities in this estuary (Fig 1e), a feature which would facilitate solid-phase adsorption reactions. Synchronous, large-scale phosphate removal also occurred in the Forth- 75% removed by 4°/oo (Fig 1f) (54). Such phosphate removal is usually regarded as one of surface adsorption onto the solid phase (55), a process which would be favoured in low ionic strength, low pH, high turbidity conditions (53). Since uranium may exist dominantly as $UO_2(HPO_4)_2^{2-}$ (56) under the prevalent pH, pCO_2 and PO_4^- levels in the Forth, then U could easily be removed in concert with the phosphate. Coincident uranium-phosphate removal has previously been suggested in water polluted by phosphate derived from a fertiliser plant (14). In the Forth, however, the excess phosphate levels are primarily sewage-derived. The degree of localised uranium removal in the Forth can be estimated (17, 57) from the intercept I on the ordinate defined by the lower straight line segment of the dashed mixing curve (Fig 1d), using the equation

$$I = Ur + \frac{Ru}{Qr}$$

where Ur is the dissolved uranium concentration of the Forth river water (0.09 $\pm$ 0.01μ gl^{-1}), Qr is the river discharge rate (taken here as 2.764 $m^3 s^{-1}$) and Ru is the removal rate of uranium from solution. With I = -0.40 $\mu g\ l^{-1}$U, Ru is calculated to be 4.4 x 10^4 g U yr^{-1}. An annual average U supply rate from the Forth would be around 1.3 x 10^5 g U yr^{-1}. Close examination of Fig 1d, however, indicates a change of curvature at around 10°/oo salinity and not a linear increase from the most deviant U data point to the extrapolated seawater end-member (3.50 $\mu g l^{-1}$ at 35°/oo) and this trend is clearly consistent with desorption of uranium as salinity increases. Thus net uranium removal from the Forth is considerably less than suggested by the maximum value of 4.4 x 10^4 g yr^{-1} calculated above.

Estuarine Porewaters

Uranium was shown above to behave conservatively in the water column of the Tamar estuary, however from the dissolved uranium data (Fig. 2 a-c) for the porewaters from three cores in St John's Lake, an embayment 2.5 km from the mouth of this estuary , it is observed that there is removal of uranium from solution onto the sediment at shallow depth. The uranium concentration decreases from about 3 $\mu g l^{-1}$ at the 0-1cm depth interval to a subsurface minimum of between 0.25 and 0.81μ gl^{-1} by 4cm depth. Beyond about 14cm depth the uranium porewater concentration increases again to values between 1.4 and 5.6 μgl^{-1}. All the cores sampled from this area become anoxic at less than 1cm below the sediment/seawater interface, semi-quanititative Eh measurements in core 053 giving values of +32.2mV at 0-1cm and -117mV at 1-2cm (44). Redox conditions thus change rapidly, and large Fe (up to 1962μ M) and Mn (up to 120 μM) porewater maxima occur just below the interface, decreasing rapidly with depth, with the Fe maxima occurring slightly deeper than those of Mn, as would be expected from the relative order of use as oxidants during microbial metabolism (58). Thus, since the Fe^{3+}/Fe^{2+} and U^{6+}/U^{4+} redox couples are similar it is probable that uranium is being reduced and fixed to the sediment just as Fe and Mn are being reduced and released to solution. From thermodynamic considerations (36) $UO_2(CO_3)_3^{4-}$ in the porewater can be used as organic matter oxidant within the usual sequence (58) between Mn and Fe, being either reduced to UO_2^+ (1) or to an insoluble U^{4+} species possibly adsorbed or complexed by an organic phase (31). Sediment tank and incubation experiments (36) suggest that the rate for such uranium reduction reactions are of the order of weeks.

An argument against the alternative removal mechanism of U^{6+} adsorption/complexation onto the sediment without reduction is observed increases in the porewaters of potential uranyl complexing ligands such as phosphate (regular increases in 053 and 075 from ~ 20μ M to ~ 190 μM over 20cm) and bicarbonate (a moderate increase in alkalinity from 4.3mM at 1cm to 9mM at 13cm measured in one core only). Consistent with uranium fixation onto the estuarine sediment is the observation (59) that ammonium carbonate leaches of Tamar estuary mud has a $^{234}U/^{238}U$ activity ratio of 1.14 ± 0.05.

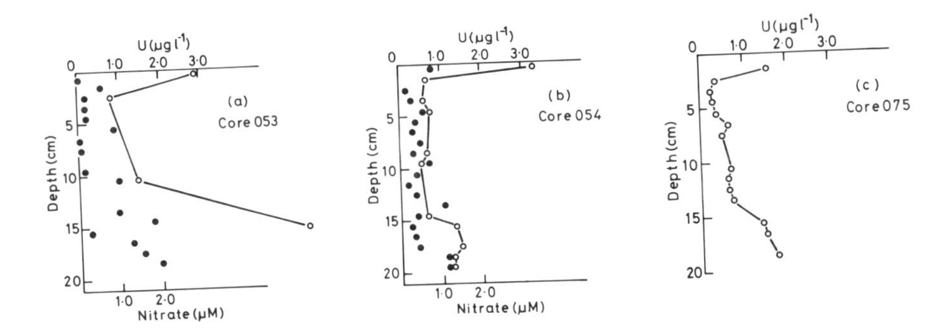

Figure 2 Porewater uranium concentration-depth profiles (o) in three cores from St John's Lake, Tamar estuary: (a) core 053, (b) core 054 and (c) core 075. Porewater nitrate profiles (●) are included for (a) and (b).

From the top two porewater uranium data points in core 054 (Fig. 2b) i.e. 3.28 μ gl^{-1} at 0-1cm and 0.63 μgl^{-1} at 1-2cm respectively, and using a diffusion coefficient of 1 x 10^{-6} $cm^2 s^{-1}$ for U (2, 60), a uranium flux of ~0.08μg cm^{-2} yr^{-1} into the sediment is calculated.

Such a flux would not be detected in water column studies of the estuary (see Fig. 1c) where flushing times are of the order of a week at average runoff (61). For a mean spring tide of 4.7m (61) and a water column concentration of ~3μ gl^{-1} near St John's Lake, the U inventory would be

13.1 $\mu g\ cm^{-2}$, giving a residence time for U of ~165 yrs (i.e. much greater than water renewal).

Figures 2 a-c reveal quite sudden increases of porewater uranium at depths > 14cm to values between 1.4 and 5.6 μgl^{-1}. The explanation for such increases we believe could be linked to bioirrigation in the cores at this site. There is abundant epifauna in the sediment, numerous polychaete worms and molluscs being found. Worm burrows and shell debris were evident in all cores extending down to 18 or 19cm (43). Mid core porewater maxima in Fe^{2+}, Mn^{2+}, NH_4^+ and PO_4^{3-} were found in other cores studied at this site (44) consistent with the emplacement of relatively reactive organic material at depth in the sediment column. For cores 053 and 054 here, evidence for disturbance comes from porewater nitrate profiles (Figs 2a, b) where nitrate is present below the maximum depths of nitrite penetration (~ 9-12cm). Nitrate can be produced by the action of autotrophic bacteria which are normally surface-dwelling and which derive their energy from the aerobic oxidation of NH_4^+ (62) but which can survive long periods in anoxic sediment, being carried down by bioperturbation (63) or infaunal pumping (64). Some reduced uranium at depth may therefore be released from the sediment by irrigated oxygen, increasing the porewater U concentration at depth to a level in core 053 (Fig 2a) which is greater than the overlying water value of ~3$\mu\ gl^{-1}$. A proportion of the enhanced dissolved uranium at depth, however, will probably be derived from water brought down from the surface.

Deep-sea Porewaters

The first example of deep sea porewaters to be discussed here are from the Cape Verde abyssal plain in the eastern Atlantic (station 10552, Fig. 3a) where slowly-accumulating (0.4 cm kyr^{-1}) calcareous marl/ooze is oxic throughout the sampled depth (170 cm). Solid-phase data shows non-authigenic uranium quite constant with depth at ~1ppm. The porewater uranium results from the in-situ sampler are similar to the accepted uranium seawater value, with a weighted mean of 3.20 $\pm$ 0.5 μgl^{-1} with no systematic trend discernable with depth. It is therefore concluded that uranium behaves conservatively in the porewaters of this oxic sediment. The systematic differences between the uranium levels found in the porewaters squeezed from the box and Kastenlot cores (Fig 3a) and those in the in-situ sampler at this station is almost certainly due to a sampling artefact (41) whereby deep sea cores on retrieval decompress, causing calcium carbonate to precipitate resulting in alkalinity losses relative to in-situ samples (65-67). Here the relative losses of U and alkalinity between the squeezed and in-situ porewaters were 44% and 27% respectively. The extent of alkalinity loss should depend on the state of $CaCO_3$ saturation of the porewaters at depth (66) such that thermodynamic corrections may not always be applicable (67, 68). Porewater uranium profiles from the box and Kastenlot cores here (Fig 3a) indicate uranium removal from porewater into the precipitated calcite.

We now consider deep-sea cores which have anoxia at depth, in which it is usually found that the porewater uranium concentrations are elevated, sometimes very markedly, above the seawater uranium value.

Station 10554, again in the eastern Atlantic, was an area where pelagic sedimentation is interrupted by the rapid deposition of turbidites of variable thicknesses (69) which already contain elevated organic carbon and authigenic uranium contents(32). Dissolved oxygen data show that the porewaters here become anoxic by 40cm. Core variability makes

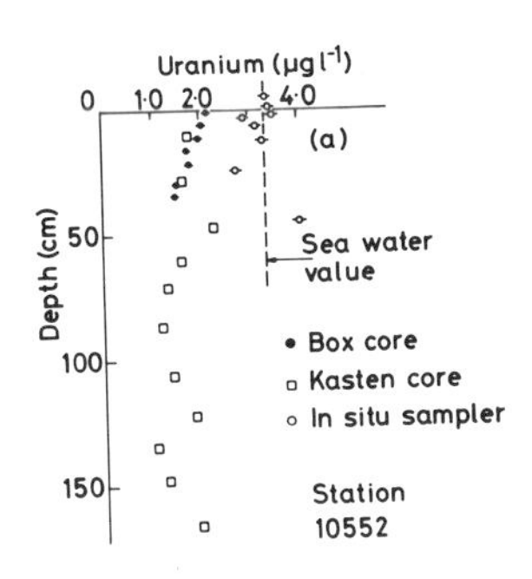

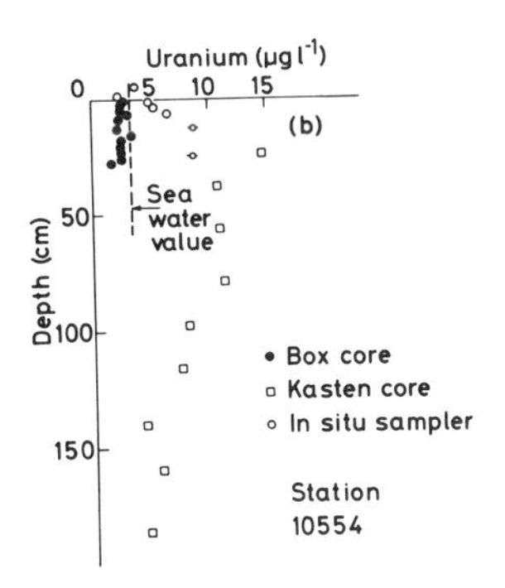

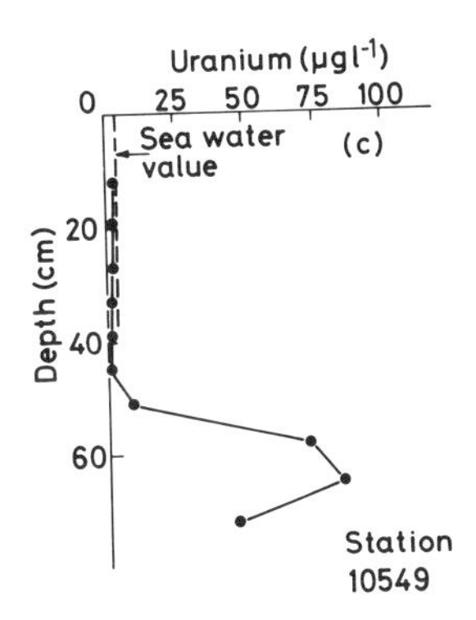

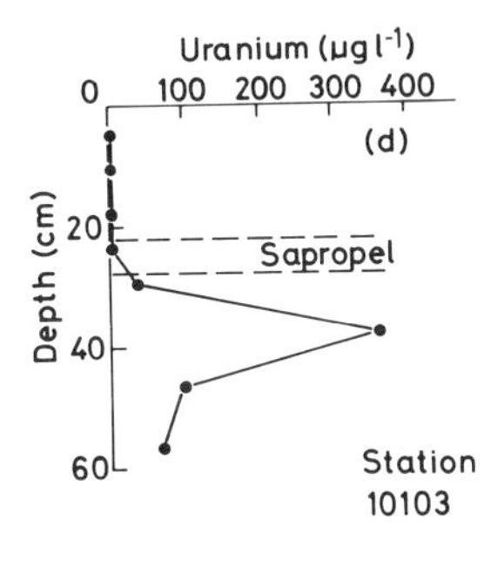

Figure 3 Porewater uranium concentration-depth profiles in deep-sea cores from three different eastern Atlantic stations (a), (b) and (c), and one eastern Mediterranean station (d).

intercomparisons and interpretation difficult here. The main observations are (1) that the in-situ porewater data indicate a sediment to seawater U flux from a sub-surface turbidite unit. This flux is estimated, using a diffusion coefficient for U of 1×10^{-6} cm^2s^{-1} (2, 60), as 11µ gcm^{-2} kyr^{-1}. (2) The pressure artefact is evident for the box core porewater samples which show systematically lower alkalinity values than the in-situ samples (41), and the uranium values are depressed (weighted mean 2.32 ± 0.03 μgl^{-1}) relative to seawater.

The second example of porewater uranium from a core with anoxic sediment is shown in Fig 3c. This Kastenlot core from station 10549 consists of glacial-period silty-marl (0.9% organic C, 60-76% $CaCO_3$) overlain by Holocene calcareous ooze (~0.5% organic C, 70-85% $CaCO_3$) (70). The solid-phase uranium content in the upper ooze was ~ 0.4ppm (0-51cm), and in the lower glacial sediment ~4.5ppm (64-136cm), the latter being authigenic. The increases in both the solid phase and porewater uranium levels were coincident. Porewater manganese peaked at ~30cm in the core and the Fe (iii)/Fe(ii) redox boundary, inferred from a colour change, occurred at ~48 cm. Porewater uranium might therefore be expected to be low in the reducing porewaters at depth. However levels of up to 85 $\mu g l^{-1}$ are present at 65cm (Fig 3c) and furthermore, negligible upward diffusion appears to be occurring from the peak to seawater. Such high uranium concentrations in solution could be due to the presence of iron-bearing colloids (71) or to uranium-organic complexes, both of sufficiently small size to pass through the 0.4 μM Nuclepore filters in the squeezer units. However, we have never observed fission stars in the fission track method used to determine the uranium contents of porewaters, which we might expect if colloidal aggregates were present. A planned comparison of the uranium contents for regular and ultrafiltered porewaters may allow us to suggest the presence of organic complexes, which the porewater uranium profile for station 10549 imply must be of high molecular weight to prevent significant diffusion.

Finally, a somewhat similar situation for an east Mediterranean 'sapropel' core is shown in fig 3d. These sapropel cores can have very high organic (up to 7.2%) and high uranium (up to 50 ppm) contents, with a good correlation between these parameters (33). The sapropel cores taken at station 10103 contained calcareous ooze overlying sapropels, both sediment types varying in thickness from core to core due to complex topography. Auxiliary data for the core shown is lacking, but from results on a neighbouring Kastenlot core, with a thick sapropel section from 28-101cm, organic carbon and solid-phase uranium in the sapropel were 2-3% and 10-18ppm respectively, the uranium being authigenic (34). The sapropel overlies a grey calcareous lutite with high U content (5-6 ppm, $^{234}U/^{238}U$ from 1.06 to 1.12) relative to the ooze above the sapropel ($\sim$1.5ppm, $^{234}U/^{238}U \sim 0.98$) but with similar low organic C (0.3%). There is clearly authigenic uranium in the lutite and it is within this layer where the massive porewater uranium peak of 360 $\mu g l^{-1}$ occurs (Fig 3d), and not within the sapropel itself. It would thus be difficult to invoke organo-uranium complexes as the reason for the peak in this case. We obviously need to know more about the geochemistry of uranium (particularly with respect to organic matter) before we can begin to generalise about is behaviour in porewater.

Conclusions

1. Uranium isotopes generally behave conservatively during estuarine mixing, but removal can occur with phosphate during high particulate loadings (Forth estuary). Coastal water can show small depletions in their uranium content, and evidence for such a coastal removal occurs in the Clyde estuary.
2. In the estuarine porewaters studied here, uranium is removed from solution probably by a redox process as the sediment rapidly becomes reducing. It is also likely that uranium release at depth in the estuarine cores is due to surficial water irrigation caused by bioturbational processes.
3. In an oxic deep-sea sediment, uranium behaves conservatively in the porewaters. There is an important pressure-related artefact affecting the uranium content of deep-sea porewaters.
4. In deep-sea sediments containing anoxic layers, porewater uranium levels are high (up to 365 $\mu g l^{-1}$ here). These high values can result in a sediment to seawater flux of uranium. However, the absence of uranium diffusion from other high porewater uranium peaks might signify the presence of high molecular weight organo-uranium complexes.

Acknowledgement

The authors wish to thank NERC and UKAEA for supporting various parts of this work.

References

1. Langmuir, D., Uranium solution-mineral equilibria at low temperatures with applications to sedimentary ore deposits. Geochim. Cosmochim. Acta., 1978, 42, 547-569.
2. Ku, T-L., Knauss, K.G. and Mathieu, G.G., Uranium in open ocean: concentration and isotopic composition. Deep Sea Res., 1977, 24, 1005-1017.
3. Chen, J.H., Edwards, R.L. and Wasserburg, G.J., ^{238}U, ^{234}U and ^{232}Th in seawater. Earth Planet. Sci. Lett., 1986, 80, 241-251.
4. Burton, J.D., Radioactive nuclides in the marine environment. In chemical Oceanography, eds. J.P. Riley and G. Skirrow, Academic Press, London, 1975, 3, pp. 91-191.
5. Anderson, R.F., Concentration, vertical flux and remineralization of particulate uranium in seawater. Geochim. Cosmochim. Acta, 1982, 46, 1293-1299.
6. Cochran, J.K., The oceanic chemistry of the U- and Th- series nuclides. In Uranium Series Disequilibrium: Applications to Environmental Problems, eds. M. Ivanovich and R.S. Harmon, Clarendon Press, Oxford, 1982, pp. 384-430.
7. Scott, M.R., The Chemistry of U- and Th- series nuclides in rivers. In Uranium Series Disequilibrium: Applications to Environmental Problems, eds. M. Ivanovich and R.S. Harmon, Clarendon Press, Oxford, 1982, pp. 181-201.
8. Turekian, K.K. and Cochran, J.K., Determination of marine chronologies using natural radionuclides. In Chemical Oceanography, eds. J.P. Riley and R. Chester, Academic Press, London, 1978, 7, pp. 313-360.
9. Bhat, S.G. and Krishnaswami, S., Isotopes of uranium and radium in Indian rivers. Proc. Ind. Acad. Sci., 1969, 70, 1-17.
10. Turekian, K.K. and Chan, L.H., The marine geochemistry of the uranium isotopes, ^{230}Th and ^{231}Pa. In Activation Analysis in Geochemistry and Cosmochemistry, eds. A.O. Brunfeld and E. Steinnes, Universitetsforlaget, Oslo, 1971 pp. 311-320.
11. Mangini, A., Sonntag, C., Bertsch, G. and Muller, E., Evidence for a higher natural uranium content in world rivers. Nature, 1979, 278, 337-339.
12. Borole, D.V., Krishnaswami, S. and Somayajulu, B.L.K., Uranium isotopes in rivers, estuaries and adjacent coastal sediments of western India: Their weathering, transport and oceanic budget. Geochim. Cosmochim. Acta, 1982, 46, 125-137.
13. Borole, D.V., Krishnaswami, S. and Somayajulu, B.L.K., Investigations on dissolved uranium, silicon and particulate trace elements in estuaries. Estuar. Coast. Mar. Sci., 1977, 5, 743-754.
14. Martin, J-M, Nijampurkar, V. and Salvadori, F., Uranium and thorium isotope behaviour in estuarine systems. In Biogeochemistry of Estuarine Sediments, ed. E.D. Goldberg, UNESCO, 1978, pp. 111-127.
15. Martin, J-M., Meybeck, M. and Pusset, M., Uranium behaviour in the Zaire estuary. Neth. J. Sea Res., 1978, 12, 338-344.
16. Martin, J-M. and Meybeck, M., Elemental mass-balance of material carried by major world rivers. Mar. Chem., 1979, 7, 173-206.
17. Maeda, M. and Windom, H.L. , Behaviour of uranium in two estuaries of the south-eastern United States. Mar. Chem., 1982, 11, 427-436.

18. Toole, J., Baxter, M.S. and Thomson, J., The behaviour of uranium isotopes with salinity change in three UK estuaries . Estuar. Coast Shelf Sci., in press.
19. Lewis, D.M., The Geochemistry of Manganese, Iron, Uranium, ^{210}Pb and Major Ions in the Susquehanna River. Ph.D. Thesis, Yale University, 1976.
20. Figueres, G., Martin, J-M. and Thomas, A.J., River input of dissolved uranium to the oceans: the Zaire river and estuary. Oceanol. Acta, 1982, 5, 141-147.
21 Bonatti, E., Fisher, D.E., Joensuu, O. and Rydell, H.S., Post depositional mobility of some transition elements, phosphorus and thorium in deep sea sediments. Geochim Cosmochim. Acta, 1971, 35, 189-201.
22. Cowart, J.B., The relationship of uranium isotopes to oxidation/reduction in the Edwards Carbonate Aquifer of Texas. Earth Planet. Sci. Lett., 1980, 48, 277-283.
23. Maynard, J.B., Geochemistry of Sedimentary Ore Deposits, Springer-Verlag., 1983.
24. Deutsch, W.J. and Serne, R.J., Uranium mobility in the natural environment: evidence from sedimentary roll-front deposits. In Geochemical Behaviour of Disposed Radioactive Waste, eds. G.S. Barney, J.D. Navratil and W.W. Schulz, A.C.S. Symposium Series No 246, American Chemical Society, 1984, pp. 287-302.
25. Thomson, J., Turekian, K.K. and McCaffrey, R.J., The accumulaton of metals in and release from sediments of Long Island Sound. In Estuarine Research vol 1, Academic Press, Inc., New York, 1975, pp. 28-43.
26. Aller, R.C. and Cochran, J.K., ^{234}Th/^{238}U disequilibrium in near-shore sediment: particle reworking and diagenetic timescales. Earth Planet Sci. Lett., 1976, 29, 37-50.
27. Carpenter, R., Peterson, M.L., Bennett, J.T. and Somayajulu, B.L.K., Mixing and cycling of uranium, thorium and ^{210}Pb in Puget Sound sediments. Geochim. Cosmochim. Acta, 1984, 48, 1949-1964.
28. Boulard, A.P. and Michard, G., Etude de l'uranium, du thorium et de leurs isotopes dans quelques carottes du bassin Angolais (Atlantique Sud-Est). Earth Planet. Sci. Lett., 1976, 32, 77-83.
29. Yamada, M. and Tsunogai, S., Post depositional enrichment of uranium in sediment from the Bering Sea. Mar. Geol., 1984, 54, 263-276.
30. Veeh, H.H., Deposition of uranium from the ocean. Earth Planet. Sci. Lett., 1967, 3, 145-150.
31. Kolodny, Y. and Kaplan, I.R., Deposition of uranium in the sediment and interstitial water of an anoxic fjord. In Proceedings of symposium on Hydrogeochemistry and Biogeochemistry vol 1, ed. E. Ingerson, Clark Co., Washington D.C., 1973, pp. 418-442.
32. Colley, S. and Thomson, J., Recurrent uranium relocations in distal turbidites emplaced in pelagic conditions. Geochim. Cosmochim. Acta, 1985, 49, 2339-2348.
33. Mangini, A. and Dominik, J., Late quaternary sapropel on the Mediterranean ridge: U-budget and evidence for low sedimentation rates. Sed. Geol., 1979, 23, 113-125.
34. Thomson, J., Holocene sedimentation rates on the Hellenic outer ridge: a comparison by ^{14}C and ^{230}Th excess methods. Sed. Geol., 1982, 32, 99-110.
35. Degens, E.T., Khoo, F. and Michaelis, W., Uranium anomaly in Black Sea sediments. Nature, 1977, 269, 566-569.
36. Cochran, J.K., Carey, A., Sholkovitz, E.R. and Surprenant, L.D., The Geochemistry of uranium and thorium in coastal marine sediments and sediment pore waters. Geochim. Cosmochim. Acta, 1986, 50, 663-680.
37. Baturin, G.N., Uranium in oceanic ooze solutions of the southeastern

Atlantic. Proc. Nat. Acad. Sci. USSR, 1971, 198, 224-226.
38. Baturin, G.N. and Kochenov, A.V., Uranium in the interstitial waters of marine and oceanic sediments. Geokhimiya, 1973, 10, 1529-1536.
39. Dysart, J.E. and Osmond, J.K., Uranium in pore waters of two southern ocean cores. Antartic J. United States, 1975, 10, 255.
40. Zhorov, V.A., Boguslavskiy, S.G., Babinets, A.Ye., Solov'yeva, L.V., Kirchanova, A.I. and Kir'yakov, P.A., Geochemistry of uranium in the Black Sea. Geokhimiya, 1982, 8, 1195-1203.
41. Toole, J., Thomson, J., Wilson, T.R.S. and Baxter, M.S., A sampling artefact affecting the uranium content of deep-sea porewaters obtained from cores. Nature, 1984, 308, 263-266.
42. Cochran, J.K. The fates of U and Th decay series nuclides in the estuarine environment. In The Estuary as a Filter, ed. V.S. Kennedy, Academic Press, 1984.
43. Upstill-Goddard, R.C., Geochemistry of the Halogens in the Tamar Estuary, Ph.D. thesis, Univ. of Leeds, 1985.
44. Alexander, W.R., Inorganic Diagenesis in Anoxic Sediments of the Tamar Estuary, Ph.D. thesis, Univ. of Leeds, 1985.
45. Sayles, F.L., Mangelsdorf, P.C. Jr., Wilson, T.R.S. and Hume, D.N., A sampler for the in-situ collection of marine sedimentary pore waters. Deep-Sea Res., 1976, 23, 259-264.
46. Peters, R.D., Timmins, N.T., Calvert, S.E. and Morris, R.J., The I.O.S. box corer: Its design, development, operation and sampling. Report no. 106, Institute of Oceanographic Sciences, Surrey, 1980.
47. Toole, J., Marine Chemistry of Uranium: Studies by Particle Track Analysis and Alpha Spectrometry, Ph.D. thesis, Univ. of Glasgow, 1984.
48. Blanchard, R.L., $^{234}U/^{238}U$ ratios in coastal marine waters and calcium carbonates. J. Geophys. Res., 1965, 70, 4055-4061.
49. Blanchard, R.L. and Oakes, D., Relationships between uranium and radium in coastal marine shells and their environment. J. Geophys. Res., 1965, 70, 2911-2921.
50. Bhat, S.G., Kirshnaswami, S., Lal, D., Rama and Moore, W.S., $^{234}Th/^{238}U$ ratios in the ocean. Earth Planet. Sci. Lett., 1969, 5, 483-491.
51. Scholkovitz, E.R., Boyle, E.A. and Price, N.B., Removal of dissolved humic acid and iron during estuarine mixing. Earth Planet. Sci. Lett., 1978, 40, 130-136.
52. Das, N., Geochemical Studies of Caithness Flags, Ph.D. thesis, Univ. of Glasgow, 1985.
53. Morris, A.W., Bale, A.J. and Howland, R.J.M., Nutrient distributions in an estuary: evidence of chemical precipitation of dissolved silicate and phosphate. Est. Coast. Shelf Sci., 1981, 12, 205-215.
54. Leatherland, T.M., Forth River Purification Board, personal communication.
55. Liss, P.S., Conservative and non-conservative behaviour of dissolved constituents during estuarine mixing. In Estuarine Chemistry, eds. J.D. Burton and P.S. Liss, Academic Press, London, 1976, 93-130.
56. Dongarra, G. and Langmuir, D., The stability of UO_2OH^+ and $UO_2(HPO_4)_2^{2-}$ complexes at 25°C. Geochim. Cosmochim. Acta, 1980, 44, 1747-1751.
57. Li, Y.H. and Chan, L.H., Desorption of Ba and ^{226}Ra from river-borne sediments in the Hudson estuary. Earth Planet. Sci. Lett., 1979, 43, 343-350.
58. Froelich, P.N., Klinkhammer, G.P., Bender, M.L., Luedtke, N.A., Heath, G.R., Cullen, D., Dauphin, P., Hammond, D., Hartman, B. and Maynard, V., Early oxidation of organic matter in pelagic sediments of the eastern equatorial Atlantic: suboxic diagenesis. Geochim. Cosmochim. Acta, 1979,

43, 1075-1090.
59. Hamilton, E.I. and Stevens, H.E., Some observations on the geochemistry and isotopic composition of uranium in relation to the reprocessing of nuclear fuels. J. Environ. Rad., 1985, 2, 23-40.
60. Li, Y.H. and Gregory, S., Diffusion of ions in sea water and in deep sea sediments. Geochim. Cosmochim. Acta, 1974, 38, 703-714.
61. Uncles, P.J., Bale, A.J., Howland, R.J.M., Morris, A.W. and Elliot, R.C.A., Salinity of surface water in a partially-mixed estuary, and its dispersion at low run-off. Oceanol. Acta, 1983, 6, 289-296.
62. Vanderborght, J.P. and Billon G., Vertical distribution of nitrate concentration in interstitial water of marine sediments with nitrification and denitrification. Limnol. Oceanogr., 1975, 20, 953-961.
63. Sorensen, J., Capacity for denitrification and reduction of nitrate to ammonia in coastal marine sediment. Appl. Environ. Microb., 1978, 35, 301-305.
64. Van der Loeff, M.M.R., van Es, F.B., Helder, W. and De Vries, R.T.P. Sediment-water exchanges of nutrients and oxygen on tidal flats in the Ems-Dollard Estuary. Neth. J. Sea. Res., 1981, 15, 113-129.
65. Emerson, S., Jahnke, R., Bender, M., Froelich, P.N., Klinkhammer, G.P., Bowser, C. and Setlock, G., Early diagenesis in sediments from the eastern equatorial Pacific. 1. Pore water nutrient and carbonate results. Earth Planet. Sci. Lett., 1980, 49, 57-80.
66. Murray, J.W., Emerson, S. and Jahnke, R., Carbonate saturation and the effect of pressure on the alkalinity of interstitial waters from the Guatemala Basin. Geochim. Cosmochim. Acta, 1980, 44, 963-972.
67. Emerson, S., Grundmanis, B. and Graham, D., Carbonate chemistry in marine porewaters: MANOP sites C and S. Earth Planet. Sci. Lett., 1982, 61, 220-232.
68. Jahnke, R., Emerson, S. and Murray, J.W., A model of oxygen reduction, denitrification and organic matter mineralisation in marine sediments. Limnol. Oceanogr., 1982, 27, 610-623.
69. Weaver, P.P.E. and Kuijpers, A., Climatic control of turbidite deposition on the Madeira abyssal plain. Nature, 1983, 306, 360-363.
70. Thomson, J., Wilson, T.R.S., Culkin, F. and Hydes, D.J., Non-steady state diagenetic record in eastern equatorial Atlantic sediments. Earth Planet Sci. Lett., 1984, 71, 23-30.
71. Giblin, A.M., Batts, B.D. and Swaine, D.J., Laboratory simulation studies of uranium mobility in natural waters. Geochim. Cosmochim. Acta, 1981, 45, 699-709.

KVI-683

NATURAL RADIOACTIVITY OF HEAVY MINERAL SANDS: A TOOL FOR COASTAL SEDIMENTOLOGY?

R.J. de Meijer, L.W. Put
Kernfysisch Versneller Instituut, Rijksuniversiteit Groningen
Zernikelaan 25, 9747 AA Groningen, The Netherlands

R.D. Schuiling
Instituut voor Aardwetenschappen, Rijksuniversiteit Utrecht
Postbus 80021, 3508 TA Utrecht, The Netherlands

J. de Reus
Dienst Getijdewateren, Rijkswaterstaat, Postbus 207
9750 AE Haren, The Netherlands

J. Wiersma
Directie Noordzee, Rijkswaterstaat, Postbus 5807
2280 HV Rijswijk, The Netherlands

ABSTRACT

The radioactivity of heavy mineral sands, due to the inclusion in their crystals of U and Th, has proven to be a fast and reliable method for mapping surface occurrences of these minerals in dune and beach sands along the Dutch coast. Gamma-spectrometric and granulometric analyses revealed that the radioactivity can be correlated with the smaller grain sizes. The absolute and relative magnitude of the U and Th activities varies strongly with magnetic susceptibility. A correlation was found between U and Th abundance and elements like Zr, P and e.g. La. All these results indicate a correlation between U, K and Th and the mineralogical composition, thus leading to "radiometric finger printing" of the sands. Concentrations of heavy mineral sands were not only found at the foot of dunes but also in the dunes and at places where islands merged, indicating that also during periods of netto accretion heavy mineral concentration occurs.

INTRODUCTION

Heavy mineral sand-deposits are the main sources for the world supply of rare metals such as zirconium and titanium. Concentrations of heavy minerals are known to occur in beach and dune sands. They are found e.g along the coasts of Kerala in India, Australia, Malaysia, Brasil, the Gulf of Mexico and the North Sea. In these sands minerals are present as ilmenite, magnetite, garnet, epidote, monazite, and rutile. Some of these deposits may visually be recognized by a darker colouring of the sands.

Since heavy mineral concentrations are often found in beach ridges it is generally accepted that the concentration took place by the wash out of the lighter minerals such as quartz in these high energy features. The presence of heavy minerals in beaches and dunes is an indicator for coastal development processes.

In the search for heavy mineral concentrations usually samples are taken at various locations and the abundance of the heavy minerals is determined in the laboratory by floating off the lighter components of the sand in heavy liquids such as bromoform. Investigations[1-5] in and along the North Sea have been carried out in this way and represent a data base obtained with elaborate work and only rather coarse sampling.

It is known that the radionuclides ^{238}U, ^{235}U and ^{232}Th may become incorporated in igneous materials when they are originally formed from the molten state. A mineral is likely to be enhanced in radioactive nuclide content if one of its predominant ions is of similar size to that of U and Th. Based on isomorphism, activity is expected in minerals containing zirconium, calcium, and rare earth elements like cerium and lanthanum. Thorium is usually associated with zirconium and cerium; uranium and thorium appear as replacements of calcium in minerals like apatite.

The radioactivity of the minerals allows the identification of deposits by radiometric means. Fast and detailed information on surface deposits was obtained in the southeast part of the USA by registrating the emitted gamma radiation by detectors mounted onto an aircraft[6,7]. Discrete sources of ^{232}Th were used by Kamel and Johnson[8] to determine the drift along the Californian coast. Also along the North Sea coast heavy minerals were located by radiometric means. Bonka[9] observed dark radioactive sands along the beaches of some German Frisian Islands.

The aim of this investigation was to find out whether heavy mineral sands also occur along the Dutch coast, to study the correlation of the radioactivity with physical parameters such as grain size and magnetic susceptibility, and to investigate the feasibility of a detailed radiometric mapping of the coast in order to obtain information on the physical processes that play a role in transport of sand along the coast. This latter topic may help to obtain knowledge and better understanding on the protection of the Dutch coast and may contribute to the present research program "Kustgenese".

Part of the results have already been published[10,11]; in this paper we only present some of the observations including some recent ones and more preliminary conclusions and we refer for details to these papers.

EXPERIMENTAL TECHNIQUE

Radiometric mapping of the beaches and adjacent dune formations have been made in the Netherlands along the North Sea coast from IJmuiden north-east till the German border. The measurements were carried out with a portable 4×4×5 cm^3 NaI crystal, mounted on a photomultiplier tube and connected to accessory electronics, including a single-channel analyser and a scaler display (Scintrex GIS5). Measurements were usually made every 100 m along the foot of the dunes; at every kilometer a transect was made from the waterline into the dunes in steps of about 25 m. Occasionally, mainly in areas with higher activities, more detailed measurements were made. In these more detailed studies contact readings were made in addition to the measurement at the usual height of about 1 m (hip height).

At a few locations profiles upto a depth of about 1 m were made using a radiation monitor with a GM-tube mounted on a stick (Wallac GMP-256).

Samples of sand were collected from the surface, from greater depth, and from the seabottom to be investigated in more detail in the laboratory. Samples were dried at 100°C. Bulk materials, sieved materials and samples separated in a Frantz isodynamic separator were investigated for U, Th and K contents by gamma-ray spectrometry with a Ge detector.

The conversion of total count rate to exposure rate was deduced from a series of measurements at various locations on Ameland with both the Scintrex and the Reuter Stokes detectors at the same spot and the same height above the ground. The readings were corrected for the contribution of cosmic ray back ground. The magnitude of these corrections determined in measurements over (fresh) water with a depth of more than 5 m.

The ^{238}U activity was deduced from the γ-decay of ^{214}Bi and ^{214}Po assuming equilibrium in all preceeding decay stages. A more direct determination of the ^{238}U activity can only be made from the intensity of the E_γ = 1.001 MeV γ-ray in the decay of ^{234m}Pa. Since this transition is weak the activities are only deducible with a rather large uncertainty. Using the 1.001 MeV γ-ray a ^{238}U concentration is found which is consistently almost a factor of two higher than deduced from the ^{214}Bi and ^{214}Po activities (results not presented). This discrepancy indicates that the assumption on equilibrium is not valid. The ^{235}U/^{238}U activity ratios tend to support such a disequilibrium.

The mineralogical composition of the sands was determined visually under a binocular microscope and/or by Guinier X-ray diffraction.

RESULTS

The results of radiometric mapping indicate rather low concentrations on the mainland, except for a spot near Bergen, and on the islands of Schiermonnikoog, Rottumerplaat and Rottumeroog, see fig. 1. High values were found on the islands Texel and Ameland. In general it can be said that each island has its own characteristic pattern. Moreover it was observed that the concentrations were lowest at the waterfront and increased across the beach, often reaching a maximum at the foot of the dunes. At various locations even higher values were found in the dunes. Since most detailed information has been gathered on the adjacent islands Schiermonnikoog and Ameland the results presented in this paper will mainly concern these two islands.

Fig. 2 shows the generalized results of the measurements of total counts on these two islands. The exposure rates are obtained from the conversion procedure described above. One notices that the values for Ameland are much higher than those found on Schiermonnikoog. On both islands the highest count rates were measured in the dunes rather than at the dune foot. On Ameland the high concentrations occur a few hundred meters inland in dunes that were formed in the last century at locations where three islands merged into the present island[12]. At Schiermonnikoog an elevated count-rate is observed in the new dunes at the east side of the island. These islands are being formed by catching drifting sands in the vegetation

Figure 1. The Netherlands

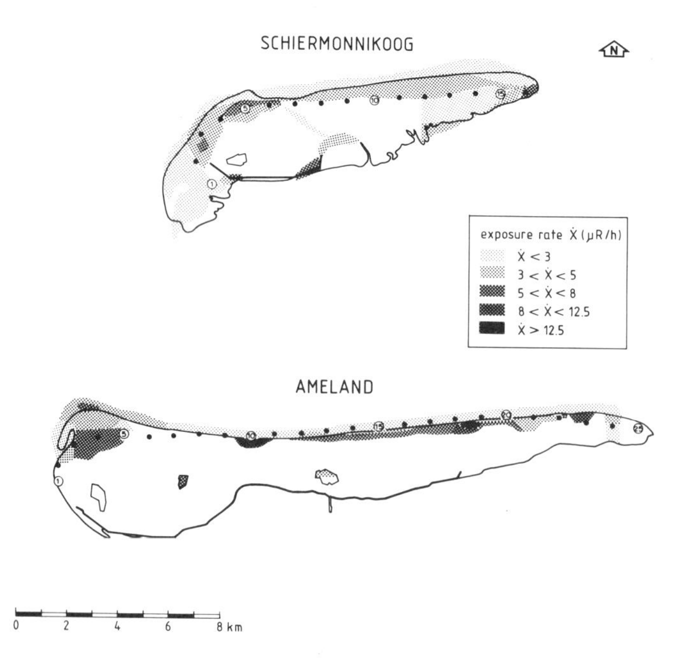

Figure 2. Gamma-ray exposure (μR/h) on the beach and in the dunes of the islands Ameland and Schiermonnikoog.

on a man-made sand dike.

From comparison measurements between the Scintrex and Reuter Stokes detectors the highest count rate of 620 cps, observed in the Zwanewaterduinen at Ameland, corresponds to an exposure rate of 25 μR/h. This spot has the highest exposure rate due to natural radiation in the Netherlands known to us. At beaches and dunes of the other islands and along the coast of the mainland maximum count rates of 80-250 cps were obtained. A full year stay at that spot would result in a total-body equivalent dose of 1.3 mSv/a, corresponding to about 50% of the average annual dose of the Dutch population. It should be noted that on the dikes protecting the southern part of the island count rates were obtained which were two times higher than at the "hot spot" in the Zwanewaterduinen. The high exposure rate is attributed to slags embedded in the concrete blocks as part of the dike construction.

Depth profiles on the beach show bands of dark minerals varying in thickness from a few millimeters to a few centimeters. The bands get more diffuse at larger depths and are usually unobservable after 0.5 m. In the older dunes no clear band structure was observed. Fig. 3 shows the count

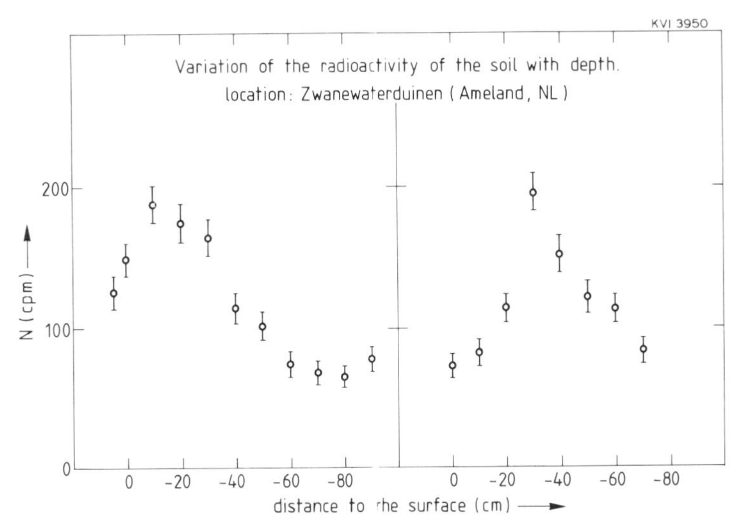

Figure 3. Variation of radioactivity of the soil with depth at two locations in the dunes of Ameland. The measurements were made with a Wallac radiation monitor.

rate variation with depth measured with the Wallac radiation monitor at two locations in the Zwanewaterduinen at Ameland. The left part of the figure shows a broad maximum 10-40 cm below the surface; at the right-hand part of the figure the maximum occurs at a depth of 30-50 cm.

Gamma-ray spectra indicate that the radioactivity in the sands is predominantly due to the natural radionuclides U, Th and K and their radioactive decay products. The results are presented in table 1 for samples from dunes and beaches collected at places with high activity, grab samples from two locations elsewhere along the coast (Domburg and Bennebroek, see fig. 1), and grab samples obtained from the sea bottom between the islands Ameland and Terschelling. From the table one sees that the U/Th ratio of the samples from Borkum, Ameland and various locations in the tidal inlet

Table 1: U, Th and K contents in Bq/kg of various sand samples collected on beaches and in dunes and of samples at various depths collected from the seabottom of the tidal inlet between Terschelling and Ameland.

Location	^{238}U	^{235}U	^{232}Th	^{40}K	$^{238}U/^{232}Th$
Beaches and dunes					
Borkum (FRG)	238 ± 8	22 ± 8	238 ± 8	340 ±30	1.00±0.05
Ameland P19	167 ± 8	23 ± 3	174 ± 8	360 ±30	0.96±0.06
Ameland P23*)	13.3± 1.5	3.0± 1.5	20.0± 1.9	380 ±30	0.67±0.10
Bergen P32	384 ±10	20 ± 2	314 ±15	11 ±11	1.22±0.07
Bergen P32.6	249 ± 3	13 ± 2	205 ± 7	45 ± 8	1.21±0.03
Domburg P15.1*)	6.5± 0.7	0.5± 0.4	4.9± 0.5	109 ± 8	1.3 ±0.2
Bennebroek*)	7.9± 1.4	1.2± 1.3	8.8± 1.4	534 ±14	0.9 ±0.2
Sea bottom*) depth (m)					
20.1	5 ± 2	0.5± 0.7	7.3± 0.6	373 ±11	0.7 ±0.3
8.8	5.4± 0.6	0.6± 0.2	4.4± 0.5	250 ± 9	1.2 ±0.2
5.4	5.6± 0.9	ND	5.7± 0.9	267 ±12	1.0 ±0.2
4.4	9 ± 2	0.5± 0.3	15.3± 1.2	264 ±11	0.59±0.14
6.8	4.8± 0.4	0.4± 0.3	5.0± 0.8	367 ±11	0.96±0.17
7.4	6.8± 0.7	0.2± 0.5	5.6± 0.8	321 ±11	1.21±0.17
3.4	9.7± 2.0	1.0± 0.2	11.7± 1.4	364 ± 8	0.8 ±0.2

*) bulk samples

between Terschelling and Ameland is consistent with a value near or slightly smaller than unity; the ^{40}K values vary between 250 and 400 Bq/kg. The samples from Bergen have a U/Th ratio of about 1.2 and a low ^{40}K content. The samples of Domburg and Bennebroek are grab samples from one of the islands in Zeeland and an old dune formation south of Haarlem, respectively. The Domburg sample has a U/Th ratio of 1.3 and a ^{40}K content of about 100 Bq/kg; the Bennebroek sample has characteristics comparable to the samples in the northern part of the country.

The samples collected from the sea bottom of the tidal inlet between Terschelling and Ameland all have a similar ^{40}K content but show variations in the U and Th activities up to a factor of four. From these scattered data it is not yet clear whether these differences are due to present or ancient selection processes.

Fig. 4 shows concentrations of various radioactive and stable elements obtained from a chemical analysis. In this figure La is chosen as a representative for the rare earth elements. From the figure it follows that in the smaller fractions Zr dominates (400,000 ppm) pointing to the occurrence of zircon in the fine fractions, and that K is mainly present in the larger grains, obviously indicating the occurrence of light minerals as K-feldspar and muscovite (Schuiling et al., 1985). Moreover one notices that the distribution of U follows that of Zr, whereas Th and La follow P reflecting the replacement in zircon and monazite respectively. The second maximum of P near the grain size of 0.2 mm corresponds with the maximum of the Sr concentration. The similarity of Sr and P near ϕ = 0.2 mm may arise from inclusion in e.g. apatite.

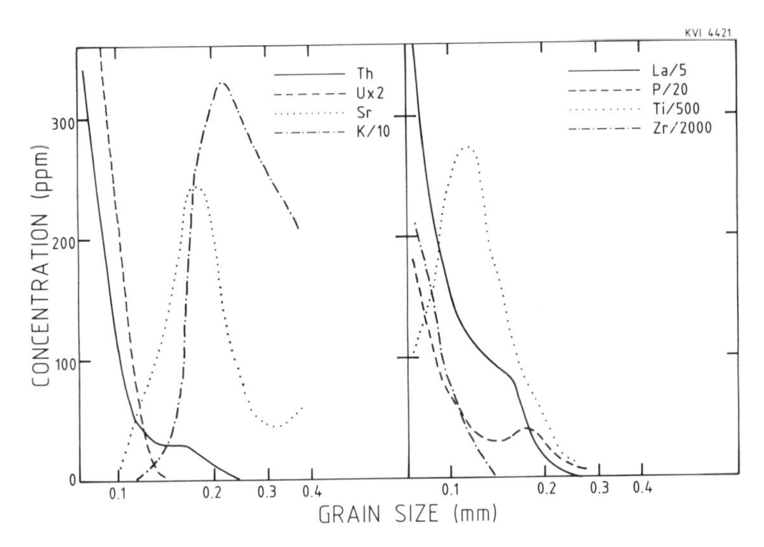

Figure 4. Concentration of various elements as determined by chemical analysis in the Ameland P19 sample. The multiplicative numbers indicate the factor by which the original concentration was multiplied or divided to bring them to the presented scale.

CONCLUSIONS AND DISCUSSION

The relation between the radioactivity and parameters such as grain size, magnetic susceptibility and chemical composition indicate that the U and Th activity reflects the mineralogical composition. The ratio of U and Th and the activity of ^{40}K may probably be used to identify the origin of the sands. Such a possibility was recently found in heavy mineral sands collected at Florida beaches along the Gulf of Mexico[16]. Here variations of an order of magnitude were observed apparently reflecting the differences in mineral composition of the sands of the nearby rivers.

In the previous paragraph it was indicated that sands may be qualified by means of their U/Th ratio and their K concentration ("radiometric finger printing"). In retrospect this result should not be too surprising. U and Th are mainly present in the heavy minerals whereas K is associated with the light minerals. Heavy minerals are usually a smaller fraction of the sand; their concentration varies depending on the selectivity in the transport processes. Thus the absolute concentration is strongly dependent on the degree of enrichment of heavy minerals, whereas the K-concentration only varies slowly with the enrichment factor. The U/Th ratio reflects the heavy-mineral composition and of course is more or less independent of the enrichment factor.

It is interesting to point out that near Bergen a borderline exists between northern sand deposits of Scandinavian and southern deposits of Middle European origin (see ref. 5). From the present results it seems that the samples from Bergen and from Domburg are of Middle European origin and that the others including possibly the sample from Bennebroek contain sand of predominantly Scandinavian origin.

The present results indicate an empirical method to identify the origin of sands in a rather fast and sensitive way. From this rather limited number of samples, however, it is too early to draw definite conclusions. A more systematic study of this aspect is presently considered.

At first sight the difference in heavy-mineral concentrations between the islands Ameland and Schiermonnikoog seems to be at first sight in agreement with the general hypothesis (see e.g. refs. 2,3,4,17) that eroding coasts show higher concentration factors for heavy minerals than stable coasts: nowadays Ameland is eroding, Schiermonnikoog is stable to slightly prograding at its western part. A closer look at the situation shows, however, that the highest concentrations occur in the dunes at places where former islands have merged. Also on Schiermonnikoog higher activities are found in new dune formations. Aside from the "hot spots" it can be concluded that the bulk material on Ameland has a higher content than that on Schiermonnikoog. This is surprising since the two islands are only separated by a narrow tidal inlet. Also Veenstra and Winkelmolen[18] stressed this difference, mainly based on higher percentages of garnet and higher rollability of beach sands from Ameland. Similarly, in our mapping we find that the bulk sands on Texel and Vlieland have a higher radioactivity than on the mainland or on the islands Terschelling, Schiermonnikoog, Rottumerplaat and Rottumeroog. At Terschelling, sandwiched between Vlieland and Ameland the two ends show areas with higher activity. In front of the islands, at a depth of about 6 m below sea level, variations in the percentage of the heavy minerals[19] is observed along the coast. It is conceivable that these variations are related to the variations in U and Th concentrations observed in the present investigation (table 1).

In the chemical analysis of an Ameland sand sample a correlation was found between the grain-size distribution and the concentration of La and P and U and Zr. The first correlation corroborates with the presence of monazite $(Ce,La,Y,Th)PO_4$ in the sample; the latter result agrees with the known inclusion of U in zircon. The results may be used to estimate the heavy-mineral concentrations in sands from their radioactivity (see also ref. 11).

So far this project has raised more questions than it has given answers. In our opinion the heavy mineral deposits have the potential of revealing information on the origin of the islands, the origin of the sands and the transport mechanisms in the sand regime around the islands and along the coast in general. Plans are being developed to investigate the possible sources of heavy minerals in front of the island and to obtain more insight in the dynamics of the transport processes by following the time dependence of the heavy mineral concentrations at certain areas of the beach by radiometric means. Also in this way we try to contribute to a better insight in and to a more effective defense of the coast of a country of which the sea is both its best ally and its strongest foe.

We like to acknowledge the help of Drs. J.G. Ackers and A. van der Wijk in earlier stages of this investigation. The assistence in the analysis of various samples by F.J. Aldenkamp, R. van der Made and O. Voorwinde and of P. Guinee, T. Top and D. Verburgh in the mapping is greatly appreciated.

REFERENCES

1. Baak, J.A., Regional petrology of the southern North Sea. Diss. Wageningen (1936).
2. Crommelin, R.D., Slotboom, G., Een voorkomen van granaatzandlagen op het strand van Goeree. Tijdschr.Kon.Ned.Aardr.Gen., 1945, 62, 143-147.
3. Lamcke, K., Natrliche Anzeicherungen von Schwermineralen in Ksten gebieten. Geologie der Meere und Binnengewässer, 1937, 1, 106-125.
4. Wasmund, E., Die Schwermineral lagerstätten der deutsche Küsten. Geol. Rundschau, 1938, 29, 237-300.
5. Eisma, D., Composition, origin and distribution of Dutch Coastal sands between Hoek van Holland and the island Vlieland, thesis Groningen, 1968, and Netherlands Journal of Sea Research, 1968, 4, 123-267.
6. Force, E.R., Grosz, A.E., Loferski, P.J., Maybin, A.H., Aeroradioactivity Maps in heavy-mineral exploration Charleston, Sourth Carolina. Prof.Pap.U.S.Geol.Surv.,(1982), 1218, 19p.
7. Grosz, A., Kosanke, K.L., Application of total-count aeroradiometric maps to the exploration for heavy-mineral deposits in the coastal plain of Virginia. Prof.Pap.U.S.Geol.Surv., 1983, 1263, 20p.
8. Kamel, A.M. Johnson, J.W., Tracing coastal sediment movement by naturally radioactive minerals. Proc. 8th Coast.Eng.Conf., 1962, 324-330.
9. Bonka, H., Erhöhte natürliche Strahlenexposition durch Schwermineral-anreicherung an der Küste Norddeutschlands. Atomkernenergie-Kerntechnik 1980, 35, 1, 5-11.
10. De Meijer, R.J., Put, L.W., Bergman, R., Landeweer, G., Riezebos, H.J., Schuiling, R.D., Scholten, M.J., Veldhuizen, A., Local variations of outdoor radon concentrations in the Netherlands and physical properties of sand with enhanced natural radioactivity. The Science of the Total Environment, 1985, 45, 101-109.
11. Schuiling, R.D., de Meijer, R.J., Riezebos, H.J., Scholten, M.J., Grain size distribution of different minerals in a sediment as a function of their specific density. Geol.Mijnb., 1985, 64, 199-203.
12. Isbary, G., Das Inselgebiet von Ameland bis Rottumeroog. Archiv der Deutschen Seewarte, 1936, 56, 3.
13. Stolk, A., Koncentratie van zware mineralen in strandafzettingen. Inst.Earth Sc., State Univ. of Utrecht, internal report.
14. Slingerland, R.L., The effects of entrainment on the hydraulic equivalence relationship of light and heavy minerals in sand. J.Sed.Petr., 1977, 47, 2, 753-770.
15. Stapor, F.W., Heavy mineral concentrating processes and density/shape/size equilibria in the marine and coastal dune sands of the Apalachicola, Florida, region. J.Sed.Petr., 1973, 43, 2, 396-407.
16. De Meijer, R.J., Aldenkamp, F.J., Greenfield, M.B., Donoghue, J., private communication.
17. Koning, A., Sedimenten met een hoog gehalte aan zware mineralen op het strand van Vlieland. Natuurwet.Tijdschr., 1947, 29, 197-202.
18. Veenstra, H.J., Winkelmolen, A.M., Size, shape and density sorting around two barrier islands along the north coast of Holland. Geol.Mijnb., 1976, 55, 87-104.
19. Vakgroep Mijntechnologie, Project Noordzee-Memorandum 82-01, 1982, Delft, University of Technology, 56pp.

CHARACTERISTICS OF CHERNOBYL FALLOUT IN THE SOUTHERN BLACK SEA

Hugh D. Livingston and Ken O. Buesseler
Woods Hole Oceanographic Institution
Woods Hole, Massachusetts 02543
USA

and

E. Izdar and T. Konuk
Institute of Marine Sciences and Technology
Dokuz Eylül Üniversitesi
Izmir
TURKEY

ABSTRACT

The input to the Black Sea of fallout radioisotopes from the 26 April 1986 Chernobyl nuclear accident may be used to study both physical and biogeochemical processes in this unusual oceanic setting. We describe measurements of the concentrations, distributions, and particle associations of ^{134}Cs, ^{137}Cs, ^{144}Ce, ^{106}Ru, and transuranic elements in surface waters of the southern Black Sea sampled in June and September 1986. Our measurements indicate that the Chernobyl tracer signal in surface waters of the southern Black Sea is both substantial and widespread - ^{137}Cs concentrations in the range 40-250 Bq/m^3 characterized waters from the Bosporus to the Caucasus. In addition, several vertical profiles of Cs isotopes down to the upper boundary of the anoxic bottom waters are used to address the question of the extent of physical mixing of the Chernobyl tracer signal after delivery to the sea surface. The vertical profile data permit an evaluation of the total Chernobyl Cs isotope deposition to these Black Sea water masses.

A comparison of the relative nuclide composition of surface waters sampled at different locations over time permits preliminary conclusions be drawn on the extent and relative rates of removal of reactive radionuclides from Black Sea surface waters. Other data relevant to these questions come from measurements of the partition of the Chernobyl radiotracers between the filterable suspended particulate and dissolved phases of surface waters.

INTRODUCTION

The Black Sea and the Baltic Sea are the two marine bodies of water geographically closest to the site of the Chernobyl nuclear reactor accident in the U.S.S.R. The accident, starting on 26 April 1986, led to the contamination of much of Europe with a wide variety of radionuclides released to the atmosphere (1,2). In addition to direct deposition, the Black Sea may receive additional input via river run-off to its northwestern corner (Fig. 1) from the heavily contaminated watersheds of the Danube and Dnepr Rivers. In this paper we present data which address the question of the amount and composition of Chernobyl fallout initally delivered to the Black Sea. It is our belief that the data discussed represent fallout delivered directly and that subsequent measurements will permit an assessment of riverine inputs. As a first attempt to characterize the geochemical behavior of Chernobyl-derived radionuclides in the Black Sea, we describe our observed measurements of the distributions of various radionuclides between the filterable and filtered phases of Black Sea surface waters and of changes in the relative composition of the various radionuclides during the first year following the accident.

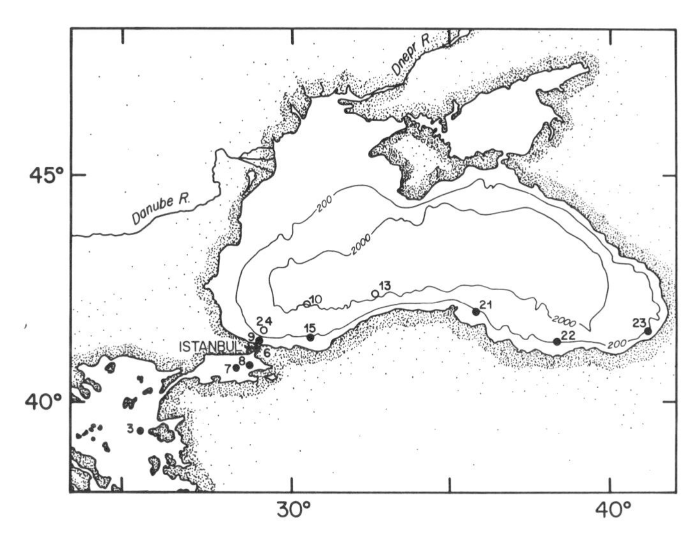

Figure 1. Location map of Black Sea sampling stations. The open circles are off-shelf stations (>200 m), while the filled circles are shelf stations (<200 m). Depth contours are in meters.

The Black Sea has unique properties set by the physical characteristics of its location and water balances. The major features include a shallow brackish surface layer overlying a more saline and anoxic deep water mass. The upper layer derives its freshwater component mostly from the large riverine inputs in the Northwest. The surface circulation is thought to consist of two cyclonic gyres, and the only outflow of surface water occurs through the shallow Bosporos straits. This outflow overrides a subsurface influx of Mediterranean derived water which provides a major ventilation source for the slow renewal of the deeper waters of the Black Sea basin.

Magnitude of Chernobyl Fallout in the Black Sea

^{137}Cs was the major long-lived radionuclide released from the Chernobyl nuclear reactor accident (1). Also, the shorter lived ^{134}Cs ($T_{1/2}$ = 2.06 years) accompanied ^{137}Cs ($T_{1/2}$ = 30.2 years) at activity levels slightly greater than 50% of those of ^{137}Cs. The total deposition of Chernobyl radioactivity to a specific location is to a first approximation assessable through measurement of Cs isotope inventories in soils. In such determinations it is necessary to subtract pre-existing ^{137}Cs present from atmospheric nuclear weapons testing fallout - making use of the absence of ^{134}Cs in the latter and the known and constant $^{134}Cs/^{137}Cs$ ratio (= 0.53) in the Chernobyl release. A number of such soil measurements made throughout Europe (but not including Turkey) were reported by Krey et al. (3).

As we were not aware of any comprehensive assessment of the scale of Chernobyl fallout to the Black Sea, we attempted to evaluate this through Cs isotope measurements in the southern Black Sea. Sampling was carried out in June and September of 1986 from the R/V K. PIRI REIS of the University of Izmir, Turkey. Preliminary measurements and sampling details have already been reported from the June 1986 cruise (4) and from the September 1986 cruise (5). In addition, measurements of material collected in time-series sediment traps have been described (6-8).

Samples for measurement of surface radionuclide concentrations were collected only in the June cruise but, in addition, several vertical profiles were collected and measured in September. The latter permit the determination of the total Cs isotope inventory in the water column at a given sampling location. The former provide an approximate scaling of the total Cs isotope delivery when viewed in the context of hydrographic measurements and of other vertical concentration profiles. The Cs isotopes were collected and measured using two different approaches. The first approach was used in surface water sampling and involved absorption of Cs isotopes from large volumes of water pumped through cotton fiber filter cartridges impregnated with cupric ferrocyanide (4). These filters have high affinities for Cs absorption and the total Cs absorbed is calculated from comparison of activities collected on two serial filters (9). The activities on each filter were measured by direct Ge(Li) gamma-spectrometry using a coaxial detector. Calibration was achieved using filters loaded with standardized amounts of ^{137}Cs and ^{134}Cs. The second approach, used for analyses of 10-liter Niskin bottle samples for vertical profiles, involved radiochemical separation of Cs isotopes and activity measurement by both gamma-spectrometry and low-level beta counting - using standard techniques (10,11).

Data for additional isotopes were obtained by either the direct gamma-spectrometric analysis of ashed MnO_2 cartridges which were placed in line after the Cs absorbers (for ^{106}Ru and ^{144}Ce) or by their wet acid digestion followed by radiochemical transuranic analysis (11,12).

The concentrations of ^{137}Cs, ^{134}Cs, and other nuclides (decay corrected to 1 May 1986) measured in surface water samples in the southern Black Sea in June and September are provided in Table 1. The positions of the sampling stations are shown in Figure 1. The levels of ^{137}Cs found are broadly an order of magnitude higher than the reported pre-Chernobyl range of values (12-16 Bq/m^3) (13). We believe that these surface values are

Table 1. Radionuclides in Black Sea surface water - large volume filtration/absorption data

Table 1a. Total Activities[(a)]

Sta. No.	------- (Bq/m^3) -------				------- (mBq/m^3) -------			
	^{137}Cs	^{134}Cs	^{106}Ru	^{144}Ce	$^{239,240}Pu$	^{238}Pu	^{241}Am	^{242}Cm
14-18 June 1986								
3	45±21	19±8	l.e.	1.6±0.8	22±1	0.3±0.2	0.5±0.1	2.0±0.5
7	41±4	17±2	9.5±1.5	1.6±0.7	8±1	0.3±0.2	0.1±0.3	3.8±1.6
6	64±2	20±1	n.c.	n.c.	n.c.	n.c.	n.c.	n.c.
9	78±2	37±1	l.e.	0.5±0.4	7±3	1.1±0.6	2±9	0.2±0.4
10	193±4	93±2	l.e.	23.8±0.6	17±3	3.9±0.9	2.1±0.7	44±3
13	41±4	19±2	30±6	6.3±0.4	10±2	0.9±0.3	1±5	10±3
15	56±1	27±1	19±4	3.9±0.5	7±1	0.7±0.3	1±4	6±4
7-13 September 1986								
21	118±3	56±1	n.c.	n.c.	n.c.	n.c.	n.c.	n.c.
22	119±2	60±1	n.c.	n.c.	n.c.	n.c.	n.c.	n.c.
23	148±4	70±2	n.c.	n.c.	n.c.	n.c.	n.c.	n.c.
19-23 September 1986								
8	60±2	29±1	16±9	1.7±0.6	8±5	0.5±0.5	0.5±0.5	1±1
10	183±2	92±1	28±1	15.3±0.4	9±1	2.2±0.4	2±1	26±2
24	250±2	126±1	47±2	21±1	12±1	2.8±0.5	1.6±0.5	34±3

Table 1b. Percent particulate (filter activity/total activity)

Sta. No.	^{137}Cs	^{134}Cs	^{106}Ru	^{144}Ce	$^{239,240}Pu$	^{238}Pu	^{241}Am	^{242}Cm
14-18 June 1986[(b)]								
3	0.02±0.01	(c)	l.e.	11±6	9±1	(d)	0±15	0±10
7	0.05±0.01	(c)	6±2	18±18	7±2	(d)	11±22	13±1
6	0.12±0.01	(c)	n.c.	n.c.	n.c.	(d)	n.c.	n.c.
9	0.05±0.01	(c)	l.e.	10±17	3±2	(d)	1±8	0±10
10	0.03±0.01	(c)	l.e.	11±1	3±1	(d)	11±5	19±2
13	0.06±0.01	(c)	2±1	9±1	2±1	(d)	2±5	3±1
15	0.05±0.01	(c)	3±1	9±2	3±1	(d)	0±15	16±10
7-13 September 1986								
21	0.06±0.005	(c)	n.c.	n.c.	n.c.	(d)	n.c.	n.c.
22	0.06±0.005	(c)	n.c.	n.c.	n.c.	(d)	n.c.	n.c.
23	0.07±0.005	(c)	n.c.	n.c.	n.c.	(d)	n.c.	n.c.
19-23 September 1986								
8	0.17±0.03	(c)	0±3	20±16	3±3	(d)	0±15	0±10
10	0.06±0.01	(c)	4±1	11±1	3±1	(d)	17±13	22±4
24	0.07±0.01	(c)	0.9±0.5	9±1	2±1	(d)	3±16	3±1

(a) = Data reported as either Bq/m^3 or mBq/m^3 (= 10^{-3} Bq/m^3), and decay corrected to 1 May 1986. All data reported with a propagated error which includes one sigma counting uncertainties and an uncertainty associated with the large volume cartridge collection efficiencies as described in (12).

(b) = Particulate data corrected to account for differences between June (cotton) and September (polyethylene) wound cartridge filters due to direct adsorption of activity onto the cotton filters (23).

(c) = See ^{137}Cs.

(d) = See $^{239,240}Pu$.

n.c. = Not collected; i.e., at these stations, only Cs cartridge data are available.

l.e. = Poor quality data due to low absorption of Ru by MnO_2 cartridge.

representative of the values throughout the surface mixed layer. Even the June sampling was sufficiently long after the input to have allowed substantial dilution and mixing in this layer. An indication of the extent of vertical penetration through physical mixing and transport can be seen from the profiles of ^{137}Cs concentrations, temperature, and salinity at a station north of the Bosporos (station 24) in September 1986 (Fig. 2). Below the mixed layer, defined by the break in the temperature profile, the levels of ^{137}Cs are consistent with pre-Chernobyl bomb fallout values - a conclusion supported by the absence of ^{134}Cs in any of these deeper samples. No Chernobyl Cs isotopes are therefore detectable in the higher salinity deeper layers representing water entering by the Mediterranean inflow.

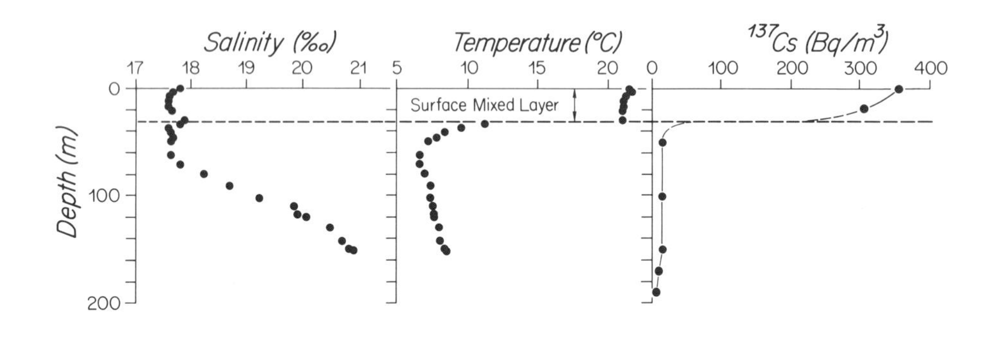

Figure 2. Vertical profiles of salinity (°/₀₀), temperature (°C) and ^{137}Cs (Bq/m^3) at station 24 in September 1986. ^{134}Cs (not shown) levels are zero below the surface mixed layer.

The measured Chernobyl ^{137}Cs inventory at station 24 was 11.2 kBq/m^2 (304 mCi/km^2) in September 1986. The total Chernobyl ^{137}Cs inventory in the upper mixed layer may be compared with the total bomb fallout ^{137}Cs deposition to the latitude of the Black Sea - estimated by conversion of ^{90}Sr deposition data (14) (by multiplication by 1.5) to be 2.69 kBq/m^2 (72.5 mCi/km^2) as of 1 May 1986. Our preliminary assessment, based on the assumption that the surface measurements at all stations can be scaled to mixed layer values (as in Fig. 2), would suggest that ^{137}Cs deposition from Chernobyl in the southern Black Sea was in range of 1-4 times that delivered from fallout from nuclear weapons testing. This estimate also assumes that all the measurements made on the June/September 1986 samples represent direct deposition of Chernobyl fallout to the Black Sea, as opposed to any subsequent additions via riverine input.

The trend with time in surface Cs isotope concentration at some of the sampling locations is also relevant to physical mixing. At most of the southwestern Black Sea stations sampled in both June and September, the concentrations of Cs isotopes were found to have risen sharply and significantly between the two sampling intervals. Our inclination at present is to interpret this change in the context of variability in deposition to the Black Sea. Due to the rapidity of the surface circulation, the water sampled in September is likely to have been at a more northerly location in the Black Sea in June and would have been transported to the sampling positions in the cyclonic gyral circulation. This scenario would imply that deposition in the northern Black Sea may have been higher than in the south. This situation is supported by meteorological models which show higher deposition of Chernobyl fallout to the northern vs. the southern Black Sea (15).

Composition of Chernobyl Fallout in the Black Sea

It has been widely recognized that not only was the depositional pattern of Chernobyl radioactivity highly variable throughout Europe but so also was the isotopic composition of the release over the time course of the accident (1,3,16). Krey et al. (3) addressed this question using data for the isotopic composition of Chernobyl fallout in soil, vegetation and other materials from various sampling locations throughout Europe. These authors divided the isotopic compositional types observed on the basis of whether the nuclide mixture was enriched in refractory nuclides (represented by ^{103}Ru) or volatile nuclides (represented by ^{131}I).

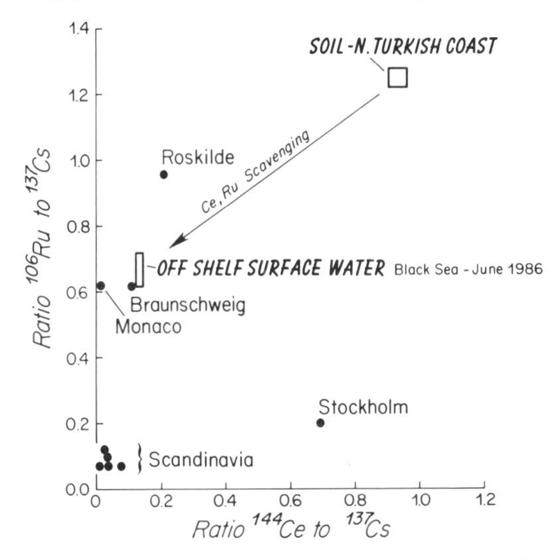

Figure 3. Composition of Chernobyl fallout in the Black Sea region - comparisons with other European locations. See text for discussion.

We attempted to describe the nature of the isotopic composition of Chernobyl radioactivity deposition in the Black Sea using a similar approach. In our approach we plotted in Figure 3 ^{106}Ru and ^{144}Ce data at various points in Europe from Krey et al. (3) normalized as their activity ratio to ^{137}Cs. This approach revealed that differing degrees of enrichment of ^{106}Ru or ^{144}Ce were found at different points, e.g. Stockholm (relatively ^{144}Ce rich), Roskilde, Denmark (relatively ^{106}Ru rich).

We only have Chernobyl radioactivity deposition data from one location in the vicinity of the Black Sea - from Kefken Island on the northwestern Turkish coast (17). These data are plotted in Figure 3 and represent fallout rich in both ^{144}Ce and ^{106}Ru. If this sample is representative of Chernobyl fallout delivered to the Black Sea, it would imply a refractory isotope-rich deposition.

It is not possible to compare this soil sample radionuclide composition directly to fallout detected in the surface waters of the Black Sea. The difficulty lies in the fact that the particle reactive nuclides ^{106}Ru and ^{144}Ce have been exposed to preferential removal processes through particle scavenging. The effect of this is to lower the ratios of ^{144}Ce and ^{106}Ru to ^{137}Cs relative to their deposition ratios. Our ^{106}Ru and ^{144}Ce data are from large volume surface samples pumped through particle filters and manganese dioxide impregnated filter absorbers (4,11). As will be mentioned later, the least affected surface waters sampled with respect to their scavenging history are the offshelf (i.e., from stations at water depths >200 m) samples collected in June. The data for total ^{144}Ce and ^{106}Ru relative to ^{137}Cs in the surface waters at offshelf stations 10 and 13 are plotted in Figure 3 and are found to be depleted in both ^{106}Ru and ^{144}Ce, relative to the soil sample. If the composition of the soil sample was representative of that delivered to the Black Sea, one could conclude that substantial removal of both ^{144}Ce and ^{106}Ru had occurred by the June cruise. The alternate explanation is that these water masses had received less ^{144}Ce and ^{106}Ru-rich fallout than that delivered to the soil of the Northwest Turkish coast. Recently reported data (18) would tend to favor the former hypothesis. Also, data to be presented subsequently on nuclide particle association and removal rates would be more consistent with this scavenging explanation rather than that of input variability.

Since transuranics were not determined in the Kefken Island soil sample, it is not possible to make a similar comparison. However, the high $^{238}Pu/^{239,240}Pu$ ratios and readily measurable ^{242}Cm in several surface water samples in Table 1 are clear indications that transuranics originating from the Chernobyl source had reached the Black Sea.

Reactive Nuclide Particle Association and Removal Rates

The particle reactivity of any element in seawater is reflected by the extent to which it associates with particulate suspended material which in turn is thought to be correlated with the residence time or removal rate of the element from surface waters (19). Our large volume surface samples were first filtered through a 0.5 micrometer wound fiber filter cartridge. This filter cartridge has been used in fallout studies in the Atlantic to measure actinide partition between particulate and dissolved phases (20). The fractions of each nuclide in each phase are measured and calculated using the same approach described by Livingston and Cochran (20). The particulate data obtained for ^{137}Cs, ^{106}Ru, ^{144}Ce, $^{239,240}Pu$, ^{241}Am, and ^{242}Cm are shown in Table 1(b) as the particulate nuclide percentages of the surface water activity found on particles relative to the total activity. In both cruises, and including stations located both on and off the shelf, the median fraction of the total surface radionuclide found on filtered particles increased in order: ^{137}Cs - 0.1%; ^{106}Ru - 3%; $^{239,240}Pu$ - 3%; ^{144}Ce - 9%; ^{242}Cm - 22% (or 3% for ^{242}Cm; see later discussion). This partition pattern indicates that the fallout from

Chernobyl in the Black Sea which we observed had entered into the system and become highly solubilized in a manner consistent with the chemistries of each radioelement. ^{137}Cs, for example, was almost entirely in the soluble phase and, as with the case of bomb fallout ^{137}Cs, will act as a tracer of pure physical mixing and transport processes. At the other extreme, ^{144}Ce and ^{242}Cm exhibited relatively strong particle associations and their geochemistries will therefore be strongly controlled by particle cycling and sinking processes. For comparison, particulate ^{230}Th in much of the Northwest Atlantic Ocean water columns, represents about 15% of the total ^{230}Th concentration (20). Certainly Ce, with redox chemistry involving both ^{III}Ce and ^{IV}Ce, appears to resemble Th in its particle affinity.

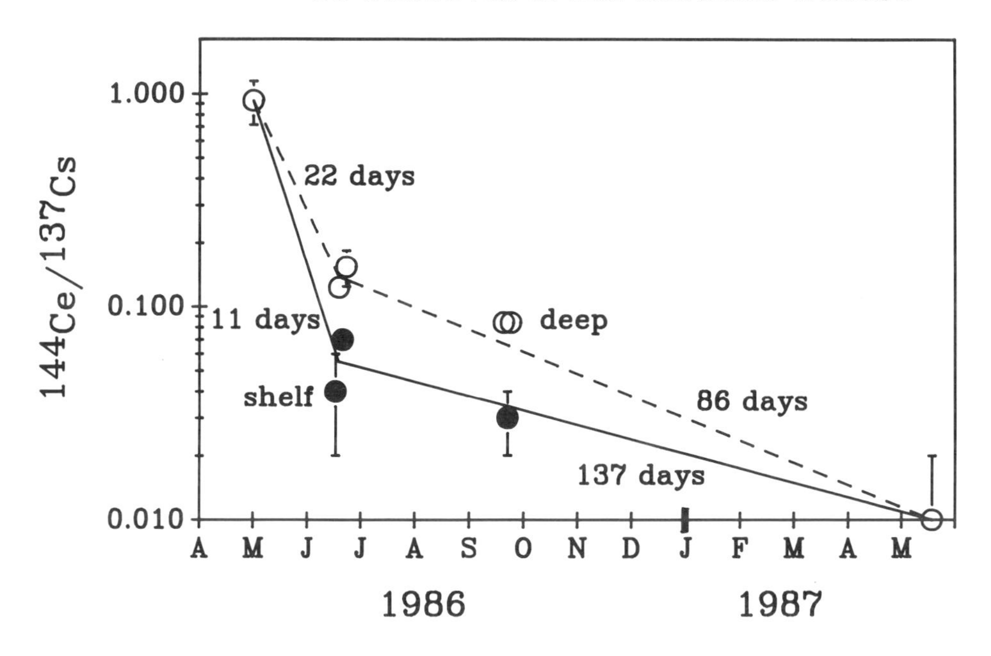

Figure 4. Plot of $^{144}Ce/^{137}Cs$ activity ratios in surface water with time. Open circles represent off-shelf and filled circles shelf samples, respectively. The value for 1 May represents a soil $^{144}Ce/^{137}Cs$ ratio. Half-removal time for ^{144}Ce is calculated and shown for the different time periods and sample types.

A final clue to the differing reactivities of the various Chernobyl radionuclides follows from their associations with sinking particle fluxes (6,7) and the temporal pattern of their removal from the surface of the Black Sea. The latter question is first addressed through data on $^{144}Ce/^{137}Cs$ ratios in surface water - plotted in Figure 4. Shown in this figure are the ratios of these nuclides to ^{137}Cs measured in the soil sample from the Northwest Turkish coast (17) and in surface waters in the southwestern

Black Sea sampled in June and September 1986, and preliminary data from May 1987. The soil ^{137}Cs data were adjusted to subtract pre-Chernobyl fallout ^{137}Cs - using a $^{134}Cs/^{137}Cs$ ratio of 0.53, and all data are decay corrected to 1 May 1986. The ratios in the soil are taken to represent the input ratios to the Black Sea from Chernobyl fallout. The use of the nuclide ratio of ^{144}Ce normalized to ^{137}Cs, removes the effects of physical mixing with respect to isotope concentration changes. The resultant ratio changes thus represent either removal through scavenging or input compositional variability. For the purposes of this discussion, we assume that the input isotopic composition to the Black Sea is constant. Using a vertical scavenging model with first-order removal kinetics, the removal rates can be calculated from the slope of the $^{144}Ce/^{137}Cs$ ratio change with time ($\partial N/\partial t = -kt$, where k =removal rate and ln2/k = half-removal time).

The first point to note is the pattern of lower ratios on samples collected at stations on the shelf of the southern Black Sea. This trend is in keeping with the open ocean/coastal ocean trend reported for many other reactive species such as ^{210}Pb, ^{234}Th, ^{238}Th, Pu, etc. (e.g., for ^{228}Th see Li et al. (21)). The trend is caused by the combination of higher productivity and suspended particle concentrations on the continental shelves which results in the enhanced scavenging of particle reactive elements in coastal regions.

Secondly, during the five months following input to the Black Sea of fallout from Chernobyl, substantial removal of the reactive ^{144}Ce is implied by the observed reduction in its ratio to ^{137}Cs in these surface waters. The ratios for $^{144}Ce/^{137}Cs$ over the 150-day interval between May and September 1986 vary by a factor of roughly 10 and 30 for the deep and shelf locations, respectively. If the assumptions above are correct, ^{144}Ce removal would appear to be a two component process - an initially rapid step with half-removal times of 11 and 22 days for shelf and deep stations respectively, followed by slower removal rates with about three to four month half-removal times. There are too little data to permit elucidation of the nature of this apparent two stage removal process. Obvious candidates to consider are the effects of spring coccolith blooms (8) or differences in the rate of ^{144}Ce removal related to redox chemistry, i.e. faster initial removal of ^{IV}Ce, for example.

The continued scavenging of Ce with time is supported by our most recent ^{144}Ce surface water data from May of 1987, approximately one year after the Chernobyl fallout inputs. Using the MnO_2 cartridge technique and gamma spectrometry, essentially no ^{144}Ce could be detected ($^{144}Ce/^{137}Cs = 0.1 \pm 0.1$). This fits the 1986 summer trends and results in a calculated half-removal time of 87-137 days at all sites.

The corresponding change for $^{106}Ru/^{137}Cs$ is a little less than a factor of 2 between May and September 1986. The behavior of the ^{106}Ru in the Black Sea will obviously be closely dependent on its oxidation state and chemical speciation. Our data do not speak directly to this question, but clearly the chemical form cannot be such as would cause particle association and removal as rapidly as would be found for trivalent or tetravalent actinides or lanthanides. Preliminary data for May 1987 suggest that the ratio change $^{106}Ru/^{137}Cs$ since September 1986 is not more than a further factor of 2.

The few transuranic data in Table 1 may be viewed in like manner for their removal patterns from surface water. Only the Pu and Cm data had analytical uncertainties which were of satisfactory quality. As no transuranic analyses were made on the Kefken Island soil sample (17), changes in observed ratios relative to soil values could not be made. $^{242}Cm/^{137}Cs$ ratios declined from 0.23×10^{-3} in June (stations 10,13) to 0.14×10^{-3} in September (stations 10,24). The direction and magnitude of this change is not unlike that noted above for ^{144}Ce and would result in a calculated half-removal time of 124 days. Curiously, ^{242}Cm partitioning between dissolved and particulate phases of surface water varied more than was found for ^{144}Ce. Although total $^{242}Cm/^{137}Cs$ ratios in surface water at stations 13 and 24 were similar to those found at station 10, much smaller fractions of ^{242}Cm were associated with filterable particulates at stations 13 and 24 - ~3% at stations 13 and 24 as opposed to 19% and 22% at the two occupations of station 10. Particulate data for ^{242}Cm at the other stations were mostly accompanied by rather large analytical uncertainties. These low values appear anomalous and we presently do not have an explanation to offer. It is worth noting that a low value for ^{106}Ru partition was also observed at station 24.

The fraction of surface $^{239,240}Pu$ associated with a filterable phase was observed to range from 2-3% for the samples in the Black Sea. The higher values at stations 3 and 7 lie in the Aegean and Sea of Marmara. These values are very similar to the oceanic partition observed for Pu derived from atmospheric weapons testing fallout in the western North Atlantic Ocean (20). The higher particle association observed for ^{242}Cm over Pu in Black Sea water parallels the higher values for ^{241}Am relative to Pu in the western North Atlantic (20). This trend is in accord with the strong chemical similarity between Am and Cm and their stronger particle associations relative to Pu.

A significant fraction of the observed surface Pu concentrations in the Black Sea can be calculated to derive from Chernobyl fallout on the basis of the observed $^{238}Pu/^{239,240}Pu$ ratios of 0.09-0.24. This can be estimated assuming a two source model of Pu input, with a $^{238}Pu/^{239,240}Pu$ end member ratio of 0.036 and 0.55 (22) for global weapons testing and Chernobyl fallout, respectively. At station 10 for example, in September 1986 one calculates that approximately 40% of the Pu in the surface waters originated from the Chernobyl souce. All of the observed ^{242}Cm has to derive to from Chernobyl as it is not present in atmospheric weapons testing fallout (due to its short half-life of 0.4 years).

On the basis of the observed distributions of these transuranic elements between particle and solution phases of Black Sea surface water it may be concluded that the transuranics deposited there from the Chernobyl accident have entered the biogeochemical cycle in a similar fashion to transuranics from global fallout. This rapid entry of Chernobyl nuclides into biogeochemical cycles is in fact noted for all the nuclides we have observed and hence allows them to be used as tracers of their stable elemental analogues in those cycles.

SUMMARY

Measurements of ^{137}Cs, ^{134}Cs, ^{144}Ce, ^{106}Ru, $^{239,240}Pu$, ^{241}Am, and ^{242}Cm introduced to waters of the southern Black Sea are described. Surface ^{137}Cs concentrations in June and September 1986 are in the range 40 to 250 Bq/m^3 - between one or two orders of magnitude greater than the pre-existing bomb fallout ^{137}Cs signal. At that time, Chernobyl Cs isotopes were confined to the surface mixed layer. Inventory estimates suggest that the Chernobyl ^{137}Cs input may range from one to four times that from nuclear weapons testing fallout to the region. The composition of the Chernobyl fallout delivered to the Black Sea appears relatively enriched in the refractory radionuclides, such as ^{144}Ce and ^{106}Ru. The behavior of the various radionuclides appears predictable on the basis of their known elemental geochemical behavior. The reactivities with respect to surface water particle association increase in order ^{137}Cs (0.1%), ^{106}Ru (3%), $^{239,240}Pu$ (3%), ^{144}Ce (9%), and ^{242}Cm (20%). Rates of removal from surface water would be in the reverse order to this sequence, with higher removal rates being found in shelf as opposed to deep water locations.

ACKNOWLEDGEMENTS

We would like to thank S. A. Casso, M. C. Hartman, W. R. Clarke and L. D. Surprenant for their help in the laboratory analyses, and M. R. Hess for her assistance in typing this manuscript. Samples were collected in the Black Sea aboard the R/V K. PIRI REIS from the University of Izmir in Turkey; we thank the officers and crew for their assistance.

Our Chernobyl Black Sea program has involved cooperation and assistance from E. Izdar's lab at the University of Izmir, E. Degen's lab at the University of Hamburg, and S. Honjo's lab at the Woods Hole Oceanographic Institution. Financial support for our work came from the Coastal Research Center in Woods Hole, the U.S. Office of Naval Research (contract N00014-85-C-0715), and primarily from the National Science Foundation (contract OCE-8700715).

This is Woods Hole Oceanographic Institution Contribution No. 6800.

REFERENCES

1. Hohenemser, C., Deicher, M., Ernst, A., Hofsäss, H., Lindner, G. and Recknagel, E., Chernobyl: An early report. Environment, 1986, 28(5), 6-13 and 30-43.

2. Buesseler, K.O., Chernobyl: Oceanographic studies in the Black Sea. Oceanus, 1987, 30(3), 23-30.

3. Krey, P.W., Klusek, C.S., Sanderson, C., Miller, K. and Helfer, M., Radiochemical characterization of Chernobyl fallout in Europe. Report EML-460, Environmental Measurements Laboratory, U.S. Dept. of Energy, New York, 1986, pp. 155-213.

4. Livingston, H.D., Clarke, W.R., Honjo, S., Ezdar, E., Konuk, T., Degens, E. and Ittekot, V., Chernobyl fallout studies in the Black Sea and other ocean areas. Report EML-460, Environmental Measurements Laboratory, U.S. Dept. of Energy, New York, 1986, pp. 214-223.

5. Livingston, H.D. and Buesseler, K.O., Chernobyl tracers in the Black Sea. Eos, 1986, 67(44), 1070.

6. Buesseler, K.O., Livingston, H.D., Honjo, S., Hay, B.J., Manganini, S.J., Degens, E.T., Ittekkot, V., Izdar, E. and Konuk, T., Chernobyl radionuclides in a Black Sea sediment trap. Nature, 1987, 329. 825-828.

7. Buesseler, K.O., Livingston, H.D., Honjo, S., Hay, B.J., Degens, E.T. and Izdar, E., The annual record of Chernobyl radionuclides in a deep Black Sea sediment trap. Eos, 1987, 68(50), 1771-1772.

8. Honjo, S., Hay, B.J., Manganini, S.J., Asper, V.A., Degens, E.T., Kempe, S., Ittekkot, V., Izdar, E., Konule, Y.T. and Benli, H.A., Seasonal cyclicity of lithogenic particulate fluxes at a southern Black Sea sediment trap station. In Particle Flux in the Ocean, eds. E.T. Degens, E. Izdar and S. Honjo, SCOPE UNEP Sonderband, Vol. 62, Mitt. Geol.-Palaont. Inst., Univ. Hamburg, Hamburg, Germany, 1987, pp. 19-40.

9. Mann, D.R. and Casso, S.A., In situ chemisorption of radiocesium from seawater. Marine Chem., 1984, 14, 307-318.

10. Wong, K.M., Radiochemical procedures for analyses of Sr, Sb, lanthanide, cesium and plutonium isotopes in seawater. In Reference Methods for Marine Radioactivity Studies I, Technical Report Series No. 118, IAEA, Vienna, 1970, pp. 119-127.

11. Buesseler, K.O., Casso, S.A., Hartman, M.C. and Livingston, H.D., Determination of fission-products and actinides in the Black Sea following the Chernobyl accident. Second Intern'l. Conf. on Low Level Measurements of Actinides and Long-Lived Radionuclides in Biological and Environmental Samples, Akita City, Japan, 16-20 May 1988 (in preparation).

12. Livingston, H.D. and Cochran, J.K., Determination of transuranic and thorium isotopes in ocean water: in solution and in filterable particles. J. Radioanal. Nucl. Chem., 1987, 115(2), 299-308.

13. Vakulovskiy, S.M., Krasnopevtsev, Y.U., Nikitin, A.I. and Chumichev, V.B., Distribution of ^{137}Cs and ^{90}Sr between water and bottom sediments in the Black Sea, 1977. Oceanology., 1982, 22, 712-715.

14. Larsen, R.S., Worldwide deposition of ^{90}Sr through 1982. Report EML-430, Environmental Measurement Laboratory, U.S. Dept. of Energy, New York, 1984, Table 7.

15. World Health Organization, Assessment of Radiation Dose Commitment in Europe due to the Chernobyl Accident, eds. E. Wirth, N.D. van Egmond and M.J. Suess, ISH-Heft 108, Institut für Strahlenhygiene des Bundesgesundheitsamtes, München, 1987.

16. IAEA Board of Governors, INSAG Summary Report on the Post-Accident Review Meeting on the Chernobyl Accident. GOV/2268, IAEA, Vienna, 1986, 140 pp.

17. Krey, P. W. Personal communication, 1987.

18. Livingston, H.D., Buesseler, K.O. and Casso, S.A., Biogeochemical scavenging of Chernobyl ^{106}Ru in oxic waters of the Black Sea. Eos, 1987, 68(50), 1771.

19. Santschi, P.H., Adler, D., Amdurer, M., Li, Y.-H. and Bell, J.J., Thorium isotopes as analogues for "particle-reactive" pollutants in coastal marine environments. Earth Planet. Sci. Lett., 1980, 47, 327-335.

20. Cochran, J.K., Livingston, H.D., Hirschberg, D.J. and Surprenant, L.D., Natural and anthropogenic radionuclide distributions in the Northwest Atlantic Ocean. Earth Planet. Sci. Lett., 1987, 84, 135-152.

21. Li., Y.-H., Santschi, C.H., Kaufman, A., Benninger, L.K. and Feely, H.W., Natural radionuclides in waters of the New York Bight. Earth Planet. Sci. Lett., 1981, 55, 217-228.

22. Aarkrog, A., The radiological impact of the Chernobyl debris compared with that from nuclear weapons fallout. J. Environ. Radioactivity., 1988, 6, 151-162.

23. Livingston, H.D., Mann, D.R., Casso, S.A., Schneider, D.L., Surprenant, L.D. and Bowen, V.T., Particle and solution phase depth distributions of transuranics and ^{55}Fe in the North Pacific. J. Environ. Radioactivity, 1987, 5, 1-24.

SPATIAL AND TEMPORAL VARIATIONS IN THE LEVELS OF CESIUM IN THE NORTH WESTERN MEDITERRANEAN SEAWATER (1985-1986)

Calmet D., Fernandez J.M., Maunier P., Baron Y.

Commissariat à l'Energie Atomique
Institut de Protection et de Sureté Nucléaire
Département d'Etudes et de Recherches en Sécurité
Service d'Etudes et de Recherches sur l'Environnement
Station Marine de TOULON
BP n°330 83507 LA SEYNE sur MER
France

ABSTRACT

The explosion at the Chernobyl 4 reactor led to the discharge into the atmosphere of ^{137}Cs, the fallout was clearly discernible in the surface waters of the north western part of the Mediterranean coast, although there was no significant health hazard. This radioelement was used as a tracer to plot the stratification and dispersal of water from the Rhône at local and regional levels. It appears to cross the Golfe du Lion along a north east - south west axis.

Vertical salinity and temperature profiles plotted with a CTD probe enabled a distinction to be made between the various masses of water present in the Golfe du Lion. There is a difference in the levels of ^{137}Cs activity in the surface and deep water, as measured at reference points. In September 1986, the surface water only showed traces to a depth of 50 m, whereas the whole of the continental shelf water showed traces in December 1986. Outside the continental shelf at the same period, there was no noticeable increase in the ^{137}Cs levels in water at greater depths.

INTRODUCTION

Since 1977, the Commissariat à l'Energie Atomique's Environmental Studies and Research Service at the TOULON Marine Station has been working on the processes involved in the dispersal of natural or anthropogenic radioelements in the Mediterranean sea, as part of a general research programme (**RADMED** programme , from **RAD**ioactivity and **MED**iterranean sea).

Samples of seawater are taken annually from the north west Mediterranean basin. Until the first three months of 1986, the results were largely used to monitor the changes in ^{137}Cs isotopes levels in the water from atmospheric fallout as a result of weapons testing between 1955 and 1979, in the northern hemisphere (1). Unlike the latter type of fallout, which is referred to as global, the explosion at the Chernobyl reactor on 26 April 1986 led to the release of radioelements at

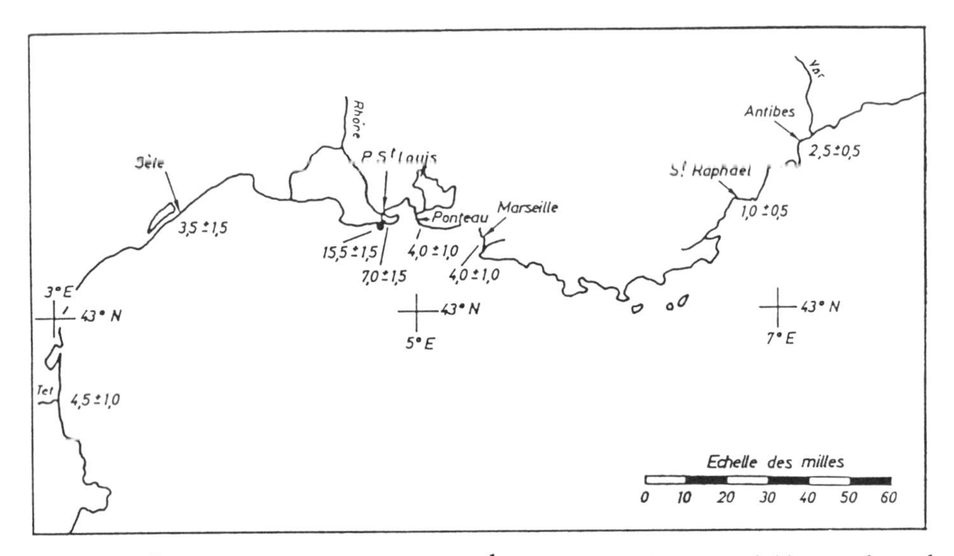

Figure 1A :^{137}Cs activity expressed in mBq.l^{-1} in sea water filtered to 0.45 um along the Mediterranean coast in February and April 1986 (LITTORAL86A).

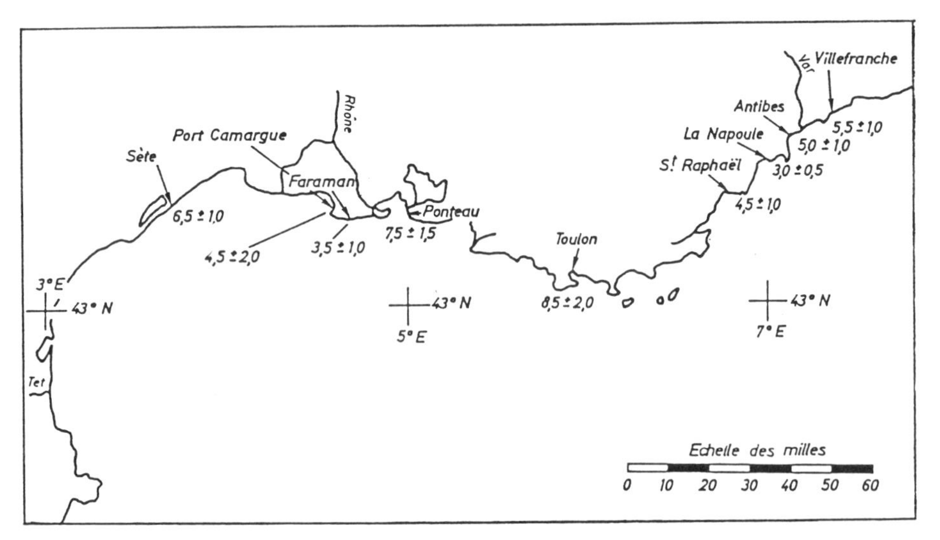

Figure 1B : ^{137}Cs activity expressed in mBq.l^{-1} in sea filtered to 0.45 um along the Mediterranean coast in November 1986 (LITTORAL86B).

a specific point in time, with a fraction of the fallout reaching the eastern part of the north-west Mediterranean basin (2). Chcrnobyl fallout deposits assesment were led from aerosols and soils measurements and from atmospheric models results. These cesium latest deposits, which can be pinpointed to a specific time and place, can therefore be regarded as tracer elements, with a specific $^{137}Cs/^{134}Cs$ close to 2.0, which can be plotted to determine the dynamics of the waters of the north-west Mediterranean basin.

The radioelements released by the Chernobyl accident reached the marine environment via two routes: a direct route, whereby the atmospheric fallout fell to surface of the sea, and an indirect route via the rivers, after surface deposits on the soil had been washed away by rainwater. On the French section of the Mediterranean coast, two rivers of unequal importance flow out to sea: the Var and the Rhône. However, the waters of the Rhône carry not only the radioelements arising from the Chernobyl accident but also those associated with the discharge of low-level radioactive liquid effluents from various nuclear plants such as cesium isotopes, with a $^{137}Cs/^{134}Cs$ ratio greater than 4.

The monitoring of radioelements arising from atmospheric fallout on the surface of the sea enables seasonal vertical changes to be detected, at regional level, in the various bodies of water whereas the monitoring of radioelements discharged by rivers leads to a definition of the limits of geographical areas under the influence of the deposits of coastal rivers.

EQUIPMENT AND METHOD

Radioactive tracers have been monitored during 7 oceanographic cruises in collaboration with research teams from the CNRS and the University. The cruises enabled us to cover a network of sampling points all over the French Mediterranean coast. After establishing the coastal area by means of 25 sampling points in the course of the CORSE86 (86.09.23-86.10.06), LITTORAL86A (86.02.03-10, 86.04.03-11, 86.04.21-29) and LITTORAL86B (86.11.05-13) cruises, efforts were concentrated in particular on the area reached by the waters of the Rhône, as defined by 15 sampling points during the DYPOL01 (86.08.28-86.09.07) and DYPOL02 (86.11.19-30) cruises, and the area of the Golfe du Lion covered by 24 sampling points as part of the PELAGOLION01 (86.09.02-11) and PELAGOLION02 (86.12.04-12) cruises.

Over the network of sampling points, volumes of water between 90 and 120 litres were systematically pumped from a depth of 0.5 m. At some points selected on the basis of results of vertical contours of salinity, temperature or turbidity obtained in real time with a bathymeter, the various bodies of water identified between the surface and 1700 m were sampled using 30 l Niskin bottles.

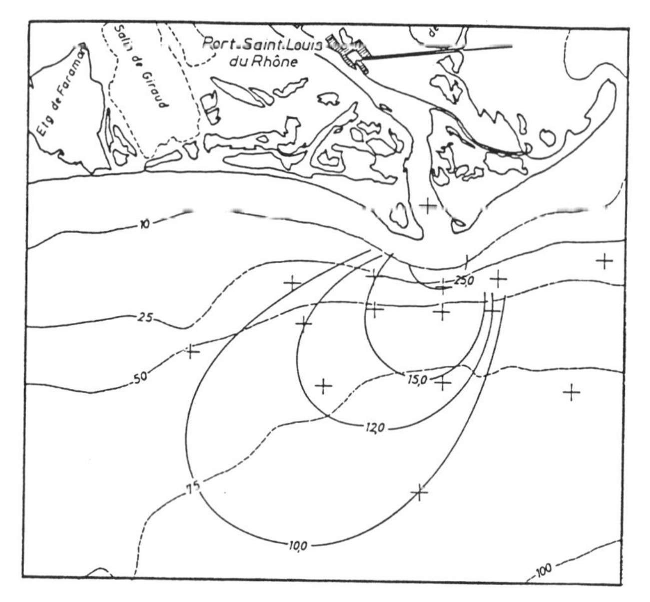

- Figure 2 : Diagram plotting lines of ^{137}Cs isoactivity expressed in mBq.l^{-1} in sea water filtered to 0.45 um within the area affected by water from the Rhône, for the period of August and September 1986 (DYPOL01).

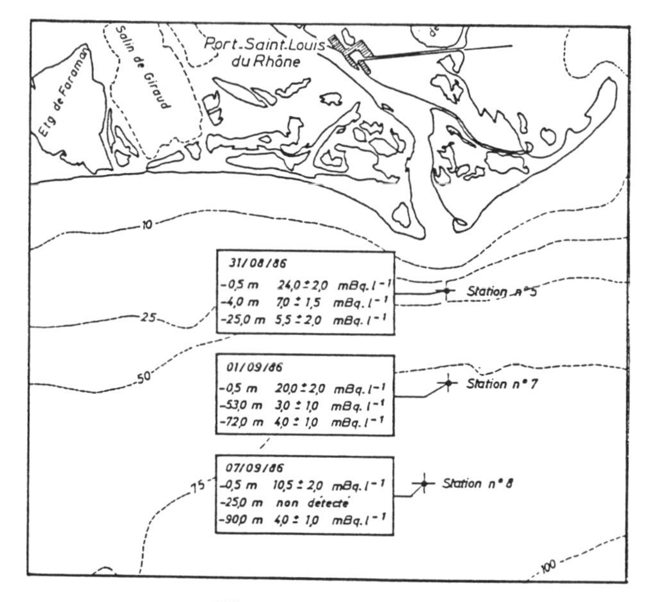

- Figure 3 : Vertical distribution of ^{137}Cs activity expressed in mBq.l^{-1} of sea water filtered to 0.45 um on the basis of 3 sample levels taken at 3 sampling points within the area affected by water from the Rhône (DYPOL01).

All of the seawater samples were filtered (0.45 μm) in order to remove radioelements associated with the particles, the quantity of which varied according to the area studied, especially in the Rhône estuary area where cesium isotopes are mainly linked to suspended matter.

Gamma-emitting radioelements were measured by radiochemical methods described elsewhere (3,4).

RESULTS AND DISCUSSION

The results of our research are presented according to two areas: the coastal fringe subjected to river pollution, and the area defined by the continental shelf of the Golfe du Lion, where the river deposits are diluted.

The coastal fringe

The results of seawater sampling conducted between February and April 1986 (LITTORAL86A), at less than 2 nautical miles from the coast, reveal the presence of a single artificial radioelement: ^{137}Cs. ^{134}Cs was not detected. Levels in the region of 1 mBq/l are typical of the eastern part of the coast whereas the west has slightly higher readings in the region of 4 mBq/l emphasing the action of the waters of the Rhône in this part of the coast (Figure 1A) where ^{134}Cs could also be detected ($^{137}Cs/^{134}Cs$ between 4 to 9).

In November 1986 (LITTORAL86B), ^{137}Cs activity levels were higher all along the coast, varying from 3 to 9 mBq/l with a $^{137}Cs/^{134}Cs$ from 2.3 to 4.0 , eliminating the differences between east and west as found in previous surveys (Figure 1B). Thus the Chernobyl fallout is clearly discernible, especially in the eastern part of the French Mediterranean coast. These observations are in line with those results obtained from atmospheric and land sampling (2).

During these two cruises, the coastal area directly under the influence of water from the Rhône appears to have a higher ^{137}Cs content. A detailed study of the area in August, September and November 1986 during the DYPOL 01 and 02 cruises confirmed the presence of a decreasing spatial gradient of ^{137}Cs levels from the estuary towards the open sea (Figure 2). The Rhône water area of dilution corresponds to a situation classically described as a fundamental current flow, under the influence of the Ligurian-Provenal current (5).

The vertical distribution of ^{137}Cs at 3 levels defined as a function of turbidity contours, at 3 sampling points distributed along a radial perpendicular to the coast, reveals the predominance of ^{137}Cs in the surface waters (Figure 3). The 3 sampling points display a marked discontinuity between the surface level of activity and the intermediate and deep levels. The discontinuity corresponds to the surface dispersal of the waters of the Rhône, which are less dense than the receiving seawater.

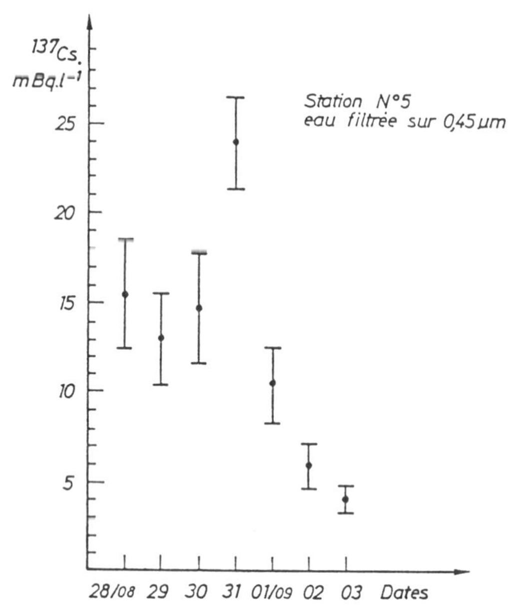

- Figure 4 : Changes over a period of time in the ^{137}Cs activity expressed in mBq.l^{-1} of sea water filtered to 0.45 um sampled at station 5 located within the area affected by water from the Rhône (DYPOL01).

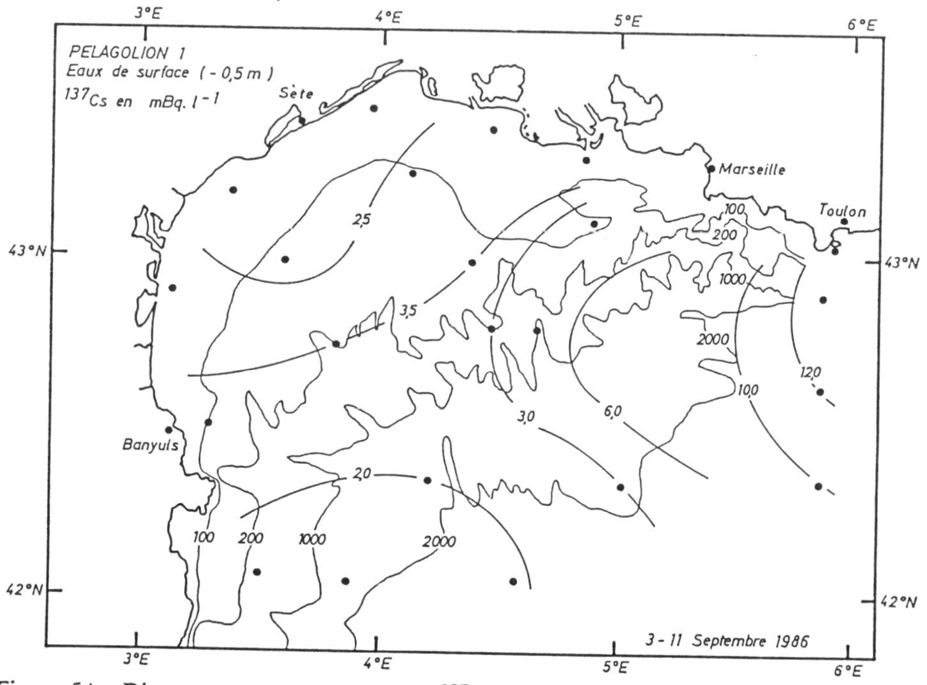

- Figure 5A : Diagram plotting the lines of ^{137}Cs isoactivity expressed in mBq.l^{-1} in surface sea water filtered to 0.45 um sampled in the Gulf of Lion in September 1986 (PELAGOLION01).

Local spatial heterogeneity, both horizontal and vertical, is accompanied by daily temporal variations which we detected at sampling point 5. Between 28.08.86 and 03.09.86, the ^{137}Cs concentration of the surface waters, under the permanent influence of Rhône deposits, fluctuated between 4 and 24 mBq/l (Figure 4). This spread of values reflects both the variations in deposits over the period of time in question and the variable rates at which freshwater and seawater mix.

The Golfe du Lion

In September 1986, levels of ^{137}Cs in the surface waters of the continental shelf of the Golfe du Lion varied from 1.5 to 6.5 mBq/l with an average of 3.0 mBq/l and a $^{137}Cs/^{134}Cs$ between 2.5 to 5.0. These figures were much lower than those obtained at sampling points at the eastern boundary of the gulf, on average 12 mBq/l (4.5 mBq/l of ^{134}Cs), but slightly higher than those at the southern boundary which were equal to 2.5 mBq/l (1.0 mBq/l of ^{134}Cs) on average (Figure 5A).

The decreasing spatial gradient from east to west seems to correspond to a stock of cesium carried by waters in the Ligurian-Provenal geostrophic current coming from the north-west Mediterranean basin. The absence of major rainfall prior to the dates of the sampling definitely restricted the amount of soil washed away into the Rhône and thus the quantity of ^{137}Cs the latter was carrying. The levels recorded in the Gulf must therefore have been chiefly due to direct atmospheric fallout onto the surface waters in the eastern part of the north-west basin.

At the time of the sampling operation in December 1986 (Figure 5B), the mean ^{137}Cs and ^{134}Cs activities of the surface waters of the continental shelf were higher: 7.0 mBq/l, for activity levels between 6.0 and 8.0 mBq/l of ^{137}Cs and 1.0 to 2.5 mBq/l for ^{134}Cs. A central strip running from north east to south west, joining the Rhône delta to the Côte Vermeil, had the highest ^{137}Cs readings which were constantly in the region of 8.0 mBq/l (Figure 5B). These findings are to be compared with those of ALAIN (6), MINAS (7) and TOURNIER (8). The higher volume of deposits in the waters of the Rhône after the autumn rains and the increase of low level industrial releases made it possible to track the Rhône water across the gulf. Sampling points at the east and south boundaries of the continental shelf recorded the lowest values, between 2 and 6 mBq/l of ^{137}Cs.

As regards sampling points on the Golfe du Lion continental shelf, the vertical distribution of ^{137}Cs and ^{134}Cs readings gave the highest values from the surface to a depth of 70 m in September 1986 and to a depth of 200 m in December 1986. The increase over a period of time in the depth of the layer of surface water characterised by high ^{137}Cs levels and appearance of ^{134}Cs reflects the gradual spread of these elements towards the sea bed.

Outside the continental shelf, at sampling points located at depths of 1700 m, ^{137}Cs was not detected in the deepest bodies of water (Figure 6). The dispersal of dissolved ^{137}Cs therefore seems to be relatively slow as, 8 months after the introduction of ^{137}Cs into the surface waters, the fraction containing dissolved ^{137}Cs had not noticeably increased in the deepest layers.

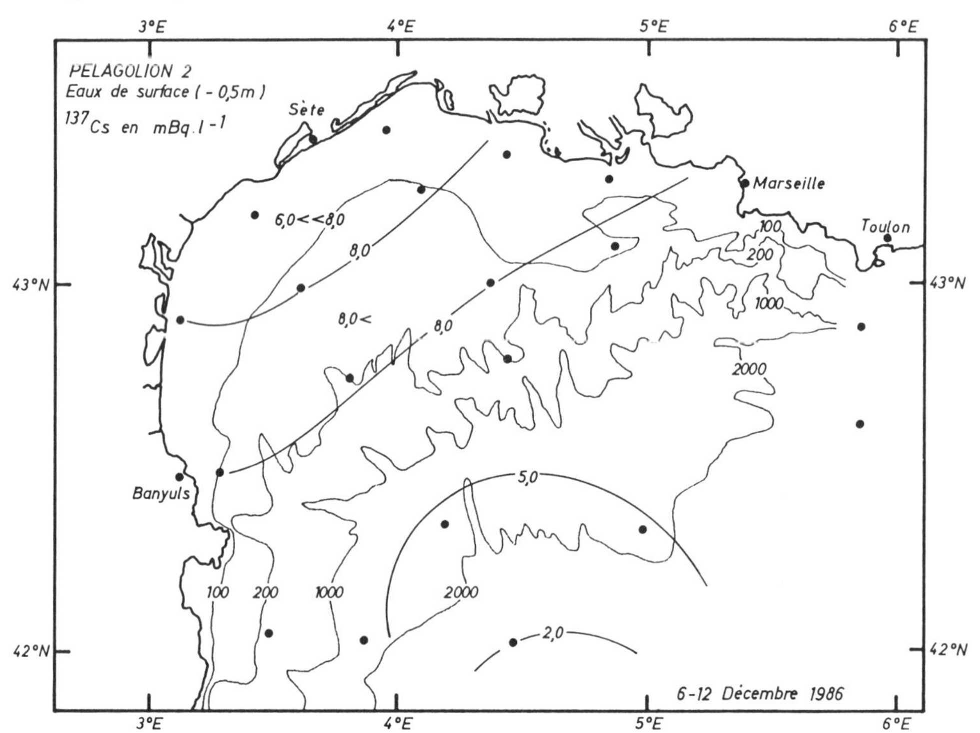

Figure 5B : Diagram plotting the lines of ^{137}Cs isoactivity in mBq/l of surface sea water filtered to 0.45 μm sampled in the Golfe du Lion in December 1986 (PELAGOLION 02).

CONCLUSION

On 26 April 1986, the explosion at the Chernobyl 4 reactor led to the injection of cesium isotopes into the atmosphere of the northern hemisphere in sufficient quantities to result in atmospheric concentrations of that element reaching a peak of 10 Bq/m^3 in Provence between 1 and 5 May 1986 (2). Atmospheric fallout to the surface waters of the eastern part of the north-west Mediterranean basin increased the ^{137}Cs levels in the water nearest to the coast in that area by a factor of 3. Initially, this led to homogenization of ^{137}Cs levels in the waters of the French Mediterranean coast, excluding the area affected by waters from the Rhône.

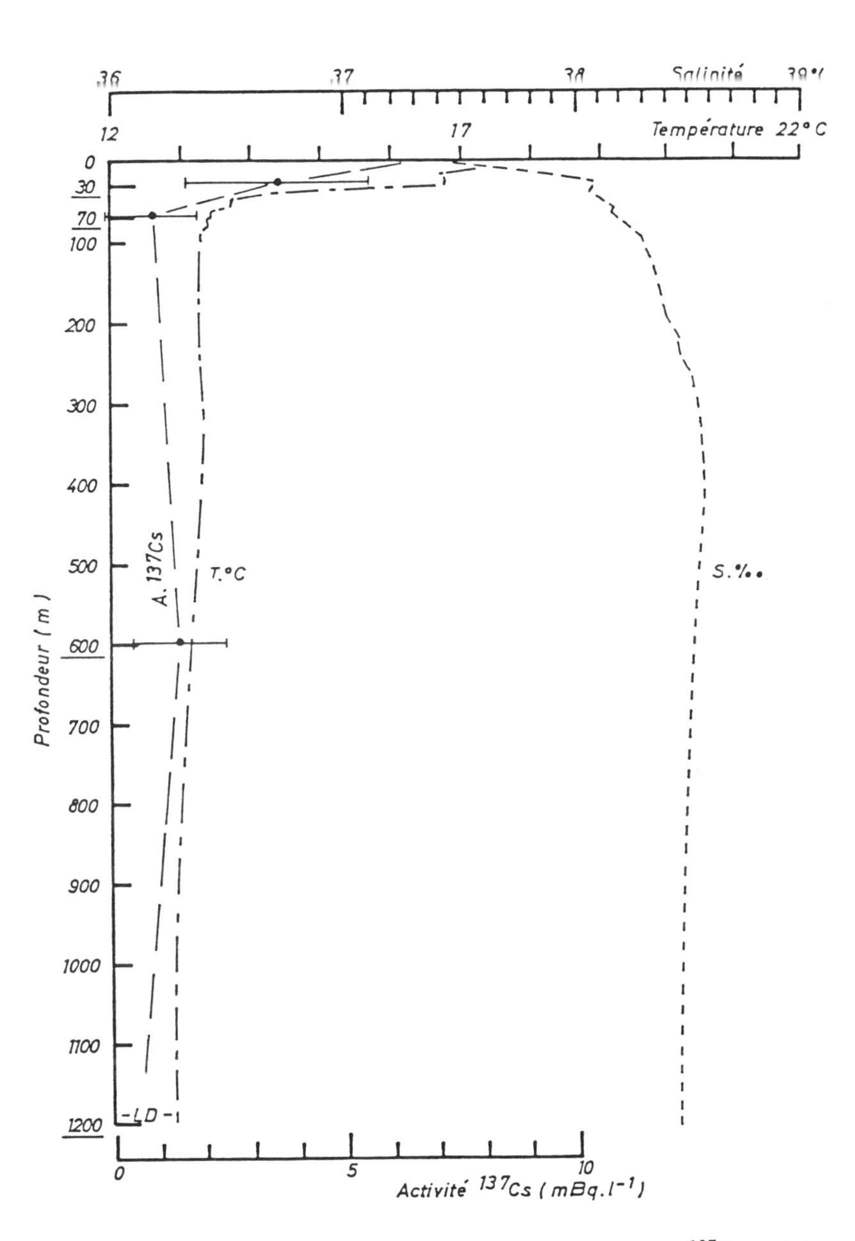

- Figure 6 : Vertical change in the salinity, temperature and ^{137}Cs activity expressed in $mBq.l^{-1}$ of sea water filtered to 0.45 um at sampling point n° 101 in the Gulf of Lion on 86/09/03 (PELAGOLION02).

A study of the area under the direct influence of deposits from the waters of the Rhône showed major local stratification together with a ^{137}Cs surface dilution gradient from the coast towards the open sea. Sampling of Rhône water enabled us to produce a map of how the water was dispersed in the Golfe du Lion, which it crosses along a north east - south west axis.

Atmospheric and earth contamination provided a reservoir of cesium in the surface waters, vertical penetration being progressive. The layer of water nearest the surface was in fact contaminated to a depth of 50 m in September 1986, and 200 m in December after the seasonal thermocline has disappeared.

We would like to express our special thanks to Mr. G. JACQUES and Mr J.C.ALOISI, who headed the PELAGOLION and DYPOL cruises, and to all members of crews of the R.V. Catherine LAURENCE, Le NOROIT and Le SUROIT.

REFERENCES

1. **United Nations**, Sources and effects of ionising radiation, UNSCEAR,1984, 781.

2. **Commissariat à l'Energie Atomique**, Institut de Protection et de Sureté Nucléaire, L'accident de Tchernobyl, Rappport IPSN 2/86 rev. 3, Octobre 1986, 163.

3. **Foulquier L., Philippot J.C., Baudin-Jaulent Y.**, Métrologie de l'environnement. Echantillonnage et préparation d'organismes d'eau douce. Mesure des radionucléides émetteurs gamma. Rapport CEA/R-5164, 1982.

4. **Calmet D.**, Synthèse radioécologique des différents compartiments de l'environnement marin du Cotentin. Thèse de Doctorat d'Etat Aix-Marseille II, 1986, 254.

5. **Blanc F., Leveau M.**, Plancton et eutrophie: aire d'épandage rhodanienne et Golfe de Fos (Traitement mathématique des données). Thèse de Doctorat d'Etat, Université d'Aix-Marseille II, 1973, 681.

6. **Allain C.**, Topographie dynamique et courants généraux dans le bassin occidental de la Méditerranée (Golfe du Lion, Mer Catalane, Mer d'Alboran et ses abords, secteur de la Corse), Rev. Trav. Inst. Pêches Marit., 24, 1960, 121-145.

7. **Minas H.J.**, Recherche sur la production organique primaire dans le bassin méditerranéen nord-occidental. Rapport avec les phénomènes hydrologiques. Thèse de Doctorat d'Etat, Université d'Aix-Marseille II, 1968, 227.

8. **Tournier H.**, Hydrologie saisonnière du Golfe du Lion en hiver, Rev. Trav. Inst. Pêches Marit., 33, 1969, 265-300.

THE CONTAMINATION OF THE NORTH SEA AND BALTIC SEA BY THE CHERNOBYL FALLOUT

H. Nies, Ch. Wedekind
Deutsches Hydrographisches Institut
Bernhard-Nocht-Str. 78
D-2000 Hamburg 4
Federal Republic of Germany

ABSTRACT

The German Hydrographic Institute (DHI) investigated the Chernobyl Fallout in the North- and the Baltic Sea. After the radioactive cloud has reached the southern North Sea at 3 May and the western Baltic at 5 May, a great deal of different radionuclides could be determined in the marine environment, such as Cs134, Cs 137, Ru 103, Ru 106, Zr/Nb 95, Ba/La 140, I131, Te/I 132. Nuclides with lower activity levels have been Sr 90, Pu 239/240, Pu 238, Am 241, Cm 242, Mo/Tc 99, Ce 141, Ce144, and Ag 110m.

During several cruises to the North Sea and to the Baltic Sea the levels of the contamination by this Fallout could be detected. The Chernobyl Fallout is discernable from other sources of artificial radioactivity in the North Sea and the Baltic sea byits typical Cs134/137 activity ratio.

The relatively fast water movement along the coast of Belgium, the Netherlands, and Germany renewed the Chernobyl contaminated water of the southern North Sea within 4 months after the accident by uncontaminated water from the Channel. However, in the German Bight itself Chernobyl Fallout is still to be identify in December 1986, which is due to runoff of the river Elbe.

During an internationally co-ordinated monitoring programme in October/November 1986 the different levels of the Chernobyl Fallout could be determined in the entire Baltic Sea. The highest contamination levels were measured in the southern part of the Bothnian Sea with up to 800 mBq/l Cs 137. In the western Baltic and in the Baltic Proper the contamination of the Sea is relatively low with about 40 mBq/l Cs 137. In the Baltic Proper the Chernobyl contamination has not yet reached the water below the halocline. However, in the Bothnian Bay an almost homogenious mixing down the water column can be established.

INTRODUCTION

Prior to the nuclear reactor accident at Chernobyl the concentrations of artificial radionuclides in the North Sea were mainly influenced by the discharges from the nuclear fuel reprocessing plants at La Hague (France) and Sellafield Works (United Kingdom). In the Baltic Sea, these radionuclides were transported only to a slight extent by particular weather and hydrographic conditions, so that the radioactive inventory of the Baltic Sea was determined by nuclear weapon Fallout of the sixties.

The Chernobyl accident has considerably increased and changed the inventory of artificial radionuclides in the marine environment of the North Sea and the Baltic Sea. In the following paper, the essential results of the measurements of Chernobyl Fallout nuclides, carried out by the Deutsches Hydrographisches Institut in 1986, are to be presented and discussed.

RESULTS AND DISCUSSION

The input of the Fallout from Chernobyl into the Sea took place via the atmosphere, whereby a large-scale and time synoptic monitoring became necessary. After the accident had become known, the DHI therefore took water samples immediately as a precaution at several positions in the North Sea and the Baltic Sea, in order to investigate them as soon as possible for radioactive contamination. Aerosol investigations carried out in Hamburg simultaneously provided data about the time of the arrival of the radioactive cloud and about its nuclide composition. The nuclide pattern was ascertained by Gamma spectroscopy. Hereby, the relative activity of the most important nuclides in the analyzed aerosols, in relation to Cs 137 on the 5th/6th May, are given in the following table:

Nuclide	Rel. activity	Half-life
Cs 137	1.00	30 years
Cs 134	0.52	754 days
Ru 103	2.00	39.3 days
Ru 106	0.75	368 days
Ba/La 140	0.67	12.8 days
I 131	6.00	8.0 days
Te/I 132	2.30	3.2 days

Some further nuclides, such as Sr90, Pu239/240, Pu238, Am241, Np239, Cm 242, Te 129m, Mo/Tc 99, Cs 136, Ce 141, Ce/Pr 144, Zr/Nb 95, Ag 110m, were detectable in lesser concentration in different samples from the region of the sea.

The DHI's radiological measurement network (positions of the measuring network station are given in Fig 1.) delivered the first information about radioactive precipitation into the sea at 3 May at the station Light Vessel "Borkumriff" at 21.00 hrs. after the set in of thunder showers. The input into the southern North Sea occurred from west to east direction by the movement of a thunder front. The input into the western Baltic was detected at the 5 May at station Fehmarnbelt. The input into the Baltic Proper was registered at 29 May by the travelling station RV GAUSS, which was in the area south of Gotland. At that time no direct input by rain of the Fallout into the water took place in this area. It merely concerned precipitation from aerosols and fog.

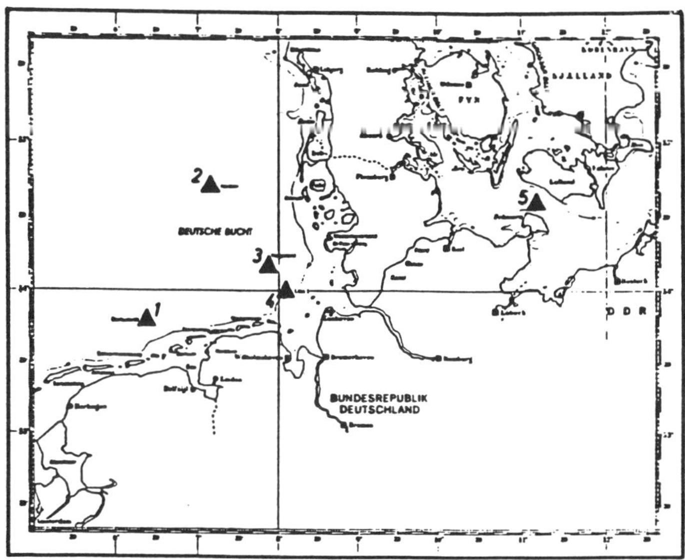

Fig. 1
Radiological measuring network of the DHI at Sea
1 Light Vessel "Borkumriff", 2 Research Platform "Nordsee"
3 Island "Helgoland", 4 LV "Elbe-1", 5 LV "Fehmarnbelt"
Travelling stations: RV "GAUSS", SWV "ATAIR", SV "SÜDEROOG"

The temporal coarse of the contamination of the sea water in the coastal region (Fig. 2 and 3) of Germany shows that the activity concentration in the surface water decreased very rapidly after the peak values directly prosecuting the input between 3 and 10 May. This is an effect of the dilution and distribution within the water column. In the western Baltic already in August the Chernobyl derived Caesium has been down-mixed almost to the bottom layers (1). In the long run, the activity decreased in the German Bight more quickly than in the western Baltic Sea. This effect is due to the well known water mass transport in the North Sea.

The Chernobyl Fallout can unequivocally identified by the activity ratio of Cs 134/Cs 137, which amounted in May 1986 to about 0.5 . The differentiation between the Chernobyl Fallout and the contamination already previously present in the North Sea from Sellafield or, to a lesser extent, from LaHague can be conducted by this ratio, because the activity ratio of this Caesium-isotopes in the water of the North Sea during the previous years amounts to maximum 0.1 (2).

The decrease of the activity concentration of Cs 137 in the Baltic Sea between May and June 1986 can be seen in Fig. 4. The relatively low concentrations in the Baltic Proper are a result of only dry deposition. The higher levels in the western Baltic are due to Washout of the radioactive cloud from the atmosphere. The values at depth below 40 m show that no Chernobyl Fallout has reached deeper layers beneath the halocline in June. They are in the same range as determined in previous years (3).

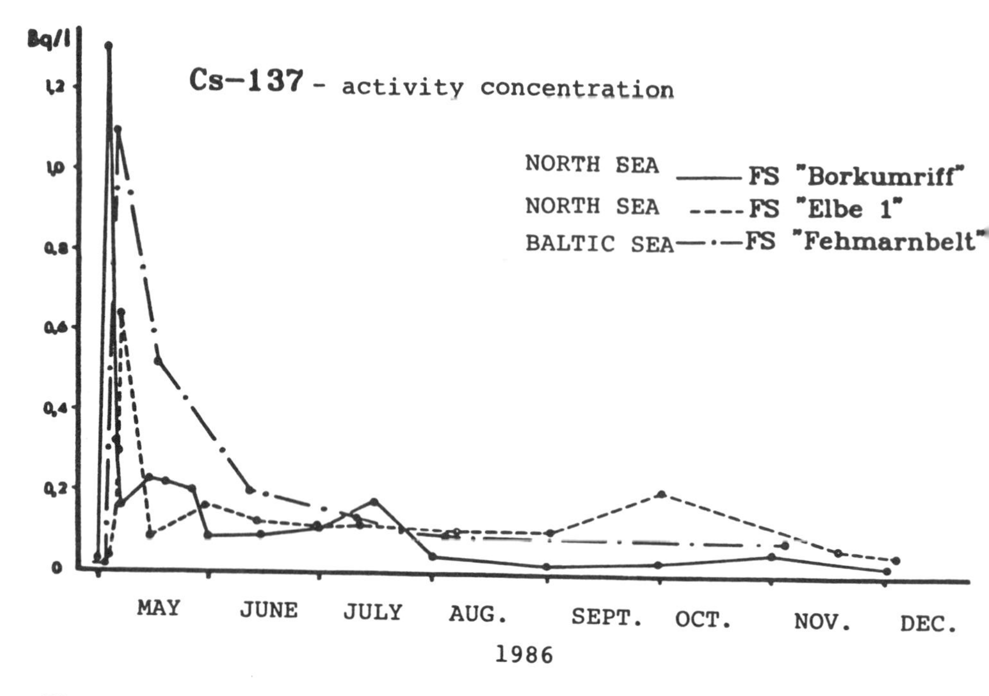

Fig. 2
Temporal coarse of the activity concentration of Cs137 at three positions in the North Sea and the Baltic Sea

On a voyage of the RV GAUSS from 23.05. to 12.06. samples were likewise taken over the larger part of the North Sea for the determination of Radiocaesium on a large scale. The results are shown in Fig. 5. The input originating from the accident in Chernobyl can be identified from the course of the isolines of the activity ratio Cs 134/Cs 137. Areas with almost pure Chernobyl Fallout are to be recognized in the inner German Bight and one spot in the Skagerrak. The Chernobyl derived contamination weakens from the German Bight towards the direction of the Northwest. Along the British east coast almost pure Caesium-isotopes discharged from Sellafield are detected, though the activity concentration is in the same order of magnitude as in areas influenced by Chernobyl. The discharges from the reprocessing plant at LaHague give no considerable contribution to the Caesium contamination along the coast line of the southern North Sea.

The Cs134 depth distribution indicates that the surface measurement data are to be viewed as being representative for the mixed layer. The thickness of the homogenious mixed layer is known at the time of investigation owing to the accompanying oceanographic measurements. In the southern part of the North Sea the mixed layer reached down to the sea floor and in the northern part of the North Sea down to a depth of 40 to 60 meters. In the vicinity of the south Norwegian coast a thickness of 100 meters can be assessed. Proceeding from that, one can calculate the radioactive inventory of the surface layer areally and represent it as areal deposition. Fig. 6 shows the Cs 137 distribution pattern of the Chernobyl Fallout deposition in kBq/m^2 resulting from the mean radionuclide ratio Cs134/Cs137.

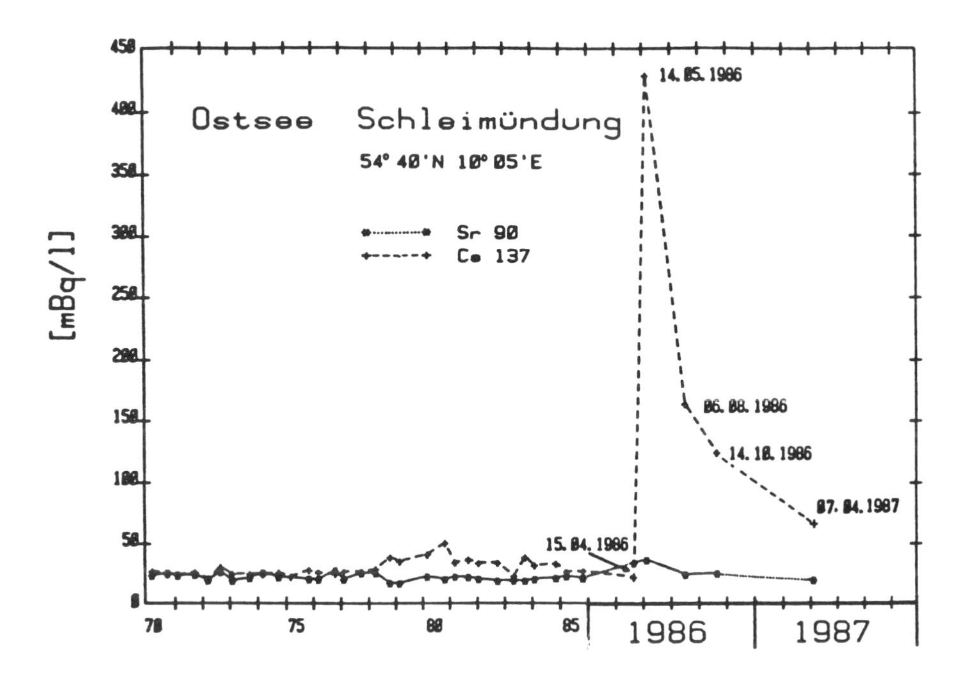

Fig. 3
The Cs 137 and Sr 90 surface activity concentration at the position Schleimünde in the western Baltic Sea

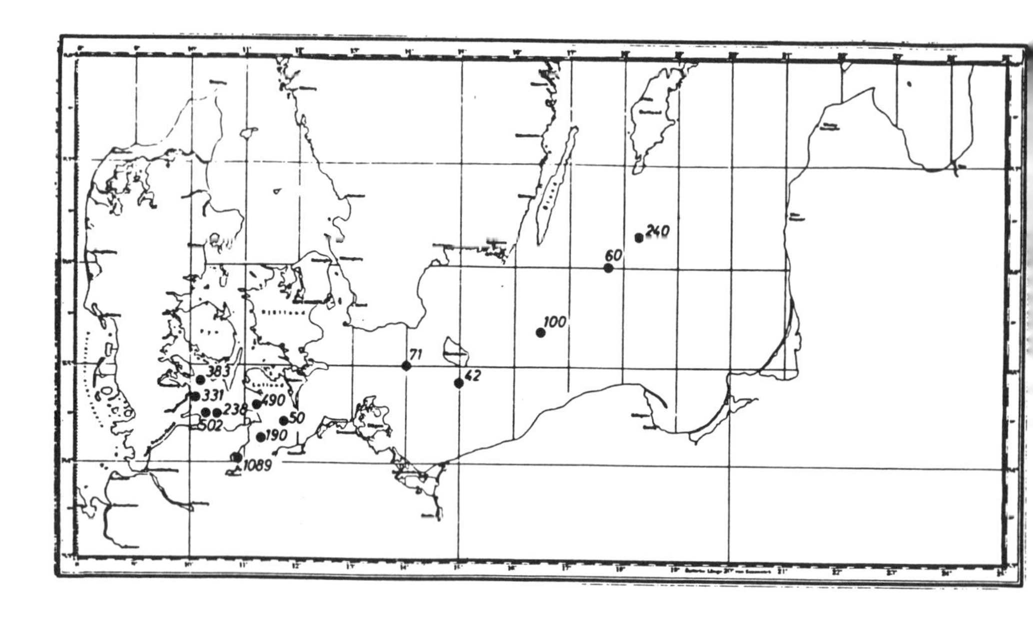

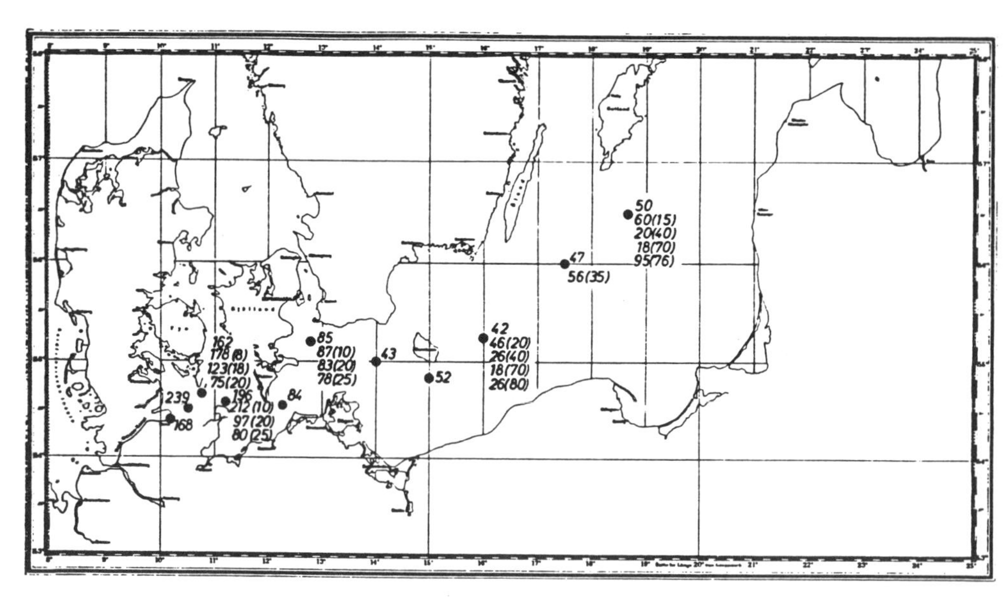

Fig. 4 Baltic Sea
upper map: Cs137 activity concentration (mBq/l) in the surface water; 29 April to 11 May 1986,

lower map: Cs 137 activity concentration (mBq/l); (Depth of the samples are given in brackets); 4 to 11 June 1986

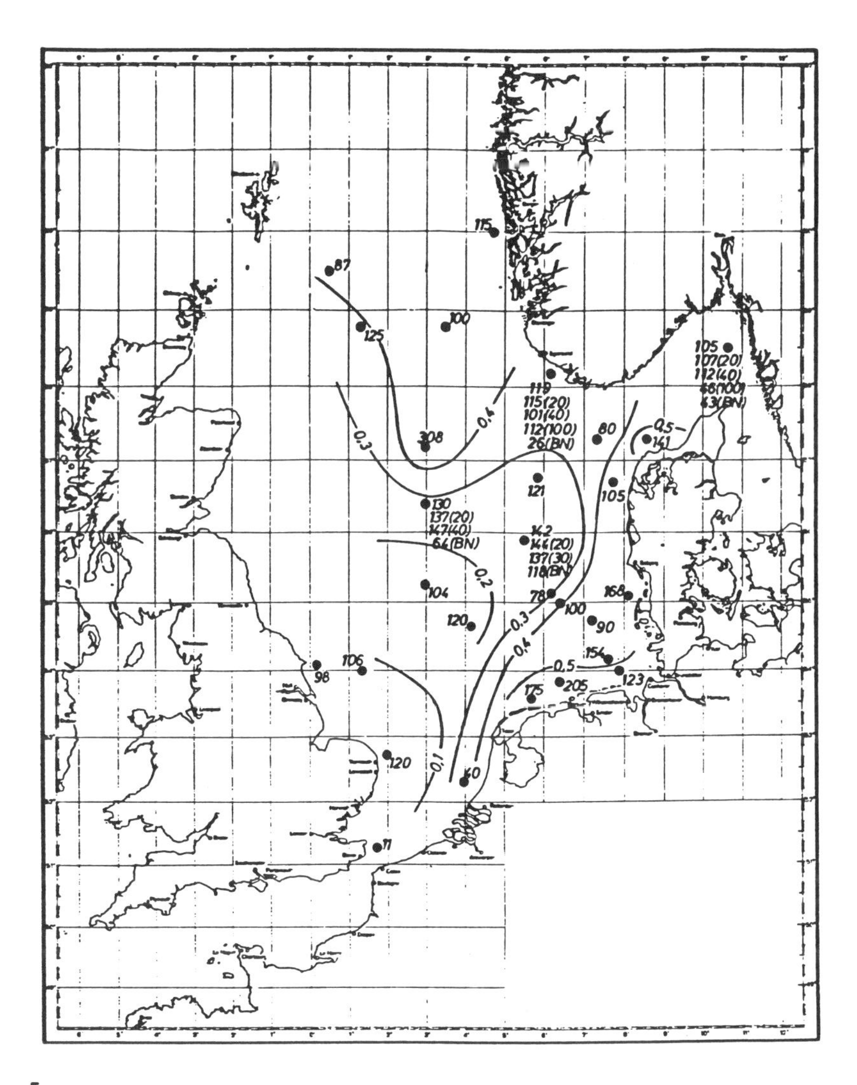

Fig. 5
Cs 137 activity concentration in the North Sea
Values in mBq/l; ( ) depth in m
Isolines are representing the Cs 134/Cs 137 activity ratio
RV GAUSS 23.05. to 12.06.1986; SWV ATAIR 11.06. to 13.06.1986

Down to greater depth, the contamination, owing to mixing of the water column after the Chernobyl input, is only partly recognizable. As shown on analyses of sediment trap material of a sediment trap about 60 nmiles west from Bergen, which was deployed at a depth of about 200 m from 24 April to 20 September 1986, the vertical transport of radionuclides into depth is accelerated considerably by the adsorption of activity on particulate matter (4).

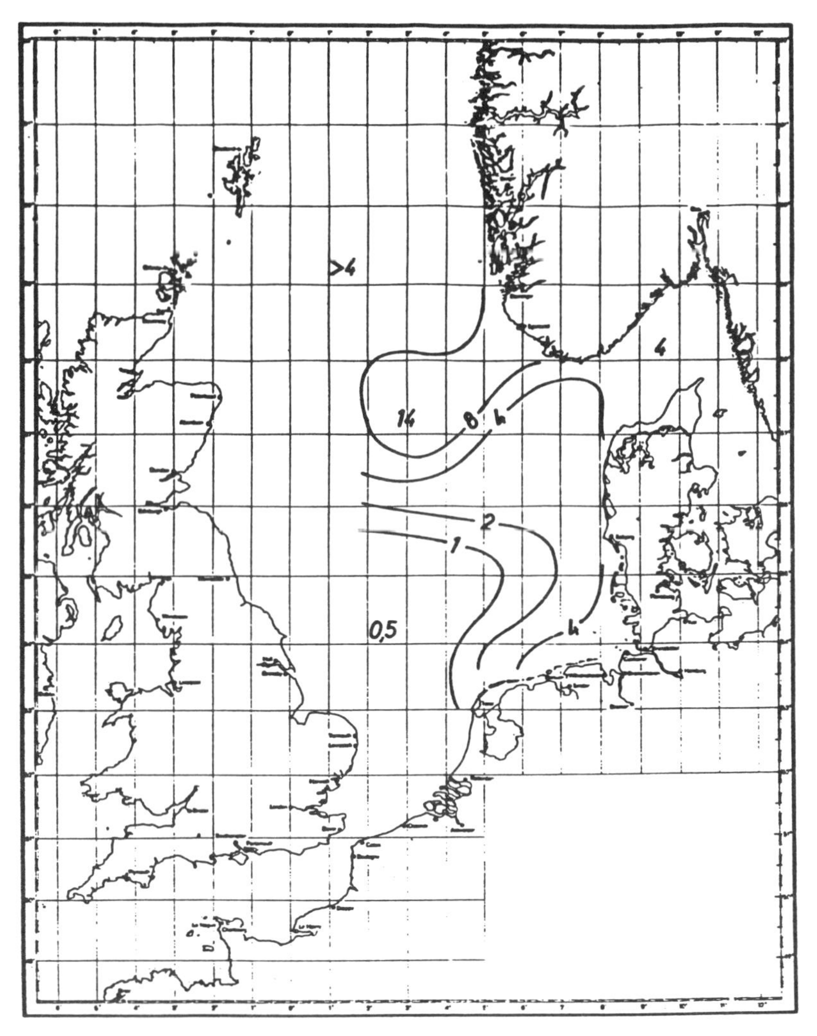

Fig. 6
Cs 137 areal deposition at the sea surface of the North Sea in kBq/m²; calculated by means of the Cs 134 activity concentration on the basis of the Cs 134/Cs 137 activity ratio 0.52
RV GAUSS 23.05. to 12.06.1986; SWV ATAIR 11.06. to 13.06.1986

The investigation of the large scale distribution of the contamination of the Baltic Sea was carried out in an international co-operation during a RV GAUSS cruise in October/November 1986. As can be seen from figs.7 and 8 the level of the contamination in the Baltic Sea is quite different in different regions of the Baltic Sea. The highest contamination levels are detected in the southern part of the Bothnian Sea with 887 mBq/l Cs137.

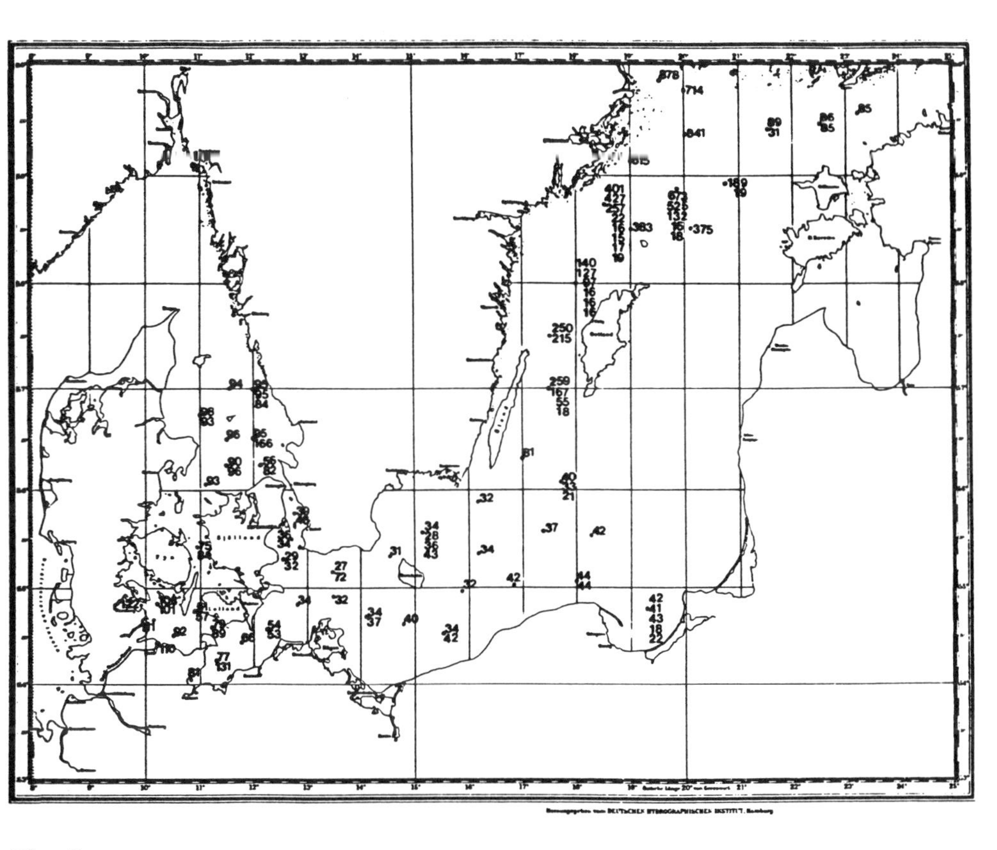

Fig. 7

Western Baltic and Baltic Proper
activity concentration of Cs 137 (mBq/l)
Depth profiles:
1. value: surface, 2. value: 20 m, 3. value: 40 m
4. value: 80 m ,5. value: 120 m, 6. value: 160 m
lowest value: sample near to the seabed

RV GAUSS 14.10. to 03.11.1986

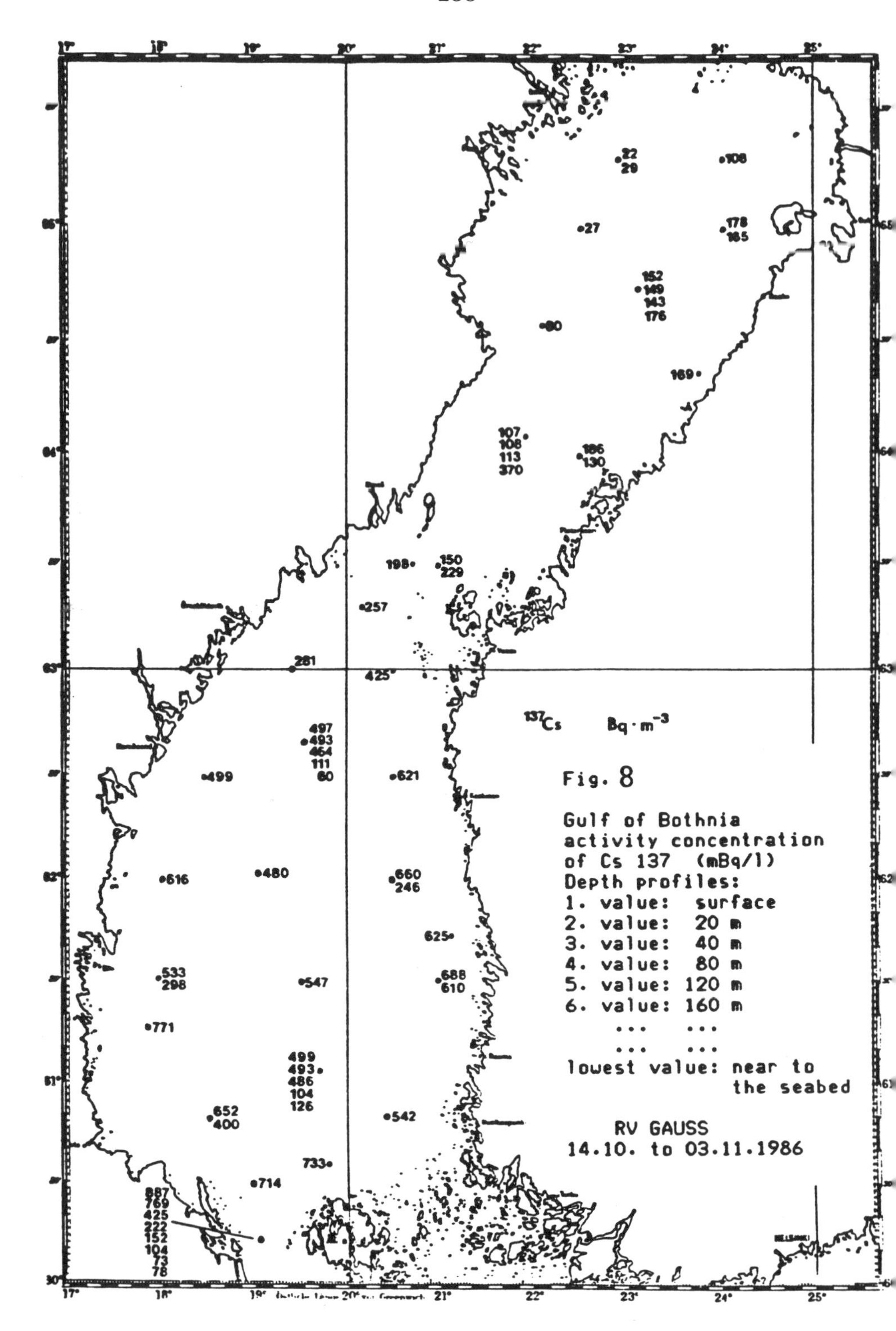

^{137}Cs Bq·m^{-3}
Fig. 8
Gulf of Bothnia
activity concentration
of Cs 137 (mBq/l)
Depth profiles:
1. value: surface
2. value: 20 m
3. value: 40 m
4. value: 80 m
5. value: 120 m
6. value: 160 m
... ...
... ...
lowest value: near to the seabed
RV GAUSS
14.10. to 03.11.1986

Coming further north in the Gulf of Bothnia the activity concentration decreases and reaches a very low value at the western part of the northernmost stations in the Bay of Bothnia with about 25 mBq/l Cs 137. The contamination of the Baltic Proper is also remarkable low with values of around 30 to 40 mBq/l Cs 137. Only in the western Baltic Sea, sligthly higher values are found.

Regarding the depth penetration of the Chernobyl Fallout, one can recognize that in the western Baltic the Fallout has reached the bottom layers, while in the Baltic Proper almost no Cs 134 is detectable below the halocline, which is found to lay at a depth between 30 and 50 m. In the Bothnian Sea the depth profile of the Chernobyl Fallout reveals a continuous decrease, though the halocline in this Sea area was still stable during the time of investigation. In the Bay of Bothnia the Chernobyl Fallout is almost homogeniously mixed down to the sediment. The salinity and temperature profiles in this sea area show that the whole water column seems to be homogeniously mixed down to the bottom. However, one reason may also be due to the fact that the Caesium portion, which is bound to particulate matter, is in the range of 5 %, while in regions further south the particulate bound Caesium does not exceed 1 % of the total amount of Cs 137 (5).

If one calculates the apparent deposition of Cs 137 in the whole Baltic Sea area by using the Cs 134/Cs 137 activity ratio, a distribution as given in fig. 9 is obtained. The highest deposition seems to have occurred in the Aaland Sea, though by swedish monitoring programmes after the accident, the highest deposition on land was in the area of the swedish town Gävle with about 120 kBq/m^2 (6). Principally, the pattern of the deposition in the Baltic Sea fits well with the data found in different countries surrounding the Baltic Sea. However, at the time of the investigation - 1/2 year after the deposition - the highly contaminated water masses have already moved anticlockwise in the Gulf of Bothnia and from the southern part of the Bothnian Sea through the Aaland Sea from north to south. This leads to a tongue shaped pattern of contaminated water masses in the area around Gotland.

In spite of the time-consuming methods of analysis, a series of Sr 90, Pu 239, Pu 238, Am 241, and Cm 242 determinations were carried out from seawater samples. It is remarkable that the activity contribution of Cm 242 surmounts that one of Pu239.

If the activity concentration of Sr 90 from previous years (3) are compared with the activity in samples, which are analyzed after the accident in Chernobyl, the Sr 90 Fallout contribution in the Baltic Sea area is very low. The activity ratio between Sr90 and Cs 137 can thus be determined to lay between 0.006 and 0.008 in those water samples, which show Cs137 values of more than 500 mBq/l.

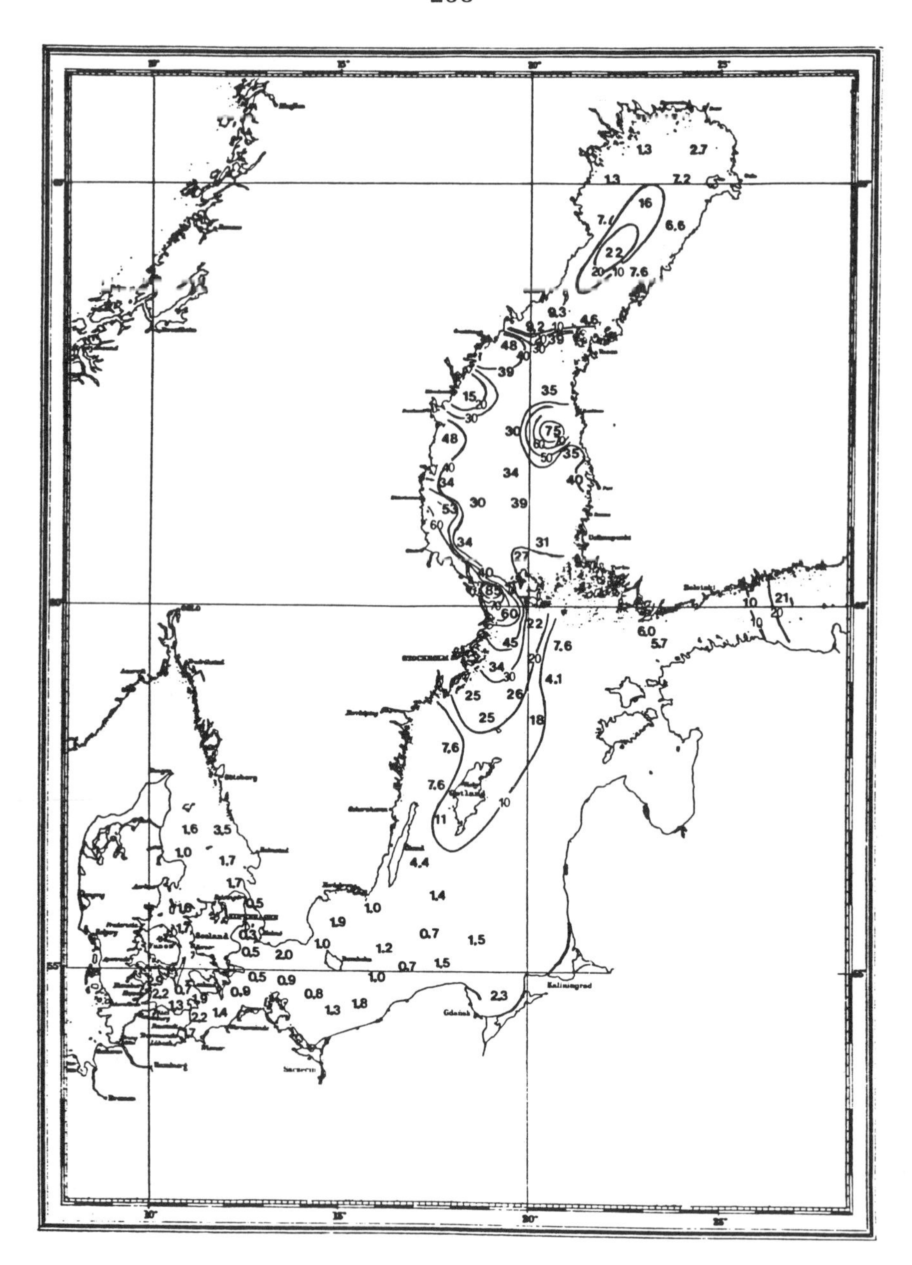

Fig. 9
The apparent areal deposition of Cs 137 (kBq/m^2) at the sea surface of the Baltic Sea; calculated by means of the Cs 134 concentration on the basis of the Cs 134/Cs 137 activity ratio in the Chernobyl Fallout; the values are decay corrected to the 1 May 1986; RV GAUSS 14.10. to 03.11.1986

CONCLUSION

Since all the above presented results have been obtained, further investigation programmes of the DHI continued during 1986 and 1987. In December 1986 the southern North Sea region shows only in the German Bight still higher Chernobyl derived activity. This must be due to the runoff of the river Elbe, which still has higher contamination levels than the river Rhein, Ems, or Weser. The same effect can also be recognized on samples of suspended matter taken by centrifugation in December 1986 in the German Bight.

At the most positions of the western Baltic Sea, since October/November to April 1987, a decrease of the activity concentration of Cs 137 by a factor of 50 % can be established.

Investigations in the future will reveal the essential depository of radionuclides in the sediment.

ACKNOWLEDGEMENTS

We are indepted to all persons, who supported our work with taking samples at sea or assisting our work on the research vessels. We thank especially G. Främcke, H. Gabriel, I. Goroncy, T. Mikkelsen, V. Rechenberg, and M. Santi for carrying out the analyses in the laboratory.

REFERENCES

1. Dr. A. Aarkrog, RISØ National Laboratory, Denmark, 1987 pers. communication

2. Kautsky, H., Dt. Hydrogr. Z., 1986, 39, 139 - 159

3. IAEA-TECDOC-362 Study of Radioactive Materials in the Baltic Sea, Vienna, 1986

4. Kempe, S., Nies, H., Chernobyl nuclide record from North Sea sediment trap; Nature, 1987, Vol. 329, 828-831

5. Dr. H. Dahlgaard, RISØ National Laboratory, Denmark, 1987 pers. communication

6. Grimås, U., Neumann, G., Notter, M., Tidiga erfarenheter av nedfallet från Tjernobyl; Radioekologiska studier i svenska kustvatten Naturvårdsverket, Rapport 3264, 1986

Cs-137 AND OTHER RADIONUCLIDES IN THE BENTHIC FAUNA IN THE BALTIC SEA BEFORE AND AFTER THE CHERNOBYL ACCIDENT

P-O Agnedal
STUDSVIK NUCLEAR
S-611 82 NYKÖPING
SWEDEN

ABSTRACT

The results of the material collected during three cruises with the R/V Gauss FRG in the Baltic Sea are presented. The purpose of the cruises was to collect water, sediment and benthic fauna and to determine the radioactivity in the samples and calculate the inventory of radionuclides in the Baltic Sea, especially Cs-137.

The results from the two first cruises show a tendency of decreasing Cs-137 concentration in some species in the benthic fauna with increasing salinity in the water, which has not been seen regarding stable cesium. The results of the third cruise performed in Oct/Nov 1986 after the Chernobyl accident are compared with those from the first two cruises and also with the Cs-137 level in the water and the fall-out situation in the adjacent land areas.

The preliminary results of the water analyses performed by DHI on board the ship during the cruise show great variations in activity level depending on the geographical site.

INTRODUCTION

In May 1986 Helcom (Baltic Marine Environmental Protection Commission - Helsinki Commission) and those who had participated in the previous Gauss-cruises in the Baltic Sea in 1983 and 1984 received an invitation from DHI to take part in a new cruise in 1986. The cruise took place in October/November.

The route was nearly the same as in 1983 and partly in 1984, which has resulted in good possibilities of studying changes in the radioactivity levels in the sediment and in the benthic fauna after the fall out from the Chernobyl accident.

WORKING PROGRAMMES

Besides the participants from Deutsches Hydrographishes Institut (DHI) there were working groups from Risø, Denmark, University of Lund and from Studsvik, Sweden, as well as from Finnish Center for Radiation and Nuclear Safety (STUK), Helsinki, Finland. The intention of the cruise was to investigate the distribution and concentration of the fall-out from Chernobyl in the Baltic Sea area and also to promote progress in the analytical procedures for determining the wide range of new nuclides present in the seawater after the accident. Most effort was therefore put into water sampling, surface and bottom water, but samples were also taken from depths inbetween. Sediment samples were taken at some stations. The Studsvik group collected benthic fauna and sediment from selected station but took no water samples.

The route with all the stations is marked on the map in Figure 1 and the benthic fauna stations are specially marked and the positions are given in Table 1. In Table 1 the number of the corresponding stations in 1983 and 1984 are also given.

METHODS

Field sampling

The modified Ockelman dredge was also used in 1986 (1). One dredge had a net with 4 mm meshes and one with 2 mm meshes. The latter one was used in the Bothnian Bay to collect small crustaceans but was unfortunately lost after two stations which limited the catch of the small crustaceans. As earlier, two to four dredgings sufficed to collect enough animals for the analyses. The samples were sorted immediately and deep-frozen. The sediment samples were taken with the Danish HAPS, a box-corer sampler. The samples were divided into two layers 0-3 and 3-6 cm and deep-frozen.

Activity measurements

In the laboratory the deep-frozen samples, after thawing and weighing (wet weight) were dried at a temperature of 105 °C to constant weight.

After homogenizing either the whole sample or a sub-sample was used for gammaspectroscopy. The measurements were performed with a 30 % Ge(Li) detector which has a resolution of 1.9 keV at 1.33 MEV. The detector is shielded with 15 cm lead and connected to a multichannel analyzer ND4410, Nuclear Data. Measuring times varied from 500 to 3 000 minutes.

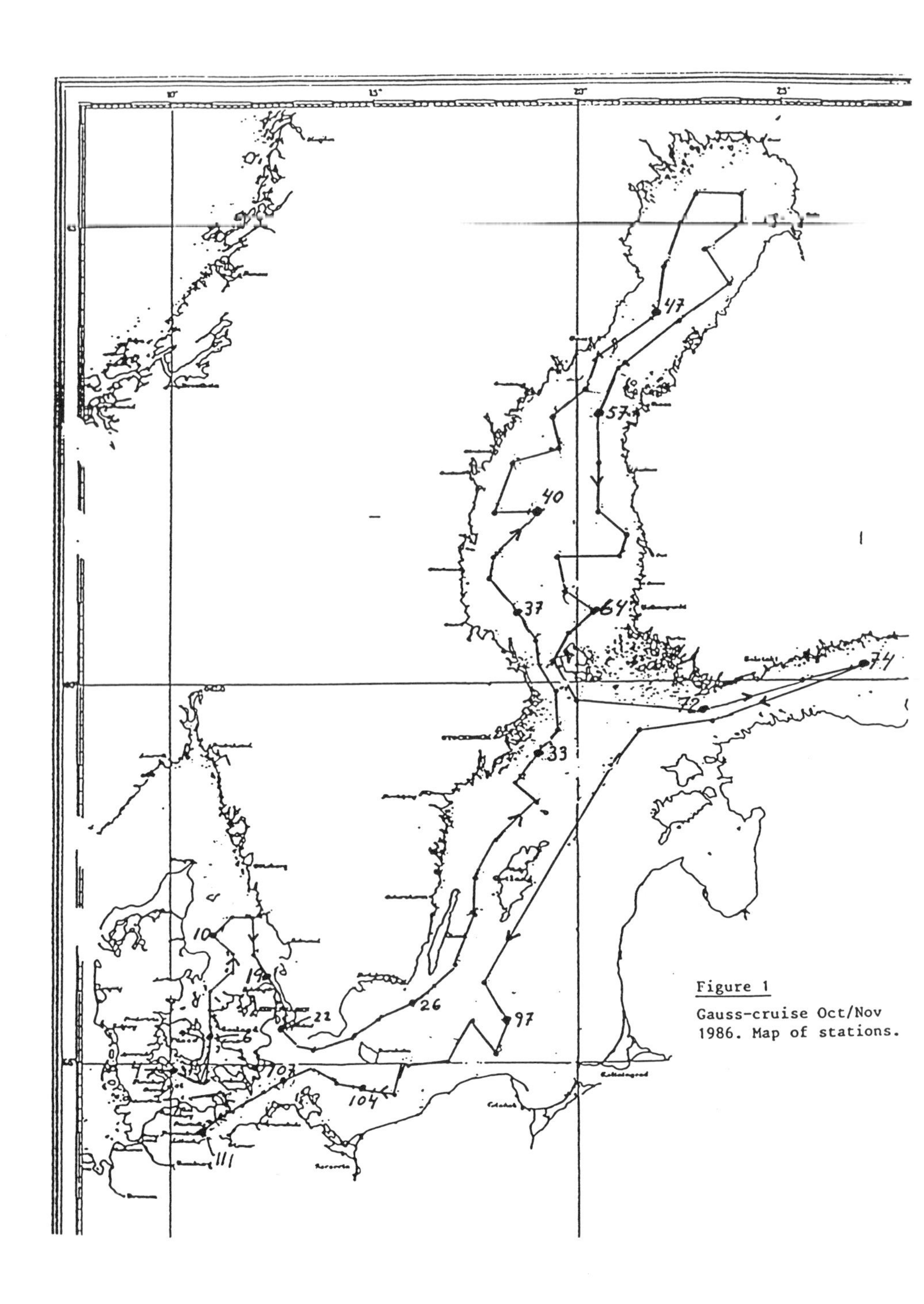

Figure 1

Gauss-cruise Oct/Nov 1986. Map of stations.

Table 1 Position of sampling stations. Radionuclide concentration values in sediment

Station no.			Position		Sediment samples Bq/kg dry weight			Cs-137	
1983	1984	1986	N	E	Cs-134	Cs137	Ag-110	1983	1984
9	2	4	54°50'	10°12'		52.5±2.4		2.2	6.1
24	10	6	55°25'	10°57'		37.3±2.4			
28	12	10	56°45'	11°00'				0.2	9.3
33	17	19	56°15'	12°07'		5.3±1.4		29.8	18.3
37	-	22	55°17'	12°33'				1.0	
119	32	26	55°53'	16°10'		3.9±1.4		5.4	5.3
110	-	33	59°06'	19°00'		9.0±0.5		4.6	
107	-	37	60°49'	18°32'	56.7 ±5	123 ±5			
104	-	40	62°01'	19°03'	125 ±5	274 ±5	5.4±2.2	7.0	
		47	64°05'	21°56'					
81	-	57	63°00'	20°30'	184 ±5	452 ±5	11 ±2.8	51.4	
73A	-	64	60°50'	20°24'	30*	170		13.6	
66A*	-	72	59°35'	23°10'				6.3	
63B*	-	74	60°09'	26°35'				7.4	
47*	-	97	55°30'	18°16'		24		1.4	
41*		104	54°38'	14°47'					
41	21	107	54°50'	12°48'					
-	-	111	54°03'	10°51'					

* Ref 5 0 - 5 cm.

Table 2:1 Radionuclide concentration values in *Saduria entomon* and *Pontoporeia affinis*

Station		Bq/kg dry weight *Saduria entomon*						
1983/84	1986	Mn-54	Ru-103	Ru-106	Ag-110m	Cs-134	Cs-137	Cs-137 (1983-84)
119	26				11.7±3.3	10.9±1.6	15.4± 2.8	4.3±0.8 3.3±1.7
110	33				54.7± 2.3	10.2±2.0	23.8± 2.7	4.9±0.9
107	37	50 ±5	44 ±13	-	620 ± 7	122 ±5	250 ± 7	5.3±0.8
104	40	13.4±3.5	-	-	31.9± 5	53 ±2.7	102 ± 5	
	47	-	-	-	76.8± 3.6	47 ±5	95.2± 4.6	
81	57	-	-	-	56 ± 2.5	19.4±2.5	51.0± 3.1	4.2±1.1
73A	64	7.5±2.0	8.3±4.8	32±11	170 ± 5	78.7±2.6	158 ± 5	
63B*	74	4.3±2.0	-	-	61 ± 2.6	32.5±2.2	61.3± 3.2	4.1±1.0
41*	104	-	-	-	-	-	11 ± 6	1.9±1.7
				Pontoporeia affinis				
104	40				772 ± 9	74 ±6	157 + 9	
81/88	47				412 ±13	91 ±9	229 ±14	

RESULTS

Sediment

The results of the analyses are listed in Table 1. The samples consist of the two collected layers 0-3 and 3-6 cm. The results from 1983 are given for comparison. The 1984 samples were taken with another method and the results are therefore not quite comparable.

The results show an increase of about 8 to 40 times from the 1983 values in the most exposed areas. The highest Cs-137 concentration, 450 Bq/kg d.w., was found at station 57 on the east side of the Bothnian Sea and the second highest at station 40, 275 Bq/kg d.w., southwest from station 57. At these two stations Ag-110 was also detected (11 and 5 Bq/kg d.w. respectively).

The ratio Cs-134: Cs-137 was around 0.44 back-calculated to October 14th 1986.

As can be seen from Figure 2 the concentrations in the south part of Baltic proper and in the Dansih Straits are not so high.

Benthic Fauna

All results of the biota are presented in Tables 2 and 3. Besides Cs-137 and Cs-134, nuclides as Mn-54, Ru-103, Ru-106 and Ag-110m have been detected in the samples. All samples more or less show the impact of the Chernobyl fall-out. The influence is most clear in the areas of the sea where a high level of fall-out was detected on adjacent land areas, Figure 3.

Two of the most common animals in the benthic fauna in the Baltic Sea are the mussel Macoma baltica and the crustacean Saduria (Mesidotea) entomon. Another crustacean Pontoporeia affinis, is also very common on soft bottoms. M. baltica has its northern limit of distribution where the salinity is less than 3-5 ^{o}/oo which is around 61°N. These species were collected both in 1983, 1984 and therefore "background" values are available. The mean value of Cs-137 in S. entomon in 1983 was 5.4 Bq/kg d.w. and in 1986 85.3 Bq/kg d.w. The Cs-134 mean value in 1986 is 46.7 which gives a ratio of 0.54 Cs-134: Cs-137. Ag-110m was found in all samples and the mean concentration is 170 Bq/kg d.w. with a maximum of 620. This value was found in S. entomon from station 37 situated outside the land area where high contamination occurred. Samples from stations 40 and 64 also have high concentrations of Cs-137 and somewhat lower values have been found at stations 47 and 57 and all four stations are situated between 60° and 64°N. Mn-54 has been detected in the samples from stations 37, 40, 64 and 74 with the highest values at the two first stations, Ru-103 and Ru-106 in the sample from station 64 and Ru-103 at station 37. In the samples from stations 47, 57 and 74 Mn-54 and Ru-103, Ru-106 have not been detected although the measuring time was the same as for the other samples.

On account of the loss of the finemesh dredge, only two samples of Pontoporeia affinis were collected. Both samples show, however, an increase of the Cs-137 content from 10 to 193 Bq/kg d.w. as mean values in 1983 and 1986 respectively and a ratio Cs-134: Cs-137 of 0.44.

A sample of Gammarids from station 107, southwestern Baltic Proper, contained Ru-103, Ru-106, Ag-110m in addition to Cs-137 (17.4 Bq/kg d.w.)

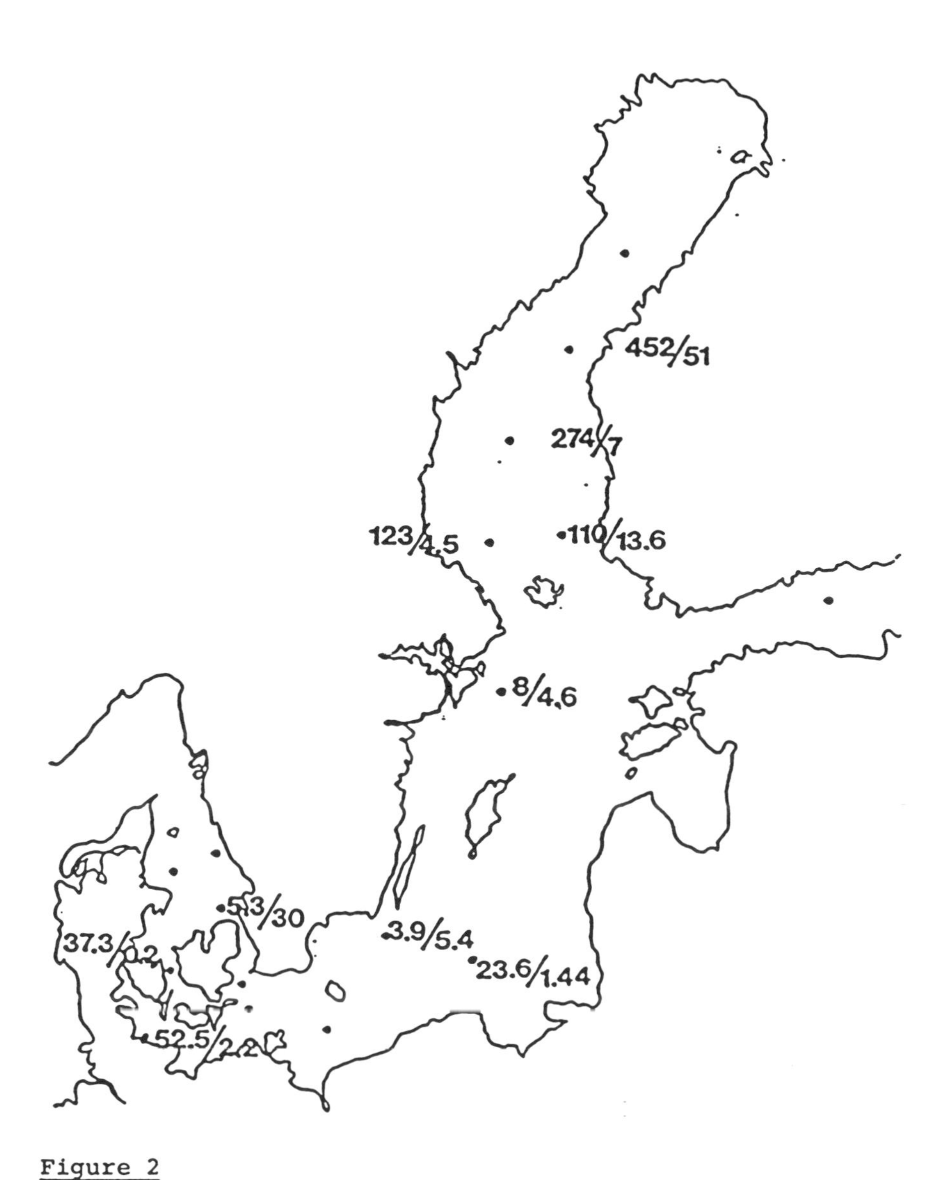

Figure 2

Cs-137 in Baltic Sea Sediment.
Bq/kg d.w. 1986/83.

Table 2:2 Radionuclide concentration values in Macoma baltica and Mytilus edulis

Station				Bq/kg dry weight		
				Macoma baltica		
1983/84	1986	Ru-103	Ag-110m	Cs-134	Cs-137	Cs-137 (1983/84+)
24*	6				15 ±6	0.70±0.42/< 5.2+
37	22				2.2±1.1	
119	26				2.7+1.1	0.54+0.25/< 0.69+
110	33		8.8±0.1	1.2±0.9	7.4±1.1	1.1 ±0.5
73A	64	8.2±2.8	51.2±1.2	13.5±0.9	30.1±1.4	
63B*	74				20.5±2.8	0.75±0.33
47*	97				3.7±6.3	
41*	104				1.2±1.0	0.88±0.35/< 3.6+
22/21	107				< 3.3	2.2 ±0.5+
		Co-60		Mytilus edulis		
37	22	3.0+0.5			2.5+0.7	
119	26				5.0±3.0	2.5/0.87+
	104				6.7±5.5	
22/21	107				2.2±0.7	0.96+

Table 3 Radionuclide concentration values in other benthic invertebrates

Station 1983	1984	1986	Species	Ag-110m	Cs-137	
24	12	10	Buccinum undatum	28.8±1.9	7.4±2.0	
			Cyprina islandica	13.4±1.8	< 2.8	
			Astarte sp	4.8±1.5	1.2±1.0	
33	17	19	Cyprina islandica	5.34±0.32	1.44±0.44	
33	17	19	Mixed mussels	5.5 ±0.7	0.9 ±0.7	
37		22	Mya sp		< 8.6	
24	10	6	Cyprina islandica		3.0±1.1	
24	10	6	Mixed mussels		< 8.8	
			Echinodermata			
24	10	6	Brittle stars		< 5.0	
24	12	10	Brittle stars		< 23	
		10	Asterias rubens	21.4±1.5	7.4±2.5	
33	17	19	Brittle stars, large		4.4±2.3	
33	17	19	Brittle stars, small		3.8±1.9	
24	12	10	Sea urchins	8.5±1.8	< 4.3	
33	17	19	Sea urchins		9.1±1.5	
			Polychaets			Co-60
24	10	6	Nephtys sp+Terebellides sp		17.5± 4.3	10.9±2.8
33	17	19	Eumenia sp	7.9±0.7	9.2±1.0	5.1±0.8
33	17	19	Aphrodite aculeata		10.3±3.9	8.4±2.4
33	17	19	Mixed polychaets		5.8±4.5	7.9±2.8
41		104	Polychaets (Terrebelides sp)		5.7± 4.1	
22	21	107	Polychaets (Terrebelides sp)		< 69	
47*		97	Polychaets (Terrebelides sp)		35 ±13	

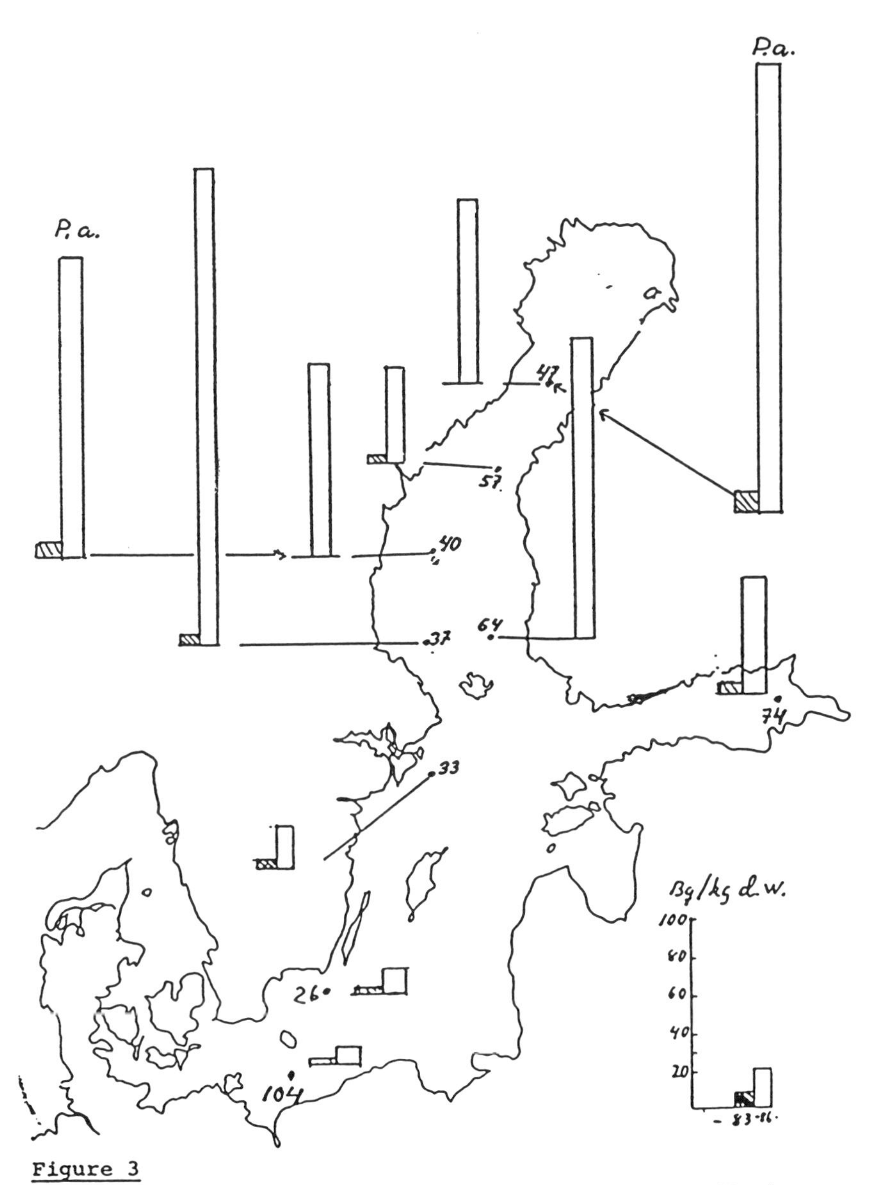

Figure 3

Cs-137 in Saduria entomon and Pontoporeia affinis (P.a) 1983/86.

as well as Cs-134, which also indicates a contamination of this part of the Baltic. A mixed sample of crustaceans from station 97 has a higher concentration of Cs-137 than was found in 1983 at stations in adjacent waters.

The molluscs represented by Macoma baltica and to some extent by Mytilus edulis show, within the same are where S. entomon has an increased concentration of Cs-137 as well as at other stations an effect of the fall-out as Ru-103 and Ag-110m have also been detected in the samples.

DISCUSSION

Aarkrog (2) has analysed water samples in the Danish waters and found that in May there was no Cs-137 from the Chernobyl fall-out in the bottom water. In August, however, the Cs-137 in surface and bottom water was 73 and 70 Bq/m^3 respectively and in October the values were 75 and 84 Bq/m^3.

The investigations from DHI (3) indicate that there was a mixing down to 20 m already in June in the Baltic Sea.

In autumn 1986 the concentration of Cs-137 in the bottom water in the Bothnian Sea varied from 29 to 400 mBq/l and in the Baltic Proper from 16 to 85 mBq/l according to the results from the measurements of DHI (4).

Finnish observations (5) show that fall-out nuclides were found in water from 100 m depth already in May and that surface sediment from 166 m depth in mid June contained fall-out nuclides.

The activity reached the bottom water/sediment fairly fast either via settling particles or by vertical mixing of the water. One reason to believe that settling particles have contributed most is that the fall-out occurred at about the same time as the phytoplankton spring maximum. The dead plankton then sink to the bottom.

The concentration of Cs-137 in the sediment from the investigated stations show an increase from 1983 of about 20 to 280 Bq/kg d.w. on average for the Bothnian Sea. In the Baltic Proper the values have hardly changed but a slight increase in the Danish Straits and in the south western part can be observed. These figures are based on few samples but the Finnish investigations have given values of the same order of magnitude (5).

CONCLUSIONS

The presence of different nuclides coming from the Chernobyl fall-out has been observed in the benthic fauna in the Danish Strait, in the northern part of Baltic Proper and in the Bothnian Sea where the highest values are found at the stations in the vicinity of those land areas which received the highest contamination.

The benthic animals, especially Saduria entomon and Pontoporeia affinis have received a high concentration in a short time. It is not possible to determine whether the uptake has occurred through water or through food.

To measure the elimination of the radioactivity in the Baltic Sea a repeated collection and analysis of the same species in the benthic fauna and of the sediment would be of great value.

ACKNOWLEDGEMENT

The author wants to express his thanks to Mr S Lampe for his assistance in the field work and to the personnel who has performed the analytical work and also to Dr S Evans for fruitful discussions. Without the contribution from the Swedish National Board of Radiation Protection the participation in the cruise had not been possible and the author is very grateful for the support.

REFERENCES

1. Agnedal, P.-O., Radioactive and stable elements in the bottom fauna and sediment in the Baltic Sea. Studsvik-NP/86/45, 1986.

2. Aarkrog, A., Slutrapportering af Risös måleprogram (Fase II) i forbindelse med Chernobylulykken. Helsefysikalisk afdelning, Januari 1987, Risö.

3. Die Auswirkungen des Kernkraftwerkenfalles von Tshernobyl auf Nord- und Ostsee. Zweichenbericht des Deutschen Hydrographischen Institutes. Zeitraum von 29 April bis 14 Juni 1986. DHI, Hamburg den 20 August 1986.

4. Nies, H,, personal communication.

5. Ilus, E. et al, Finnish studies on radioactivity in the Baltic Sea after the Chernobyl accident in 1986. Suppl 11 to Annual Report STUK-A55, April 1987.

ARTIFICIAL RADIOACTIVITY IN THE NORTHEAST ATLANTIC

Hartmut Nies

Deutsches Hydrographisches Institut
Bernhard-Nocht-Str. 78
D-2000 Hamburg 4
Federal Republic of Germany

ABSTRACT

The Deutsches Hydrographisches Institut carried out a research programme during the years 1983 to 1985 in the Northeast Atlantic. One aim of the deep sea project NOAMP (Nord-Ost-Atlantisches-Monitoring-Programm) was the determination of the radioactive burden in the water column of this area, because in earlier years low-level radioactive waste has been dumped in the Northeast Atlantic. During 5 cruises with the german RV METEOR at 21 positions large water samples have been taken and analyzed for artificial radioactivity, in order to clarify, whether the waste has lead to detectable contamination of the water masses. In the following poster some results of the determination of Cs 137, Sr 90, Tritium and Plutonium are given and discussed in oceanographic context. All the analyzed samples are due to contamination of global fallout. A difference between dumpsite and control areas cannot be established.

INTRODUCTION

On two positions of the Northeast Atlantic more than 21 000 TBq β/gamma-activity and more than 500 TBq α-activity of low-level radioactive waste have been dumped (1). The stations, at which water samples have been taken from vertical profiles down to the bottom, are shown in fig. 1. Especially the long-lived nuclides Cs 137, Sr 90, and Tritium are used as tracer nuclides, because the activity of these nuclides gave the highest contribution to the waste. Plutonium was determined in vertical profiles at three positions.

RESULTS AND DISCUSSION

The highest activity concentration of Cs 137 is found in the surface layer down to a depth of about 500 m. The values lay between 2.8 and 4.5 mBq/l (Figs. 2a - 2g). The vertical profiles show an irregular structure. The reason for this scattering, especially in the upper 250 m, are due to intrusions of different water bodies and biological processes in the euphotic layer. Below 500 m depth, the concentration of Cs 137, Sr 90, and Tritium decreases in general down to a minimum of about 1.5 mBq/l for Cs 137 at a depth of about 1000 m (centre of the Mediterranean sea water). Below this layer the concentration increases sligthly to an intermediate maximum at a depth between the layer of the Mediterranean water and the Labrador water of 1200 to 1600 m. Below 2000 m depth the activity concentration of Cs 137 and Sr 90 lies between 0.4 and 1.1 mBq/l. At a depth of about 3000 to 4000 m the Cs 137 concentration exceeds the detection limit of about 0.05 mBq/l only by a very small amount. In the layers near to the bottom the concentration seems to increase very sligthly. This effect may be due to adsorption of the nuclides to particles within the nepheloid layer, because the samples have been analyzed without filtration.

In deeper layers higher values of Cs 137 and Tritium are found at station 71 (Fig. 2c, 4c), which is situated on the westernmost edge of the investigated positions. The Θ-S relation gives the indication that this water originates from northern of the subpolar front in the North Atlantic current system (2).

The activity ratio of Cs 137/Sr 90 of the most samples lies between 1.18 and 1.68 (Fig 5). Theoretically, a ratio of 1.45 would be expected in line with that found in global fallout. The activity of Tritium is found also in almost linear relation to Cs 137 and Sr 90 (Fig. 6). However, because of its quite different chemical property compared to that one of Caesium and Strontium, Plutonium is quite differently distributed in the water column of oceanic systems. The surface activity concentration of Pu 239/240 varies between 15 and 25 µBq/l (Fig 7). The maximum values of more than 30 µBq/l are ascertained at a depth between 800 and 1000 m. In this layer the activity concentrations of the previously mentioned nuclides have a relativ minimum. At a depth of 800 m a minimum of the dissolved oxygen, and a weak, intermediate Orthophosphate-, Nitrate-, and Silicate-maximum is found (Fig. 8). These nutrient profiles are controlled by the rate of decomposition of organic matter synthesized in the euphotic layers of the oceanic water column. Therefore, the plutonium maximum at this layer can be interpreted as the transport of adsorbed plutonium on organic particles at the sea surface. These particles are remineralized in the oxygen minimum layer, in which Plutonium is removed from the particles and released into the water. In the deeper layers the concentration of Pu does not increase as it can be seen for the other artificial nuclides. Only in a few samples, Pu 238 could be determined. In the few samples, in which Pu 238 was above the detection limit, the Pu 238/Pu 239+240 activity ratio lies at 0.04. This leads to the conclusion that the origin of the determined Pu in the water samples is due to global fallout, because the ratio in the dumped waste would be expected to be significantly higher.

CONCLUSION

The measured activity concentration of the investigated radionuclides gives no indication of a contamination of the water column in the Northeast Atlantic by the dumped low-level radioactive waste. The difference between the vertical profiles of the determined nuclides of samples from the dumping areas and of the areas of comparison can not be found. Differences in the profiles can be attributed to the geographical position due to the distribution of the different water masses in the Northeast Atlantic.

REFERENCES

(1) Review of the continued suietability of the dumpsite for radioactive waste in the North-East Atlantic, Nuclear Energy Agency / OECD, Paris 1985

(2) Kupferman, S.L., Becker, G.A., Simmons, W.F., Schauer, U., Marietta, M.G., Nies, H., Nature, Vol. 319, 474-477 (1986)

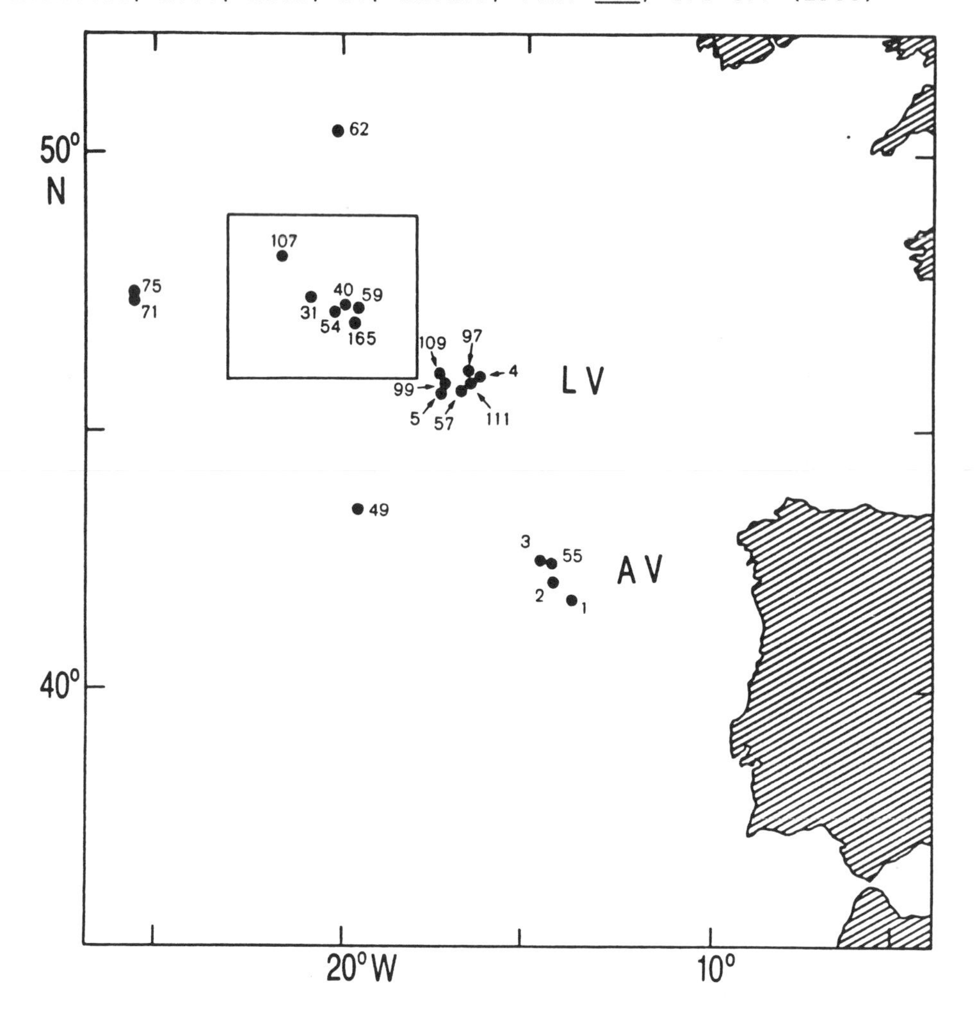

Fig. 1
Positions of the investigated stations
The NOAMP-investigation area is framed
LV: Last dumping area
AV: Previous dumping area of the Nuclear Energy Agency

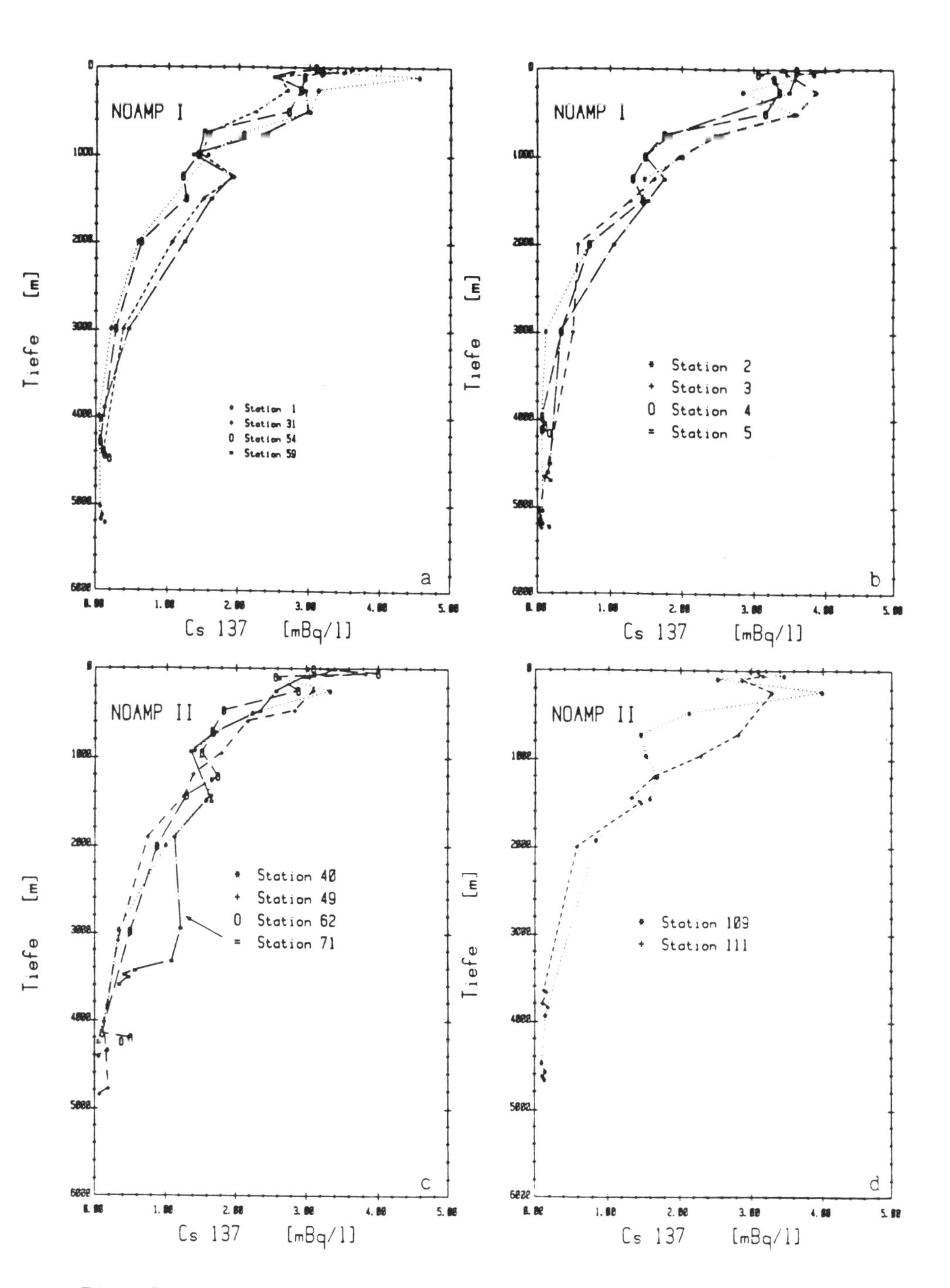

Fig. 2 a - d
Vertical profiles of the activity concentration of Cs 137

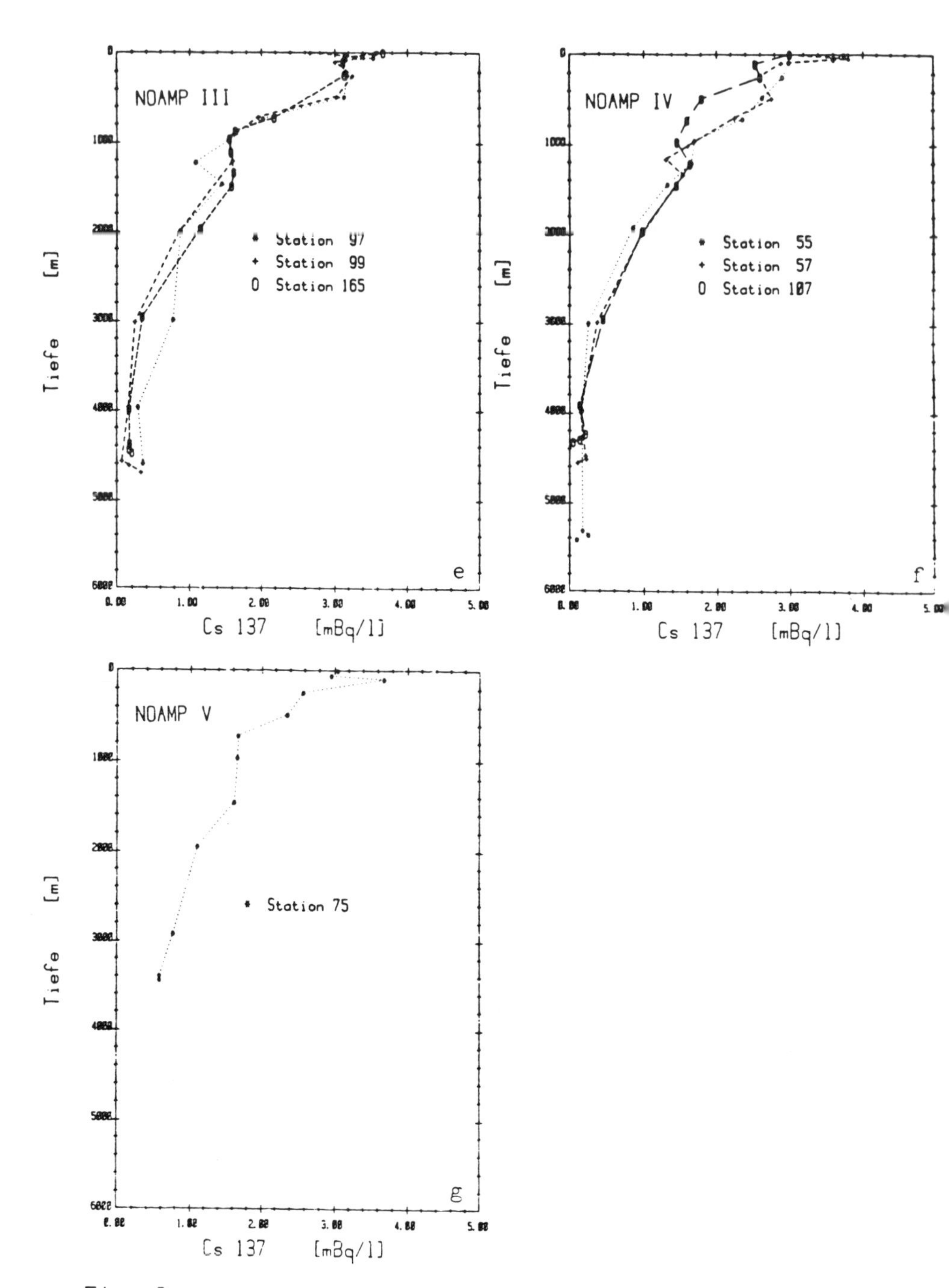

Fig. 2 e - g
Vertical profiles of the activity concentration of Cs 137

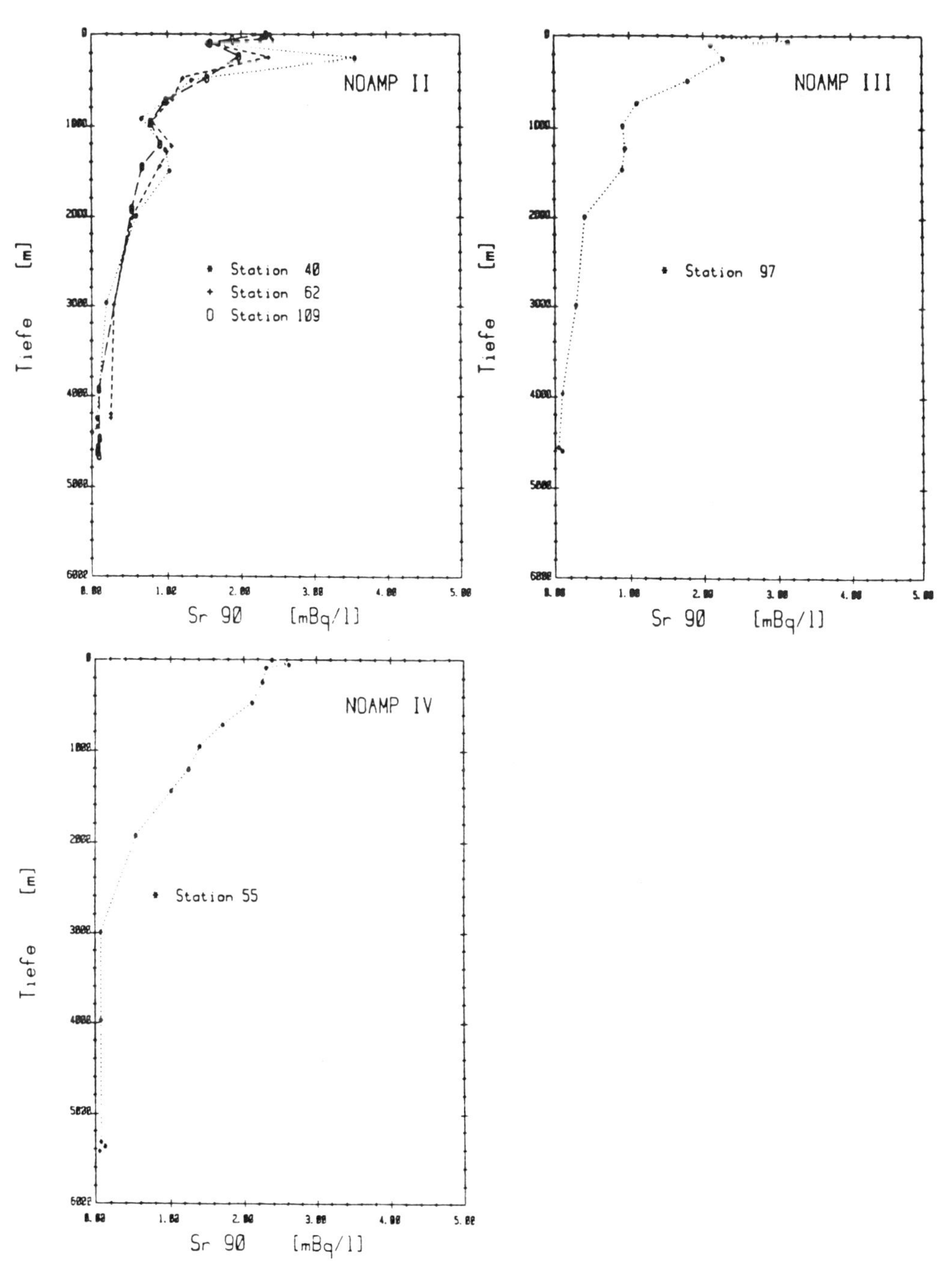

Fig. 3 a - c
Vertical profiles of the activity concentration of Sr 90

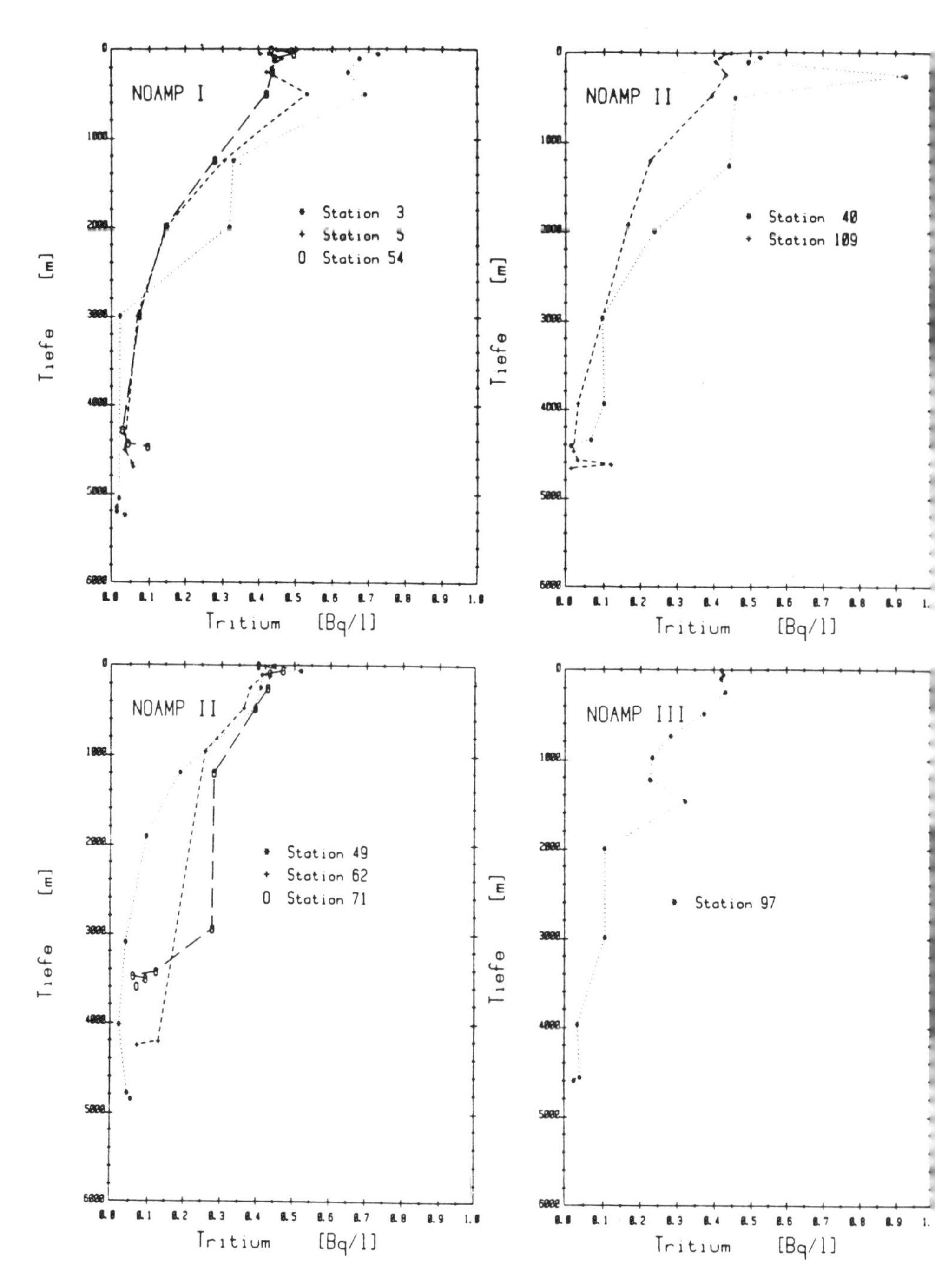

Fig. 4 a - d
Vertical profiles of the activity concentration of Tritium

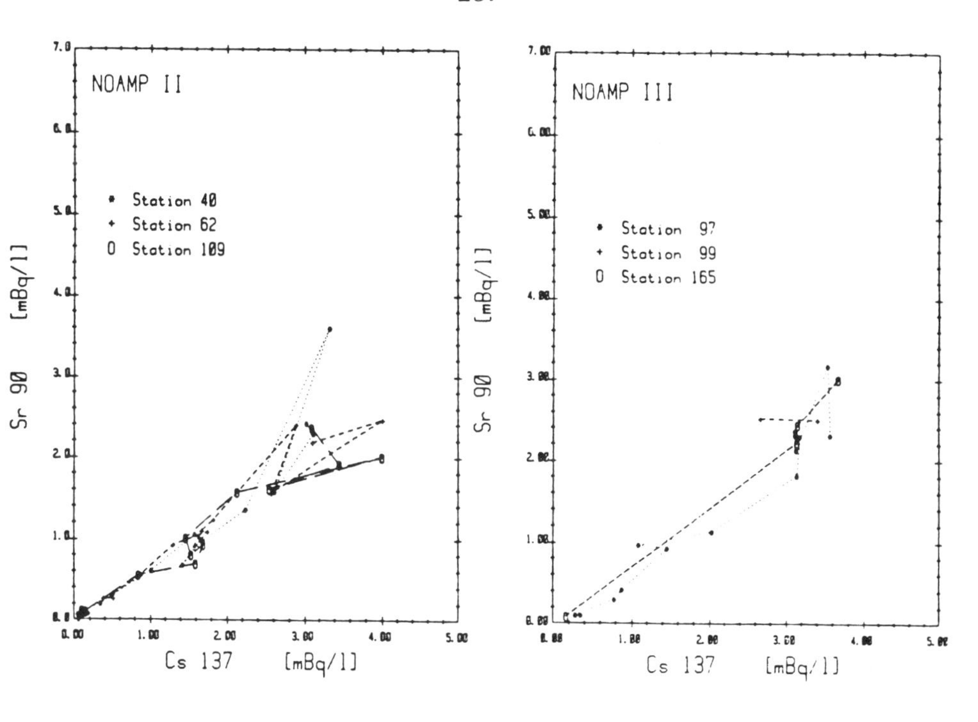

Fig. 5 a - b
Correlation between Cs 137 and Sr 90

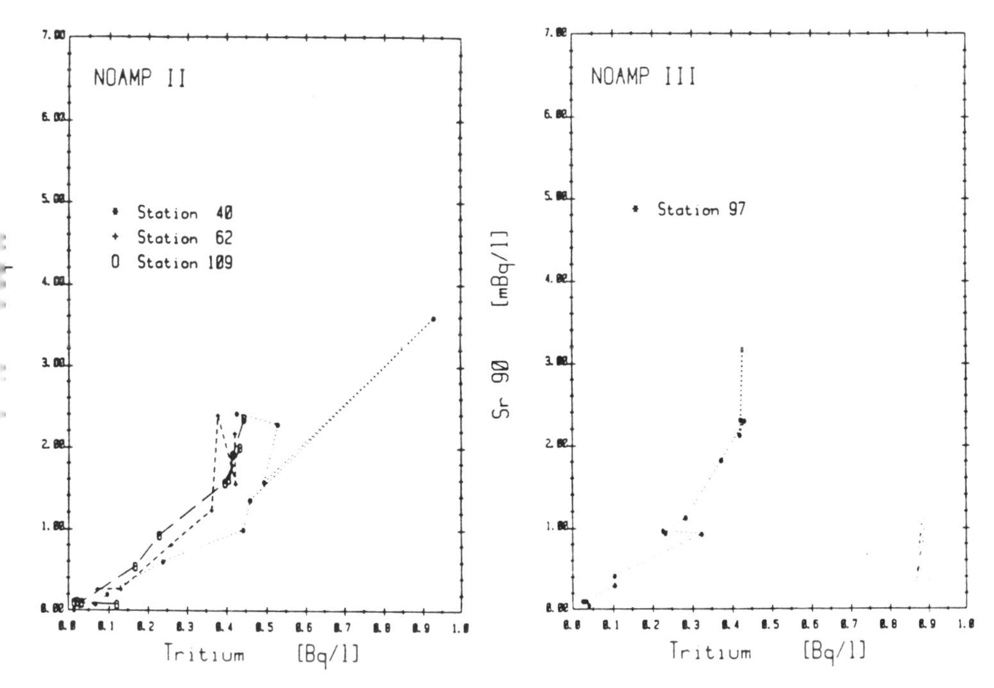

Fig. 6 a - b
Correlation between Sr 90 and Tritium

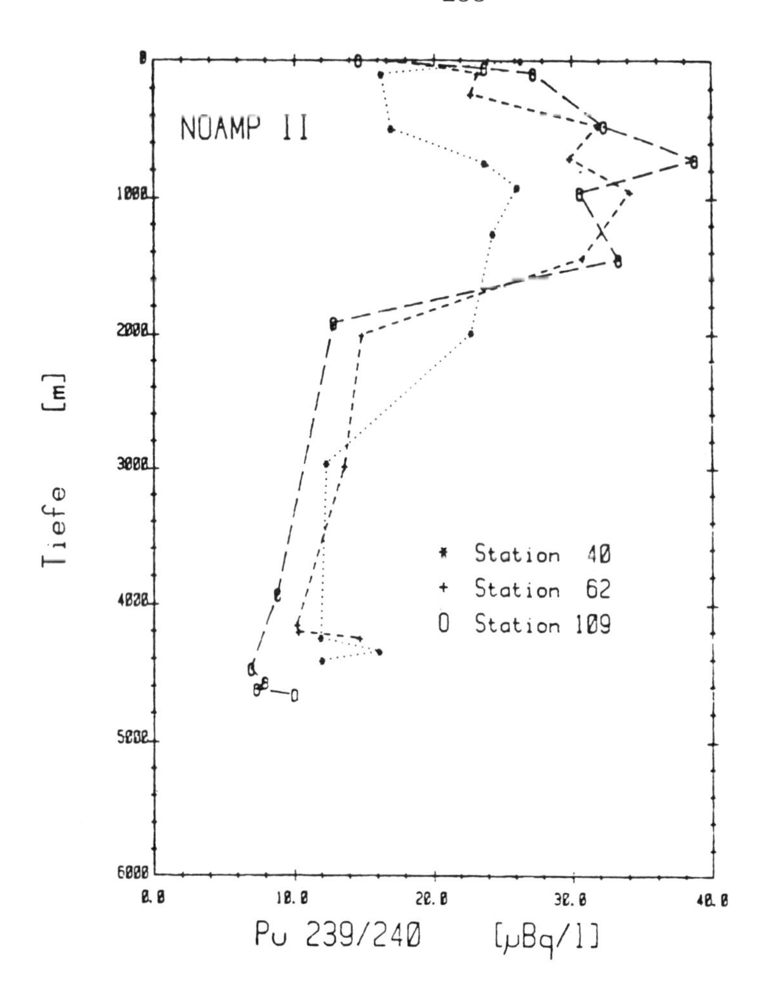

Fig. 7
Vertical profiles of the activity concentration of Pu 239/240

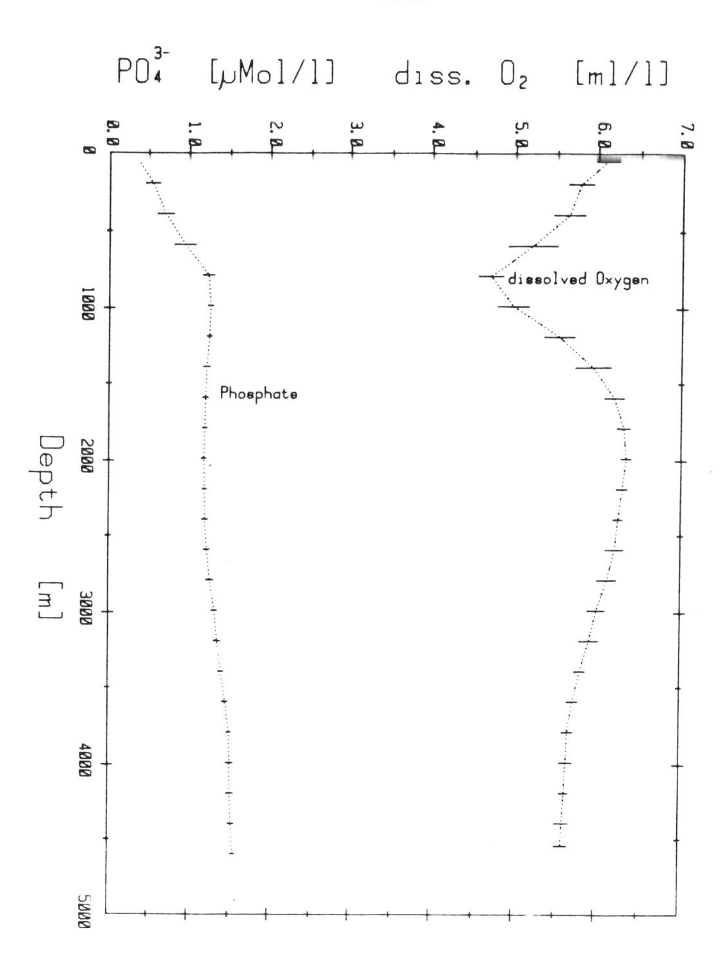

Fig. 8
Vertical profile of dissolved oxygen and orthophosphate

Utilisation de radionucléides artificiels (^{125}Sb - ^{137}Cs - ^{134}Cs) pour l'observation (1983-1986) des déplacements de masses d'eau en Manche

Auteurs

P. GUEGUENIAT, R. GANDON, Y. BARON
Laboratoire de Radioécologie marine de la Hague CEA IPSN/DERS/SERE

JC. SALOMON
Centre Océanologique de Bretagne. IFREMER BREST

J. PENTREATH
Laboratoire de Lowestoft-Ministry of Agriculture, Fisheries and Food (UK)

JM. BRYLINSKI
Station Marine de Wimereux

L. CABIOCH
Station Biologique de Roscoff

Ce travail a été réalisé avec la collaboration technique de Mrs R. LEON et J. MOISAN.

Résumé

Les radionucléides artificiels représentent d'excellents marqueurs pour l'étude des mouvements des masses d'eaux en Manche et à l'entrée de la mer du Nord à partir du moment où les impacts revenant aux diverses sources d'émission (Sellafield, La Hague, retombées atmosphériques consécutives aux tests nucléaires en atmosphère, effet Tchernobyl) sont clairement différenciés. Ceci a été réalisé dans ce travail avec 3 radionucléides ^{125}Sb - ^{137}Cs - ^{134}Cs. Dans ces conditions on a suivi simultanément la pénétration en Manche des eaux atlantiques, non soumises à des rejets industriels, et les mouvements des eaux marquées par l'établissement de la Hague caractérisés par des coefficients de mélange longitudinaux supérieurs aux coefficients de mélange transversaux. Les résultats obtenus dans trois secteurs stratégiques importants, Pas de Calais, Golfe Normand Breton, eaux côtières du Pays de Caux/Nord/Pas de Calais sont détaillés.

La Manche constitue un modèle exemplaire de mer à très fort régime de marée. Les mers mégatidales affectées par des marées du même ordre sont peu nombreuses à l'échelle planétaire. L'intensité du régime hydrodynamique, induit par la marée, a des conséquences capitales sur la nature, la distribution et la dynamique des suspensions et des sédiments, le cheminement et la transformation des espèces chimiques. Elle conditionne la structure, le fonctionnement et la production des peuplements.

Mais la Manche n'est pas seulement intéressante en tant que modèle de mer mégatidale. Elle appartient à l'un des plus vastes plateaux continentaux, incluant des mers de régimes hydrodynamiques variés, depuis les zones de faible énergie de la mer Celtique, de la majeure partie de la mer du Nord et de la Baltique, jusqu'aux fortes énergies dynamiques qui affectent l'Iroise, la Manche, l'entrée de la mer du Nord et de la mer d'Irlande. Dans cet ensemble, elle constitue globalement une zone de transit d'eaux atlantiques vers la mer du Nord à travers le Pas-de-Calais.

Elle recueille des apports côtiers, parmi lesquels ceux de la Seine sont les plus importants. On peut donc y trouver des terrains de choix pour l'étude des cheminements et des échanges aux interfaces, de substances dissoutes et de particules sédimentaires ou vivantes, d'origine côtière ou franchement marine, au sein d'un système de mers épicontinentales suffisamment vaste pour que les échanges avec l'océan n'y soient pas les seuls transferts importants.

La dualité eaux cotières-eaux extérieures apparaît nettement dans l'hydrologie du Pas de Calais où s'individualise un "fleuve côtier" caractérisé par une légère dessalure, séparé des eaux extérieures par une structure frontale (Brylinski et al. (1984), QUISTHOUDT et al. (1987)). Des espèces indicatrices montrent aussi l'existence d'une zone côtière très marquée au Nord du Pays de Caux (Cabioch - Glaçon (1977)) par laquelle transitent probablement des eaux issues de la Baie de Seine à la faveur d'un courant induit par la marée. Des résurgences d'eau douce au pied des falaises, les cours d'eau côtiers interviennent également dans la dessalure. Ce phénomène relatif aux eaux côtières a été également observé sur le modèle réduit hydraulique de la Manche construit sur la plaque tournante de Grenoble.

Ce travail aura pour objectif de montrer comment les radionucléides artificiels peuvent être utilisés pour marquer les déplacements des eaux atlantiques vers la mer du Nord à partir du moment où les impacts respectifs des diverses sources d'émissions (Sellafield, La Hague, retombées atmosphériques) sont clairement différenciées.

1 - LES RADIONUCLEIDES ARTIFICIELS EN MILIEU MARIN

1.1 - Sources

Il existe pour les radionucléides artificiels quatre principales sources d'introduction dans les eaux du Nord Est de l'Europe :

1. Usine de traitement de combustibles irradiés de La Hague dont les rejets sont effectués dans les eaux du Nord Ouest Cotentin.
2. Usine de traitement de combustibles irradiés de Sellafield dont les rejets sont effectués en mer d'Irlande.
3. Retombées atmosphériques consécutives aux tests nucléaires en atmosphère.
4. Effet Tchernobyl.

La prise en compte d'un traceur radioactif le ^{137}Cs, principalement rejeté par l'établissement de Sellafield ((Kautsky et Murray, 1981), (Jefferies et al., 1982)) a contribué à améliorer les connaissances déduites de la compilation de données de courant, de salinités, de températures, de photos par satellites (Pingree, 1980 ; Otto, 1983, Ellet et Mc Dougall, 1984)*. Ainsi dans un travail de synthèse Djenidi* et al. 1985 ont pu estimer des valeurs moyennes, pour les flux aux frontières, d'un modèle numérique couvrant la mer du Nord, le Skaggerak, la Manche, la mer Celtique et la mer d'Irlande. De son côté, Garnier (1985) a passé en revue les études théoriques et expérimentales susceptibles de concourir à la validation de modèles de transport et de dilution à longue distance des radionucléides artificiels.

La cartographie du ^{137}Cs, en mer du Nord, a pu être établie à plusieurs reprises par les scientifiques anglais et allemands avec de grandes campagnes se déroulant pendant un mois. Ces cartographies ont apporté des informations précieuses concernant la dynamique de l'ensemble de la mer du Nord et principalement les échanges à partir de la mer d'Irlande vers la mer du Nord par la voie Nord entre l'Ecosse et la Norvège. Par contre la communication Manche - mer du Nord, marquée par les rejets de l'établissement de La Hague, n'a été que peu exploitée, en particulier sur le plan du ^{137}Cs en raison de la prédominance des rejets de Sellafield (10 - 100 fois plus élevés) . . .

* Considérations empruntées à Djenidi et al. (1986).

1.2 - Stratégie utilisée

Notre stratégie est susceptible de relancer les études de mouvements d'eau basées sur les traceurs radioactifs :

Elle concilie les approches nucléaires et classiques (analyse multiparamétriques modèles numériques . . .) avec l'apport d'un groupe structuré (GRECO-Manche) et des équipes en physique, sédimentologie, biologie et chimie.

Elle introduit l'étude d'un nouveau radionucléide le ^{125}Sb.

On a reporté dans le tableau n°1 les valeurs des rejets de l'établissement de La Hague entre 1983 et 1986. Les données présentées complétent celles de Calmet et Guégueniat (1985) qui n'allaient pas au-delà de 1982. Seuls les rejets de 3 radionucléides ^{125}Sb (marqueur de La Hague), ^{137}Cs (marqueur de Sellafield), ^{134}Cs (marqueur de Tchernobyl) importent dans ce travail, les autres valeurs de rejets seront utiles pour la compréhension d'autres travaux : (Aarkrog et al. (1987), Germain et al. (1987), Masson et al. (1987)), portant sur la distribution des radionucléides artificiels chez l'algue brune Fucus et chez la moule.

Années	$^{3}_{1}H$	$^{106}Ru.Rh$	^{134}Cs	^{137}Cs	$^{90}Sr + Y$
1983	1.16×10^6	674.008	4920	22981	283314
1984	1.46×10^6	702.081	4810	29804	219061
1985	2.59×10^6	874.245	8177	29371	93696
1986		806.066	1393	9597	

Années	^{144}Ce	^{125}Sb	^{60}Co	$^{238}Pu + ^{239}Pu$
1983	4764	148903	13454	84
1984	6349	131451	24601	136
1985	4352	108607	15332	140
1986		142501	12476	

TABLEAU N° 1 : Rejets annuels de l'Etablissement de La Hague de 1983 à 1986 (en GBq).

Les conditions de rejet ont beaucoup d'importance dans la dispersion. A la Hague, avant le début 1984, les rejets commençaient deux heures et demie avant la pleine mer et se terminaient une demi-heure après. Depuis, la durée de rejet est plus courte et se déroule entre 2 heures et demie avant la pleine mer et une demi-heure avant.

1.3 - Propriétés des radionucléides artificiels mesurés

1.3.1 - L'antimoine 125

L'antimoine 125 (période 2,7 ans) existe à l'état de valence V dans l'eau de mer [Sb (OH)$_5$], [Sb (OH)$_6$]$^-$. Il est dosé après coprécipitation, sur MnO_2 à pH3 : les rendements atteignent 90 - 100 %.

Les coefficients de distribution du ^{125}Sb sur les sédiments sont compris entre 100 et 500. Dans ces conditions le pourcentage de formes dissoutes atteint plus de 99 % pour une charge en matières en suspension (MES) de 10 mg.l^{-1}, plus de 95 % pour une charge en MES de 100 mg.l^{-1}. Les teneurs des eaux de la Manche en MES étant très inférieures à 100 mg.l^{-1} on peut se dispenser de filtrer les eaux avant d'effectuer les traitements d'extraction du ^{125}Sb. La solubilité du ^{125}Sb dans les eaux du N.O. Cotentin a été régulièrement vérifiée : dans les conditions les plus défavorables rencontrées (forte charge en suspension de 40 mg.l^{-1}), le pourcentage de ^{125}Sb particulaire atteignait 0,2 %.

Par ailleurs si on établit un rapport entre les activités présentes dans les sédiments superficiels (baie du Mont St Michel par exemple) et les activités rejetées à l'émissaire on obtient en 1978 les valeurs suivantes (x 10^{-16}) ^{144}Ce = 157 ; ^{137}Cs = 6,2 ; ^{106}Ru = 4,4 ; ^{125}Sb = 1,4. Ainsi, à activités égales rejetées, le ^{125}Sb se fixe sur les sédiments considérés 3,1 fois moins que le ^{106}Ru, 4,4 fois moins que le ^{137}Cs, 112 fois moins que le ^{144}Ce. Un calcul analogue effectué sur les résultats d'analyses de radioactivité artificielle dans des constituants biologiques, recueillis aux environs immédiats de l'émissaire de rejet de l'usine de La Hague, montre qu'à activité égale rejetée, le ^{125}Sb se fixe 33 fois moins que le ^{106}Ru sur l'algue Fucus serratus, 5 fois moins sur l'algue Ascophyllum nodosum, 3 fois moins sur l'algue Pelvetia canaliculata, 20 fois moins sur les mollusques gastéropodes (Germain et al. 1979).

Les facteurs de concentration du ^{125}Sb sur algues, constituants du milieu marin (plancton en particulier) sont très faibles, l'algue Pelvetia présente toutefois la particularité de constituer un meilleur capteur que les autres algues étudiées jusqu'ici (Germain et al.1979).

L'antimoine 125 représente par conséquent un traceur conservatif au sein de la masse d'eau qui offre par ailleurs la particularité d'être un marqueur spécifique de l'établissement de La Hague. Les valeurs des rejets 149 000 GBq en 1983, 131 500 en 1984, 108 600 en 1985, 142500 en 1986, sont environ 4 fois plus élevées qu'à Sellafield. L'antimoine est absent dans les retombées atmosphériques antérieures et postérieures à Tchernobyl.

1.3.2 - Le Césium 137 (période 30 ans)

Le ^{137}Cs est un element marqueur des rejets de Sellafield à caractéristique essentiellement soluble. Il est présent dans le bruit de fond de l'eau de mer
(2 à 6 mBq.l^{-1}). En 1983 dans le cadre d'une collaboration anglo-française nous avons mesuré des rapports ^{137}Sb/ ^{125}Cs de 0.02 à 0.05 à Sellafield, de 3-10 à la Hague.

1.3.3 - Le Césium 134 (période 2 ans)

Le ^{134}Cs ou plus précisément le quotient ^{137}Cs/^{134}Cs est caractéristique des retombées de Tchernobyl. Ce rapport est voisin de 2 nettement inférieur à celui des rejets de la Hague ou de Sellafield qui se situent en moyenne entre 5 et 10. Le ^{134}Cs est absent dans les retombées atmosphériques antérieures à Tchernobyl.

1.4 - Cadre du travail

Ce travail a été financé dans le cadre d'une collaboration IPSN-COGEMA, il s'intégre dans les actions du GRECO Manche. Les moyens à la mer ont été fournis par le GRECO Manche et la Marine Nationale ; les comptages ont été effectués au Groupe d'Etudes Atomiques de la Marine Nationale.

Dans ce qui va suivre on présentera les grandes lignes des transferts obtenues en 1983 et 1986 avec l'antimoine et le cesium 137. Ensuite on détaillera les données obtenues dans le Pas-de-Calais.

2 - LES GRANDES LIGNES DES TRANSFERTS EN MANCHE

2.1 - La pénétration des eaux Atlantiques (^{137}Cs en 1983)

La topographie des zones non soumises à des rejets industriel a été établie pour la première fois en mai-juin 1983 avec le ^{137}Cs (voir figure N°1) : des eaux Atlantiques, vierges de rejets industriels, s'avancent jusqu'au large du Cap de La Hague. Dans cette situation il n'y a pas interférence en Manche Ouest entre les rejets de La Hague et ceux de Sellafield empruntant la voie Sud.

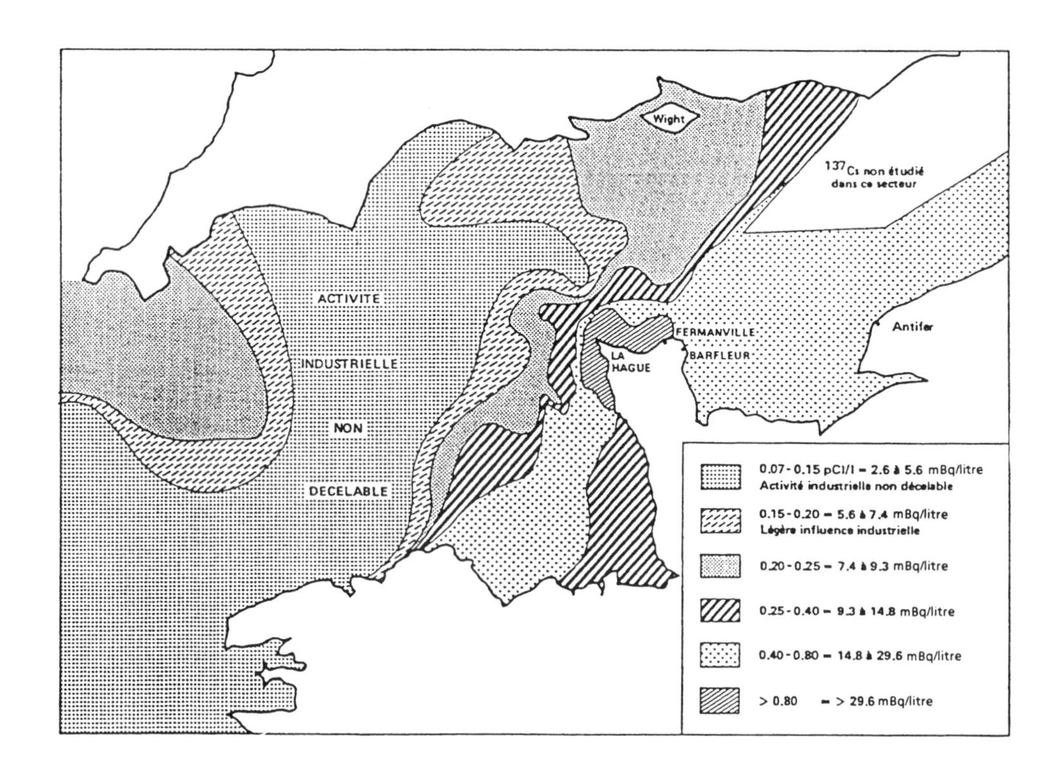

FIGURE N°1 : Distribution du ^{137}Cs en Manche Est et en manche Ouest au cours de l'année 1983

2.2 - La dispersion du ^{125}Sb en Manche (1986)

Sous la poussée des eaux Atlantiques les radionucléides rejetés par l'Etablissement de La Hague se dispersent essentiellement vers l'Est. On observe sur la figure N°2 qui représente la distribution du ^{125}Sb en juin 1986, une structure en bandes parallèles au sens du courant avec un fort gradient décroissant entre les eaux françaises et anglaises. Les côtes Sud de l'Angleterre sont en quelque sorte protégées par cette dérive : les activités des eaux comprises entre 0 et 12 $mBq.l^{-1}$ sont nettement moins élevées que celles de la Baie de Seine et des eaux côtières du Pays de Caux Nord Pas-de-Calais (30-80 $mBq.l^{-1}$), elles-mêmes inférieures à celles du N.O. Cotentin (80-300 $mBq.l^{-1}$). On retrouve sous une autre forme la prédominance en Manche des coefficients de mélanges longitudinaux sur les coefficients de mélanges transversaux qui selon Pingree (1975) atteint un facteur 10.

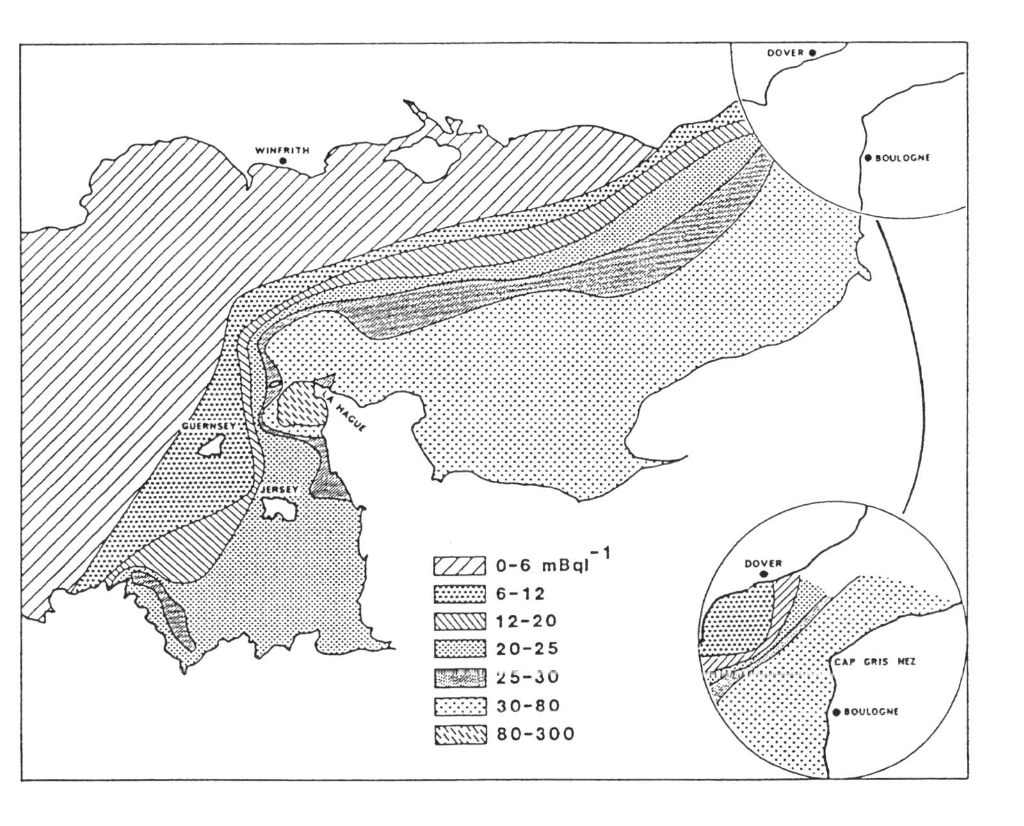

FIGURE N°2 : Dispersion du ^{125}Sb en juin 1986

2.3 - Pénétration des eaux de la Manche en Mer du Nord

Les eaux marquées par l'Etablissement de La Hague sont caractérisées par un rapport $^{125}Sb/^{137}Cs$ compris entre 3,2 et 4 qui tient compte du bruit de fond de l'eau de mer. Dès l'entrée en Mer du Nord, seule une cellule d'eau possède encore cette propriété (voir figure N°3).

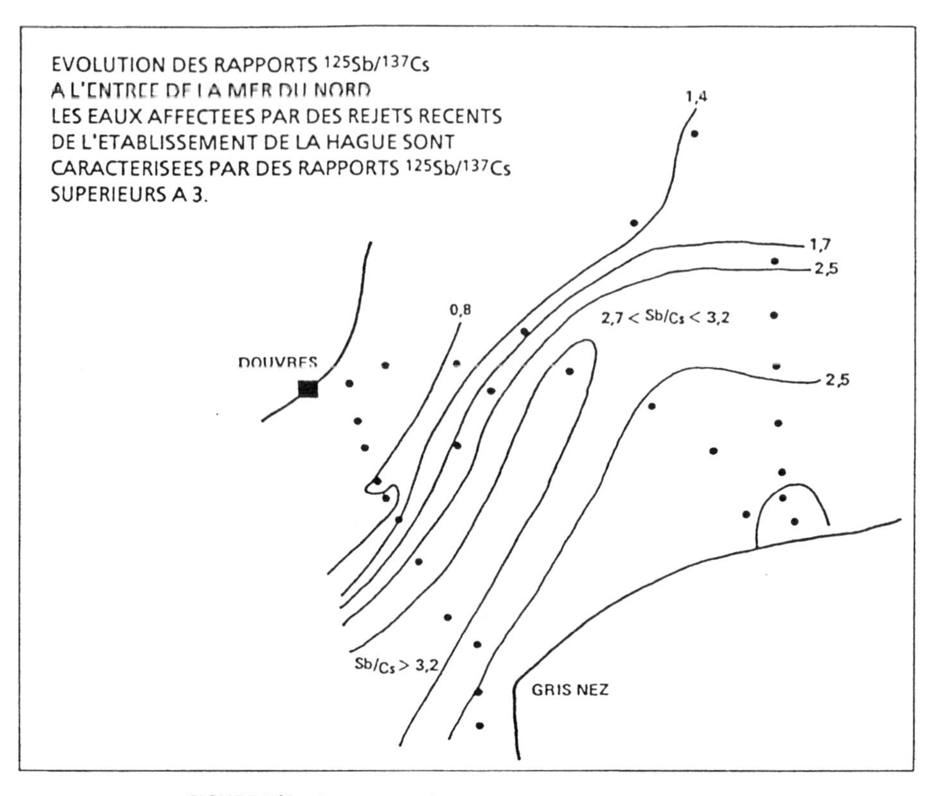

FIGURE N°3 : Campagne des mortes eaux en Juin 1986

Le Pas-de-Calais représente un secteur clef pour étudier les transferts entre la Manche et la Mer du Nord. Sur un axe cap Gris Nez - Douvres le rapport $^{125}Sb/^{137}Cs$ augmente de 2,2 dans le fleuve côtier à 3.2 puis diminue par paliers pour atteindre 0,8 dans les eaux côtières de Douvres. Dans cette dernière région la fraction de ^{137}Cs imputable aux rejets de l'Usine de la Hague ne représente que 25 %, le reste pourrait être du aux rejets de l'Usine de Sellafield et aux retombées de Tchernobyl. Près des côtes françaises la source principale de ^{137}Cs serait la Hague (pour 60 %). Le reste pourrait être attribué à des apports telluriques marqués par les retombées de Tchernobyl affectant le fleuve côtier où les eaux côtières de l'entrée de la mer du Nord, hypothèse que permet d'envisager la distribution de la radioactivité du ^{134}Cs en Manche.

Le fleuve côtier

Le césium 134, marqueur des retombées de Tchernobyl se concentre dans les eaux côtières du pays de Caux et du Nord Pas-de-Calais à des teneurs de l'ordre de 5.5 à 7 mBq par litre et des rapports $^{137}Cs/^{134}Cs$ faibles, compris entre 3.8 et 4.5. Dans les eaux marines extérieures les activités en césium 134 sont moins élevées (1,5 - 4 mBq par litre) et les rapports $^{137}Cs/^{134}Cs$ plus élevés (5 - 11).

Dans les eaux du Nord-Ouest Cotentin les valeurs de ^{134}Cs rencontrées ne dépassent pas 5.5 mBq par litre alors que les rapports $^{137}Cs/^{134}Cs$ atteignent 7 à 12.

On estime que la fraction de ^{134}Cs observée en excès dans le fleuve côtier provient des apports telluriques de la Seine et des autres rivières de moindre importance (Somme, Canche, Authie ...)

2.4 - Comparaison entre les situations du ^{125}Sb entre 1983 et 1986

Depuis 1983, les rejets en ^{125}Sb de l'Etablissement de La Hague ont été relativement constants ; jusqu'à cette date les fractions dispersées vers l'Est ou vers l'Ouest représenteraient respectivement 95 % et 5 % de ces rejets. Cette première estimation ressort d'une exploitation par modèles mathématiques, des données de radioactivité artificielle : F. Salomon et al. (1987). A l'heure actuelle, avec des rejets plus courts, la dérive vers l'Ouest est encore plus faible comme le montre la comparaison des figures N°2 (situation 1986) et N°4 (situation 1983) mais le nouvel état d'équilibre n'était pas encore atteint en 1985.

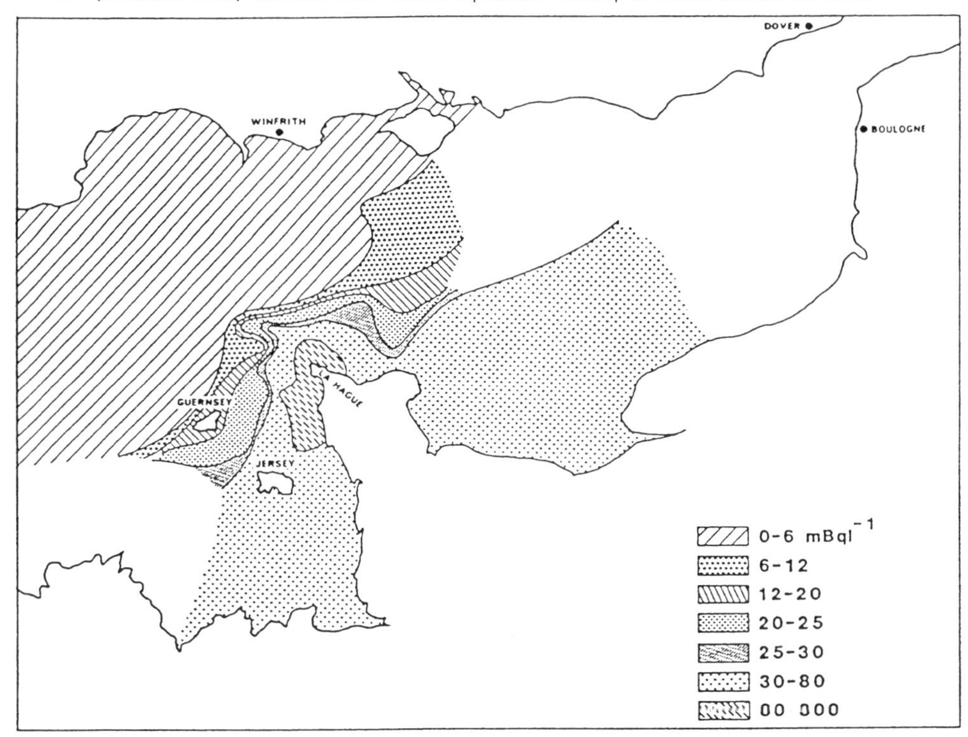

FIGURE N°4 : Dispersion du ^{125}Sb en Juin 1983

3 - LE PAS-DE-CALAIS

Le Pas-de-Calais constitue un détroit clef pour l'injection des radionucléides artificiels de l'Etablissement de La Hague en Mer du Nord. Il ne fonctionne pas de manière uniforme aussi bien en surface qu'en profondeur. La structure présentée sur la figure N° 2, n'est pas figée. En effet, lors d'une campagne réalisée en vives eaux en novembre 1985 nous avons constaté que les eaux Atlantiques ($2 < {}^{137}Cs < 6 mBq l^{-1}$) s'étaient avancées jusque dans le centre du Pas de Calais. Il en résultait une distribution différente du ^{125}Sb représentée sur la figure N°5.

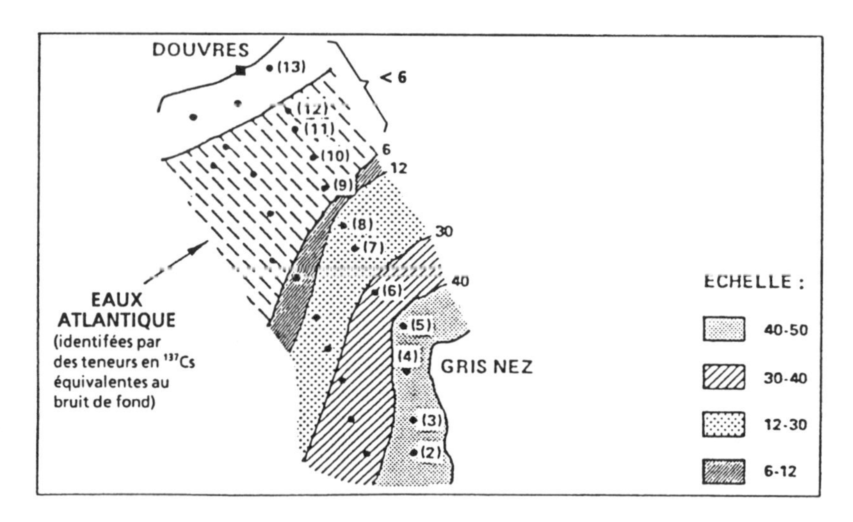

FIGURE N°5 : Evolution du ^{125}Sb le long des deux radiales transmanches parcourues en période de vives eaux, en Novembre 1985.

a) Les eaux les plus influencés par les rejets de l'Etablissement de La Hague sont plaquées sur le littoral dans le fleuve côtier.

b) Les côtes anglaises, dans cette configuration, ne sont plus affectées par les rejets de l'Etablissement de La Hague.

Cette campagne avait été effectuée avec une optique pluri-disciplinaire : les résultats (figure N°6) de température, salinité, bathymétrie Quisthoudt et al. (1987) complètent très utilement les précédents et matérialisent le fleuve côtier.

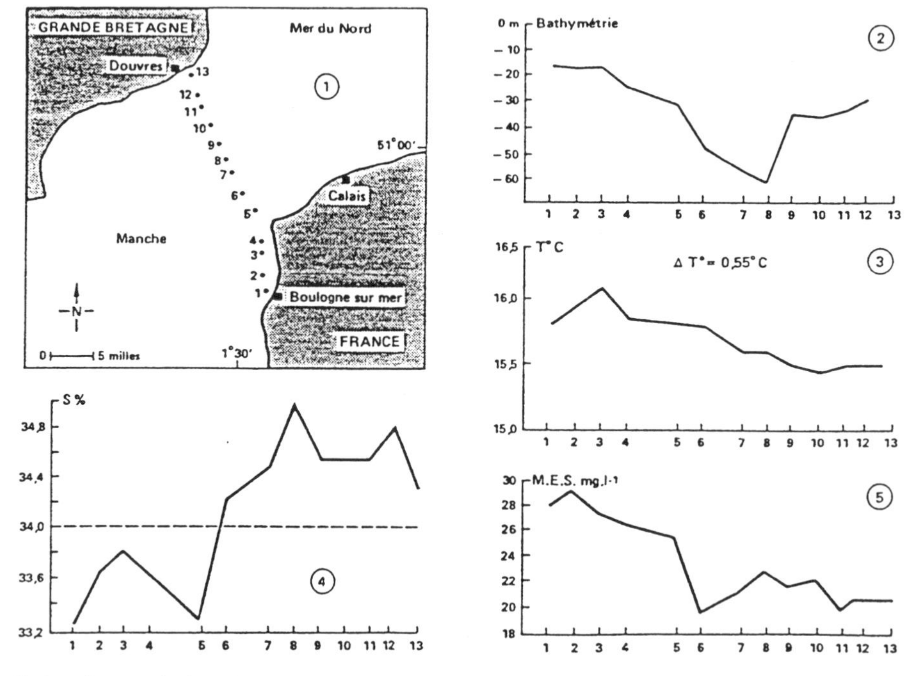

FIGURE N°6 : Evolution de la bathymétrie (2), température (3), salinité (4) et des matières en suspension (5) sur la radiale Boulogne-Douvres (1) selon Quisthoudt et al. (1987).

Conclusion

Les mesures de radioactivité artificielle effectuées par de nombreux auteurs en mer d'Irlande, mer Celtique, Manche, mer du Nord, apparaissent très prometteuses sur le plan des applications fondamentales en océanographie physique.

Pour exploiter pleinement l'utilisation des traceurs radioactifs comme outil océanographique il faudrait pouvoir coordonner les diverses actions engagées et modifier la conception de l'organisation des campagnes. Ainsi une campagne sur l'ensemble de la mer du Nord qui dure un mois avec des moyens lourds à la mer doit pouvoir s'appuyer sur des radiales répétitives effectuées dans le Pas-de-Calais et l'entrée de la mer du Nord avec des moyens plus légers. C'est à cette condition, utilisation simultanée à plusieurs navires, que l'on pourra exploiter à fond les résultats de radioactivité. Dans une même optique ces mesures doivent pouvoir s'intégrer dans un vaste contexte de collaboration entre diverses disciplines. En particulier les découvertes faites sur les propriétés du "fleuve côtier" dans le Pas-de-Calais témoignent l'intérêt d'une stratégie qui doit, dans ce cas, aboutir à une étude de transferts des polluants métalliques et organiques dissous entre la Manche et la mer du Nord.

Remerciements

Les campagnes ont été réalisées à bord du Pluteus, du "Côtes de Normandie", du SEPIA navires océanographique du CNRS et du Dahlia du Groupe d'Etudes Atomiques de la Marine Nationale de Cherbourg. Nous remerçions vivement ces équipages pour leur collaboration, leur efficacité et leur excellente ambiance de travail malgré des conditions souvent difficiles.

Bibliographie

PINGREE (RD) (1980)
Physical Oceanography of the Celtic Sea and English Channel, in the North-West European Shelf Seas : the Sea Motion II : Physical Chemical Oceanography, and Physical Resources,
Elsevier Ocean. Series, 24 B, 415 - 466

SALOMON (JC), GUEGUENIAT (P), ORBI (A), BARON (Y) (1987)
A lagrangian model for long term tidally induced transport and mixing. Verification by artificial radionuclide concentrations.
Colloque Cherbourg. Radioactivité et océanographie

AARKROG (A), DAHLGAARD (H), DUNIEC (S), GUEGUENIAT (P), HOLM (E) (1987)
The application of seaweeds as bioindicators for radioactive pollution in the Channel and Southern North Sea.
Colloque Cherbourg. Radioactivité et océanographie

BRYLINSKI (JM), DUPONT (J), BENTLEY (D) (1984)
Conditions hydrobiologiques au large du Cap Gris Nez (France) : premiers résultats.
Océanologica Acta. Vol 7, 3, pp 315 - 322

CABIOCH (L), GLACON (R) (1977)
Distribution des peuplements denthiques en Manche orientale, du Cap d'Antifer à la Baie de Somme.
C.R. Acad. Sc. Paris t 285 (18 juillet 1977)

CALMET (D), GUEGUENIAT (P) (1985)
Behaviour of radionuclides released into coastal walers.
IAEA - TECDOC 329

DJENIDI (S), RONDAY (F), BECKERS (Ph) and PIRLET (A) (1985)
Etude de la circulation résiduelle en vue de l'évaluation de la dispersion d'effluents radioactifs dans l'environnement marin (Plateau Continental Nord-Européen).
Rapport SC - 012 - B/BIAF/423 - F (SD)/DN01, Commissariat à l'Energie Atomique, 107 pp

DJENIDI (S), NIHOUL (JC), RONDAY (F), GARNIER (A) (1986)
Modèles mathématiques de courants résiduels sur le plateau continental Nord-Européen.
La Baie de Seine (GRECO-MANCHE). Université de Caen, 24-26 avril 1985 - IFREMER - Actes de colloques n°4 pp 73-84

ELLET (DJ) and Mc DOUGALL (N) (1984)
Quelques résultats d'observations effectuées à l'Ouest de la Grande Bretagne.
Vol. 1, Séries temporelles de mesures océaniques (1983), C.O.I. Unesco Paris, 21-25

GARNIER (A) (1985)
Choix de méthodes d'évaluation de l'impact à longue distance de rejets radioactifs en milieu marin.
Océanis Vol. 11, Fascicule 6, pp 597-625

GERMAIN (P), MASSON (M), BARON (Y) (1979)
Etude de la répartition de radionucléides émetteurs gamma chez des indicateurs biologiques littoraux des côtes de la Manche et de la Mer du Nord de février 1976 à février 1978.
Rapport CEA - R - 5017

GERMAIN (P), BARON (Y), CALMET (D), MASSON (M) (1987)
Répartition de deux traceurs radioactifs (^{106}Ru-Rh, ^{60}Co) chez deux espèces indicatrices (Fucus serratus L., Mytilus edulis L.) le long du littoral français de la Manche.
Colloque Cherbourg - Radioactivité et océanographie

JEFFERIES (DF), STEELE (AK) and PRESTON (A) (1982)
Further studies on the distribution of ^{137}Cs in British Coastal Waters.
Irish Sea, Deep Sea Res., Vol. 29, n°6A, 713-738

KAUTSKY (H) and MURRAY (CN) (1981)
Artificial radioactivity in the North Sea.
Revue de l'Energie Atomique, Vienne (Autriche), suppl. 2, 63-105

MASSON (M), PATTI (F), CALMET (D) (1987)
Teneurs en ^{99}Tc chez Fucus sp et dans les eaux de mer sur le littoral de la Manche : synthèse.
Colloque Cherbourg - Radioactivité et océanographie

OTTO (L) (1983)
Currents and water balance in the North Sea, in North Sea Dynamics, Springer Verlag.
Sundermann et Lenz Eds. 26 - 43

PINGREE (RD), PENNYWICK (L) and BASTIN (G.A.W) 1975
A time varying temperature model of mixing in the English channel J. Mar. Biol. Assoc. V.K Vol 55, 975-992

QUISTHOUDT (C), BENTLEY (D), and BRYLINSKI (JM) (1987)
Discontinuité hydrobiologique dans le détroit du Pas-de-Calais journal of Plankton Research Vol 9 n°5

DETERMINATION OF DISTRIBUTION PROCESSES, TRANSPORT ROUTES AND TRANSPORT TIMES IN THE NORTH SEA AND THE NORTHERN NORTH ATLANTIC USING ARTIFICIAL RADIONUCLIDES AS TRACERS

Hans Kautsky
Vogt Wellsstrasse 24a
D-2000 Hamburg 54

ABSTRACT

The fundamental distribution pattern of the Cs-137 in the North Sea, as well as the transport routes - within a certain variation width - can be viewed as being largely constant.

By following the temporal course of the Cs-137 : Sr-90 ratio at different points, a transport time can be deduced of around 3 years between the Sellafield Works and the Pentland Firth ; a further year would be needed to reach the southeastern North Sea. Measurements of the Cs-134 : Cs-137 ratio lead to corresponding transport times.

The distribution of the Cs-137 in the northern North Atlantic illustrates well the previously known current picture in this region.

The ratio Cs-134 : Cs-137 permits an estimate of the transport times from the Sellafield Works as far as the northern North Atlantic to be made. As far as Spitsbergen, the transport times are 6 to 7 years, as far as the region eastwards of Greenland they are 7 to 9 years.

GENERAL

The isotope ratios Cs-137 : Sr-90 and Cs-134 : Cs-137, and their temporal changes, are utilized for estimation of transport times. The basis for the studies presented are essentially the measurement results gained in the years 1977 to 1985 in the Deutsches Hydrographisches Institut in Hamburg (1) (2) (3). In several points, they are supplemented by corresponding results originating from Aarkrog and his collaborators (4) (5) (6).

One can look upon the nuclear fuel reprocessing plants of Sellafield Works on the Irish Sea and La Hague on the English Channel as being the principal sources of the artificial radionuclides present today in the North Sea and in the adjoining regions of the Atlantic towards the North. So far, the discharge from La Hague has been slight compared with that of the Sellafield Works.

In April 1977, the beginning of an influx appeared of larger quantities of Cs-137 into the North Sea in the region of the Pentland Firth. The activity concentration values rose to over 400 $Bq.m^{-3}$ in the surface water compared to 170 to 190 $Bq.m^{-3}$ in September 1976.

TRANSPORT PATTERN

During the years 1977 to 1985 an annual cruise was undertaken for the investigation of the distribution of artificial radionuclides in the water of the North Sea. The general distribution picture of caesium is largely consistent with the results of earlier investigations (Fig. 1).

The water flowing into the North Sea in the region of the Pentland Firth and the Orkney Islands moves along the British East coast towards the South. In the course of this route, one can recognize in 3 regions - at circa 57° to 58°N, 55° to 56°N and 53° to 54°N - transport currents directed towards the Northeast.

The water near Dover, flowing into the North Sea from the South, moves along the mainland coast, in a relatively narrow strip, towards the North. The activity concentrations of Cs-137 therein, and originating from La Hague, are essentially lower than in the water coming from the North. A steep concentration gradient of the Cs-137 therefore exists between these two water bodies. The one coming from the North and that coming from the South, flow between about 52°20'N 2°E and 56°N 6°30'E, alongside but separate from one another. Westwards of Jutland - between 54°30'N and 56°N - the boundary lies mostly at 6°30'E.

A mixing of these two water bodies, and that of the water masses on the northernmost transport route towards the East, first takes place at about 57°N and particularly in the region of the Skagerrak. The further transport then takes place mainly along the Norwegian coast towards the North ; small quantities reach the Baltic Sea.

Almost the whole of the Cs introduced into the sea with the waste waters of the Sellafield Works is transported into the northern North Atlantic and then into the North Sea towards the Norwegian coast. The vertical distribution of this Cs, in the northern North Atlantic, demonstrates that the water transport over these wide stretches takes place in the surface layer (down to about 200 m depth).

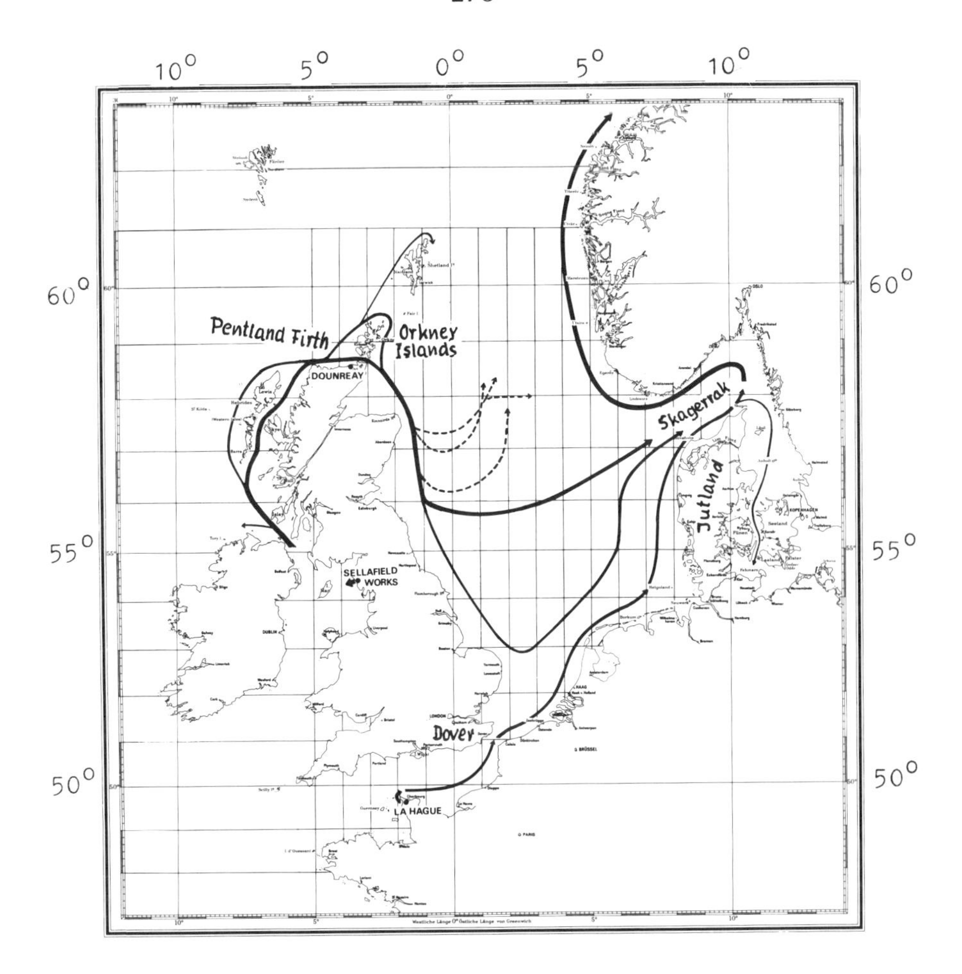

Figure 1 . Transport routes of Cs-137 deduced from the measurements of the activity concentration distribution in the years 1971 to 1984. Dotted lines indicate different temporal transport routes.

In the horizontal distribution pattern of the Caesium (Figs. 3, 4), there appears very clearly in outline the different current routes in the northern North Atlantic (9) :

a. Along the Norwegian coast, the Norwegian Atlantic Current
b. The branch of that current branching-off at about 64°N
c. The North Cape Current
d. The branching-off West Spitsbergen Current at about 70°N
e. The Bear Island Current coming from the Northeast

In addition, in 1982 - especially in connection with the measurement data of Aarkrog (4) (5) - one can recognize very well the water transport running from NNE towards SSW between about 78°N 3°W and 72°N 18°W eastwards of Greenland.

TRANSPORT TIMES

The course of the two curves of the Cs:Sr ratio in the discharge of the Sellafield Works and in the region of the Pentland Firth indicates a transport time of about 3 years from the Sellafield Works as far as the Pentland Firth (Fig. 2). If one compares further the corresponding ratio figures westwards of Jutland at 55°30'N 6°E, this results in a practically parallel curve, but offset by about a further year.

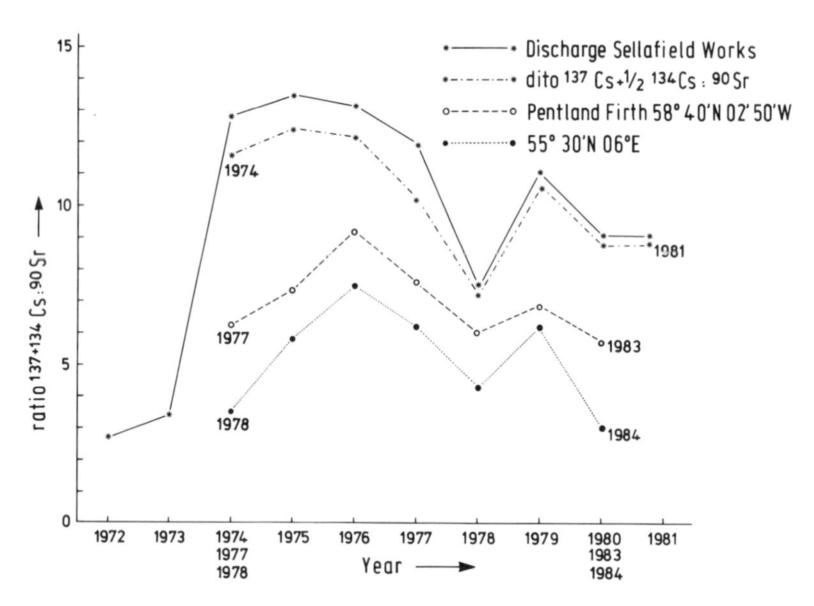

Figure 2 . Comparison of the ratio of the activity concentrations of Cs-137+134 : Sr-90 in the discharge from the Sellafield Works and at 2 positions in the North Sea

In order to estimate the transport times one can also use the temporal change of the Cs-134 : Cs-137 isotope ratio relative to the isotope mixture discharged from the nuclear fuel reprocessing plants.

In Table 1, the discharge of the Cs isotopes and their ratio in the waste waters for the Sellafield Works (7) and for La Hague (8) are given. In Table 2, the decay corrected Cs-134 : Cs-137 ratio figures - referred to the different discharge years - are summarized for the years 1982 to 1985.

If one considers the temporal changes of the ratio of the Cs-134 : Cs-137 in the water on the way from the Sellafield Works through the North Sea, then one attains transport times which are comparable with those which have been calculated before in another manner.

Especially over greater distances the exactitude of these methods depends naturally upon the relevance of the measurement data. In practice, only two factors should be viewed as being actually essential :

Table 1. Annual descharges of Cs-134 and Cs-137 (TBq.a-1) from the Sellafield Works and from La Hague.

	Sellafield Works			La Hague		
Year	Cs-134	Cs-137	Cs-134:Cs-137	Cs-134	Cs-137	Cs-134:Cs-137
1970	251	1 154	0.22	14	89	0.16
1971	236	1 325	0.18	48	242	0.20
1972	215	1 285	0.17	6.1	33	0.19
1973	166	768	0.22	8.4	69	0.12
1974	997	4 061	0.25	9.0	56	0.16
1975	1 081	5 231	0.21	4.3	34	0.12
1976	738	4 289	0.17	6.5	35	0.19
1977	594	4 478	0.13	9.5	51	0.19
1978	404	4 088	0.099	7.8	39	0.20
1979	235	2 562	0.091	3.6	23	0.16
1980	239	2 966	0.080	3.9	27	0.15
1981	168	2 357	0.071	6.0	39	0.15
1982	138	2 000	0.070	8.4	51	0.17
1983	89	1 200	0.074			
1984	35	434	0.080			

Table 2. Annual ratios of Cs-134 and Cs-137 in the discharges of the Sellafield Works, decay corrected to the years 1982, 1983, 1984 and 1985. Half lives of Cs-134 and Cs-137 are 2.07 and 30.02 years.

Release Year	1982	1983	1984	1985
1970	0.00514	0.00377	0.00276	0.00201
1971	0.00577	0.00421	0.00299	0.00225
1972	0.00738	0.00541	0.00396	0.00291
1973	0.013	0.00955	0.00699	0.00512
1974	0.020	0.015	0.011	0.0079
1975	0.023	0.017	0.012	0.0091
1976	0.026	0.019	0.014	0.010
1977	0.027	0.020	0.014	0.010
1978	0.028	0.021	0.015	0.011
1979	0.035	0.026	0.019	0.014
1980	0.042	0.031	0.022	0.016
1981	0.052	0.038	0.028	0.020
1982	0.070	0.051	0.037	0.027

a. The statistical counting error from the determination of the activity concentrations.
b. The input of Cs isotopes from other sources. Thereby, practically only 2 sources come into consideration :
a. The worldwide fallout.
b. The waste waters from La Hague.

An exact calculation of the influence of La Hague upon the Cs-134 : Cs-137 ratio is not possible. But this influence might be viewed as being very small compared with that of the Sellafield Works due to the fact that their discharge is many times that of La Hague.

During the years 1982 to 1985 water samples of 500 l to 2000 l were taken annually at certain positions in the northern North Atlantic. The separation of the Cs took place over ion exchangers. The exchanger was measured γ - spectro scopically in a Marinelli beaker with a semiconductor detector (Ge (Li)) for 250 hours.

The transport time calculations of the water from the ratio of the activity concentrations of the Cs-134 : Cs-137, were made based upon the following suppositions :

a. The only source for radiocaesium worthy of mention are the Sellafield Works.
b. In the surface water of the North Atlantic, there are still about 3 Bq.m-3 of Cs-137 present from the fallout of the atomic bomb tests (3).

If one considers the data from the different regions (Fig. 5) then one can calculate the following mean transport times of the water transport from the Sellafield Works to the respective regions :

Southern Norway	4 to 5 years
Middle Norway (Lofoten)	5 years
Northern Norway (North Cape)	5 to 6 years
Bear Island to the middle of Spitsbergen	6 to 7 years
North of Spitsbergen	7 to 8 years
Eastwards of Greenland	7 to 9 years
SSE of the southernmost point of Spitsbergen	8 to 9 years
70°N 0° Longitude	4 years

Two values appear to be of particular interest :

a. The one SSE of the southern tip of Spitsbergen, with a transport time of 8 to 9 years, could indicate the inflow of the Bear Island Current, which carries water indirectly over the Barents Sea and the Arctic Ocean again into the region of the West.
b. The transport time of 4 years calculated in the central Norwegian Sea in 1985 at 70°N 0° is certainly too short.

In general, the transport time values observed in 1985 appear to be somewhat shorter. This could be due to the fact that the background for fallout Cs-137 of 3 Bq.m-3, determined in 1982, has receded further and is only now conditionally applicable as a correction value.

The counting error of the Cs-134 values measured in the region eastwards of Greeland and at 70°N 0° lies between 20 and 30 %. The highest errors occurred in 1985.

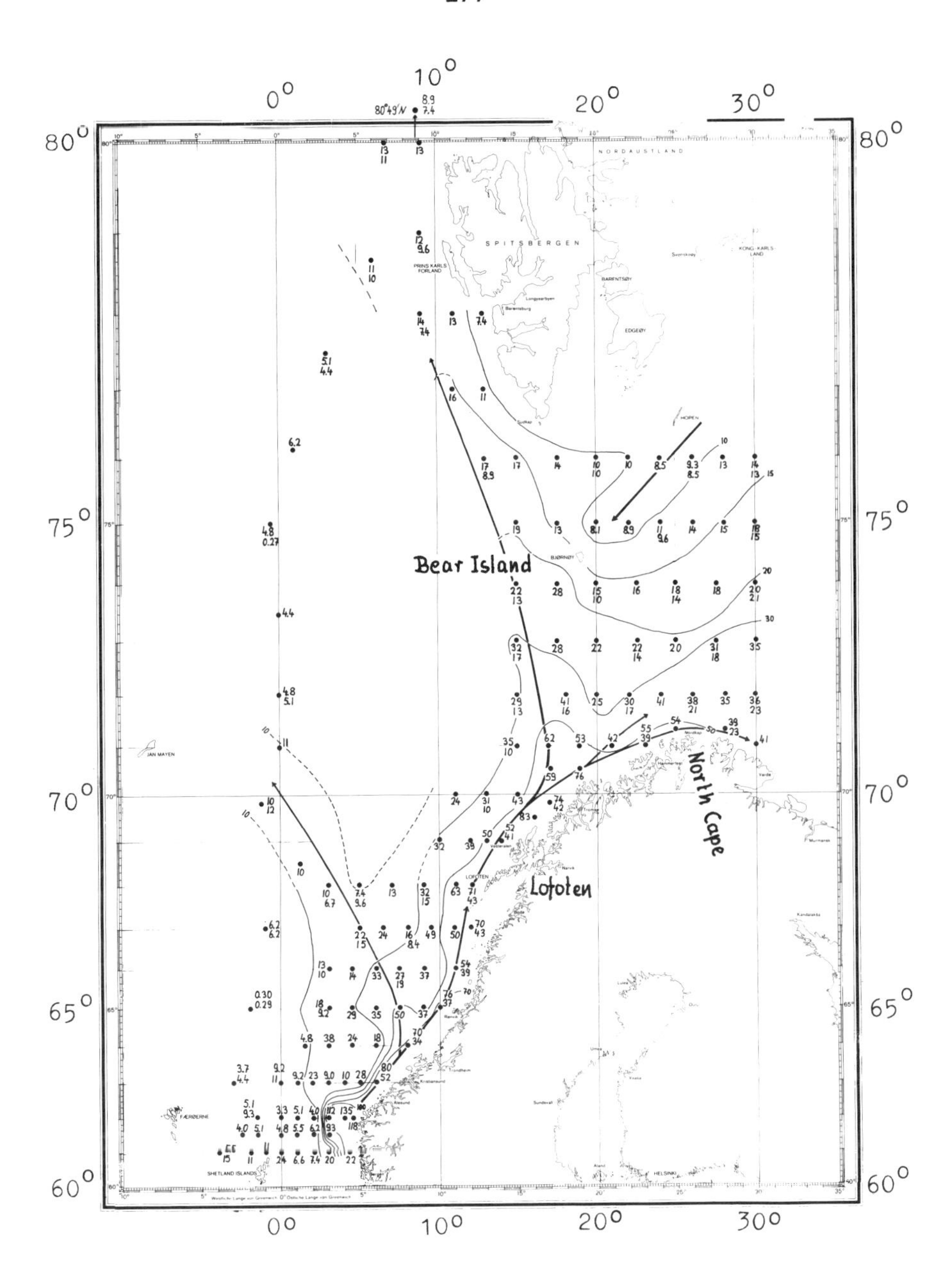

Figure 3 . Cs-137 + 134 (Bq.m-3) in surface water and 100 m depth. September 1979

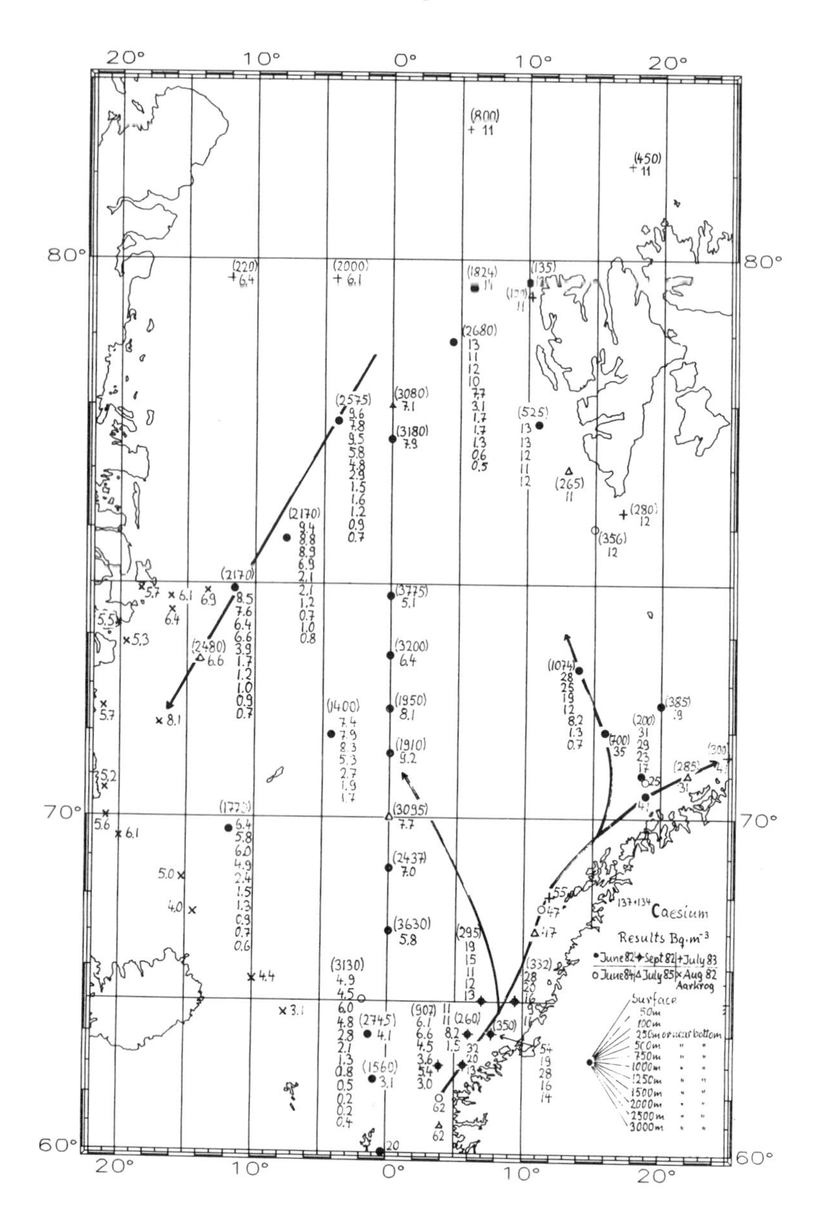

Figure 4 . Cs-137 + 134 (Bq.m-3) in different depths of the northern North Atlantic in different years

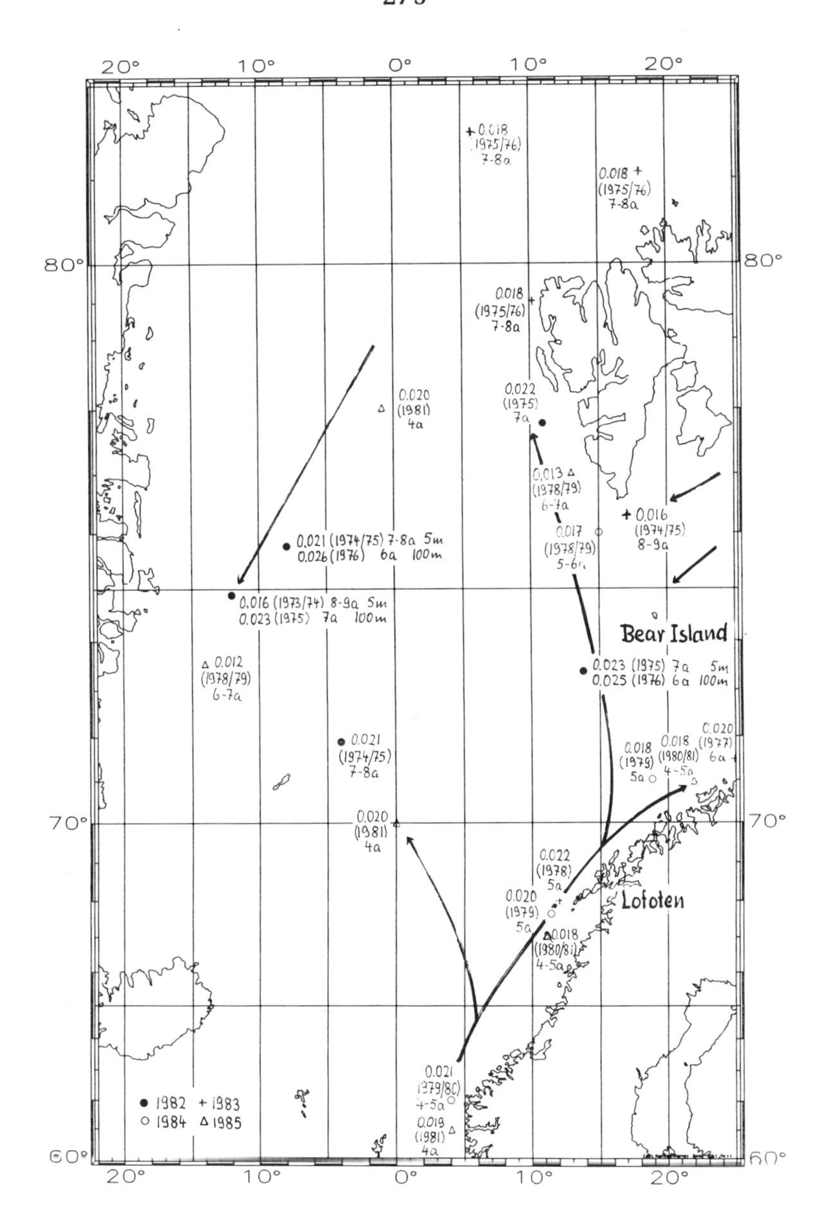

Figure 5 . Ratio of Cs-134 : Cs-137 in the water (surface and 100 m)

Figures in brackets indicate the relevant year of discharge of the Sellafield Works

Figures with "a" indicate the approximate transport time in years

REFERENCES

1. Kautsky, H., Distribution and content of different artificial radionuclides in the water of the North Sea during the years 1977 to 1981 (complemented with some results from 1982 to 1984. Dt.hydrogr.Z. 38, 1985, 193-224)

2. Kautsky, H., Distribution and content of 137 + 134Cs and 90Sr in the water of the North Sea during the years 1982 to 1984. Dt.hydrogr.Z. 39, 1986, 139-159

3. Kautsky, H., Investigations on the distribution of 137Cs, 134Cs and 90Sr and the water mass transport times in the northern North Atlantic and the North Sea. Dt.hydrogr.Z. 40, 1987, in press

4. Aarkrog, A., Dahlgaard, H., Hallstadius, L., Hansen, H., and Holm, E., Radiocaesium from Sellafield effluents in Greenland waters. Nature 304, 1983, 49-51

5. Aarkrog, A., Botter-Jensen, L., Dahlgaard, H., Hansen, H., Lippert, J., and Vilson, S.P., Environmental radioactivity in Denmark in 1982. Riso National Laboratory, Riso report R-487, 1983

6. Aarkrog, A., Boelskifte, S., Buch, E., Christensen, G.L., Dahlgaard, H., Hallstadius, L., Hansen, H., and Holm, E., Environmental radioactivity in the North Atlantic region. The Faroe Islands and Greenland included. Riso Report No. 528, 1984

7. Cambray, R.S., Annual discharge of certain long-lived radionuclides to the sea and the atmosphere from Sellafield Works, Cumbria 1957 to 1981. Harwell : United Kingdom Atomic Energy Authority, AERE-M 3269, 1982

8. Calmet, D., and Gueguenìat, P., Les rejets d'effluents liquides radioactifs du centre de traitement des combustibles irradiés de La Hague (France) et l'évolution radiologique du domaine marin. Int. Atomic Energy Agency Vienna, TECDOC - 329, 1985

9. Trangeled, S., Oceanography of the Norwegian and Greenland Seas and adjacent areas. NATO SACLANTCEN memorandum SM-47, 1974

SELLAFIELD RADIOCAESIUM AS A TRACER OF WATER MOVEMENT IN THE SCOTTISH COASTAL ZONE

P E Bradley*+, B E Economides*+, M S Baxter*+ and D J Ellett°

* Scottish Universities Research and Reactor Centre, East Kilbride, Glasgow G75 0QU, UK.
+ Department of Chemistry, University of Glasgow, Glasgow G12 8QQ, UK.
° Scottish Marine Biological Association, Oban, Argyll PA34 4AD, UK.

ABSTRACT

Radiocaesium has been discharged into the Irish Sea from the Sellafield nuclear fuel reprocessing plant in Cumbria for more than three decades and has been used here as a tracer of water movement in the western UK coastal system.

Extensive radiocaesium and hydrographic data collected during the last ten years from many cruises within the Hebridean Sea Area have helped to characterise the controlling oceanographic flow parameters and can be used to describe flow patterns. Salinity/radiocaesium plots define water sources and compositions for the Scottish coastal system. Unlike previous radiocaesium-related tracer studies in this area, the present work benefits from the past ten years of accumulated data and consequently it can assess the long-term variations in hydrographic conditions as well as the short-term changes in flow characteristics. This is particularly important as radiocaesium discharges from Sellafield have decreased radically since 1985 and it is thus essential to maximise the use of this versatile, but perhaps transient, tracer.

Fallout from the Chernobyl nuclear reactor accident of 1986 has provided a further source of radiocaesium to the study area and this component was detected in Scottish coastal waters very quickly after the Chernobyl cloud reached the UK. Chernobyl-derived radiocaesium was, of course, distinguishable from the Sellafield component by virtue of their different isotopic ratios and its distribution with depth has provided a further useful indicator of vertical mixing.

INTRODUCTION

The area of study comprises that region of the inner continental shelf between the North Channel and the Outer Hebrides of Scotland (Fig. 1). Sampling transects extend westwards from the Scottish mainland to the continental shelf, the edge of which is marked by the 200m isobath. A unique feature of this portion of the British coastal system is that, for almost its entire length, it is exposed to the open Atlantic Ocean, an important characteristic, as it provides an extra source of coastal water and increases the hydrodynamic complexity of the system.

There are three distinct sources of Scottish coastal water: Atlantic water (with high salinity and negligible radiocaesium), Irish Sea derived water (slightly lower salinity but clearly labelled with Sellafield Radiocaesium) and freshwater runoff from the mainland (near-zero salinity and negligible radiocaesium). Although, in some seasons and close to their sources, these three water types can be differentiated via their salinities alone (1, 2), at other times and in most regions the differences become much less distinct and additional tracers are desirable. Radiocaesium isotopes discharged from the Sellafield nuclear fuel reprocessing plant in Cumbria have been shown to pass through the region (3-8) and to provide a most useful tracer system for study of the rate and direction of the movement of water originally labelled within the Irish Sea.

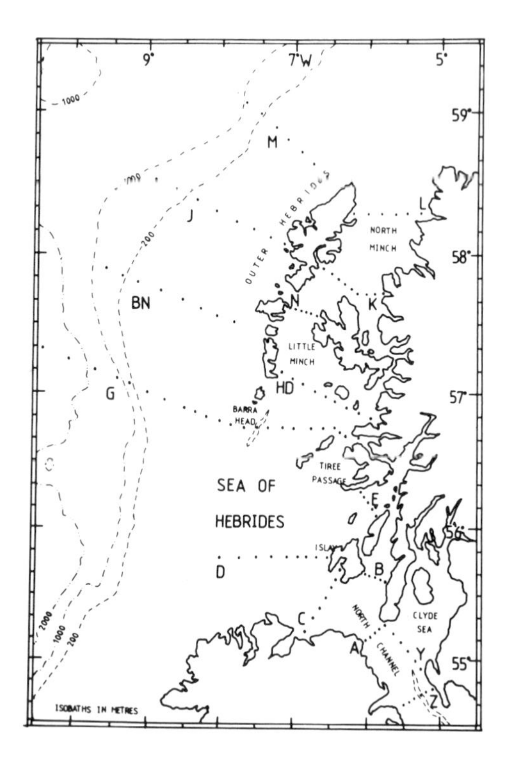

Fig 1. The Area of Study showing the positions of sampling stations

The radiocaesium isotopes of interest (^{137}Cs and ^{134}Cs) are of particular value as they exhibit relatively conservative behaviour and hence long residence times (> 10^2 years) in coastal water (9). They are present at easily detectable levels in Scottish coastal waters and their quantification is relatively straightforward using ion-exchange and - spectroscopy techniques. Radioactive isotopes have, of course, considerable advantages as water tracers as they exhibit decay during tranport and here the half-life of ^{134}Cs (2.05 y) is highly compatible with the rate of coastal current movement. Thus the ratio of ^{134}Cs to ^{137}Cs halves every 2.2 years and, under favourable circumstances, this decay effect can be used to measure transport rates from the point of input, regardless of, for example, freshwater dilution and caesium removal processes. Furthermore, as Sellafield discharges of radiocaesium have been seasonally pulsed, matching of short-term temporal trends in absolute activities along the plume flow path has enabled complementary current rate to be derived.

The discharges of Sellafield radiocaesium have recently decreased considerably, in parallel with the installation of a new effluent decontamination plant. It is appropriate therefore to review here the kinds of oceanographic studies to which this unique, and perhaps transient, tracer system has been applied during the recent decade in which levels have been highly detectable. With this in mind, we summarise here some examples of the typical usage of the Sellafield radiocaesium tracers as directed at the Scottish coastal water system in a tripartite collaborative programme between SURRC*, SMBA+ and the University of Glasgow.

METHODS

Water samples are collected at sampling stations, such as those shown in Fig 1, by SMBA on regular cruises in the Hebridean Sea Area (usually 4 to 5 times per year). Additional depth profiles of temperature and salinity are simultaneously recorded at these and other stations. Surface samples are obtained by direct pumping via a clean seawater system, whilst deeper water is collected using a messenger-triggered 25 litre sampling bottle lowered to the required depth on a weighted hydrographic wire. Monthly water samples are collected from the Clyde Sea Area in a similar manner by the Clyde River Purification Board. Samples are acidified on collection to pH2 with nitric acid and stored in 25 litre polythene bottles.

The analytical procedure for radiocaesium determination has been described in detail elsewhere (10) and involves filtration and caesium extraction steps prior to counting. The water is initially filtered through a 0.22μm cellulose nitrate membrane filter to remove particulate material, followed by caesium removal by passage through a selective ion exchanger (potassium hexacyanocobalt (II) ferate (II)). The resulting cartridge is subsequently dried, sealed and counted directly on a 3" well-type NaI(Tl) spectrometer for 23 hours.

RESULTS AND DISCUSSION

Radiocaesium distributions have now been measured during a great many cruises, involving perhaps around 2000 radiometric measurements. It is clearly impossible to present and discuss the entire data set here and our approach is therefore firstly to describe the kind of results obtained during a typical recent cruise and latterly to review some of the more significant long-term trends which become noticeable when the data from all the cruises are reviewed in retrospect. The specific cruise used to illustrate the potential of the radiocaesium tracer approach is that of August 1985.

Cruise data for August 1985

The results of the cruise are presented in Fig 2 which shows the surface distribution of ^{137}Cs in $Bq l^{-1}$. As the only significant source of radiocaesium to the area at this time was Sellafield, the dilution and dispersal of the radiocaesium plume can be regarded as representative of the processes acting upon the Irish Sea-derived coastal water body. From the ^{137}Cs distribution, there are several immediately obvious and typical points worthy of comment;

i) levels of ^{137}Cs decrease in the region from south to north and east to west. These trends are indicative of the dilution of the Irish Sea-derived water as it travels northwards and encounters and mixes with radiocaesium-free Atlantic water en route.

(ii) there is a strong concentration gradient to the west of Islay and south of Tiree, with the trend of the front lying in a SW-NE orientation. The front forms a boundary which separates high salinity, radiocaesium-free Atlantic water to the west from Irish Sea derived water to the east. The front also marks a division between stratified water to the west and vertically well-mixed coastal water to the east. The occurrence and position of such fronts in the Hebridean Sea Area have been predicted from considerations of tidal mixing and water depth by several other workers (11-14).

(iii) Further north, around Coll and Tiree, the front becomes more diffuse, although the higher radiocaesium concentrations still occur to the east, and the majority of the radiocaesium-labelled coastal water flows through the Tiree Passage. This region is one of extremely irregular bottom topography which probably controls the observed channelling of the coastal radiocaesium plume through the passage.

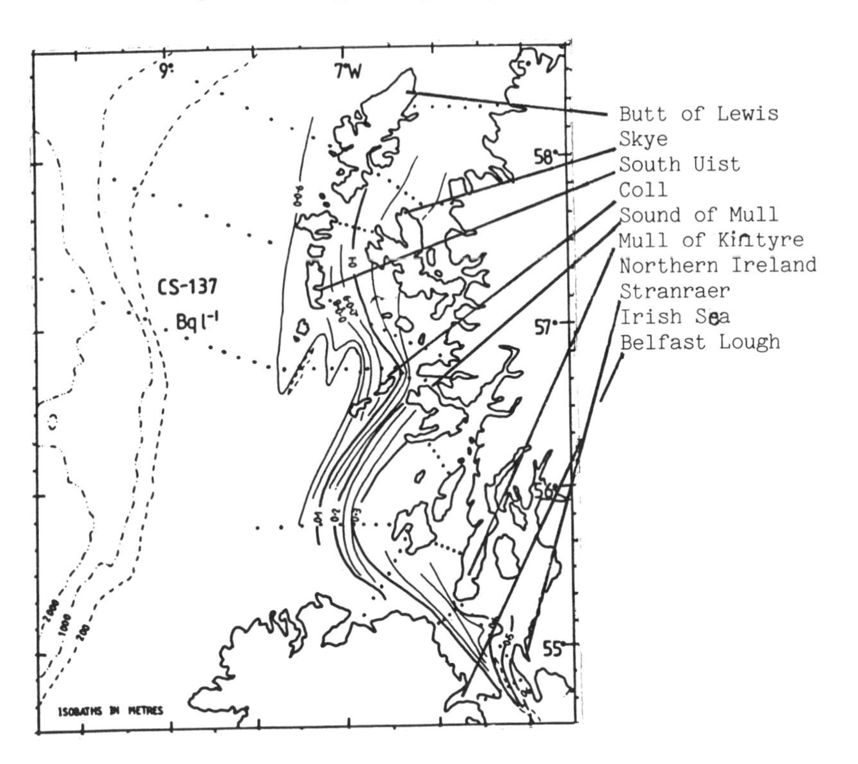

Fig 2 Surface ^{137}Cs in Bql^{-1} for RRS Challenger cruise 8/85 (August 1985)

(iv) There is an increase in radiocaesium concentration around Barra Head and South Uist, reflecting a splitting of the coastal current in the Little Minch. The ^{137}Cs distribution indicates that, although the plume continues northwards through the North Minch, a component turns anticlockwise around Barra Head and flows towards the Butt of Lewis, along the western sde of the Outer Hebrides. This splitting of the coastal flow may be attributed to the diversion of flow at the constricted passage of the Little Minch (15).

These general observations indicate the value of the radiocaesium isotopes as directional flow tracers and this kind of application can be, and has been, focussed down and quantified in localised key areas of interest (7). In essence, the radionuclide acts here like a conventional dye tracer, simply mapping the flow and indicating dilution around specific island groups and other coastal features. There is, however, considerable further potential.

Source water terms and compositions

As previously mentioned, the three contributory water sources to the Scottish coastal system can be effectively differentiated in terms of their salt and radiocaesium contents. Thus, many of the features already described can be clarified by reference to a plot of ^{137}Cs against salinity. Fig 3 illustrates part of the mixing triangle for the three water types and shows the positions of Atlantic water (salinity 35.4 ppth, negligible ^{137}Cs) and Irish Sea derived water (salinity 33.8 ppth, 0.7 Bql^{-1} ^{137}Cs), with freshwater (negligible salinity and ^{137}Cs) plotting to the left of the diagram. All the data from the August 1985 cruise have

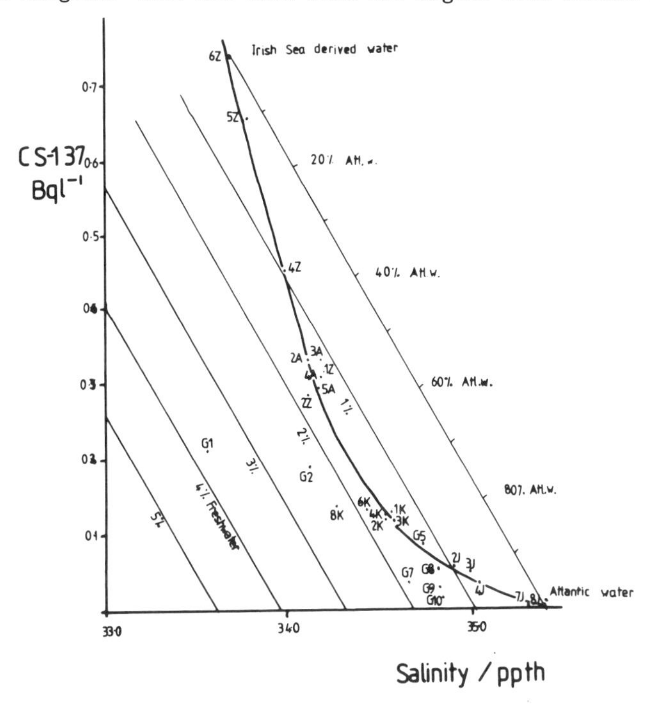

Fig 3. Radiocaesium/salinity plot for surface ^{137}Cs observations taken in August 1985

been plotted on this diagram and can typically be explained in terms of mixing between the three water sources. In other words, for any water sample collected at the time of the cruise, a measurement of salinity and ^{137}Cs is sufficient to characterise unambiguously its composition and origin. This is clearly a most valuable capability from an oceanographic viewpoint.

One obvious feature of the diagram is that most of the points lie to its right, showing that most waters sampled contained less than 2% freshwater. The exceptions were sampled close to the mainland, for example in the Sound of Mull, where the freshwater content rose to 4%. It can be seen further that all the other data lie on a curve which links Irish Sea to Atlantic water and which bows towards the freshwater end-member. The upper portion of this curve (above 0.1 Bql^{-1}) indicates the effect of the cumulative influx of freshwater to the system, although the major influence is addition of Atlantic salt water. This section of the plot also illustrates the extensive dilution of the Irish Sea derived water within the confines of the North Channel. Dilutions, by up to 60% Atlantic water, are indicated between entering the system at the southern end of the North Channel (station 6Z) and flowing through the northern exit (stations 4A and 5A).

The lower position of the curve (below 0.1 Bql^{-1}) represents stations to the west of the Outer Hebrides which are characterised by compositions closer to those of Atlantic water. Clearly the islands of the Outer Hebrides act as a barrier to freshwater flow entering the coastal system from the Scottish mainland.

Estimation of transit times for the coastal current

In addition to the dye-type modes of use, the radiocaesium isotopes can be used to assess the rate of water transport by considering the different decay characteristics of ^{137}Cs and ^{134}Cs. The preferential decline of ^{134}Cs activity means that the activity ratio of ^{137}Cs:^{134}Cs halves every 2.2 years, thus allowing transport rates in the coastal current to be estimated.

One constraint on the use of ratio data in this way is that the input values must have remained constant for a period greater than, or equal to, the duration of the transport process under study. Otherwise, ie. if the input ratio value changes, an extra source of variation is introduced and the ratio values in the region of interest can no longer be interpreted solely in terms of radioactive decay during transport.

For our purposes, the input site for the Hebridean Sea Area has been taken as the North Channel. Although the ratio values across the North Channel transect for any given cruise are variable, the deviations from the mean are not significantly larger than the errors, mainly counting-derived, on the individual data points. It is thus evident that the input ratio to the area was effectively constant at $\simeq 0.044 \pm 0.001$ for the period between June/July 1984 and August 1985.

Ratio values measured in the North Minch in August 1985 again show some deviation, although the mean is once more within the errors on individual measurements. ^{137}Cs and salinity measurements for this transect show the region to be both vertically and horizontally well-mixed and thus the ratio data for the transect have been averaged to a value of 0.039 ± 0.005.

An estimate of the transit time, t, between the North Channel and the Minch is given by;

$$t = 1/\lambda \ \ln \left(\frac{\text{input ratio at North Channel}}{\text{ratio at the Minch}} \right)$$

where, $\lambda = 0.693/2.2 \ y^{-1}$

$$\text{thus, } t = 1/0.315 \ \ln \frac{(0.044 \pm 0.001)}{(0.039 \pm 0.005)} = 4.6 \pm 0.6 \text{ months}$$

The above estimate of transit time implies for the cruise period, a mean flow rate through the Hebridean Sea Area of approximately 2.5 $kmday^{-1}$. It is of interest to examine whether this approximate value of transit time is representative of normal conditions for the coastal current. Table 1 summarises some mean flow rate values from previous cruises in this programme and by other workers. The general range of estimated flow rates is 1 to 12 kmday^{-1}, our most recent value of 2.5kmday^{-1} being clearly intermediate.

Table 1. Flow rate estimates for the Scottish coastal current

Region	Period	Flow rate (kmday^{-1})	Method	Reference
N. Channel to Skye	Feb 80-Jul 81	1.9	Ratio data	McKay et al, 1986 (7)
Hebridean Sea Area	Summer 76	1.6	Ratio data	McKinley et al 1981 (16)
Hebridean Sea Area	Summer 77	5	Ratio data	McKinley et al 1981 (16)
Tiree Passage	1976 (8 tidal periods)	11.8	Current meter	Ellett and Edwards, 1983 (17)
Tiree Passage	Aug-Dec 85	8.6	Current meter	SMBA Data
N. Channel	71-78	1.3	Cs-data	Jeffries et al 1982 (18)
N. Channel		5	temperature and salinity	Craig, 1959 (15)
Hebridean Sea Area	May 76	5	Cs-data	Livingston and Bowen, 1977 (5)
Hebridean Sea Area		5	drift bottles	Barnes and Goodley, 1961 (19)
Hebridean Sea Area	May 82	6-12	current meter	Simpson and Hill, 1986 (20)

During the long-term period of study, by far the most striking observation was the considerable increase in water flow which occurred between October 1976 and March 1977, when incursing Atlantic waters in the southern Irish Sea forced a vigourous northerly outflow through the North Channel, with flow rates reaching the upper limit of 12 kmday^{-1}. In more normal times, North Channel outflows and flow rates in the Hebridean Sea are commonly in the 1-5kmday^{-1} range, as measured by current meter, drift bottle and radiocaesium methods. In constricted areas, such as the Tiree Passage, normal flow rates are higher as the northerly flow experiences channelling.

As mentioned in the introduction, the long-term trends in Sellafield ^{137}Cs discharges and ^{137}Cs levels in the area of study have enabled complementary current rate data to be derived. Early assessments of flow rates (3, 5) assumed that water transport in the Scottish coastal current occurred by simple advection. However, a comparison of the Sellafield ^{137}Cs discharge record with ^{137}Cs levels in the Clyde Sea Area (Fig 4) shows that a more complex process is operating. Box-modelling techniques, using the Sellafield ^{137}Cs discharge data as input and comparing the resultant output with the observed Clyde Sea Area values, have enabled estimation of both transit times and residence times in the northern Irish Sea. The results of such modelling imply a transit time between Sellafield and the Clyde Sea of 4-6 months, compounded with a residence time of approximately 12 months in the northern Irish Sea (21).

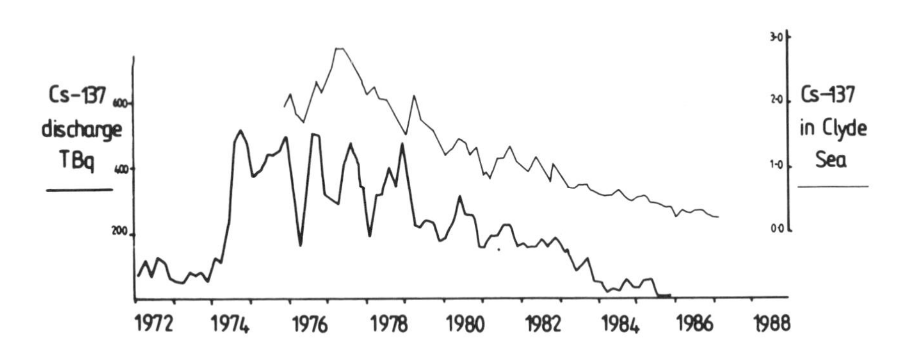

Fig 4 Sellafield ^{137}Cs discharge record and ^{137}Cs levels in the Clyde Sea Area

Long-term patterns of coastal flow

A further indication of the long-term pattern of coastal flow through the system can be gained by reference to a plot of salinity and ^{137}Cs values, across the North Channel, against time. Figure 5 shows such a diagram for the transect between Mull of Kintyre and Northern Ireland (the 'A' transect of fig 1).

From the temporal ^{137}Cs plot, it is apparent that the concentration of ^{137}Cs has been consistently higher on the eastern side of the North Channel and that levels of ^{137}Cs have gradually decreased during the period 1983-86, in parallel with the general decline in Sellafield discharges. The gradients of ^{137}Cs concentration across the channel have varied considerably with time, from periods of intense heterogeneity (eg. in November 1983 to January 1984 and November 1984) to intervening periods of more uniform ^{137}Cs concentrations (eg. in May 1984, May 1985 and May/June 1986). It is well known, of course, that surface water transport and

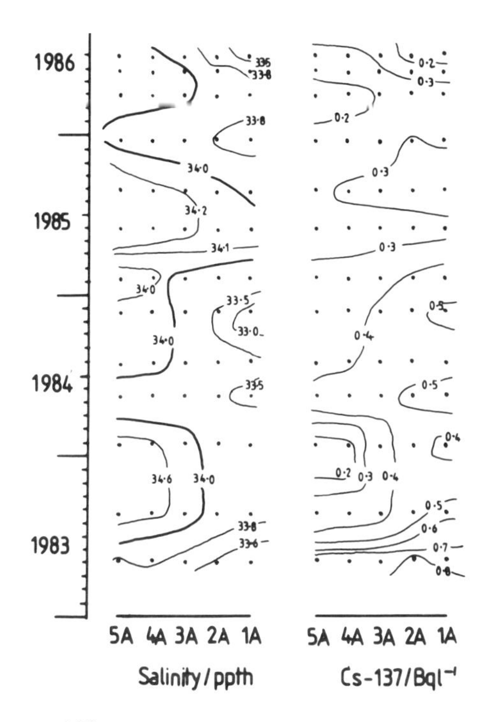

Fig 5. Isopleths of ^{137}Cs and salinity for transect A across the North Channel

resultant water mixing are strongly dependent on wind speed and direction and it has indeed been shown that this applies to the North Channel (22). Wind speed and direction are relatively constant in this region at all times of year, exceptions being during March and May when the prevailing wind is from the north-east rather than the south-west. It thus seems highly likely that the observed broadening and increased lateral homogeneity of the ^{137}Cs plume during the month of May is a reuslt of this change in local wind conditions.

Similarly, the salinity versus time plot shows the spread of lower salinity, Irish Sea derived water across the North Channel at this time of year, with higher salinity Atlantic water present on the western side of the Channel during the rest of the year. Additionally, the plot shows that salinities across the entire Channel are lower during winter months. This is a regularly observed feature in coastal waters and reflects the annual cycle of river flow, with larger influxes of freshwater during winter. The source of this freshwater is undoubtedly the Clyde Sea Area, which has its greatest freshwater input during November to January (23).

The Effects of Chernobyl fallout

The first activity attributable to the nuclear reactor accident at Chernobyl was detected in Britain on 2nd May 1986 (24) and the subsequent fallout, including ^{134}Cs and ^{137}Cs, introduced another significant source of radiocaesium to the Scottish coastal system. It was therefore important to quantify both the magnitude and the tracer potential of this new radiocaesium component.

The effects of the Chernobyl radiocaesium on the near-shore environment are illustrated here for the Clyde Sea Area (CSA) in which ^{137}Cs and ^{134}Cs have been measured continuously, on a monthly basis, over the last ten years (fig 6).

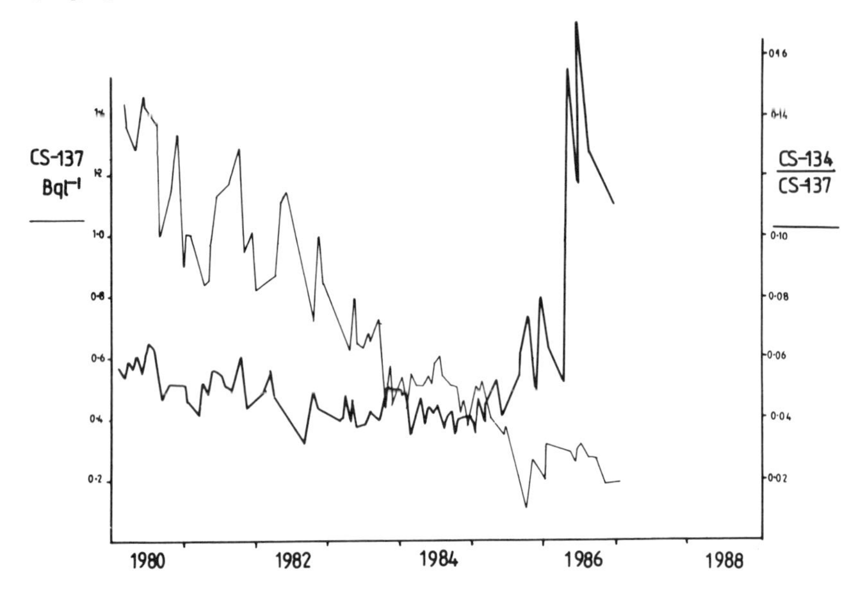

Fig 6. ^{137}Cs levels and ^{134}Cs:^{137}Cs ratio values for the Clyde Sea Area

During this period, levels of ^{137}Cs have declined steadily, in sympathy with the Sellafield discharge record, and it is evident that ^{137}Cs activites before and after the Chernobyl accident do not differ significantly. However, fig 6 also shows the corresponding variation in the ^{134}Cs:^{137}Cs ratio in the CSA during the same period - and here there is a very marked increase in the ratio following the accident. Prior to the Chernobyl event, the ratio was stable at 0.04 to 0.05, increasing slightly during 1985 in response to earlier Sellafield emissions. Following the accident (by 6th June), the ratio had increased to 0.155. The ^{134}Cs:^{137}Cs activity ratio in Chernobyl fallout has been measured at ~0.5 and it would thus appear that, although the Chernobyl component of ^{137}Cs was small relative to the Sellafield ^{137}Cs inventory of the area, its enrichment in ^{134}Cs was sufficient to triple the ^{134}Cs:^{137}Cs ratio in Scottish surface coastal waters. For example, if we assume that our post-Chernobyl data apply to a well-mixed North Channel, then then ^{134}Cs and ^{137}Cs inventories were increased by respectively, 300% and 30% as a result of the Chernobyl input.

Chernobyl-derived radiocaesium was also detected in the North Channel in surface and deep samples collected between Stranraer and Belfast though (the 'Z' line of Fig 1) on June 6th 1986. As the Chernobyl input occurred across the sea surface, the vertical distributions of ^{134}Cs and ^{137}Cs can be used to assess rates of vertical mixing. The ^{134}Cs:^{137}Cs ratios through the water column range from 0.127 to 0.170, indicating a significant contribution of Chernobyl radiocaesium, and show no general decrease with depth. This distribution indicates that vertical mixing in the North

Channel is extensive and rapid, reaching completion within 4 weeks. Tidal currents in this area are amongst the fastest along the Scottish coastline (17), reaching a maximum of over $2ms^{-1}$ south-west of Kintyre, and vertical mixing in the channel is consequently expected to be intense (25). The Chernobyl pulse has provided a useful lower limit for its rate.

Radiocaesium in sediments

In addition to the above use of Sellafield radiocaesium as a tracer of water movement, the vicinity of the Sellafield source-term has provided the ideal opportunity to investigate the uptake and partitioning of radiocaesium on coastal marine sediments under real field conditions.

It is generally assumed that, in the marine environment, radiocaesium behaves as a conservative hydrated monovalent cation which once in sediments, has an affinity for illitic clays. However, work involving sequential chemical extraction of sedimentary phases suggests that, in Scottish coastal sediments at least, the partitioning of ^{137}Cs is related, to a larger than expected degree (12-50%), to the concentrations of sedimentary oxides, organics and, less importantly, carbonates (26). McKay and Baxter have thus suggested, and found some evidence, that these components act to prevent ^{137}Cs release from clay mineral exchange sites and therefore that diagentic remobilisation of the nuclide is possible. Thus even the small component of Sellafield radiocaesium which is scavenged by coastal sediments remains potentially mobile in the long-term.

CONCLUSIONS

This paper has attempted to highlight the various ways in which Sellafield radiocaesium has, as a tracer, contributed to the understanding of coastal oceanography in the northern UK. It has mapped flow directions, monitored flow rates, quantified dilution factors, characterised water compositions and given some insight into geochemical associations in sediments. In parallel with the research programme, measurements on marine organisms and foodstuffs have shown that the radiological implications to the Scottish population are vanishingly small and it is therefore gratifying to find that, from a scientific viewpoint, the Sellafield radiocaesium releases have been of positive value.

ACKNOWLEDGEMENTS

We wish to thank the officers and crew of R.R.S. Challenger for their assistance at sea. Staff of the Clyde River Purification Board were also instrumental in collecting regular samples from the Clyde Firth. The considerable technical help of Drs. W. Jack, E. M. Scott and G. T. Cook is gratefully acknowledged. Two of us (P.E.B. and B.E.E.) further acknowledge the support of Natural Environment Research Council studentships. The Scottish Marine Biological Association is supported by The Natural Environment Research Council.

REFERENCES

1. Ellett, D.J. Temperature and salinity conditions in the Sea of Hebrides 25-29 May 1976. Annales Biologiques 1977 33 28-30.
2. Ellett, D.J. Some oceanographic features of Hebridean waters. Proceedings of the Royal Society of Edinburgh 1979. Series B77 61-74.
3. Jefferies, D.F., Preston, A. and Steele, D.K. Distribution of Cs in British coastal waters. Marine Pollution Bulletin 1973 4, 118-122.
4. Kautsky, M. The distribution of the radionuclide caesium-137 as an

indicator for North Sea watermass transport. Deutsche Hydrographische Zeitschrift.
5. Livingston, H.I. and Bowen, V.T. Windscale effluent in the waters and sediments of the Minch. Nature 1977 269, 586-588.
6. Baxter, M.S. and McKinley, I.G. Radioactive species in sea water. Proceedings of the Royal Society of Edinburgh 1978. Series B76, 17-35.
7. McKay, W.A., Baxter, M.S., Ellett, D.J. and Meldrum, D.T. Radiocaesium and circulation patterns west of Scotland. J. Environmental Radioactivity, 1986, 4 205-232.
8. Economides, B.E. PhD Thesis, 1986, Glasgow University.
9. Swan, D.S., Baxter, M.S., McKinley, I.G. and Jack, W. Radiocaesium and Pb-210 in Clyde Sea Loch Sediments. Est. Coast. Shelf Sci. 1982 15 pp 515-36.
10. MacKenzie, A.B., Swan, D.S., McKinley, J.G., Baxter, M.S. and Jack, W. The determination of Cs-134, Cs-137, Pb-210, Ra-226, Ra-228 concentrations in nearshore marine sediments and sea water. Journal of Radioanalytical Chemistry 1979 48 29-47.
11. Fernhead, P. On the formation of fronts by tidal mixing around the British Isles. Deep Sea Research 1975 22 311-321.
12. Hughes, D. A simple method for predicting the occurrence of seasonal stratification and fronts in the North Sea and around the British Isles. ICES. CM1976 (C:1) 13pp.
13. Pingree, R.D., Hooligan, P.M. and Mardel, G.T. The effects of vertical stability on phytoplankton distributions in the summer on the north-west European Shelf. Deep Sea Research 1978 25 1011-1028.
14. Simpson, J.H., Edelston, D.J., Edwards, A., Morris, N.G.G. and Tett, P.B. The Islay Front: physical structure and phytoplankton distribution. Est. Coast. Mar. Sci. 1979 9 713-26.
15. Craig, R.E. Hydrography of Scottish Coastal Waters. Marine Research 1958 1959 (2) 30pp.
16. McKinley, I.G., Baxter, M.S., Ellett, D.J. and Jack, W. Tracer applications of radiocaesium in the Sea of Hebrides. Est. Coast Mar. Sci. 1981 13 69-82.
17. Ellett, D.J. and Edwards, A. Oceanography and inshore hydrography of the Inner Hebrides. Proc Roy Soc Edinburgh 1983 83B 143-160.
18. Jefferies, D.F., Preston, A. and Steele, A.K. Further studies on the distribution of Cs in British coastal waters-Irish Sea. Deep Sea Res. 1982 Vol. 29 No. 6A pp713-738.
19. Barnes, H. and Goodley, E.F.W. The general hydrography of the Clyde Sea Area, Scotland Part 2. Description of the area; drift bottles and salinity data. Bulletin of Marine Geology 1961 5 112-150.
20. Simpson, J.H. and Hill, A.E. The Scottish Coastal Current. In The Role of Freshwater Outflow in Coastal Marine Ecosystems, ed. S. Skreslef, Springer-Verlag, 1986, Berlin Heidelberg.
21. McKay, W.A. and Baxter, M.S. Water transport from the N.E. Irish Sea to western Scottish coastal waters: further observations from time-trend matching of Sellafield radiocaesium. Est. Coast Shelf. Sci. 1985 21 471-480.
22. Howarth, M.J. Non-tidal flow in the North Channel of the Irish Sea. In Hydrodynamics of semi-enclosed shelf seas, ed. J.C.J. Nihoul, Elsevier, 1982. Amsterdam, Oxford and New York. pp205-241.
23. Poodle, T. Freshwater inflows to the Firth of Clyde. Proceedings of the Royal Society of Edinburgh, 1986 90B, 55-66.
24. Camplin, W.C., Mitchell, N.T., Leonard, D.R.P. and Jefferies, D.F. Radioactivity in surface and coastal waters of the British Isles: Monitoring of fallout from the Chernobyl reactor accident. Aquat. Environ. Monit. Rep. MAFF Direct. Fish. Res. 1986 Lowestoft (15) 1-49.

25. Edwards, A., Baxter, M.S., Ellett, D.J., Martin, J.H.A., Meldrum, D.T. and Griffiths, C.R. Clyde Sea hydrography. Proceedings of the Royal Society of Edinburgh, 1986 B77 61-74.
26. McKay, W.A. and Baxter, M.S. The Partitioning of Sellafield-derived radiocaesium in Scottish Coastal sediments. J. Environ. Radioactivity 1985 2 93-114.

THE APPLICATION OF SEAWEEDS AS BIOINDICATORS FOR RADIOACTIVE POLLUTION IN THE CHANNEL AND SOUTHERN NORTH SEA

Asker Aarkrog[1], Henning Dahlgaard[1], Simon Duniec[2], Pierre Guegueniat[3] and Elis Holm[4]

1) Risø National Laboratory, DK 4000 Roskilde, Denmark
2) Lund University, Sweden
3) CEA, Laboratorie de radioecologie marine, 50107 Cherbourg
4) International Laboratory for Marine Radioactivity, Monaco

ABSTRACT

Radioactive discharges from nuclear installations may be applied as tracers for water masses in the oceans. At a distance from the source the water concentrations of the various radiotracers may however be so low that the method becomes inpractical. In the case of tracing coastal currents this problem may be partially overcome by using bioindicators. Such organisms have the ability to accumulate radionuclides (and other substances) from the water. Brown algaes are well suited as bioindicators as they show biomagnification factors typically in the order of 10 000 for several nuclides.

The authorized discharges of radionuclides from the reprocessing plant at Cap de la Hague in France have been used for tracing the water masses from the central part of the English Channel to the German Bight where the Jutland Current starts. In a joint French-Swedish-Danish effort samples were collected from the European and British sides of the Channel coast in the first half of 1985. The samples consisted of brown algae (mostly _Fucus vesiculosus_) and a number of seawater samples. The seaweed was analysed for ^{40}K, ^{60}Co, ^{90}Sr, ^{99}Tc, ^{106}Ru, ^{125}Sb, ^{134}Cs, ^{137}Cs, ^{238}Pu, $^{239,240}Pu$ and ^{241}Am. Sr-90 and ^{99}Tc were determined in the seawater. From the mutual determinations of seaweed and seawater the observed ratios between bioindicator and water were calculated. Furthermore, the ratios of the radionuclide concentrations in the various species were estimated. From this it was possible to normalize all seaweed data to _Fucus vesiculosus_ concentrations. These concentrations were then related to the sea distance from Cap de la Hague, and over a distance of more than one thousand kilometers the following regressions were found: Bq ^{99}Tc kg^{-1} dry weight = 19000 $km^{-0.7}$; ^{60}Co: 2000 km^{-1}; ^{106}Ru: 300 $km^{-0.6}$ and $^{239,240}Pu$: 3 $km^{-0.6}$. Hence the activity decreased approximately in inverse proportion to the distance or square root of the distance from the source. Similar distance relations have earlier been observed in British coastal waters for discharges from Sellafield in the UK. The paper shows how these observations may be used to estimate unreported discharges.

INTRODUCTION

For several years the authorized radioactive discharges from nuclear reprocessing plants in Western Europe have been used for tracer studies in the North Atlantic region (Kautsky (1973)[9], Livingston et al. (1982)[10], Jefferies et al. (1982)[8] Aarkrog et al. (1983)[1] and others (Casso et al., (1984))[5]. The study of the various radionuclides discharged has made it possible to identify transport routes for waterborne pollution, and to estimate transit times and dilution factors.

The BNFL Sellafield reprocessing plant in the U.K. was by far the most important source of new radioactive pollution of the North-East Atlantic region in the seventies and the early eighties.

Calmet and Gueguениat (1985)[4] and Patti et al. (1984)[11] have shown however that the French reprocessing plant at La Hague has contributed significantly to the radioactivity levels in the English Channel in recent years. As the discharges from Sellafield have been rapidly decreasing (BNFL, 1979-86), the relative importance of the discharges from La Hague is increasing.

In seawater from the German Bight and from the Jutland Current Aarkrog et al. (1983-1985)[2] observed in recent years radionuclide ratios: $^{90}Sr/^{137}Cs$, $^{134}Cs/^{137}Cs$ and $^{99}Tc/^{137}Cs$ that were definitely higher than those expected in effluents from Sellafield. We have assumed that this indicated a contribution of activity from La Hague.

As part of the Jutland Current enters the Danish Straits (cf. Fig. 1) and as it is of general environmental interest to know how much pollution is transported from the German Bight to the Straits, we presume that the radioactive discharges from La Hague may be applicable for such a study.

METHODS

In a joint French, Swedish and Danish effort samples of seawater, seaweed and mussels were collected in the first half of 1985 from the continental as well as British side of the English Channel.

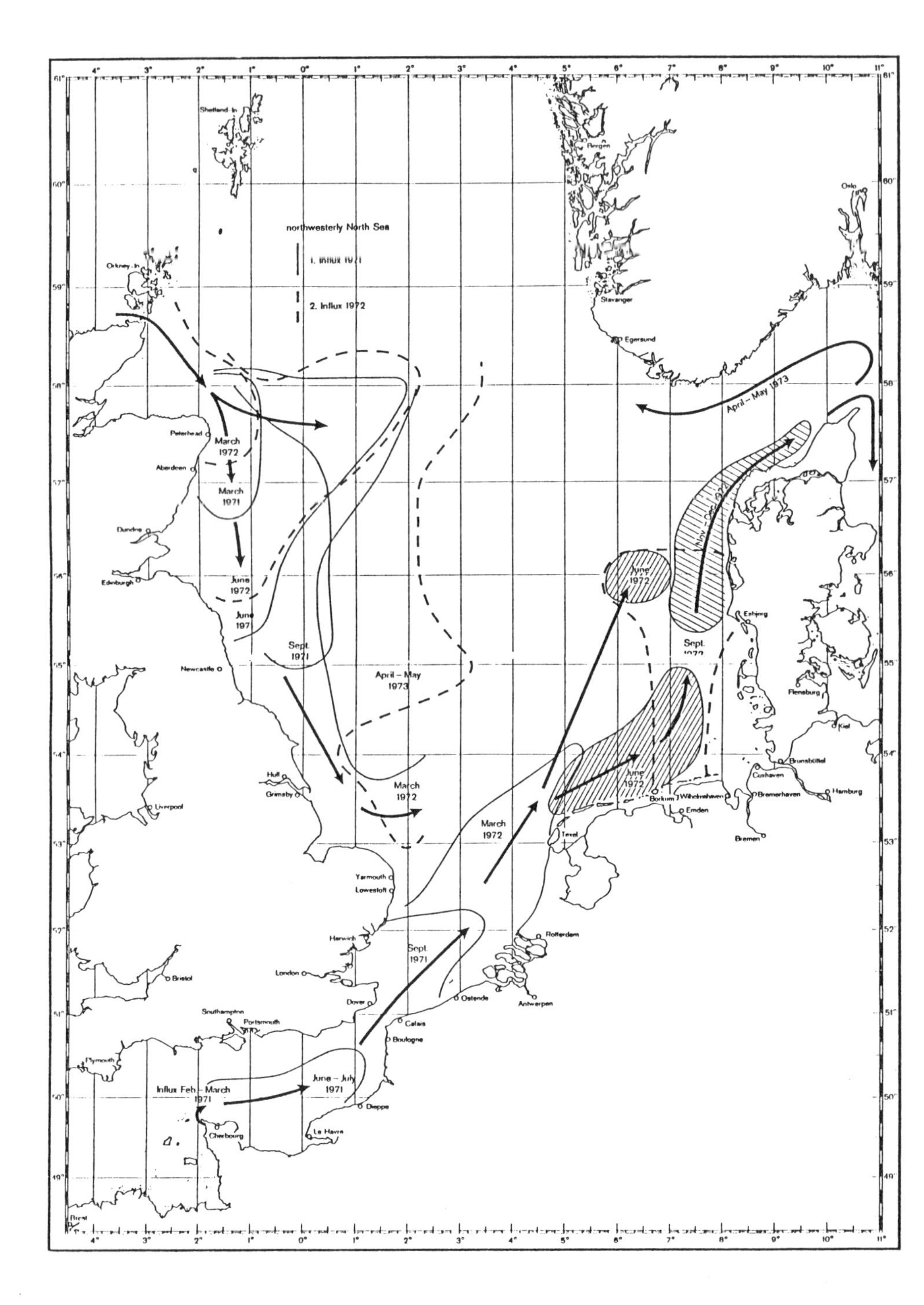

Fig. 1. Water mass transport in the North Sea based upon ^{137}Cs measurements (Kautsky, 1973).

The samples have been analysed for γ-emitters by Ge(Li) spectroscopy, for ^{90}Sr, ^{99}Tc and transuranics by radiochemistry (Harley (1972)[6], Holm et al. (1984), Talvitie (1971))[12] at Lund University and Risø National Laboratory. The water samples were analysed as unfiltered water and the seaweed samples were measured as dried samples; the ratio of fresh weight/dry weight ~ 5. The relatively high ^{106}Ru content compared with ^{99}Tc in the water samples made it necessary to recount the ^{99}Tc samples in order to correct for any ^{106}Ru present.

RESULTS AND DISCUSSION

The use of bioindicators for oceanic tracer studies implies that the radionuclide concentrations in the bioindicator are proportional to those in the water. In Table 1 water samples have been analysed for ^{99}Tc and it is possible to compare these levels with those in mutually collected seaweed samples. The observed mean ratio between ^{99}Tc in *Fucus vesiculosus* (dry weight) and seawater becomes: $(1.5 \pm 0.33) \cdot 10^5$ ($\pm$ 1 S.E; N = 13) (salinity $\geq$ 30 o/oo). This may be compared with the ratios found by Patti et al. (1984)[11] who for Fucoids in general reported: $(1.4 \pm 0.26) \cdot 10^5$ ($\pm$ 1 S.E; N = 6). These samples were collected between Wimereux and Cancale and the fresh weight was assumed to be 5 times higher than the dry weight.

The ratio between ^{99}Tc in Fucus and water may however also be obtained from an integration of the distance relations obtained from the data in Table 1 (see below)

$$\frac{\left[\int_1^{1000} 2\pi \, x 18\ 800\ x^{-0.673}\, dx\right] \text{ Fucus ves.}}{10^{-3} \cdot \left[\cdot \int_1^{1000} 2\pi \, x 46 \cdot x^{-0.43}\, dx\right] \text{ water}} = 0.9 \cdot 10^5$$

We may conclude that the observed ^{99}Tc concentration ratio of *Fucus vesiculosus* (dry weight)/seawater in the English Channel is on the order of 10^5, which is in agreement with earlier estimates obtained from the Danish Straits (Aarkrog et al. 1983 1985)[2].

Clockwise around Scotland we have seen in June 1982 that the radionuclide concentrations arising from Sellafield discharges to the Irish Sea decrease with the distance (X km) from Sellafield according to a power function:

$$\text{Bq kg}^{-1}_{\text{(dry weight Fucus vesiculosus)}} = B \cdot X^{-A} \qquad \text{(Eqv 1)}$$

Similar distance relationships have been found for discharges from La Hague to the English Channel for the north-easterly direction and Table 2 summarize values of B and A.

Table 1. Radionuclides in seaweed and surface sea water collected along the English Channel in 1985 (Unit: Bq kg^{-1} dry weight for seaweed and Bq m^{-3} for sea water)

Station number	Species	Date	Position N	Position E or W	Location	Km*	% dry matter	Salinity in o/oo
85501	Fu.ve.	9/4	53°52'	8°43'E	Cuxhaven (D)	940		
85502	Seawater	"	"	"	"	"		11.0
85503	Fu.ve.	"	53°37'	7°10'E	Norddeich (D)	835	18.7	
85504	My.ed.	"	"	"	"	"	13.3	
85505	Fu.ve.	10/4	53°10'	5°24'E	Harlingen (NL)	720		
85506	Seawater	"	"	"	"	"		11.0
85507	Fu.ve.	"	52°28'	4°36'E	IJmuiden (NL)	595	23.8	
85508	Seawater	"	"	"	"	"		15.2
85509	Fu.ve.	"	51°27'	3°36'E	Vlissingen (NL)	435		
85510	Fu.ve.	11/4	51°14'	2°55'E	Oestende (B)	390	19.6	
85511	Seawater	"	"	"	"	"		30.7
85512	Fu.ve.	"	50°58'	1°51'E	Calais (F)	305		
85594	Fu.ve.	24/6	50°46'	1°37'E	Wimereux (F)	280	14	
85513	Fu.ve.	11/4	50°52'	1°35'E	Cap Gris-Nez (F)	290	11.9	
85514	Seawater	"	"	"	"	"		31.0
85515	Fu.ve.	"	50°04'	1°22'E	Le Treport (F)	240		
85516	Seawater	12/4	"	"	"	"		5.2
85593	Fu.ve.	4/7	"	"	"	"	20	
85517	Fu.ve.	12/4	49°52'	0°42'E	St.Valerey-en-Caux (F)	200	18.1	
85518	Fu.se.	"	"	"	"	"		
85519	Seawater	"	"	"	"	"		31.1
85592	Fu.ve.	4/7	"	"	"	"	19	
85520	Fu.ve.	12/4	49°46'	0°22'E	Fécamp (F)	170	19.6	
85521	Fu.se.	"	"	"	"	"		
85522	Seawater	"	"	"	"	"		32.3
85591	Fu.ve.	4/7	49°30'	0°06'E	Le Havre (F)	150	20	
85590	Fu.ve.	5/7	49°17'	0°18'W	Luc sur Mer (F)	125	17	
85523	Fu.ve.	12/4	49°21'	0°45'W	Port-en-Bessin (F)	105	18.4	
85524	Seawater	"	"	"	"	"		32.9
85525	Fu.ve.	"	49°34'	1°16'W	St.Vaast-la-Hague (F)	75		
85526	As.no.	"	"	"	"	"	22.1	
85527	Fu.sp.	12/4	49°34'	1°16'W	St.Vaast-la-Hague (F)	75		
85528	Pe.ca.	"	"	"	"	"	52.9	
85529	Seawater	"	"	"	"	"		
85589	Fu.ve.	27/6	"	"	"	"	20	33.4
85530	Fu.ve.	13/4	49°42'N	1°16'W	Pte.de Barfleur (F)	55		
85531	Fu.se.	"	"	"	"	"	17.1	
85532	As.no.	"	"	"	"	"		
85533	Pe.ca.	"	"	"	"	"	24.1	
85534	Pa.vu.	"	"	"	"	"	18.0	
85535	Seawater	"	"	"	"	"		
85536	Fu.ve.	"	49°42'N	1°28'W	Cap Lévy (F)	40		34.1
85588	Fu.ve.	27/6	49°41'N	1°28'W	Fermanvill (F)	"	18	
85537	Fu.se.	13/4	49°43'N	1°52'W	Le Hable (F)	11	19.8	
85538	Fu.sp.	"	"	"	"	"		
85539	Seawater	"	"	"	"	"		34.9
85540	Fu.sp.	13/4	49°43'	1°56'W	Goury, Cap de la Hauge (F)	6	23.4	
85587	Fu.ve.	24/6	"	"	"	"	18	
85585	Fu.se.	11/4	49°40'	1°56'W	Herquemoulin (F)	-6	17	
85586	Fu.se.	2/7	"	"	"	"	21	
85584	Fu.ve.	1/7	49°22'	1°48'W	Carteret (F)	-38	30	
85583	Fu.ve.	"	48°50'	1°35'W	Granville (F)	-101	20	
85582	Fu.ve.	4/7	48°41'	1°51'W	Cancale (F)	-114	24	
85545	Fu.ve.	14/4	48°38'	2°02'W	Saint Malo (F)	-120		
85546	Fu.se.	"	"	"	"	"	27.2	
85547	As.no.	"	"	"	"	"		
85548	Fu.sp.	"	"	"	"	"	35.8	
85549	Pe.ca.	"	"	"	"	"		
85581	Fu.ve.	4/7	48°31'	2°45'W	St. Brieuc (F)	-134	24	
85544	Fu.ve.	14/4	48°50'	3°28'W	Perros Guirec (F)	-145	19.8	
85541	Fu.ve.	13/4	48°43'	3°58'W	Roscoff (F)	-185		
85542	Fu.se.	"	"	"	"	"	21.8	

^{40}K**	^{60}Co	^{90}Sr	^{99}Tc	^{106}Ru	^{125}Sb	^{137}Cs	^{238}Pu	$^{239,240}Pu$	^{241}Am
26.1			70			2.3 .			
		24	1.6						
35.5	2.4	5.2	200		1.6A	2.1	0.013	0.030	
17.4	1.05	0.056		6.5A		1.89			
32.3	8.1		280			1.4			
			11.7						
31.6	3.2	4.1	124	4.3A	1.55	2.3		0.022	0.011
		33	2.2						
32.4			360			1.4			
35.6	5.4	4.7	200	3.8B	2.0	1.97		0.035	0.023
		40	4.4						
29.0	6.9		250			1.8			
40.8	17.6	4.3	780	13.9A	4.7	3.9	0.046	0.096	
45.9	7.6	5.1	670	8.2A	2.4	2.9	0.043	0.088	0.029
			3.6						
29.2	9.4		350	11.3		2.0			
		10.7	0.48						
31.8	32	4.7	670	23	3.8	3.1	0.077	0.169	
41.1	14.3	5.6	1350	12B	3.0A	3.0	0.083	0.20	0.026
33.3	10.8		620	21.7		1.0			
			4.4						
41.8	28	4.8	1370	17.3	3.2	4.0	0.076	0.150	
38.7	12.2	5.5	906	16.9	2.8	2.6	0.093	0.199	0.050
32.7	13.4		540	22		1.9			
			6.1						
48.8	14.8	4.5	380		2.9A	4.8			
40.3	17.1	6.1	730	13.6A	4.8	3.9	0.050	0.054	
34.6	13.7	6.5	450	23	3.3A	2.8	0.107	0.184	0.088
			6.4						
33.7	14.7	6.9	790	10.7B	3.4A	2.6	0.080	0.27	
30.2	7.9	6.4	1180	13.2A	5.0	2.8	0.092	0.136	0.073
30.2	13.4	5.6	470	27A	4.8	2.0	0.064	0.125	
22.8	9.9	4.9	760	14.8A	43	2.0	0.057	0.125	0.042
			3.8						
37.0	23	6.4	790	13B	4.9	4.0	0.068	0.174	
35.3	46	8.0	1500	30	4.2A	3.6	0.125	0.35	
37.1	52	7.7	1100	53	6.2	4.1			0.24
30.1	16.5	6.9	2100	48	6.5	2.7	0.22	0.43	
23.9	18.3	1.14	980		12.9	2.3	0.096	0.175	
13.8	14.5	8.3		119	10.4	5.9			
			12.1						
33.7	88		1640	53		4.1			
42.2	56		1450	36A	4.5B	3.5	0.194	0.172	
42.7	200	10.0	2400	153	6.7	4.8	0.44	0.79	0.45
32.5	123		2200	65		3.4			
		89	17.1						
36.8	200	20	2300	184	11.8	12.0	0.42	0.67	
33.4	260	6.3	4700	127		5.2	0.51	0.55	
49.6	410	15	4300	250	8.8A	7.2	0.55	0.94	2.25
47.2	270	3.8	3600	184	4.4B	5.2	0.42	0.48	
40.1	48	3.5	1700	24A	3.8B	4.1			
36.5	26	3.7	1210	17A	2.1A	2.3			
38.5	18.6	3.5	750	15.3	2.0A	3.2			
31.7	9.2		800	6.7		0.7			
38.9	15.1	2.6	460	9.5A	1.8A	1.63	0.098	0.24	0.064
38.2	7.3		1260			1.8			
32.0	12.8	3.5	500	5.6B	1.7A	1.35		0.120	B.D.L.
22.9	6.1		650			0.9			
30.8	36	3.0	1020	23	1.8A	1.66	0.087	0.146	0.027
42.7	1.36	0.39	46			0.74A		0.059	
42.2		0.27	12.6			0.89		0.102	
49.9		0.44	3.3			1.10A		0.089	0.011

Table 1. (continued)

Station number	Species	Date	Position N	E or W	Location	Km*	% dry matter	Salinity in o/oo
85543	As.no.	13/4	48°43'	3°58'W	Roscoff (F)	-185		
85580	Fu.ve.	26/6	"	"	"	"	20	
85550	Fu.ve.	20/4	50°04'	5°42'W	Sennen Cove (GB)	275	20.6	
85551	Seawater	"	"	"	"	"		35.1
85552	Fu.ve.	"	50°01'	5°05'W	Coverack (GB)	230		
85553	Fu.ve.	"	50°14'	3°51'W	Hope Cove (GB)	150	31.0	
85554	Fu.se.	"	"	"	"	"		
85555	Seawater	"	"	"	"	"		34.9
85556	Fu.sp.	21/4	50°34'	2°26'W	Bill of Portland (GB)	105	18.3	
85557	Seawater	"	"	"	"	"		35.0
85558	Fu.sp.	"	50°37'	1°57'W	Swanage (GB)	105		
85559	Fu.sp.	"	50°43'	0°47'W	Selsey (GB)	145	20.5	
85560	Seawater	"	"	"	"	"		33.8
85561	Fu.sp.	"	50°44'	0°12'E	Birling Gap (GB)	195		
85562	Fu.ve.	"	"	"	"	"	18.7	
85563	Fu.se.	"	"	"	"	"		
85564	Fu.sp.	22/4	51°06'	1°13'E	Dover (GB)	275	16.5	
85565	Seawater	"	"	"	"	"		34.9
85566	Fu.ve.	"	51°21'	1°27'E	Broadstairs (GB)	305		
85567	Fu.se.	"	"	"	"	"	20.3	
85568	Fu.sp.	23/4	51°56'	1°17'E	Harwich (GB)	340		
85569	Fu.se.	"	"	"	"	"	19.8	
85570	Seawater	"	"	"	"	"		34.2
85571	Fu.ve.	"	55°29'	8°25'E	Esbjerg N (DK)	1140	23.7	
85572	Seawater	"	"	"	"	"		26.6
1193	Fu.ve.	19/4	55°28'	8°24'E	Esbjerg S (DK)	"	21.6	
1194	Seawater	"	"	"	"	"		22.9
85573	Fu.ve.	25/4	55°05'	8°34'E	Rømø (DK)	1100	19.3	
85574	Seawater	"	"	"	"	"		27.1
1189	Fu.ve.	18/4	55°09'	8°34'E	"	"	19.6	
1190	Fu.ve.	18/4	55°05'	8°34'E	Rømø	1100	21.2	
1191	My.ed.	"	"	"	"	"	14.6	
1192	Seawater	"	"	"	"	"		25.1
1183	Fu.ve.	17/4	54°08'	8°52'E	Büsumhafen (D)	990	17.9	
1184	Seawater	"	"	"	"	"		24.5
1185	Fu.ve.	"	54°31'	8°50'E	Norderhafen (D)	1030	20.7	
1186	Seawater	"	"	"	"	"		28.1
1187	Fu.ve.	18/4	54°44'	8°43'E	Dagebüllhafen (D)	1060	23.8	
1188	Seawater	"	"	"	"	"		28.5
1184 1186 1188	Seawater	17/4 ~	54°31'	8°50'E	Büsumhafen/ Norderhafen/ (D) Dagebüllhafen	~ 990		~ 27

*Shortest sea distance from Cap de la Hague in Km
**Unit: g K kg^{-1} dry weight

Fu.ve.: Fucus vesiculosus, Fu.se.: Fucus serratus, Fu.sp.: Fucus spiralis,
As.no.: Ascophyllum nodosum, Pe.ca.: Pelvetia canaliculata, Pa.vu.: Patella vulgata,
My.ed.: Mytilus edulis

85507: 0.12 B
85510: 0.11 B
85534: 0.34
85540: 0.24

In these four samples it was possible to determine ^{134}Cs, and the $^{134}Cs/^{137}Cs$ ratios were calculated. In the samples collected close to Cap de la Hague the background was too high for a reliable ^{134}Cs.

A: counting error 20-33%
B: counting error > 33%

^{40}K**	^{60}Co	^{90}Sr	^{99}Tc	^{106}Ru	^{125}Sb	^{137}Cs	^{238}Pu	$^{239,240}Pu$	^{241}Am
32.1		0.29	14.9			0.46A	0.041	0.138	
45.8	7.6	0.25	116			0.7B		0.094	0.98
42.9	1.84	0.27	54			0.68		0.053	0.011
			2.0						
36.9	2.3		81						
35.1	7.7	0.41	145	3.6B		0.74	0.017	0.105	0.008
38.3	6.8		115			1.2			
			0.5						
35.7	46	0.46	134			1.1A	0.0174	0.073	
			2.1						
32.8	224		119						
30.5	42	0.98	219			1.3A	0.0116	0.031	0.013
			0.83						
51.8	55	0.77	310						
41.8	77		700			1.6A	0.025	0.079	0.017
55.4	228		320						
33.6	24	0.65	156			1.65	0.0105	0.041	0.009
		21	2.3						
44.8	13.8		300±20			1.8			
34.7	15.0	0.91	156			2.6	0.0165	0.044	0.020
29.8	4.4		110			7.4			
35.6	7.6	1.53	104			11.3		0.038	0.044
			0.76						
24.9	1.15	4.3	200		1.4A	2.4			
			2.5						
29.3	1.38	4.2	320		2.1A	3.1			0.030
			2.0						
30.6	1.0A	3.3	190			3.4			
			1.27						
31.3	0.78A	3.6	102			3.1		0.027	0.040
28.6	0.71A	3.9	240		1.1A	3.2			0.0167
8.42	0.78	0.019B		4.9A		1.34			
		31	2.9						
26.0		3.4	81			2.2			0.043
			2.6						
29.2	0.84	3.5	230		1.5A	2.3		0.027	0.0063
			3.5						
26.4	0.76A	3.5	118		1.7A	2.8		0.091	0.049
			3.5						
						16.7			

Table 2. Values of A and B in Eqv. 1 from liquid discharges of various radionuclides to the sea from the nuclear reprocessing plants Sellafield and La Hague.

Radionuclide	La Hague (Table 1)		Sellafield (Aarkrog et al. 1983-85)	
	A	B	A	B
^{60}Co	1.02	2100	0.84	216
^{90}Sr	-	-	1.12	7040
^{99}Tc	0.67	18800	0.62	46000
^{106}Ru	0.63	315	-	-
^{137}Cs	-	-	0.94	23000
$^{239,240}Pu$	0.61	2.7	1.04	1520

The equations were based on samples collected out to a distance of approximately 1000 km from the source. The backgrounds (from Sellafield and global fallout) of ^{90}Sr and ^{137}Cs in the English Channel do not make it possible to calculate meaningful distance equations for these radionuclides for La Hague.

In the calculation of the distance relations we normalized all seaweed data to Fucus vesiculosus. We used the following ratios for ^{99}Tc (calculated from Table 1): Ascophyllum/ves.: 1.41; serratus/ves.: 0.64; spiralis/ves.: 0.55 and pelvetia/ves. = 0.81. In a similar way the other radionuclides studied were normalized to Fucus vesiculosus.

The distance relations may vary with time. If the discharges of a nuclide increase from year to year, A in Table 2 will increase, and if the annual discharges show a decreasing tendency A will decrease. It is however remarkable that the A-values in Table 2 are so similar; they all lie between 0.6 and 1.1. This shows that the coastal currents in the North Sea region run over relatively long distance without any appreciable mixing with the open sea.

From Table 2 we may estimate the discharges from La Hague relative to those from Sellafield. We assume that the Fucus samples collected at the Channel coast in the first half of 1985 represented La Hague releases from 1983 and 1984, and the samples from the Scottish coastline in June 1982 were related to Sellafield discharges in 1980-81. If we divide the distance integrals:

$\int_1^{1000} 2\pi \, x \cdot B \cdot x^{-A} dx$ for La Hague (H) with those from Sellafield(S)

we find for ^{99}Tc: H/S = 0.30, for ^{60}Co H/S = 3.3 and for

$^{239,240}Pu$ H/S = 0.024. The annual mean discharges in 1980-81 from Sellafield (BNFL 1979-1986)[3] were: 31 TBq ^{99}Tc, 0.76 TBq ^{60}Co, and 17.5 TBq $^{239,240}Pu$. We would thus predict the annual mean discharges from La Hague in 1983-84 to 9 TBq ^{99}Tc, 2.5 TBq ^{60}Co, and 0.4 TBq $^{239,240}Pu$. We have no complete data from 1983-84 from La Hague, but for 1983 Patti et al. (1984)[11] reported 11.7 TBq ^{99}Tc and for 1982 Calmet & Guegeniat (1985)[4] reported annual discharges of 3.1 TBq ^{60}Co and 0.12 TBq $^{239,240}Pu$. If we assume that the discharges from La Hague in the later years were relatively constant, we may conclude that the above estimates from Fucus measurements seem reasonable.

CONCLUSION

The radionuclide concentrations in seaweed along the continental side of the English Channel are inversely proportional to the distance (in km) or square root of the distance from La Hague. Similar distance relations have been observed earlier in British coastal water for liquid discharges from Sellafield These relations may be applied for estimates of the annual discharges from the reprocessing plants in Western Europe.

REFERENCES

1. Aarkrog A., Dahlgaard H., Hallstadius L., Hansen H. and Holm E. (1983). Nature 304, 49-51.
2. Aarkrog A. et al. (1983-1985). Risø Reports, Nos. 488, 509, 510, 527 and 528. Risø National Laboratory, Roskilde, Denmark.
3. BNFL (1979-1986). Annual report on radioactive discharges and monitoring of the environment 1978-1985. British Nuclear Fuels Limited, Warrington, Cheshire, U.K.
4. Calmet D. & Gueguenlat P. (1985), in: IAEA-TECDOC-329 pp. 111-144 International Atomic Energy Agency, Vienna.
5. Casso S.A. & Livingson H.D. (1984). WHO-84-40. Woods Hole Oceanographic Inst.
6. Harley J.H. (editor) (1972). HASL-300. Environmental measurements Laboratory, New York.
7. Holm E., Rioseco J., & Garcia-Leon M. (1984). Nucl. Instr. and Meth. in Phys. Res. 223, 204-207.
8. Jefferies D.F., Steele A.K. and Preston A. (1982). Deep-Sea Res. 29, 713-738.
9. Kautsky H. (1973). Deutschen Hydrographischen Zeitshrift, 26, 242-246.
10. Livingston H.D., Bowen V.T. & Kupferman S.L. (1982). J. mar. Res. 40, 1227-1258.
11. Patti F., Masson M., Vergnaud G. & Jeanmaire L. (1984), in: Technetium in the Environment (Desmet G & Myttenaere C. editors) pp. 37-51. Elsevier Applied Science Publishers, London & New York.
12. Talvitie N.A. (1972) Analyt. Chem. 43, 1827.

FUCUS VESICULOSUS AS AN INDICATOR FOR CAESIUM ISOTOPES IN IRISH COASTAL WATERS

I.R. McAulay and D. Pollard
Department of Physics,
Trinity College,
Dublin 2, Ireland.

The discharge of liquid radioactive waste from the British Nuclear Fuels reprocessing plant at Sellafield on the Cumbrian coast has given rise over the past two decades to an increased level of artificial radioisotopes in the Irish Sea.

Samples of seawater, fish, seaweed and sediment have been taken regularly for a number of years from various parts of the Irish Sea both by the Fisheries Radiobiological Laboratory at Lowestoft in the United Kingdom and by researchers from Irish Universities. Levels of artificial radioisotopes in the Irish Sea are higher than those found in any other European coastal waters and have enabled much information to be acquired about the behaviour of such substances in the marine environment. The purpose of this study was to investigate the temporal and spatial distribution of caesium-137 along the Irish coast. The study covers the period from Spring 1983 to May 1986, at which time radioactivity levels in seaweed were seriously disturbed by fallout from the Chernobyl accident.

The radiocaesium distribution was mapped using seaweed of the genus Fucus as a bioindicator. Seaweed samples were collected regularly over the survey period from a number of fixed sampling points along the east Irish coast between latitude 54° 17' and latitude 52° 59'. Approximately one hundred samples were analysed to determine 134-Cs,

137-Cs and 40-K activity concentrations. A smaller number of sea-water samples were also collected and analysed.

Seaweed samples were dried to constant weight, ground and sieved so as to achieve a consistent matrix density and packed to a standard counting geometry. Sea-water samples were filtered through a membrane filter (pore diameter 0.45 μm) and the caesium concentrated onto ammonium molybdophosphate impregnated resin. This technique has been fully described by Baker (1975).

The activity concentrations were determined by high resolution gamma spectrometry. The detector used was an Ortec Gamma-X high purity germanium coaxial type with a beryllium entrance window.

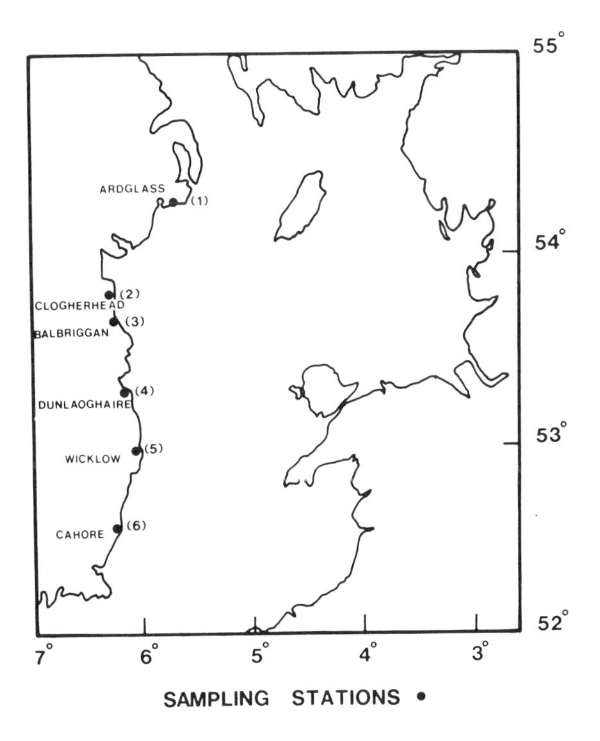

FIG. 1

Seaweed samples were collected from the stations shown in Fig. 1 on a regular basis from the beginning of 1983. During 1984 sampling was discontinued at stations 2 and 6 as they were found to provide little extra information. Water sampling was carried out at station 3 during 1985 and 1986. Figures 2, 3, and 4 show the variation in mean annual 137-caesium concentration (in Bq kg^{-1} dry weight) for the three years 1983, 1984 and 1985 at each of the sampling stations.

Figure 5 shows the variation of mean annual caesium-137 concentration at station 3 over the period and figure 6 shows the variation at station 5.

Figures 7 and 8 show the results of the individual measurements at stations 3 and 5. The data for the first four months of 1986 continued to show a decline, averaging 46 Bq/kg dry weight of 137-Cs at station 3.

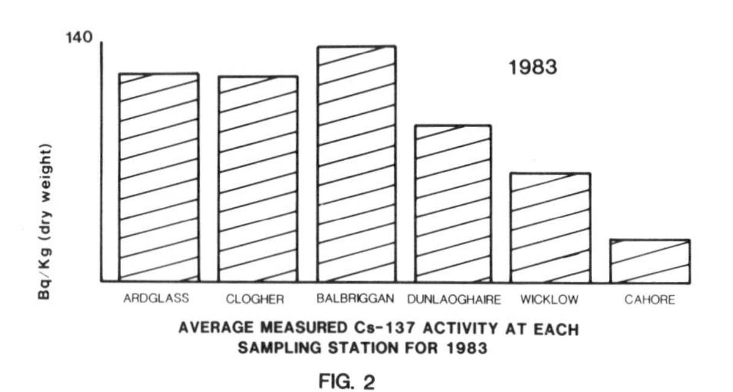

AVERAGE MEASURED Cs-137 ACTIVITY AT EACH SAMPLING STATION FOR 1983

FIG. 2

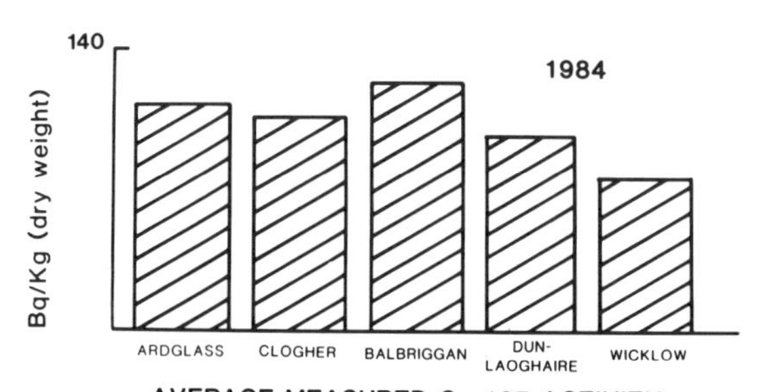

AVERAGE MEASURED Cs 137 ACTIVITY AT EACH SAMPLING STATION FOR 1984

FIG.3

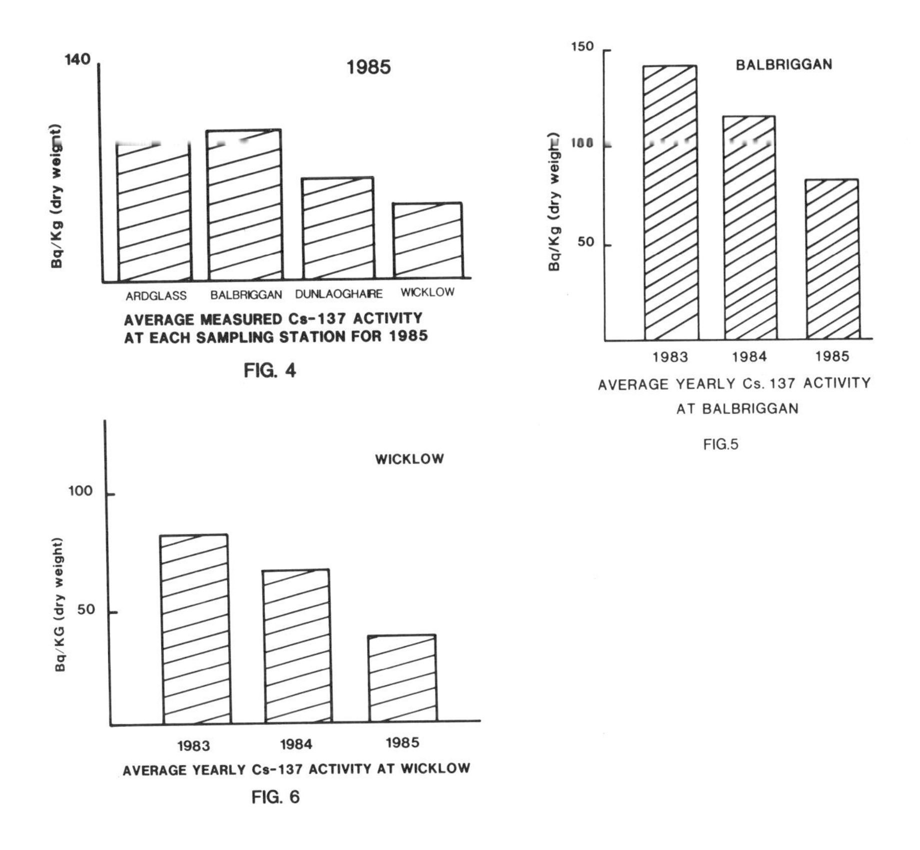

AVERAGE MEASURED Cs-137 ACTIVITY AT EACH SAMPLING STATION FOR 1985

FIG. 4

AVERAGE YEARLY Cs. 137 ACTIVITY AT BALBRIGGAN

FIG.5

AVERAGE YEARLY Cs-137 ACTIVITY AT WICKLOW

FIG. 6

The fallout from Chernobyl caused a very sharp increase in caesium-137 levels in Fucus vesiculosus, rising to 148 Bq/kg at station 3 in June 1986 and then declining slowly to 50 Bq/kg by December 1986. This sharp increase was due to direct contamination of the seaweed by rainfall and land runoff on 3rd and 4th May 1986, when the Chernobyl cloud crossed Ireland.

Caesium-134 was also measured in the seaweed samples and shows a pattern broadly similar to that shown by the 137 isotope. As the minimum detectable activity for 134 caesium was approximately 1.5 Bq/kg^{-1} dry weight with the sample geometry and counting time used, the fluctuations are less easily determined. Figure 9 shows the mean caesium-134 variation

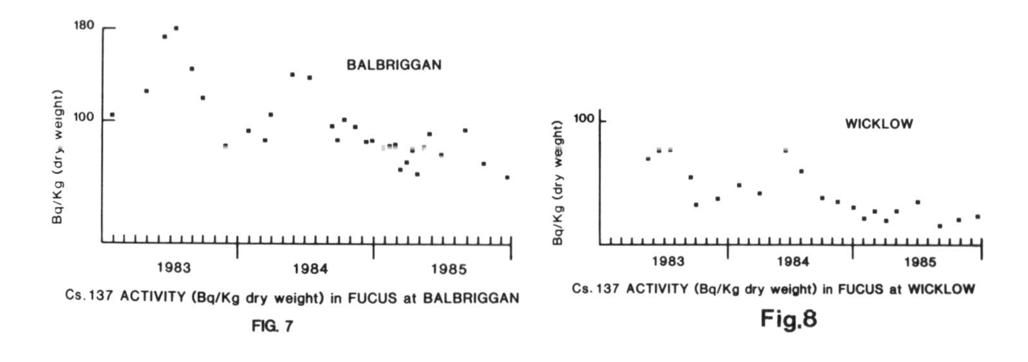

Cs. 137 ACTIVITY (Bq/Kg dry weight) in FUCUS at BALBRIGGAN

FIG. 7

Cs. 137 ACTIVITY (Bq/Kg dry weight) in FUCUS at WICKLOW

Fig.8

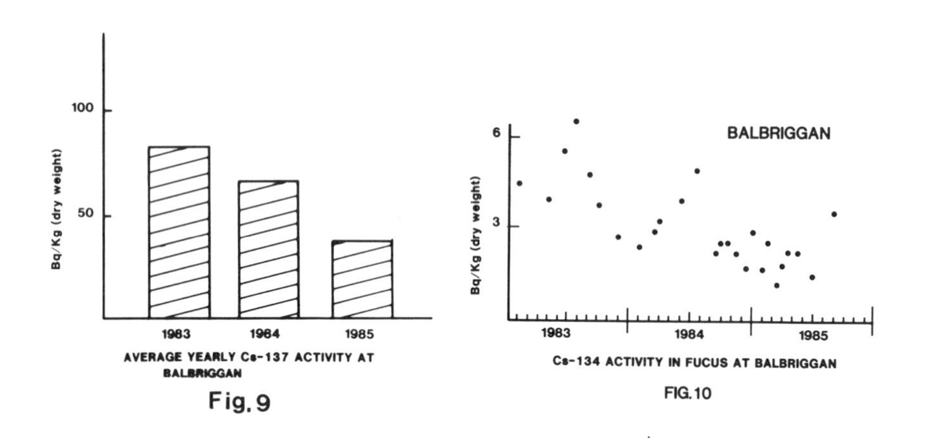

AVERAGE YEARLY Cs-137 ACTIVITY AT BALBRIGGAN

Fig. 9

Cs-134 ACTIVITY IN FUCUS AT BALBRIGGAN

FIG.10

at station 3 and Figure 10 the pattern of the individual measurements.

The average value of the caesium-137:caesium-134 ratio in the North Irish sea was 32, though this figure was becoming difficult to estimate at the end of 1985 as the caesium-134 levels were approaching the minimum detectable activity. The caesium-137 activities measured in sea water samples taken at station 3 during 1985 and 1986 are shown in Table 1.

TABLE I

Caesium 137 in seawater at sampling station 3

SAMPLE DATE	ACTIVITY (m Bq/per litre)
11th April 1985	300
30th April 1985	330
13th June 1985	350
22nd July 1985	310
24th Sept. 1985	260
29th Nov. 1985	330
30th Jan. 1986	270
June 1986	170
Oct. 1986	180

Error in above values is typically 8%.

DISCUSSION

The general conclusions are (1) that there is, at any given time, a decrease in the caesium activity in the Irish Sea south of Dublin, but a reasonably uniform concentration in the North Irish Sea; and (2) there has been a steady decrease with time of radiocaesium activity in the Irish Sea.

On the basis that the ratio of 137-Cs to 134-Cs in the Sellafield discharges is 14, it appears that the age of the caesium at the Irish coast is about 2½ years. The M.A.F.F. compartment model of the Irish Sea suggests that the transit time of Sellafield discharges to the Irish coast is about one year. However, the ratio of the caesium isotopes in seaweed in the vicinity of Sellafield was about 20 in 1985, due to the presence of 'older' caesium in the coastal water which mixes with the caesium in the discharges. A transit time of one year to the eastern coast of Ireland would imply that the ratio of caesium isotopes should be about 28 at sampling stations 1, 2 and 3. The difference between this value and the measured value is at any rate partly due to the existence of older caesium-137 at the Irish Coast.

The figures obtained in this study therefore confirm the main predictions of existing models for the behaviour of radioactivity in the Irish Sea. The individual measurements as presented in figures 7, 8 and 10 show an obvious annual cycle in the caesium concentrations in fucus seaweed with the peak values obtained in samples taken in the summer months. No such cycle was found in the potassium values obtained and it is not yet clear whether this periodicity in the caesium values is a biological phenomenon or whether it reflects a variation in the caesium concentration in the coastal waters of Ireland.

Direct measurements of the caesium content in Irish Sea fish had enabled an estimate of the collective dose to the Irish population in 1981 to be made (1985, McAulay and Doyle) The value of 2 man-Sievert was based on the mean of a wide range of activities measured, due in turn to difficulty in locating precisely the area in which each sample was caught and the uncertainty in relating spawning grounds to catch location if that could be determined. Fucus vesiculosus provides a more precisely specified sample for radioactivity measurement and enables the variation in caesium activity to be followed more easily from year to year. This in turn enables a more reliable estimate to be made of the collective dose from fish consumption in the region, provided that the fish consumed comes from the coastal regions. This is true for the species of fish which make up the bulk of consumption in Ireland. Thus, it may be estimated from the figures presented here that the caesium activity in the Irish Sea has decreased by approximately 20% per annum from 1983 to 1986. The collective dose from fish consumption will have decreased at the same rate and is estimated to have reduced to 0.7 man-Sievert to the Irish population in 1986.

The authors would like to thank Mr. T. Lane for his careful work in preparation of the samples; they also wish to acknowledge that this work was carried out with the assistance of research contract B16-B-043-IRL granted by the Commission of the European Communities.

References

Baker, C.W., (1975). The determination of radiocaesium in sea and fresh waters. M.A.F.F. Fisheries Laboratory Technical Report No. 16, M.A.F.F. Lowestoft, Suffolk, U.K.

McAulay, I.R. and Doyle, C., (1985). Radiocaesium levels in Irish Sea fish and the resulting dose to the population of the Irish Republic. Health Physics 48 (3); 333-337.

Répartition de deux traceurs radioactifs (^{106}Ru-Rh, ^{60}Co) chez deux espèces indicatrices (Fucus serratus, L., Mytilus edulis, L.) le long du littoral français de la Manche

P. GERMAIN, Y. BARON, M. MASSON
CEA, IPSN, SERE, Laboratoire de Radioécologie Marine de La Hague
B.P. 270 - 50107 CHERBOURG - France

et

D. CALMET
CEA - Station Marine de Toulon. IFREMER/CT
B.P. 330 - 83507 LA SEYNE SUR MER CEDEX - France

RESUME

Les mesures de ^{106}Ru-Rh, ^{60}Co chez l'algue Fucus serratus et le mollusque bivalve Mytilus edulis, prélevés le long du littoral de la Manche (en 1985 et 1986), indiquent que les radioéléments rejetés par l'usine de La Hague marquent principalement le littoral du Cotentin, mais qu'ils sont décelables depuis une zone proche de Roscoff à l'Ouest jusqu'en mer du Nord à l'Est. Les niveaux de radioactivité décroissent en fonction de l'éloignement du Cap de La Hague ; cependant, dans le champ proche, certains cas montrent un gradient inverse. A l'échelle de la Manche, le marquage des espèces est plus limité vers l'Ouest que vers l'Est.

INTRODUCTION

La mesure de radioéléments artificiels est effectuée dans les divers compartiments de la mer de la Manche à des fins radioécologiques. Dans ce cadre, l'étude de la distribution spatiale des radioéléments artificiels dans des organismes marins (algues et animaux) se poursuit depuis plusieurs années le long du littoral de la Manche (Fraizier, Guary, 1977 ; Germain et al, 1979 ; Masson et al, 1983 ; Germain, Miramand, 1984 ; Calmet et al, 1987). Les résultats obtenus fournissent entre autre des informations relatives à la dispersion dans l'eau de mer des éléments artificiels rejetés par l'industrie nucléaire. Deux espèces indicatrices ont été retenues : une algue Fucus serratus et un mollusque bivalve Mytilus edulis. Des échantillons de ces espèces ont été prélevés en plusieurs stations le long du littoral de la Manche, de part et d'autre de l'émissaire de l'usine de La Hague, de la Bretagne jusqu'au Pas de Calais, lors de campagnes menées en 1985 et 1986. Une spectrométrie γ a été réalisée sur chaque échantillon. A partir de l'ensemble des radioéléments détectés, nous avons retenu pour le présent travail, ^{106}Ru-Rh et ^{60}Co dont les niveaux, sur l'ensemble du littoral, sont significativement supérieurs au bruit de fond.

MATERIEL ET METHODES

Les deux espèces Fucus serratus et Mytilus edulis ont été prélevées sur l'estran des stations indiquées sur la Figure 1, en Juin-Juillet 1985, Avril-Mai 1986 et en Octobre 1986. Chaque échantillon correspond à un prélèvement réalisé depuis le plus haut niveau occupé par l'espèce jusqu'à son plus bas niveau. L'ensemble des prélèvements de chaque campagne a été effectué en trois semaines. Après élimination des sédiments et des épiphytes, séparation de la chair et de la coquille pour les moules, les échantillons ont été séchés à 90°C, broyés et conditionnés dans des boîtes de comptage de volume déterminé. Les spectrométries γ ont été réalisées sur des détecteurs Ge (Li) d'efficacité 15 % et ayant une résolution de 2 KeV à 1,33 MeV. La distribution spatiale des radioéléments dans les espèces est exprimée suivant une fonction puissance du type $y(d) = \alpha d^{-\beta}$ où y (d) est le niveau d'activité du radioélément au sein du bioindicateur à la distance d de la station à l'émissaire. Les coefficients d'ajustement α et β sont deux constantes déterminées par la méthode des moindres carrés. Le coefficient r de Bravais - Pearson est retenu pour traduire la qualité de l'ajustement des points.

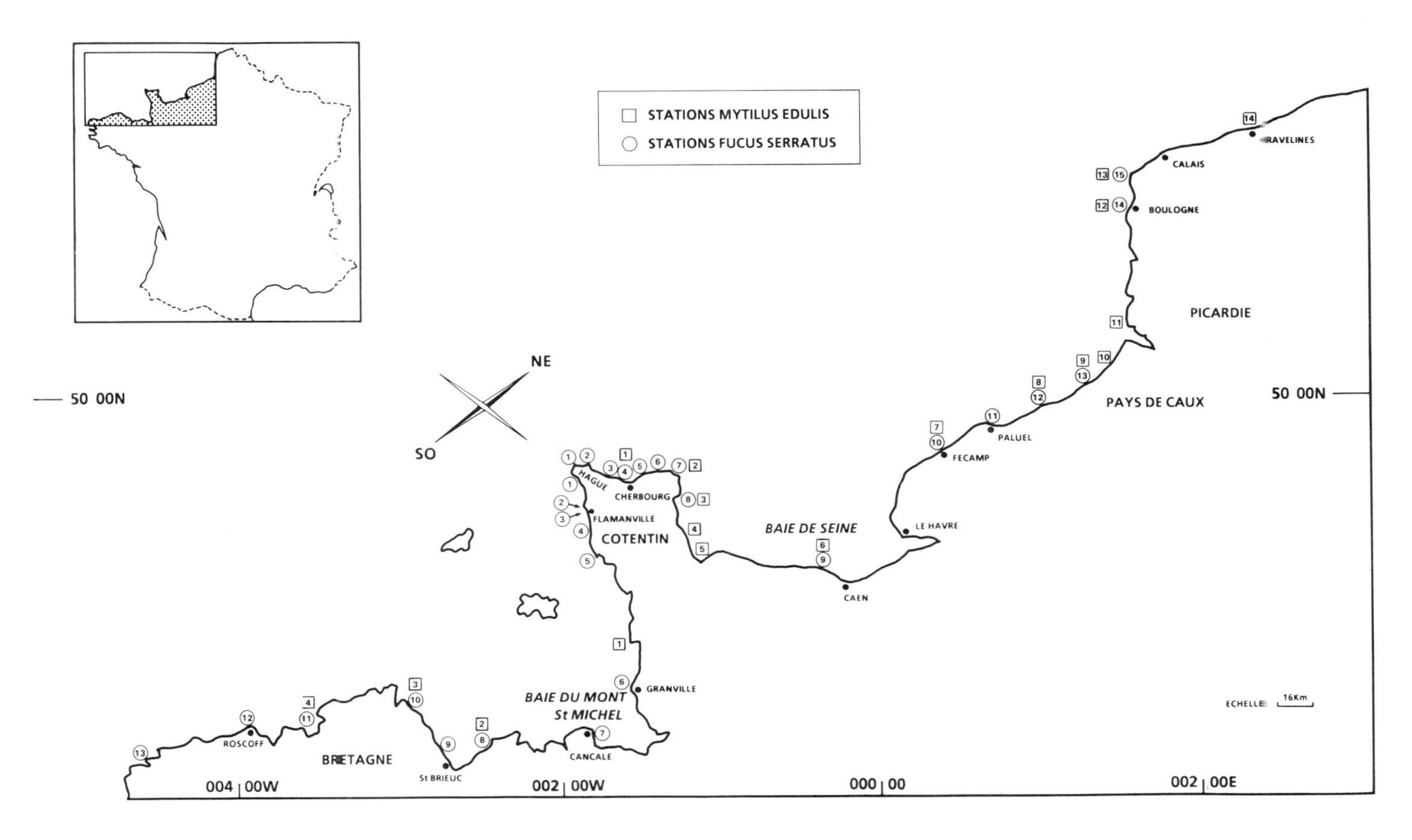

Figure 1 Localisation géographique des stations de prélèvement

RESULTATS ET DISCUSSION

La distribution des radioéléments d'origine industrielle dans les espèces côtières en une station donnée, est fonction de plusieurs facteurs tels que la présence de plusieurs sources, les quantités rejetées, les caractéristiques hydrodynamiques des sites, la position sur l'estran des espèces, les capacités différentes entre les espèces de fixer les éléments, la décroissance radioactive, la qualité physicochimique des radioéléments... En ce qui concerne ce dernier point, le ruthenium rejeté par l'usine de La Hague se présente dans l'eau de mer essentiellement sous forme de composés nitro-nitrosylruthenium lentement hydrolysables. Le cobalt se présente principalement sous forme soluble cationique Co^{++} dans l'eau de mer ; il s'y ajoute d'autres formes cationiques et des formes anioniques résultant de complexes organiques ou minéraux. L'interprétation de l'évolution spatiale des radioéléments industriels dans des espèces doit tenir compte de ces facteurs. Cependant, la comparaison dans l'espace des niveaux de radioactivité enregistrés au cours de plusieurs campagnes permet déjà de dégager des remarques concernant la dispersion des radioéléments à partir d'un émissaire. L'intérêt est renforcé par le fait que les résultats des niveaux de radioactivité dans les espèces traduisent les évènements relatifs aux rejets et à leur dispersion lors des semaines précédant la récolte des échantillons.

Les radioéléments rejetés par l'usine de La Hague marquent principalement le littoral du Cotentin, mais ils sont décelables dans les algues et les animaux depuis une zone située entre St Brieuc et Roscoff à l'Ouest, et jusqu'en mer du Nord à l'Est. Les évolutions spatiales des niveaux de ^{106}Ru-Rh et ^{60}Co et les ajustements mathématiques sont représentés sur les Figures 2 à 5 et le Tableau 1. La dilution des radioéléments se traduit par une décroissance des niveaux de radioactivité au sein des espèces en fonction de l'éloignement du point de rejet. Les valeurs du coefficient r sont significativement différentes de 0 confirmant que l'activité décroît en fonction de la distance.

Une dissymétrie dans la distribution des radioéléments apparaît entre les zones situées au Nord Est et au Sud Ouest de l'émissaire. Ainsi, dans le cas des _Fucus serratus_, une rupture brutale des niveaux ^{106}Ru-Rh et ^{60}Co se manifeste entre 23 et 35 Km au Sud Ouest. Vers le Nord Est, la diminution est progressive. A plus longue distance, le marquage des espèces est plus limité vers l'Ouest que vers l'Est. Ainsi à Roscoff, 180 Km à l'Ouest de l'émissaire de l'usine de La Hague, les éléments ne sont plus détectés, alors qu'ils le sont encore au Cap Gris Nez et à Gravelines, 290 et 330 Km à l'Est. Les niveaux de radioactivité enregistrés dans les espèces sur les côtes bretonnes sont très proches de ceux du Pays de Caux et de Picardie. Lors du passage du Nord Cotentin vers la Baie de Seine, une fraction des éléments rejetés par l'usine de La Hague progresserait dans la partie Ouest de cette Baie. Ainsi, les teneurs en ^{106}Ru-Rh et ^{60}Co dans les chairs de moules prélevées sur les fonds marins au Nord Est du Cotentin, en Avril-Mai 1986, sont respectivement de 800 Bq kg^{-1} sec et 50 Bq kg^{-1} sec, alors que plus au Sud à 20 et 35 Km, les teneurs sont de 180 et 100 Bq kg^{-1} sec pour ^{106}Ru-Rh, 9 et 7 Bq kg^{-1} sec pour ^{60}Co.

D'un point de vue général la répartition assymétrique des niveaux de radioactivité dans les espèces de part et d'autre du point de rejet de l'usine de La Hague correspond à une dérive générale des masses d'eau de mer vers la mer du Nord. Ces résultats vont dans le sens d'observations antérieures (Germain _et al_, 1979 ; D. Calmet, 1986 ; D. Calmet _et al_, 1987). Cependant, quelques faits particuliers sont à noter. Ainsi, un gradient spatial inverse chez _Fucus serratus_ apparaît pour les deux radioéléments dans le champ proche du point de rejet vers le Sud Ouest. Cela tient aux résultats enregistrés en une station à 23 Km au Sud du point de rejet, et dont il n'a pas été tenu compte dans la définition des équations des courbes ajustées. Ce phénomène est, sans doute, à relier à la complexité de la courantologie dans cette région (Salomon _et al_, 1987) et à la qualité physico chimique des radionucléides étudiés. A Gravelines, en mer du Nord, les niveaux de ^{60}Co dans les chairs de moules augmentent, ce fait étant lié aux rejets de la Centrale électronucléaire. Un marquage des échantillons de la campagne du 22 avril au 27 mai 1986 par le ^{106}Ru-Rh des retombées de Tchernobyl n'est pas à exclure, cependant, en se référant aux niveaux des ^{103}Ru et ^{134}Cs, présents dans ces retombées, son intensité est faible.

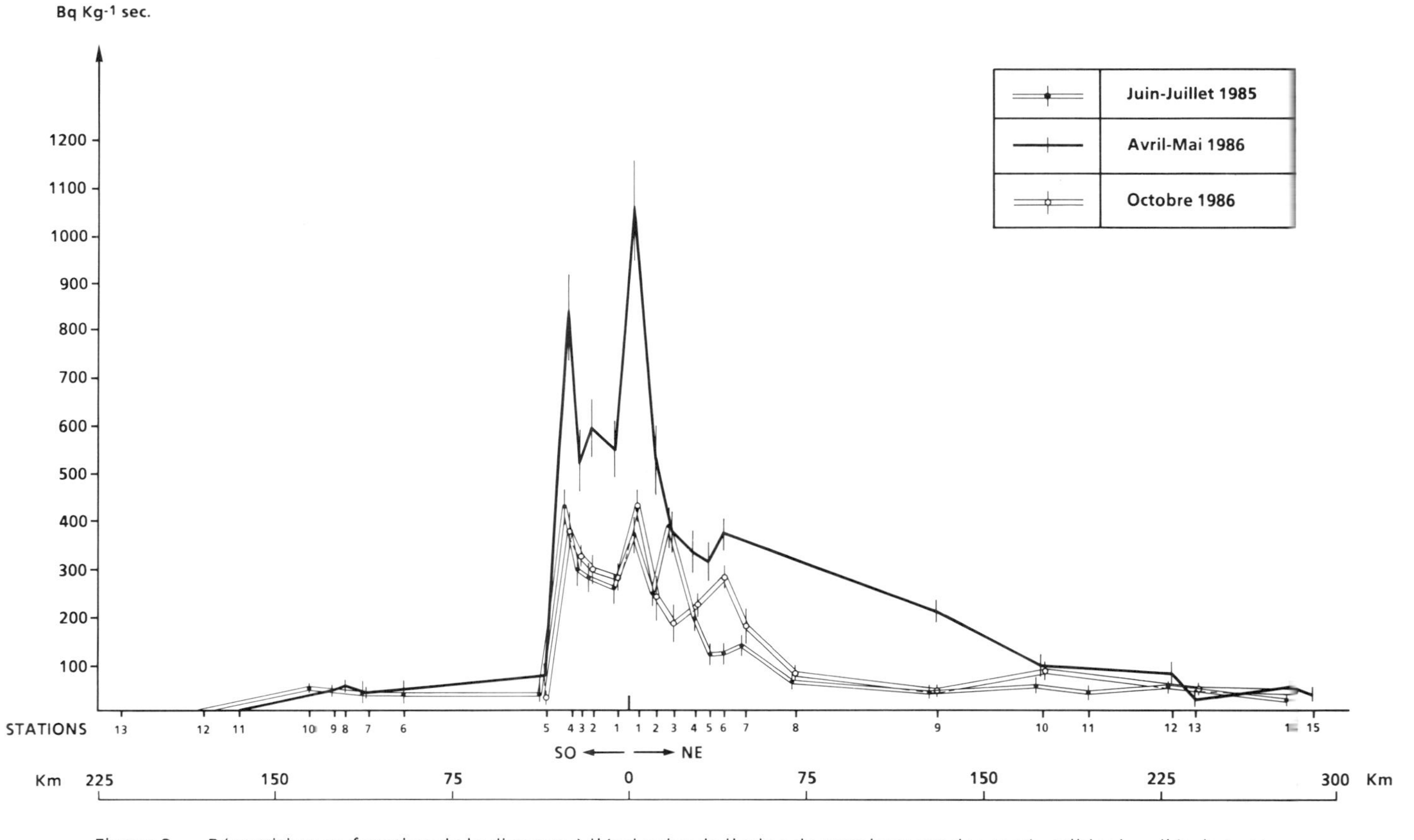

Figure 2 Répartition en fonction de la distance à l'émissaire de l'usine de retraitement des combustibles irradiés de La Hague du ^{106}Ru-Rh chez <u>Fucus serratus</u> en 1985 et 1986

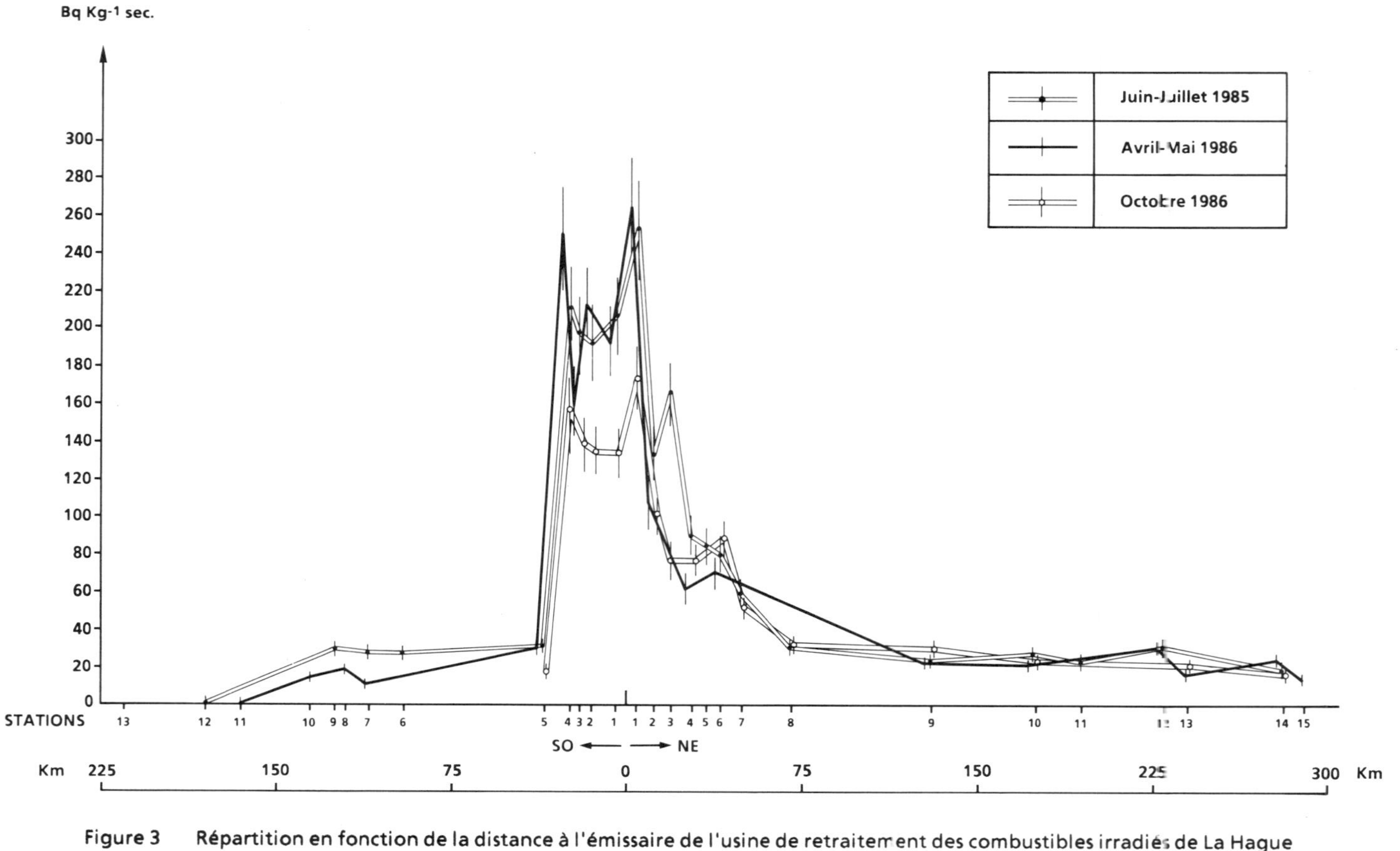

Figure 3 Répartition en fonction de la distance à l'émissaire de l'usine de retraitement des combustibles irradiés de La Hague du ^{60}Co chez Fucus serratus en 1985 et 1986

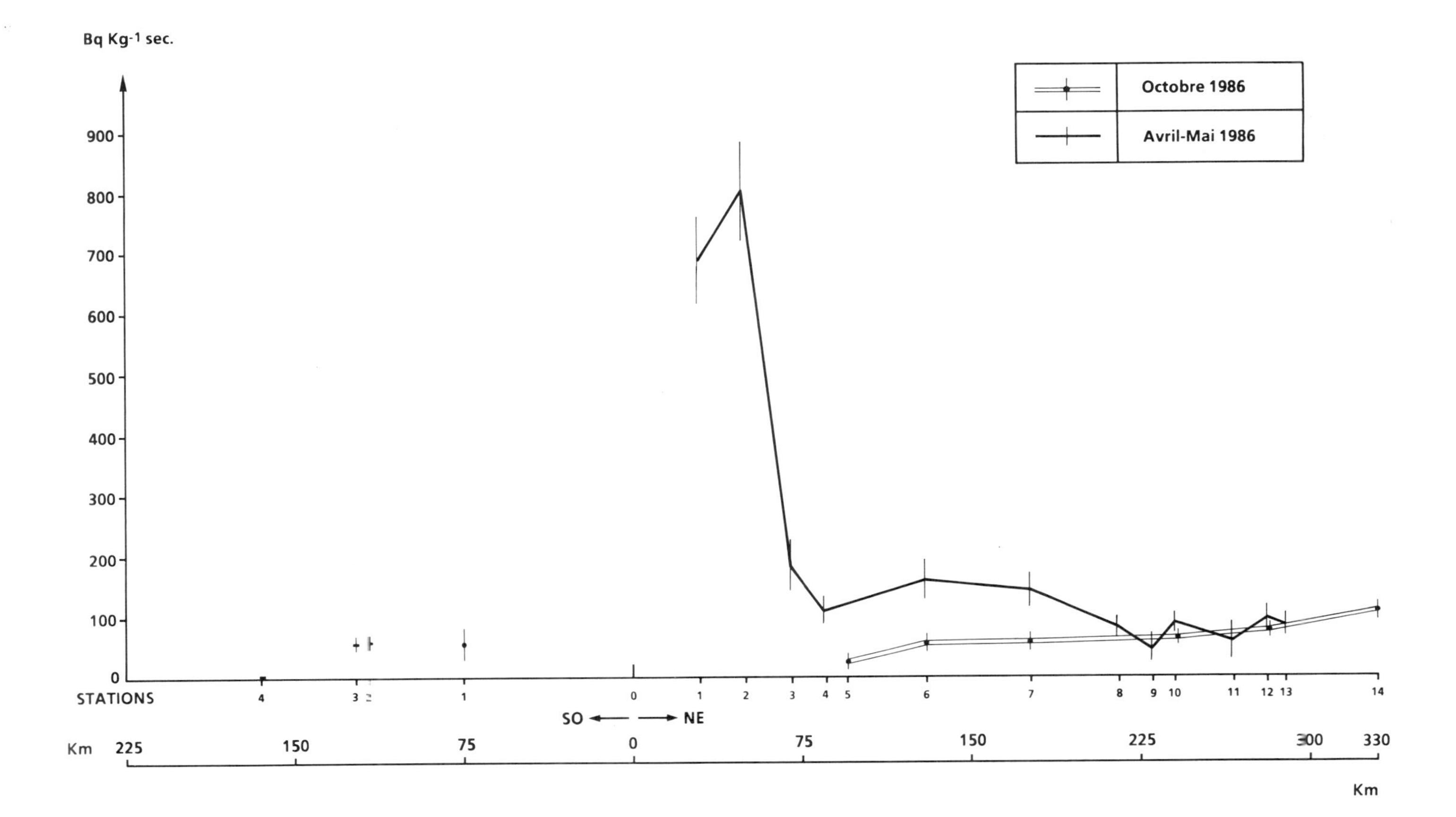

Figure 4 Répartition en fonction de la distance à l'émissaire de l'usine de retraitement des combustibles irradiés de La Hague du ^{106}Ru-Rh chez <u>Mytilus edulis</u> (chair) en 1986

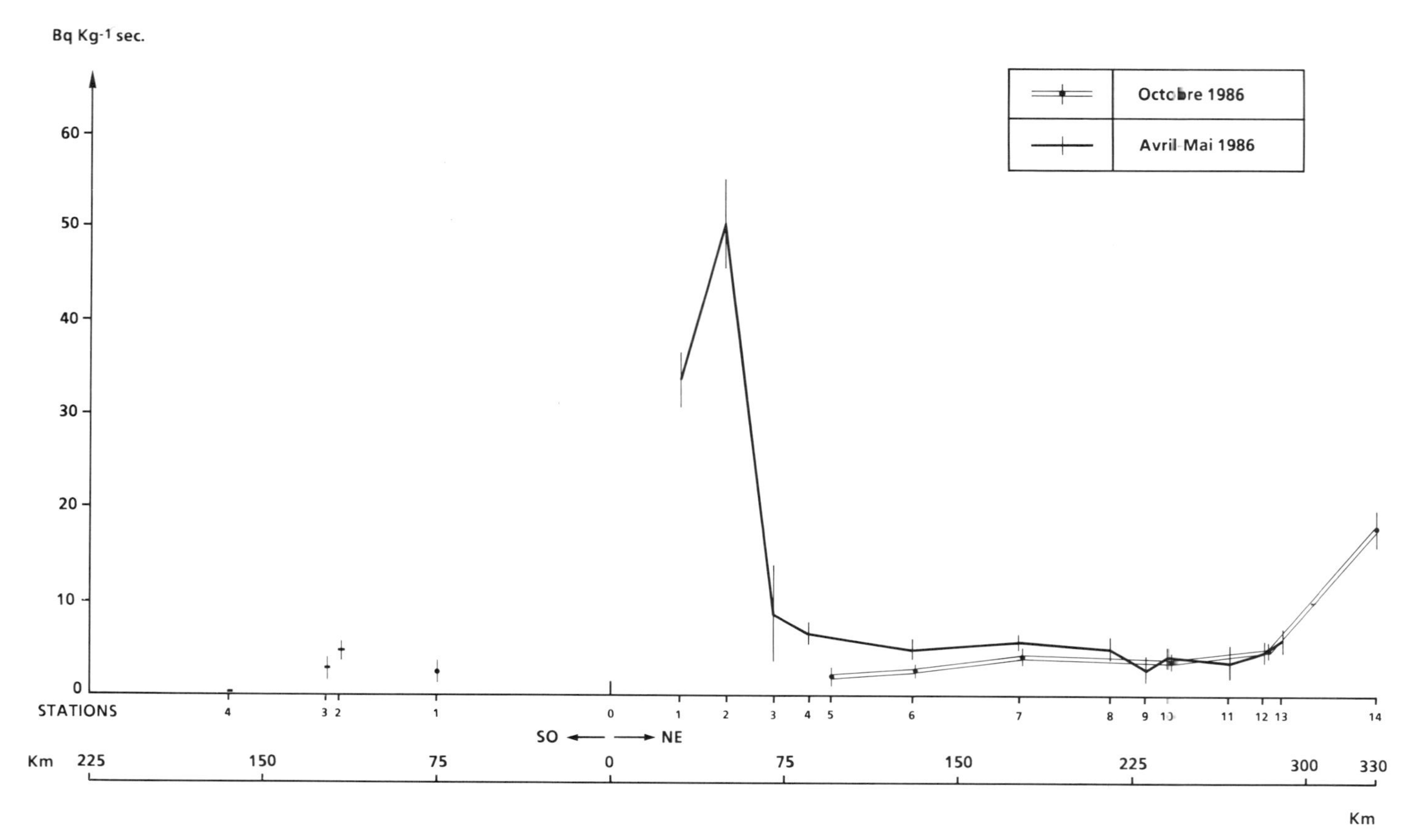

Figure 5 Répartition en fonction de la distance à l'émissaire de l'usine de retraitement des combustibles irradiés de La Hague du ^{60}Co chez Mytilus edulis (chair) en 1986

Tableau 1 - Coefficients d'ajustement α et β des fonctions puissance de la distance $y = \alpha d^{-\beta}$

Espèce	Date	^{106}Ru-Rh		^{60}Co	
		Sud Ouest	Nord Est	Sud Ouest	Nord Est
Fucus serratus	Juin-Juillet 1985	$y = 1581{,}9\, d^{-0{,}954}$ $r = -0{,}806$	$y = 1870{,}9\, d^{-0{,}751}$ $r = -0{,}935$	$y = 2984{,}2\, d^{-1{,}098}$ $r = -0{,}828$	$y = 810{,}1\, d^{-0{,}665}$ $r = -0{,}960$
Fucus serratus	Avril-Mai 1986	$y = 6102{,}0\, d^{-1{,}004}$ $r = -0{,}852$	$y = 5071{,}5\, d^{-0{,}830}$ $r = -0{,}928$	$y = 4957{,}4\, d^{-1{,}309}$ $r = -0{,}854$	$y = 600{,}9\, d^{-0{,}652}$ $r = -0{,}963$
Fucus serratus	Octobre 1986	-	$y = 1382{,}1\, d^{-0{,}625}$ $r = -0{,}903$	-	$y = 479{,}2\, d^{-0{,}580}$ $r = -0{,}973$
Mytilus edulis	Avril-Mai 1986	-	$y = 22090\, d^{-1{,}034}$ $r = -0{,}880$	-	$y = 953{,}3\, d^{-0{,}992}$ $r = -0{,}860$
Mytilus edulis	Octobre 1986	-	$y = 0{,}485\, d^{0{,}905}$ $r = 0{,}940$	-	$y = 0{,}063\, d^{0{,}759}$ $r = 0{,}914$

y = niveau de radioactivité du bioindicateur
d = distance de la station à l'émissaire
r = coefficient de Bravais - Pearson
(-) ajustement non réalisé du fait d'un nombre insuffisant de données

BIBLIOGRAPHIE

Calmet, D. - Synthèse radioécologique des différents compartiments de l'environnement marin du Cotentin. Thèse de Doctorat d'Etat ès sciences, Université Aix Marseille II, 1986, 259 p.

Calmet, D., Patti F., Charmasson S. - Spatio-temporal variations in the technetium-99 content of Fucus serratus in the English Channel during 1982 - 1984. J. Environ. Radioactivity, 1987, pp 57 - 69.

Fraizier, A., Guary, JC. - Diffusion du plutonium en milieu marin. Etude quantitative effectuée sur des espèces marines du littoral de la Manche de Brest (Pointe St Mathieu) à Honfleur. Rapport CEA-R-4822, 1977, 18 p.

Germain, P., Masson, M., Baron, Y. - Etude de la répartition de radionucléides émetteurs gamma chez des indicateurs biologiques littoraux des côtes de la Manche et de la mer du Nord de Février 1976 à Février 1978. Rapport CEA-R-5017, 1979, 55 p.

Germain, P., Miramand, P. - Distribution and behaviour of transuranic elements in the physical and biological compartments of the Channel french shore. Nucl. Instr. and Methods in Phys. Res., 223, 1984, pp. 502-509.

Masson, M., Patti, F., Cappelini, L., Germain, P., Jeanmaire, L. - Etude de la dispersion du technétium-99 sur les côtes françaises de la Manche à l'aide de deux indicateurs biologiques : Fucus sp. et Patella sp. J. radioanal. Chem., 77, 1, 1983, pp. 247-255.

Salomon, J.C., Gueguenìat, P., Orbi, A., Baron, Y.. A lagrangian model for long term tidally incluced transport and mixing. Verification by artificial radionuclide concentrations.International symposium on radioactivity and oceanography, radionuclides : a tool for oceanography, 1-5 june 1987, Cherbourg, France.

THE BEHAVIOUR OF DISSOLVED PLUTONIUM IN THE ESK ESTUARY, U.K.

M. Kelly, S. Mudge, J. Hamilton-Taylor, K. Bradshaw
Department of Environmental Science,
University of Lancaster,
Lancaster LA1 4YQ,
UK

ABSTRACT

The Esk Estuary, U.K., receives inputs of dissolved and particulate plutonium from discharges in the Irish Sea from the Sellafield nuclear fuel reprocessing plant. Both oxidation state categories for ^{238}Pu and ^{239,240}Pu (i.e. (Pu(III,IV) and Pu(V,VI)) show non-conservative behaviour in the dissolved phase in the estuary, with enrichment in low salinity waters due to desorption of plutonium from resuspended sediments. Dissolved plutonium concentrations in the estuary can exceed those of the Irish Sea waters entering the estuary because of this mechanism. Survey data show the spatial and temporal distributions of the enriched waters during the tidal cycle. The variation of dissolved plutonium concentration in the estuary, for the two oxidation states, agrees with the results of laboratory desorption experiments using contaminated Esk sediment. It is dependent primarily on suspended sediment concentration (for both oxidation states) and on the salinity/pH of the waters, especially for Pu(III,IV).

INTRODUCTION

The Esk Estuary (Cumbria, U.K.) receives inputs of plutonium (^{238}Pu, ^{239,240}Pu, ^{241}Pu) from the Irish Sea, both in solution and associated with sediments, derived originally from the authorised discharges from the Sellafield nuclear fuel reprocessing plant, which lies 10 km N of the estuary. Considerably smaller amounts enter in river water via fallout from weapons testing and discharges to the atmosphere from Sellafield. Earlier work has shown that dissolved plutonium (i.e. < 0.22 μm) behaves non-conservatively in the estuary, with relative enrichment taking place in low salinity water (1,2). Recent laboratory experimental work with Esk Estuary materials has shown that the enrichment probably is due to desorption of plutonium from the particulate phase (3,4).

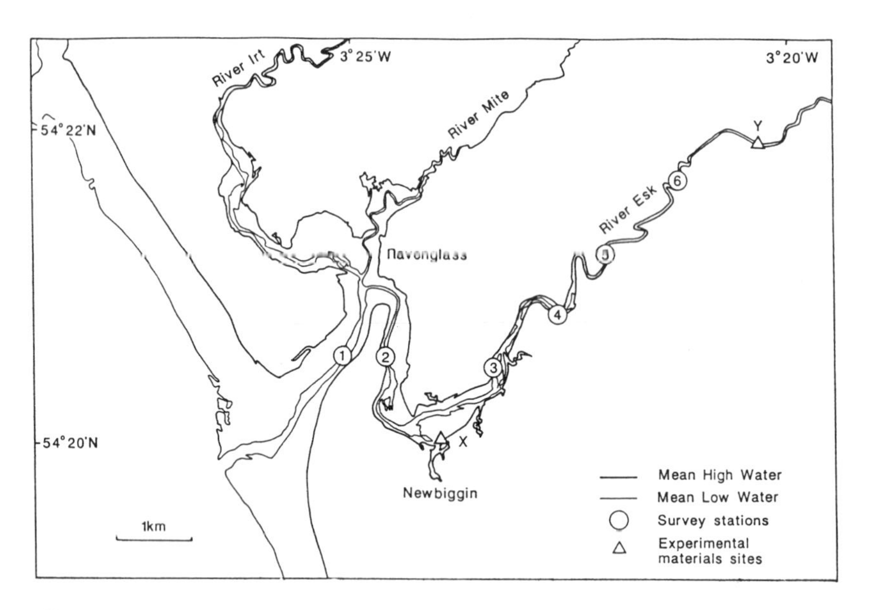

Figure 1. Location map for Esk Estuary, U.K.

Plutonium in the Irish Sea is known to exist in two oxidation states, considered to be Pu(IV) and Pu(V) (5,6), with the oxidised species dominant in the dissolved phase and the reduced predominant in the particulate. A recent survey in the Esk Estuary (25.5.86) was carried out in order to determine the separate behaviour of these co-existing oxidation states of plutonium in the estuarine environment, and to see whether their distributions are similar to those predicted by the experimental studies.

METHODS

One litre water and suspended particulate samples were taken on a spring tide, at mid depth and mid channel, at a number of stations in the estuary (Fig. 1). pH and salinity were measured on the samples immediately after being taken. The samples were filtered through 0.22 µm Millipore filters and acidified within 2 hours. These were subsequently analysed by alpha spectrometry for dissolved $^{239,240}Pu$ and ^{238}Pu in the two oxidation state categories, (III,IV) and (V,VI), according to the method of Lovett and Nelson (7).

RESULTS

The field survey showed that both oxidised and reduced states of dissolved plutonium were present in the estuary, and furthermore, that both behave non-conservatively (Fig. 2). Plutonium activities were found to be higher in low salinity waters than in the seawater entering the estuary. That is, desorption of plutonium from sediment had become a more important source of dissolved plutonium in the estuary

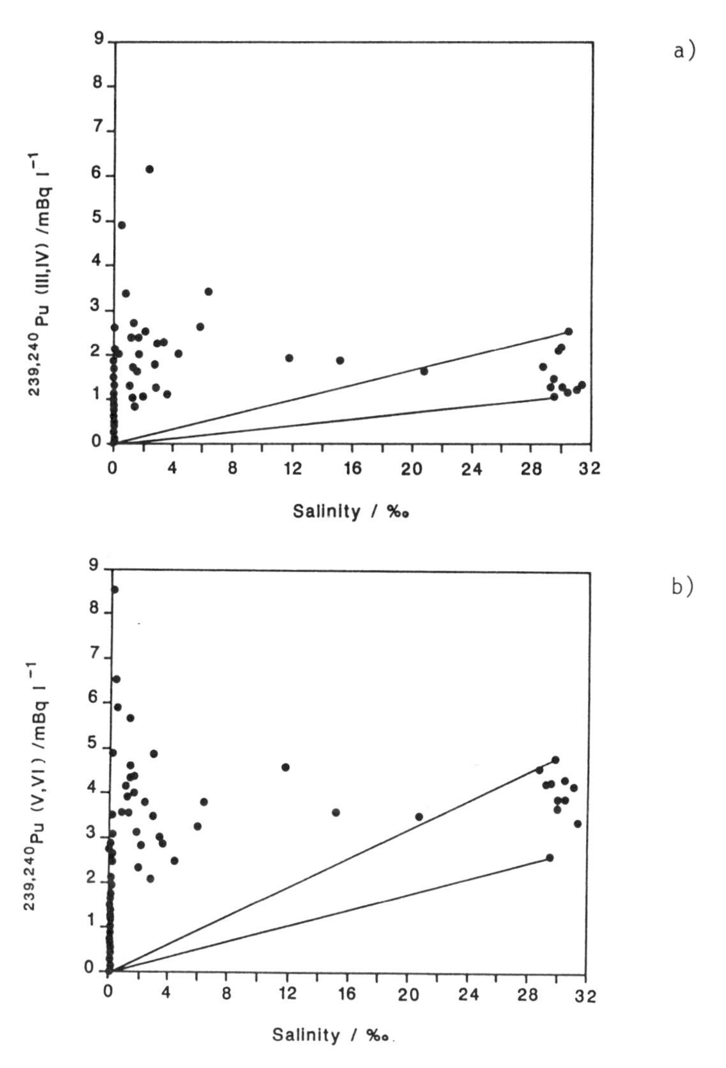

Figure 2. Dissolved plutonium variation with salinity in the Esk Estuary. One sigma errors are ∿ 7% of value.
a) ^{239,240}Pu(III,IV); b) ^{239,240}Pu(V,VI)

than the current discharges to the Irish Sea. This is the opposite of the situation found in our 1981 survey (8) and it reflects the recent decrease in the magnitude of plutonium discharges from Sellafield (9).

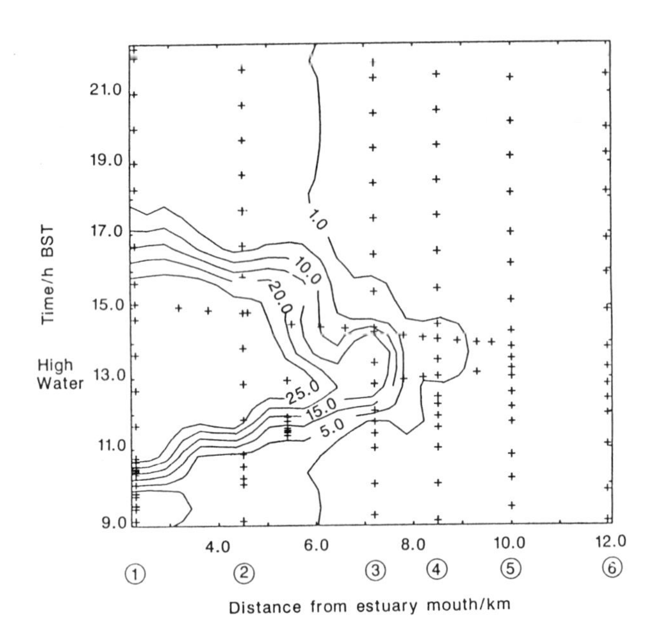

Figure 3. Distribution of salinity in space and time during a tidal cycle in the Esk Estuary. (Units of contours are ‰). Circled numbers are station locations shown in Fig. 1.

Figures 3-6 show the combined spatial and temporal distributions of variables in the Esk Estuary during a complete tidal cycle ($12\frac{1}{2}$ h), from near its mouth to the vicinity of the tidal limit. The salinity distribution (Fig. 3) is characteristic of this shallow estuary during large amplitude spring tides (4 m water level rise at Ravenglass, 4.5 km from the mouth) i.e. an asymmetric tidal curve with a prolonged ebb tide, penetration of undiluted seawater relatively far up estuary at High Water, and steep salinity gradients in space and time. The 1-2 hours of rapidly changing salinities on the flood and ebb tides are the periods of highest water velocities and suspended sediment concentrations (1). In the lower reaches, ∿ 0-6 km, the salinity in the shallow (typically $<$ 1 m) low water channel remains between 1-5‰ because of saline water drainage from intertidal sediments.

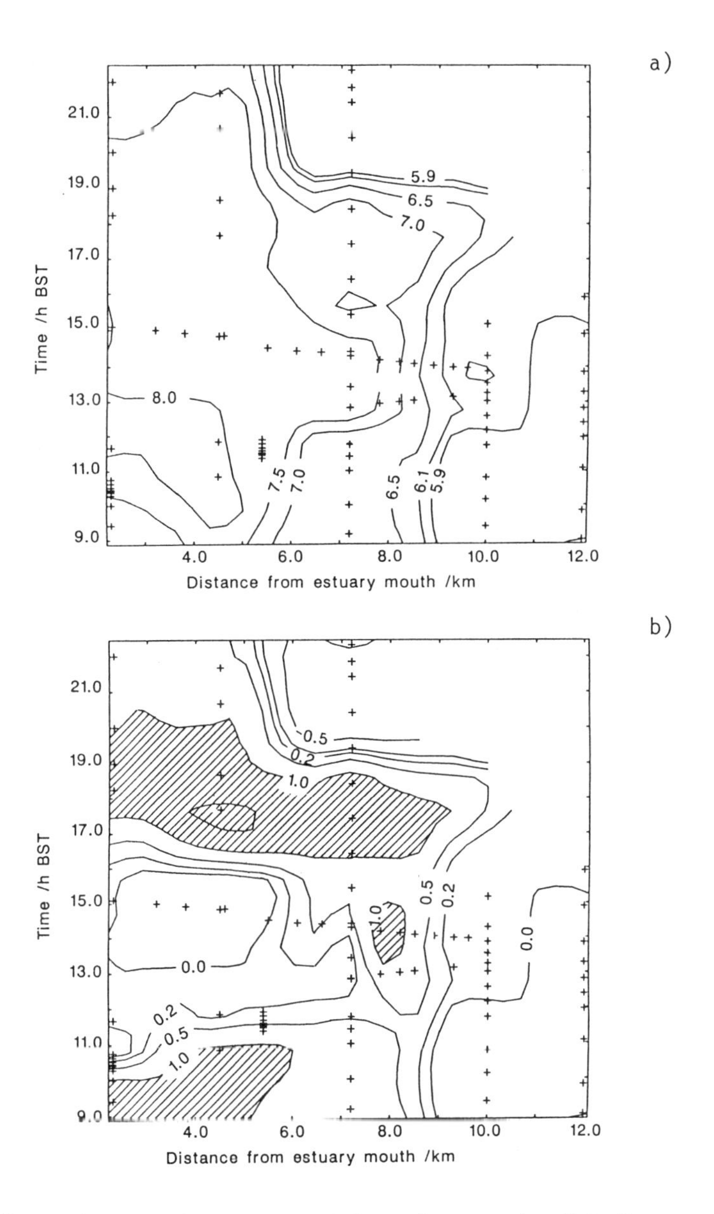

Figure 4. Distribution during a tidal cycle in the Esk Estuary of: a) measured pH; b) calculated pH anomaly, maximum anomaly shaded (> 1 pH unit) (contour units are pH units).

The measured pH distribution is shown in Fig. 4a. pH also behaves non-conservatively giving values which depart from those expected by mixing of the end member seawater and river water. Fig. 4b shows the magnitude of this anomaly obtained from a comparison of observed values

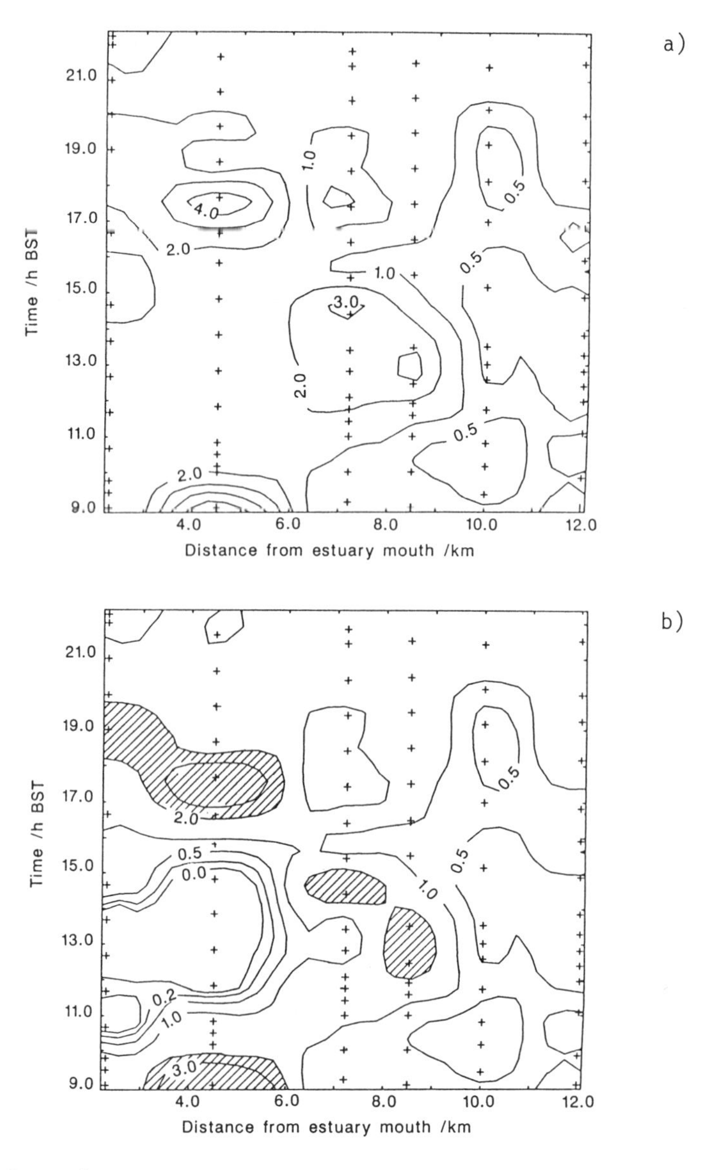

Figure 5. Distribution during a tidal cycle in the Esk Estuary of: a) measured dissolved ^{239,240}Pu(III,IV); b) calculated anomaly of dissolved ^{239,240}Pu(III,IV), maximum anomaly shaded (> 2.0 mBq l^{-1}) (contour units are mBq l^{-1}).

with experimental ones determined by mixing filtered sea and river water. The pH anomaly shows a characteristic pattern, with peak values in the lower half of the estuary during mid ebb, and early flood, and at the limit of salt intrusion at High Water.

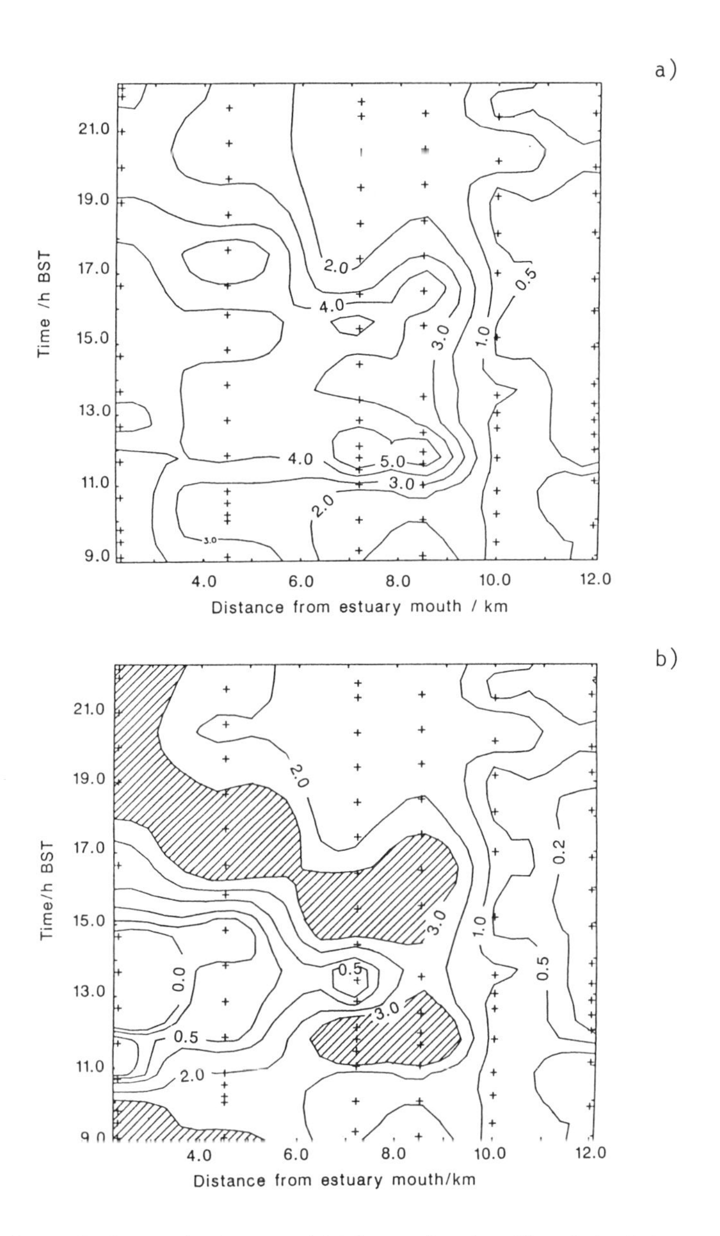

Figure 6. Distribution during a tidal cycle in the Esk Estuary of: a) measured dissolved $^{239,240}Pu(V,VI)$; b) calculated anomaly of dissolved $^{239,240}Pu(V,VI)$, maximum anomaly shaded (> 3.0 mBq l^{-1}) (contour units are mBq l^{-1}).

Observed dissolved $^{239,240}Pu(III,IV)$ and (V,VI) distributions are shown in Figs 5a & 6a and the anomalies due to their non-conservative behaviour in Figs 5b & 6b. The anomalies were calculated from the differences between the observed values and those predicted by the theoretical dilution curve for the observed salinity. There is a striking correspondence in space and time between the distributions of the peak dissolved Pu anomalies and those of pH (Fig. 4b), with Pu enriched water occurring around the limit of salt intrusion at High Water, and the middle and lower reaches in mid ebb and again in late ebb/early flood. These are mainly, but not exclusively, periods of low salinity around 1-5‰. The Pu(V,VI) anomaly typically is less restricted in space and time than that of Pu(III,IV).

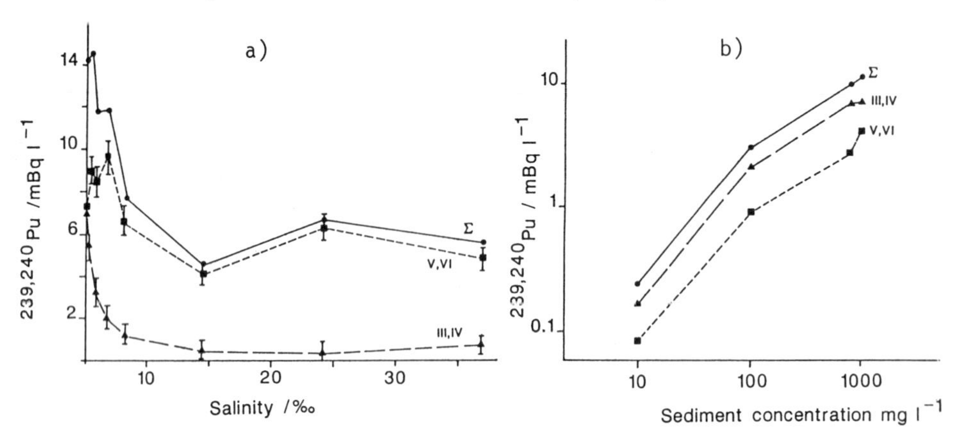

Figure 7. Experimental rapid (1 h) desorption of $^{239,240}Pu$ from Esk Estuary sediment: a) variation with salinity b) variation with sediment concentration.

DISCUSSION

The laboratory experiments investigating plutonium desorption from Esk Estuary sediments suggest that the link between pH and elevated dissolved plutonium values is a mechanistic one, with exchange reactions taking place between competing ionic species held on the sediment surface, including Pu species, and those in the solution phase, including protons (3). Desorption occurs rapidly (< 1 h), but on this time scale only a small proportion of total sediment bound plutonium is available (∿ 5%). Both Pu(III,IV) and Pu(V,VI) are apparently desorbed from the sediment, although the possibility of a post-desorption redox reaction also being involved has still to be investigated fully. The amounts desorbed are dependent on salinity and sediment concentration (Fig. 7). Finally, the reactions are reversible so that adsorption onto sediment also can occur. These experimentally obtained results enable the dissolved plutonium distribution patterns to be interpreted.

For a given tidal cycle the main source of the plutonium contaminated sediment in suspension is probably intertidal deposits in the estuary itself, with fresh particulates brought in from the sea being of lesser importance. Reworking of the finer grained higher activity mud deposit is mainly by bank collapse, which can occur at any state of the tide, although usually at periods of high velocity. These mud deposits mainly occur in the middle reaches of the estuary where they are built up by deposition from the higher stages of the tide around High Water. In these areas the muds tend to retain a saline pore water (> 20‰).

At particular states of the tide these sediments, equilibrated with proton poor, high ionic strength porewaters, can become suspended in relatively proton rich, low ionic strength fresh and brackish waters, leading to plutonium desorption as predicted by the experiments. The amount desorbed will be dependent on salinity and suspended sediment concentration. Conversely, adsorption can occur when uncontaminated sediments, which also contribute to the estuarine sediment sources, are entrained in the saline waters. Dispersion of suspended sediment between water masses of different salinities can lead to further desorption or to adsorption. The characteristic patterns shown by the peak dissolved plutonium anomalies in the Esk Estuary (Figs 5b & 6b) are considered to be due to the following sequence of events.

i) In the earliest stages of the flood tide low salinity water (< 5‰), present in the low water channel, is advected back upstream in a highly turbulent flow, entraining fine grained sediments with saline pore waters, resulting in Pu desorption and proton adsorption as a function of salinity and sediment concentration.

ii) The anomalies, together with the sediment, migrate with the tide to the upper estuary and are still clearly evident at High Water. The small difference in their magnitudes may be related to dispersion and adsorption.

iii) Sediment entrainment occurs again on the ebb as velocities increase, leading to further plutonium desorption from sediment. The scale and extent of the anomalies appear to be greater than on the flood, which may be related to the more widespread interaction between saline sediments and low salinity waters, and to the longer duration of the ebb tide.

The detailed pattern of behaviour of dissolved plutonium in the Esk Estuary will change from tide to tide, according to the salinity distribution, which varies with tidal amplitude and river discharge, and the sediment supply, which itself depends on the tide and river discharge, and the wave activity in the estuary and the Irish Sea. In addition the behaviour will be affected by variations in the plutonium specific activity of the sediment, due to grain size effects (10), initial contamination levels and differences in available labile fractions, which may be influenced by longer term diagenetic processes.

The intertidal sediments in the Esk Estuary, which are contaminated by discharges of plutonium from Sellafield, especially in the years of the peak discharge from 1968 to 1979, are proving to be an important source of plutonium for the estuarine surface waters. However, whether or not the estuary is now a net exporter of plutonium to the Irish Sea, as has been suggested (2), requires further detailed investigations.

ACKNOWLEDGEMENTS

This work was carried out as part of a contract with the U.K. Department of the Environment, and the results are published with the agreement of the Department. The field assistance of C. Allen, M. Emptage, C. Baker and L. Rosser is gratefully acknowledged.

REFERENCES

1. Assinder, D.J., Kelly, M. and Aston, S.R., Tidal variations in dissolved and particulate phase radionuclide activities in the Esk Estuary, England, and their distribution coefficients and particulate activity fractions. J. Environ. Radioactivity, 1985, 2, 1-22.

2. Eakins, J.D., Burton, P., Humphreys, D.G., and Lally, A.E., The remobilisation of actinides from contaminated intertidal sediments in the Ravenglass Estuary. In Proc. CEC Seminar, Renesse, The Netherlands, 1984. CE Publication X11/380 85 EN, 1985, 107-22.

3. Hamilton-Taylor, J., Kelly, M., Mudge, S. and Bradshaw, K., Rapid remobilisation of plutonium from estuarine sediments. J. Environ. Radioactivity, 1987, 5.

4. Burton, P.J., Laboratory studies on the remobilisation of actinides from Ravenglass estuary sediment. Sci. Total Environ., 1986, 52, 123-45.

5. Nelson, D.M. and Lovett, M.B., Oxidation state of plutonium in the Irish Sea. Nature, 1978, 275, 599-601.

6. Pentreath, R.J., Harvey, B.R. and Lovett, M.B., Chemical speciation of transuranium nuclides discharged into the marine environment. In Speciation of Fission and Activation Products in the Environment, ed. R.A. Bulman and J.R. Cooper, Elsevier Applied Science Publishers, London, 1986, pp.

7. Lovett, M.B. and Nelson, D.M., Determination of some oxidation states of plutonium in seawater and associated particulate matter. In Proc. Symp., Techniques for Identifying Transuranic Speciation in Aquatic Environments. 24-28 March 1980, Ispra, Italy, IAEA, 1981, 27-35.

8. Assinder, D.J., Kelly, M. and Aston, S.R., Conservative and non-conservative behaviour of radionuclides in an estuarine environment, with particular respect to the behaviour of plutonium isotopes. Environ. Technol. Lett., 1984, 5, 23-30.

9. British Nuclear Fuels Limited, Annual Report of radioactive discharges and monitoring of the environment, 1971-73 to 1984. 1974-1985.

10. Aston, S.R., Assinder, D.J. and Kelly, M. Plutonium in intertidal sediments in the northern Irish Sea. Estuar. Coastal Mar. Sci., 1985, 20, 761-771.

MIXING PROCESS IN NEAR-SHORE MARINE SEDIMENTS AS INFERRED FROM THE DISTRIBUTIONS OF RADIONUCLIDES DISCHARGED INTO THE NORTHEAST IRISH SEA FROM BNFL, SELLAFIELD

D S Woodhead
Ministry of Agriculture, Fisheries and Food,
Directorate of Fisheries Research,
Fisheries Laboratory,
Lowestoft, Suffolk NR33 OHT, England

ABSTRACT

A variety of radionuclides is discharged, under authorization, into the northeast Irish Sea from the nuclear fuel reprocessing plant at Sellafield, Cumbria. Depending upon their chemical characteristics, these radionuclides become associated, to a greater or lesser degree, with particulate material in the water column and variable proportions are removed to the sea bed. The distributions of a number of these radionuclides within the sea bed have been determined from a set of 43 cores taken from the coastal sediments of the northeast Irish Sea. Comparisons of the relative distributions of the different radionuclides, together with details of the history of the discharges, provide insights into the general processes which are effective in mixing them into the sea bed.

The radionuclides studied vary from those which indicate processes in the short term to those which reflect the results of three decades of discharges. In some instances the ratios of certain isotopes - such as those of Cs and Pu - have been used to indicate the relative rates at which incorporation to depth in the sediment occurs in different areas.

INTRODUCTION

The radionuclides discharged under authorization into the northeast Irish Sea from the nuclear fuel reprocessing plant at Sellafield, Cumbria have a wide range of half-lives and chemical properties which are of use in investigating the processes which govern their behaviour in the environment. Depending upon their affinity for particle surfaces, varying proportions of the nuclides become incorporated into the sediments of a relatively narrow coastal region extending from the Solway Firth in the north to Morecambe Bay in the south. For example, it has been estimated that approximately 85% of the total Irish Sea inventories of $^{239/240}Pu$ and ^{241}Am are contained in the sediments of this area (1). Radionuclide concentration profiles determined for sediment cores indicate that there is substantial, but variable, penetration of contaminant radionuclides into the sea bed (2-4) and positive evidence has been obtained for bioturbation as one mechanism which can result in their incorporation to a depth of at least 1.4 m into the soft, muddy, offshore sediments (5,6).

Clearly, the sediments have represented a significant and extensive sink for radionuclides in the past, and are important as a potential future source to the water column in an era of declining discharges. Studies of the distribution and behaviour of the radionuclides within the sea bed have provided some information on the manner and extent to which fine, settled sediments become mixed over varying time scales.

MATERIALS AND METHODS

The positions of the sampling stations (Figure 1) were chosen, on the basis of the results of the previous survey (1) and information concerning the distribution of fine sediment, to give the best possibility of making an accurate assessment of the inventories of radionuclides in the sediments of the region.

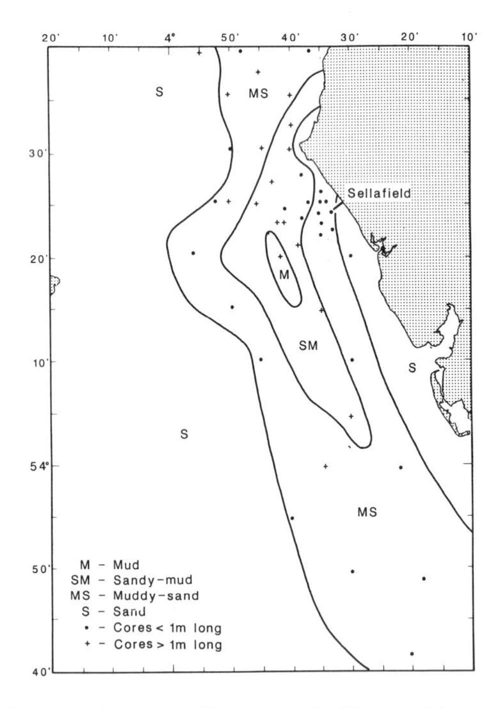

Figure 1. Distribution of fine sediment in the northeast Irish Sea, and sediment sampling positions for RV CIROLANA cruise 5/83.

Sediment samples were obtained using either a 2 m barrel gravity corer with a 10 cm internal diameter acrylic liner, or a Reineck corer with a detachable steel box, 20 x 28 x 40 cm deep. The barrel cores were sub-sampled as 5 cm thick slices at the top and close to the bottom of the core; the intervening section was sub-sampled to depth thicknesses no greater than 66 cm, separated by 5 cm slices. Thus, a typical core produced sections at 0-5, 5-71, 71-76, 76-142 and 142-147 cm depth horizons. Each core segment was then sub-sampled vertically at a constant horizontal cross-section to discard the outer surface of the core together with any attendant contamination carried down from the surface layers of the sediment. The Reineck box cores were first sub-sampled with barrel core liners, and then treated as above. The sediment samples were weighed wet, freeze dried, weighed dry, thoroughly homogenized and aliquots were taken for Ge(Li) γ-ray spectrometry and chemical processing to produce thin sources for α-spectrometry.

RESULTS

Figure 2(a) shows the derived distribution of ^{106}Ru as the integrated quantity of the nuclide per unit area of the sea bed. The contours have been interpolated between the data values on the assumption that there is an inverse square dependence on the distance separating the neighbouring sampling points; this is an arbitrary approach but it produces a reasonably smooth set of contours and provides a consistent means of displaying the results. There is a pronounced northwards displacement of the areas of greatest sediment contamination relative to the discharge point. This presumably reflects the fact that ^{106}Ru arises mainly in the effluent streams which are discharged from the sea tanks at, or about, high water when the tidal stream is ebbing to the north. It also implies a relatively rapid removal of the nuclide from the water column to the sea bed, both because of its association with the iron floc which is known to occur in these effluents (7) and through scavenging by natural particles in the water column. The latter is probably a minor contributor because, for a K_D of 10^3 and a suspended load of 10 mg l^{-1}, only 1% of the radionuclide in the water column is associated with the particles. In these circumstances the transfer of a significant fraction of the activity to the sea bed within half a tidal cycle (to maintain the northerly displacement) would, in the absence of a high sediment load, require very rapid sediment settling and resuspension rates.

There are smaller areas of relatively high contamination to the north of St Bees Head and offshore, and the contour pattern does not correlate entirely with the distribution of the fine (< 63 μm) sedimentary material; for example, the offshore patch of fine mud (> 90% particles by mass < 63 μm) is to the south of the area of greatest contamination. This rather patchy distribution is probably the consequence of a number of factors, including the variability of the monthly discharges (a factor of 16 over the previous 18 months), the local variability of scavenging processes on the time scale of the half-life of the nuclide and, possibly, local, short-term movements of fine sediment.

Figure 2(b) shows the distribution of the nuclide at depths greater than 5 cm. It can be seen that it generally represents less than 30% of the total inventory and implies that mixing of the sediment-associated radionuclide downward into the sea bed is a relatively slow process.

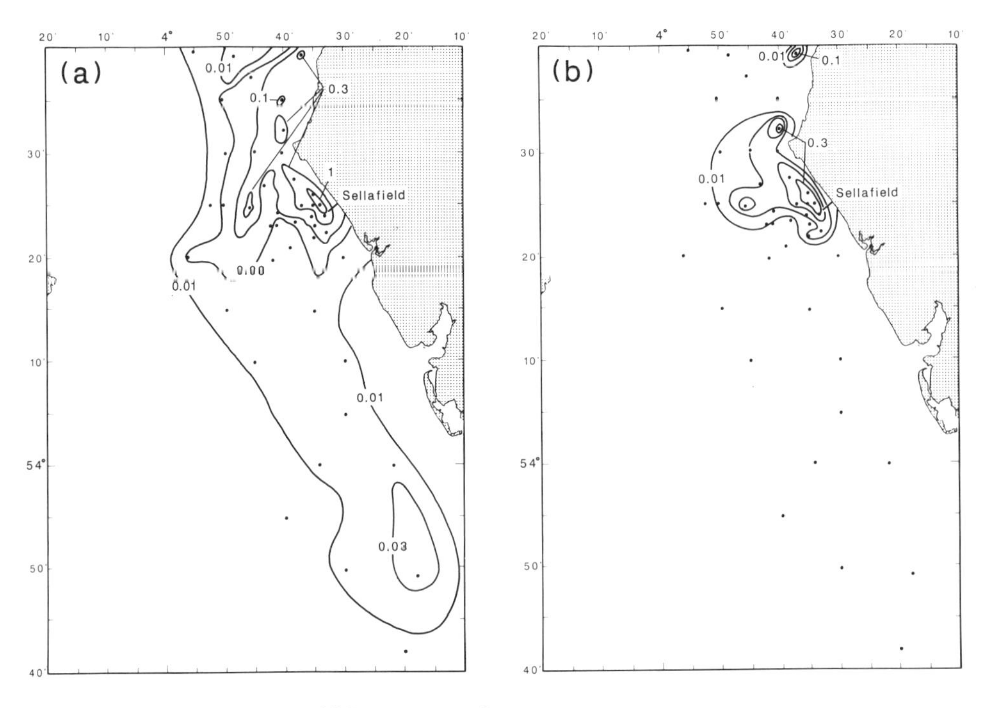

Figure 2. Inventory of ^{106}Ru, MBq m^{-2} : (a) total core; (b) at depths > 5cm.

If rather gross assumptions are made, these data can be used to obtain estimates of the magnitude of the rate of mixing of this radionuclide into the sea bed. The assumptions are: (i) that there is a constant input rate; (ii) that there is no nett sedimentation; and (iii) that the quantities of the nuclide in the 0-5 cm layer and at depths greater than 5 cm represent the partial integrals of an exponential distribution generated by the combined effects of quasi-diffusive mixing and radioactive decay (8). The diffusion coefficients derived in this way lie in the range 3-180 cm^2 a^{-1} and are of the same order as those which have been derived from ^{234}Th data (9).

The distribution of total ^{137}Cs in the sediment (Figure 3(a)) shows that, although the general pattern is similar, there is considerably less detail than was the case for ^{106}Ru. This is a consequence of the greater half-life of ^{137}Cs during which smoothing of the effects of environmental variability and discharge history can take place. Somewhat surprisingly, in view of the fact that the bulk of this nuclide is discharged continuously in the purge water arising from the fuel storage ponds, a northwards displacement, relative to the outfall, of the areas of highest sediment contamination remains apparent. The depth distribution of the nuclide is quite different however. The nuclide distribution pattern for the 0-5 cm surface layer (Figure 3(b)) shows that it accounts, on average, for less than 30% of the total, with the greater part of the activity at depths down to 76 cm (Figure 3(c)). Figure 3(d) indicates the regions in which there are detectable traces of the nuclide at depths greater than 71 cm; at one station this amounts to approximately 10% of the total with traces apparent in the sample from 142-147 cm.

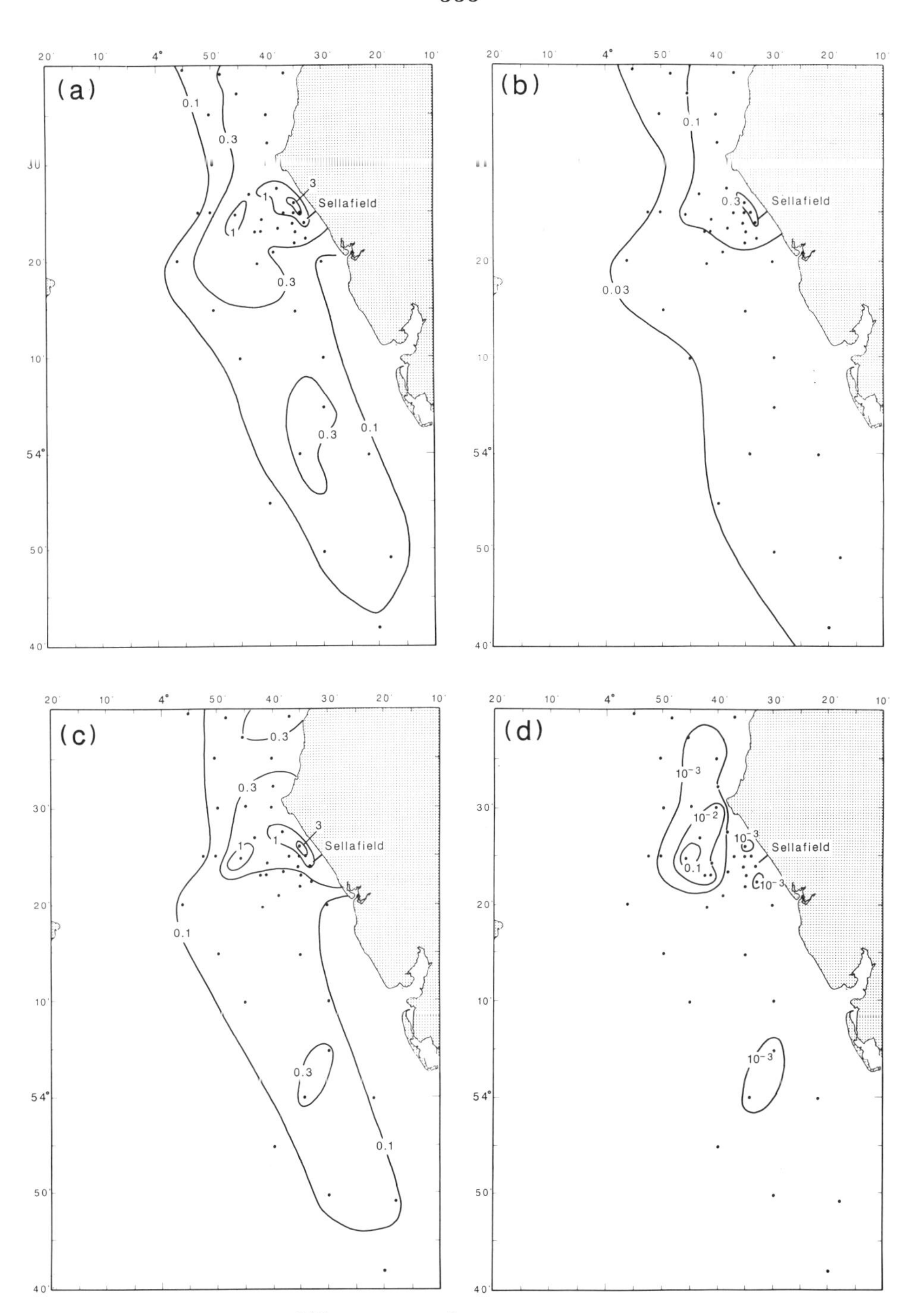

Figure 3. Inventory of ^{137}Cs, MBq m^{-2} : (a) total core; (b) surface layer, 0-5 cm; (c) subsurface layers 5-76 cm or the bottom of the core; (d) at depths > 71 cm.

In addition to ^{137}Cs, the shorter-lived ^{134}Cs is also discharged and this provides a potential means of "ageing" the contamination present in the sediment. Over the ten years prior to sampling the $^{137}Cs/^{134}Cs$ ratio in the annual discharge had increased from 4 to 14, and for the cumulative environmental inventory it had increased from 12 to 40 (10). In isolation from any source the ratio increases with a doubling time of 2.21 a. There are three offshore areas (Figure 4(a)) where the $^{137}Cs/^{134}Cs$ ratio for the total core is greater than 50; i.e. greater than the ratio for the cumulative environmental inventory. This means that there must be a significant time lag ($>$ 0.7 a) between discharge and the appearance of the caesium in association with these sediments. Examination of the ratio in the surface (0-5 cm) sediments in Figure 4(b) shows again that the caesium contamination in the offshore areas is relatively "old" as it does not reflect either the recent discharge, or the seawater for the region for which a value of 18 is typical (D. F. Jefferies, personal communication). Because the ratio is greater than that for the cumulative environmental inventory it confirms the existence of a time lag between discharge and caesium input to these sediments. This conclusion is, perhaps, unexceptionable for the area to the south into Liverpool Bay, but is rather more surprising for the area immediately to the west and north of Sellafield, particularly as it is completely enclosed by an area in which the surface sediments do appear to reflect an input from recent discharges.

Examination of the relative distributions of the two nuclides with depth in the sediment can also provide information on mixing. This is most easily achieved by comparing the ratio at depth with that in the surface sediment. This is shown in Figure 4(c) where lower values, indicative of more rapid mixing, occur mainly in the offshore area.

As was the case for ^{106}Ru, the distribution of ^{134}Cs with depth can provide some information on mixing rates. With the same basic assumptions, the derived diffusion coefficients lie in the range $10 - 2 \times 10^4$ cm^2 a^{-1}, values which are generally higher and more variable than those obtained for ^{106}Ru.

$^{239/240}Pu$, which has a higher affinity for particles and a long half-life, shows a very similar pattern to those for ^{106}Ru and ^{137}Cs for the distribution of total activity in the sediment (Figure 5(a)). Compared with ^{137}Cs a rather smaller proportion of the total activity (approximately 65%) has been mixed to depths greater than 5 cm (Figure 5(b)) and again, mainly in the offshore area, there are traces of the nuclides at depths greater than 71 cm (Figure 5(c)) with one station showing penetration to 142-147 cm.

The comparison of the behaviours of ^{137}Cs and $^{239/240}Pu$ can be made more clearly in terms of the activity ratio. In the discharge this has fluctuated substantially but generally increased over the past ten years, while the cumulative environmental value has been increasing steadily to a value of 50 at the time of sampling (10). The values for the total cores and for the surface layer of the sediment (0-5 cm) are given in Figure 6.

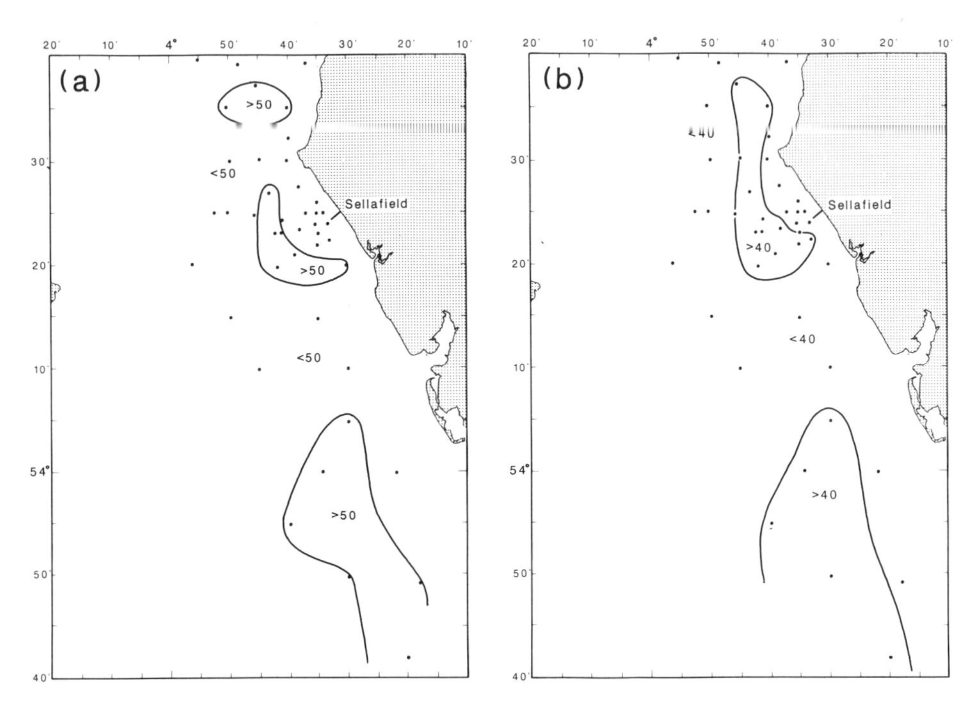

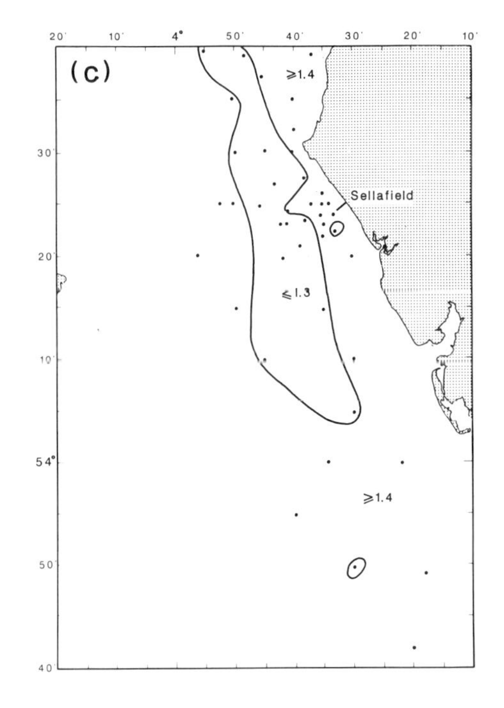

Figure 4. $^{137}Cs/^{134}Cs$ ratios :
(a) total core;
(b) surface layer, 0-5 cm;
(c) comparison of ratio at depths > 5 cm with that in the surface layer.

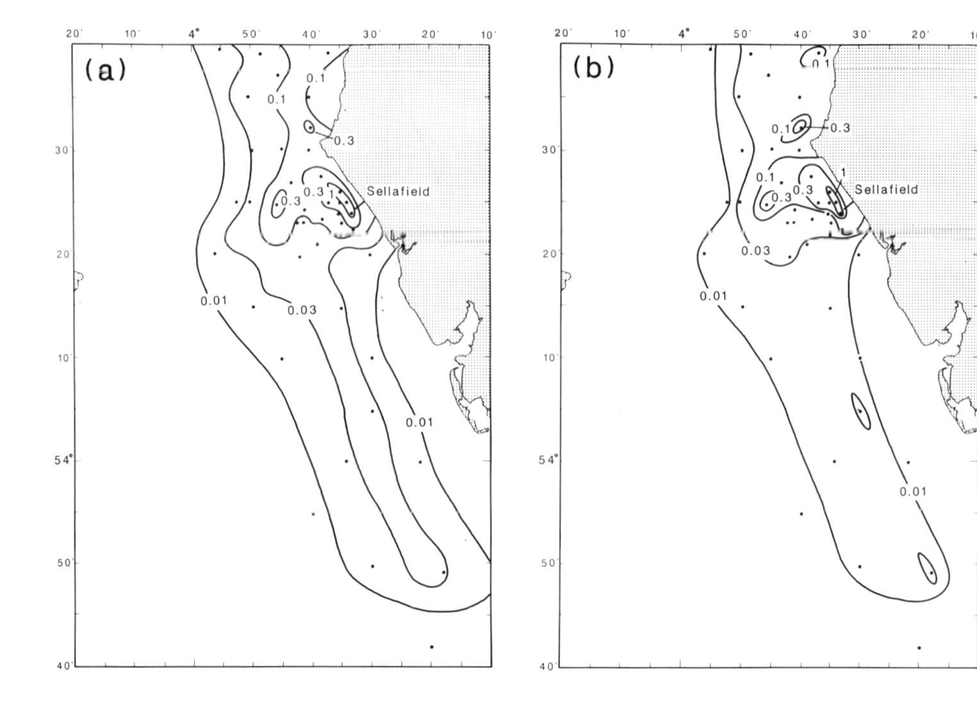

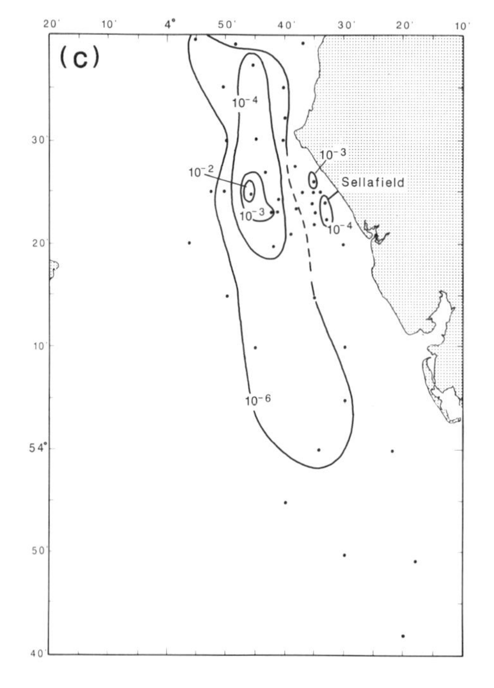

Figure 5. Inventory of $^{239/240}Pu$, MBq m^{-2} :
(a) total core;
(b) at depths > 5 cm;
(c) at depths > 71 cm.

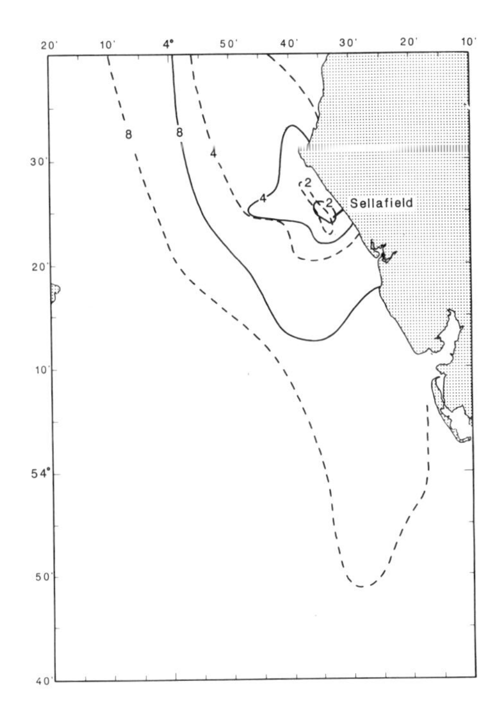

Figure 6. $^{137}Cs/^{239/240}Pu$ ratios. Full lines - total core; pecked lines - surface layer, 0-5 cm.

The distributions show the northward displacement from the discharge point and the values increase with distance. The latter is to be expected from the more soluble, less particle reactive nature of ^{137}Cs compared with $^{239/240}Pu$, but it does demonstrate quite clearly the relatively enhanced scavenging of plutonium from the water column. For the surface sediment, equivalent values occur further from the discharge point than for the total sediment, a result which is, at first sight, at variance with the fact that the ratio has increased in the discharge. However, it must be remembered that the discharges have been declining (10), and it appears quite possible that the surface sediment has now become a nett source of caesium to the water column. This is an additional process which could account, in part, for the presence of caesium contamination which, in terms of the $^{137}Cs/^{134}Cs$ ratio is "older" than the cumulative environmental inventory. As discussed above, this occurs in some offshore areas (see also Figure 4).

CONCLUSIONS

It has been shown that the Sellafield discharges to the northeast Irish Sea have provided a number of useful tracers of general sediment mixing processes in this region. The temporal variability of the source term for individual radionuclides requires that care be taken in the interpretation of the environmental data, but activity ratios are particularly useful for investigating environmental processes.

REFERENCES

1. Pentreath, R. J., Lovett, M. B., Jefferies, D. F., Woodhead, D. S., Talbot, J. W. and Mitchell, N. T. The impact on public radiation exposure of transuranium nuclides discharged in liquid wastes from fuel element reprocessing at Sellafield, U.K. In Radioactive Waste Disposal, Proc. IAEA Conf., Seattle, May 1983, 5, IAEA, Vienna, 1984, pp.315-329.

2. Hetherington, J. A., Jefferies, D. F. and Lovett, M. B. Some investigations into the behaviour of plutonium in the marine environment. In Impacts of Nuclear Releases into the Aquatic Environment, Proc. Symp. Otaniemi, June-July 1975, IAEA, Vienna, 1975, pp.193-210.

3. Pentreath, R. J., Jefferies, D. F., Lovett, M. B. and Nelson, D. M. The behaviour of transuranic and other long-lived radionuclides in the Irish Sea and its relevance to the deep sea disposal of radioactive wastes. In Marine Radioecology, Proc. 3rd NEA Seminar, Tokyo, October, 1979. OECD, Paris, 1980, pp.203-221.

4. Kirby, R., Parker, W. R., Pentreath, R. J. and Lovett, M. B. Sedimentation studies relevant to low-level radioactive effluent dispersal in the Irish Sea. Part III: An evaluation of possible mechanisms for the incorporation of radionuclides into marine sediments. Rep. Inst. Oceanogr. Sci., Wormley, Surrey, UK, 1983, 178, 66pp.

5. Kershaw, P. J., Swift, D. J., Pentreath, R. J. and Lovett, M. B. Plutonium redistribution by biological activity in Irish Sea sediments. Nature, Lond., 1983, 306, 774-775.

6. Kershaw, P. J., Swift, D. J., Pentreath, R. J. and Lovett, M. B. The incorporation of plutonium, americium and curium into the Irish Sea seabed by biological activity. Sci. Total Environ., 1984, 40, 61-81.

7. Pentreath, R. J., Harvey, B. R. and Lovett, M. B. Chemical speciation of transuranium nuclides discharged into the marine environment. In Speciation of Fission and Activation Products in the Environment, Proc. CEC/NRPB Seminar, Oxford, UK, April 1985, ed. R. A. Bulman and J. R. Cooper, Elsevier Applied Science Publishers, London, 1986, pp.312-325.

8. Kershaw, P. J. ^{14}C and ^{210}Pb in NE Atlantic sediments: evidence of biological reworking in the context of radioactive waste disposal. J. Environ. Radioact., 1985, 2, 115-134.

9. Kershaw, P. J., Gurbutt, P. A., Young, A. K. and Allington, D. J. Scavenging and bioturbation in the Irish Sea from measurements of $^{234}Th/^{238}U$ and $^{210}Pb/^{226}Ra$ disequilibria. International Symposium on Radioactivity and Oceanography: Radionuclides - a Tool for Oceanography, Cherbourg, France, June 1987 (this volume).

10. Pentreath, R. J. Sources of artificial radionuclides in the marine environment. International Symposium on Radioactivity and Oceanography: Radionuclides - a Tool for Oceanography, Cherbourg, France, June 1987 (this volume).

RADIONUCLIDE DISTRIBUTIONS IN THE SURFACE SEDIMENTS OF LOCH ETIVE

TM Williams, AB MacKenzie, RD Scott
Scottish Universities Research and Reactor Centre
East Kilbride, Glasgow, G75 OQU, Scotland

and

NB Price, IM Ridgway
Grant Institute of Geology
University of Edinburgh
West Mains Road
Edinburgh, EH9 3JW, Scotland

ABSTRACT

Loch Etive is a sea loch (fjord) lying approximately 90km to the north-west of Glasgow, Scotland. The fjord consists of two main basins separated by sills and seasonal stratification of the innermost basin results in the transient development of oxygen depleted conditions in the water. The innermost basin also receives a large freshwater input at the landward end of the fjord resulting in brackish conditions in the surface waters. The fjord thus represents a well characterised system exhibiting considerable variations in geochemical conditions within which to study radionuclide geochemistry. Soluble components of the waste discharged from the B.N.F.L. nuclear fuel reprocessing plant at Sellafield are transported by the dominant water flow along the west coast of Scotland giving rise to readily detectable levels in the sediments of Loch Etive. Gravity cores of sediment and surficial sediment samples along a transect from landward to seaward ends of the fjord have been analysed for ^{134}Cs, ^{137}Cs and ^{241}Am along with the natural radionuclides ^{210}Pb and ^{228}Th in order to assess the uptake of these nuclides under the varying conditions in the fjord and to establish a sedimentation rate. Variations in the radionuclide concentrations are considered with respect to variations in bathymetry, hydrography, redox conditions, organic content and manganese concentrations in the sediments. Apparent Pb-C, Cs-C and ^{228}Th-Mn association observed in this preliminary study are being more fully investigated.

INTRODUCTION

This study is concerned with the behaviour of natural and manmade radionuclides within the sediments and sealochs (fjords) of the west coast of Scotland. Fjords provide a setting in which there is considerable variation in the suspended particulate matter concentration, salinity and residence time of the waters. Additionally, the underlying sediments are characterised by variations in the organic matter concentration, redox conditions and sediment accumulation rates. The distributions of thorium, caesium and americium have been measured in one particular fjord, Loch Etive, and an attempt is made to relate the distributions of these radionuclides to the geochemical conditions prevailing within the fjord.

Sellafield.

The main source of supply of manmade radionuclides to the coastal environment of the West of Scotland is the liquid effluent discharge from the British Nuclear Fuels Ltd reprocessing plant at Sellafield in North-West England (Figure 1). A summary of some of the major radionuclides discharged from Sellafield, during 1985, is shown in Table 1(1). There have been considerable variations in output over the years, but these are well documented (1,2) and it has been possible to use the waste nuclides as tracers for numerous marine, physical, geochemical and biogeochemical processes (3,4,5).

Table 1. Radionuclides discharged by British Nuclear Fuels Ltd at Sellafield.

Radionuclide	Annual Discharge (1985) T Bq
^{90}Sr	52
^{106}Ru	81
^{134}Cs	30
^{137}Cs	325
^{238}Pu	0.8
$^{234,240}Pu$	2.6
^{241}Pu	81
^{241}Am	1.6

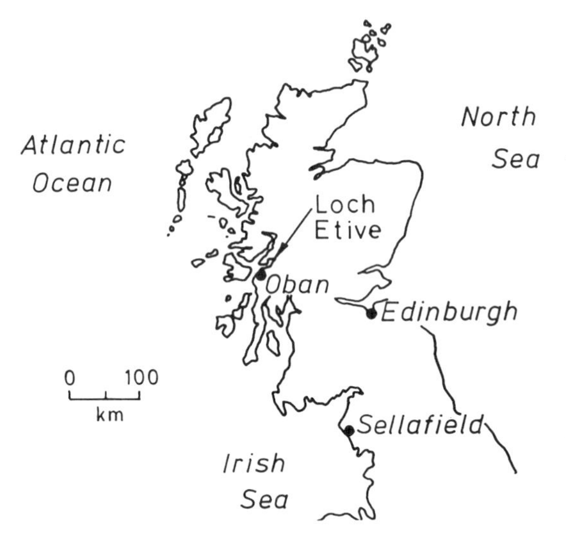

FIGURE 1. Location of Sellafield and Loch Etive.

The dispersal of these radionuclides in the marine environment is controlled initially by partioning in the seawater-sediment system and subsequently by the movement of the contaminated water or sediment. It is well established that the plutonium discharged to the Irish Sea rapidly becomes associated with particulate matter and that more than 90% is associated with the sediments close to the point of discharge (6,7). ^{241}Am has an even greater affinity for uptake by particulates and all of the discharged ^{241}Am has been incorporated in the sediments of the North East Irish Sea (8,9). In addition to this well proven affinity of plutonium and americium for uptake by particulates in the sea, recent analyses of the effluent itself have shown that most of the $^{239,240}Pu$ and ^{241}Am is in particulate form entering the Irish Sea (9,10). The presence of "hot particles" containing high concentrations of actinides in sediments close to Sellafield provides evidence that some of the discharged particulates persist in the environment (10,11). In contrast ^{137}Cs exhibits relatively conservative behaviour (12) in seawater and its dispersal around the west coast of Scotland provides a useful tracer for the study of the movement of water in this area (13,14). The small amount of plutonium which is soluble as Pu(+5)(15) exhibits similar behaviour to the ^{137}Cs and is dispersed around the Scottish coast (16); decay of the soluble fraction of ^{241}Pu gives a more widespread distribution of ^{241}Am than would otherwise be expected. The isotope activity ratios for different species discharged from Sellafield have varied and the $^{238}Pu/^{239,240}Pu$ ratio has ranged from 0.29 to 0.34(1) during the previous eight years, in comparison with the corresponding ratio of 0.04 for atmospheric fallout in the northern hemisphere (17). Similarly the $^{134}Cs/^{137}Cs$ ratio has ranged from 0.07 to 0.1. These activity ratios allow sources of radionuclides to be identified.

Loch Etive.

The mechanisms involved in radionuclide uptake by sediments and in redissolution processes are sensitive to many interacting environmental parameters such as Eh, pH and microbial activity and are consequently dependent on the unique conditions which prevail at any particular site (18). The West Coast of Scotland provides a setting in which extreme variations of such parameters occur within a short geographical distance. In addition to variation in sediment type and biological activity the waters and sediments of some of the fjords can show major variations in redox conditions often with seasonal variations. An excellent example of this is Loch Etive.

Loch Etive, situated on the West Coast of Scotland (Figure 1), is a narrow fjord 28km in length with a surface area of $26km^2$ (Figure 2). It has a large catchment area of $1400km^2$ with two main freshwater inputs, the River Etive (40%) entering at the landward end of the fjord and the River Awe (60%) entering at Bonawe. Secondary sources are the rivers Kinglass, Liver and Noe flowing into the south side of the fjord. The high freshwater input from rainfall ($200cmy^{-1}$) produces an intense halocline in the upper waters resulting in surface salinities at the landward end of the fjord between $5^{o}/oo$ and $21^{o}/oo$ in winter and summer respectively.

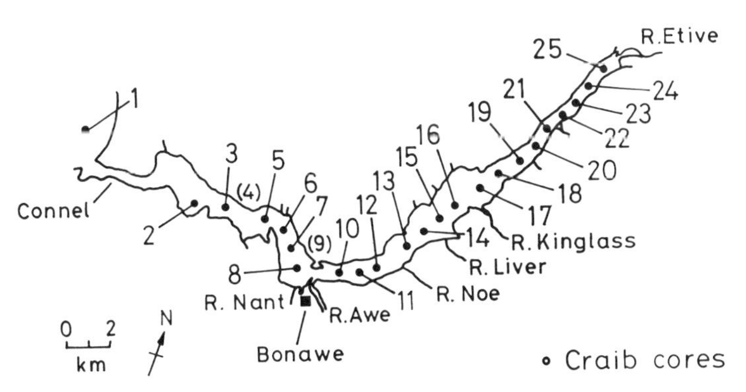

FIGURE 2. Sediment core sampling locations in Loch Etive.

The fjord consists of two main basins separated by a shallow sill at Bonawe and is connected to the open sea by a prominent sill at Connel, 300m wide, 5km long and 10m deep. The outer basin to the west of Bonawe has a maximum depth of 60m. Here circulation is typically estuarine (19) and well oxygenated waters are found at depth. In contrast, circulation of the inner basin (maximum depth 170m) is restricted causing the bottom waters to stagnate for months or years with slowly changing temperature, salinity and dissolved oxygen. Renewal is salinity controlled and occurs during periods of low freshwater run off (20).

The contrast in environmental setting between the outer and inner basins is reflected both in the character of the sediments and in the suspended particulate matter of the water column. Price and Calvert (21) have suggested that in the outer basin a considerable amount of the particulate matter is derived from outside coastal areas. In comparison they report that the composition of suspended particulate matter in the inner basin is high in iron and manganese especially in its lower waters. This was considered to be due to redox recycling about the sediment interface. Surficial sediments are dominated by organic-rich (3 to 8% organic carbon), predominantly non-carbonate, fine grained silts and muds except near the landward end of the fjord where sand is more dominant especially in proximity to the mouth of the River Etive. The influence of freshwater input is reflected by the composition of the organic matter. The higher organic carbon contents of the inner basin (C/N (atomic) ratios of 12 to 15), indicate a higher terrigenous contribution compared with that in the outer basin (C/N atomic = 9 to 10 (22,23)). The sediments of the outer basin generally show a well developed red-brown oxidising surface extending 5cm below the interface with green/grey sediment at depth. In the inner basin benthic activity is even more restricted. The often oxygen depleted water (30% of surface saturation) and higher organic carbon contents promote only a thin, 1cm thick oxidising surface sediment enriched in iron and manganese (24) below which grey/black sediments occur.

SAMPLING and ANALYSES

The samples analysed in the present work were stored subsamples of cores analysed by XRF in studies of heavy metal diagenesis (23,24). The main group of samples discussed here is a set of surface samples collected at approximately 1km intervals along the length of the fjord in November 1981 by means of a slow descent Craib corer (25) which preserves the seawater sediment interface. The sampling stations are shown in Figure 2, Station 1 represents open sea conditions and no samples could be obtained at stations 4 and 9. The cores were sectioned and dried and the 0-2cm section was taken as representative of the surface material.

Surface samples were analysed for Sc, La, Sm, Mn and Fe by instrumental neutron activation analysis. Direct gamma spectroscopy analysis was employed for ^{134}Cs, ^{137}Cs, ^{241}Am, ^{210}Pb and ^{228}Th determinations.

RESULTS and DISCUSSION

The results for analyses of the surface transect samples are presented in Table 2 and Figures 3, 4 and 5.

Before discussing the radionuclide distributions in the surface transect it is useful to consider the results for the stable elements which can indicate major variations of sediment type or geochemical conditions. Thus, as a consequence of their extremely non-conservative behaviour in seawater (26), scandium and the rare earth elements can be considered as providing an indication of any major variations in sediment type in the fjord. From the results in Table 2 it can be seen that these elements do not show any order of magnitude changes in concentration but significant trends are observed which almost certainly reflect local mineralogical variations.

Table 2. Stable element concentrations and radionuclide activities in Loch Etive surface sediments.

Station	Sc	La	Sm	^{134}Cs	^{137}Cs	^{241}Am	Station	Sc	La	Sm	^{134}Cs	^{137}Cs	^{241}Am
1	13.4	43	6.4	9.1	275	4.5	14	7.4	44	5.7	5.6	220	4.9
2	8.8	56	4.8	5.3	173	2.3	15	8.9	49	7.0	5.2	223	2.8
3	5.9	37	4.6	5.5	125	2.0	16	8.9	51	7.3	4.7	238	3.1
4	-	no sample available				-	17	6.4	48	6.1	8.2	160	3.0
5	12.9	64	8.3	12.7	353	7.0	18	8.1	49	6.4	8.4	208	2.8
6	11.5	61	7.2	7.7	272	4.1	19	9.2	59	7.5	4.0	195	3.2
7	11.0	55	7.2	8.5	292	3.8	20	8.9	54	7.2	4.5	172	2.3
8	NA	NA	NA	12.6	300	8.0	21	6.0	48	5.7	2.6	88	ND
9	-	no sample available				-	22	9.3	57	5.6	3.9	173	ND
10	8.4	38	5.2	3.5	120	3.4	23	8.2	25	3.7	4.3	212	1.9
11	10.1	48	6.9	6.2	201	4.2	24	8.1	56	6.0	2.8	195	3.2
12	10.6	71	6.6	6.9	236	4.8	25	8.3	44	5.5	4.3	196	2.9
13	8.7	48	6.3	7.9	195	3.9							

NB: The concentrations of Sc, La and Sm are reported in ppm and ^{134}Cs, ^{137}Cs, ^{241}Am activities in $BqKg^{-1}$.

N.D. = not detectable N.A. = not analysed

Typical uncertainties for stable elements ± 5%

Typical uncertainties for radionuclides ^{134}Cs and ^{241}Am ± 25%

for ^{137}Cs ± 3%

In contrast, manganese, which has an active redox chemistry in the marine environment, ranges in concentration from 0.04% in the outer basin sediment to as much as 4% in the surface deposits of the inner oxygen depleted basin as shown in Figure 3.

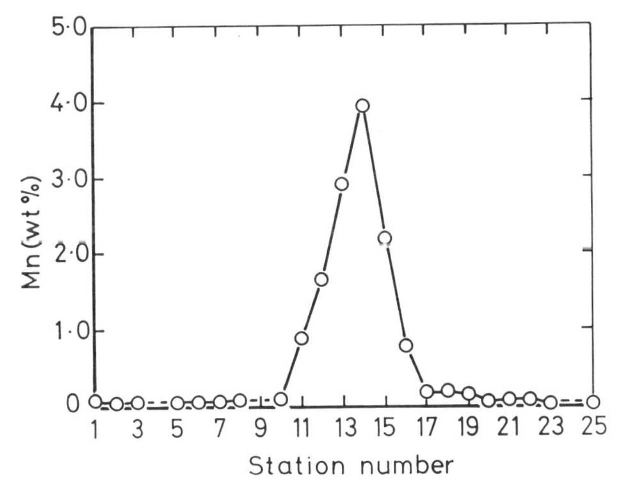

FIGURE 3. Manganese content of Loch Etive surface sediments.

Organic material also represents an important component involved in biogeochemical processes in the sediment and the organic content of the sediments shows a general decrease from the landward to seaward end of the fjord indicating a decrease in terrigenous organic input along the fjord. Superimposed on this general trend are three distinct minima occurring at stations 10, 17 and 21 (Figure 4) which are opposite major inlets and therefore represent areas of enhanced detrital terrigenous input. The stable element analyses therefore indicate that within the restrictive area

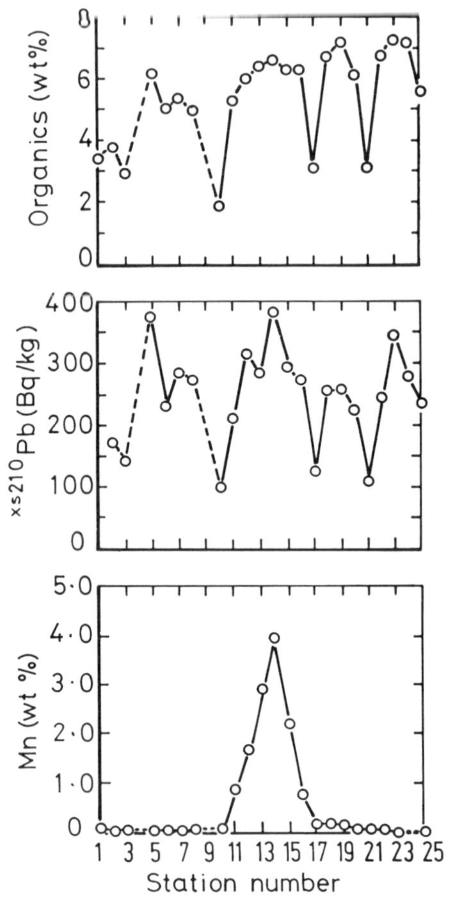

FIGURE 4. Manganese, excess ^{210}Pb and organic content of Loch Etive surface sediments.

of the fjord there are mineralogical variations, changes in the quantity (and probably the type) of organic matter and major, redox controlled, variation in manganese cycling all of which are potentially important in determining radionuclide distributions and behaviour in this system as discussed below.

Considering firstly the non-conservative natural radionuclide ^{232}Th (Figure 5) a general decrease in concentration is observed in the seaward direction with the trend following the decrease of detrital input towards the sea and the more detailed features of the distribution resembling those of the other non-conservative elements such as the rare earths. In contrast, ^{228}Th which is produced from ^{232}Th in the decay chain

$$^{232}\text{Th} \longrightarrow \; ^{228}\text{Ra} \longrightarrow \; ^{228}\text{Ac} \longrightarrow \; ^{228}\text{Th}$$

shows a pronounced maximum in concentration at the deepest part of the inner basin. In addition, systematic deviations from secular equilibrium between ^{228}Th and ^{232}Th are observed between the two basins with excess ^{228}Th in the inner basin (figure 5) but a deficiency of ^{228}Th relative to ^{232}Th in the outer basin.

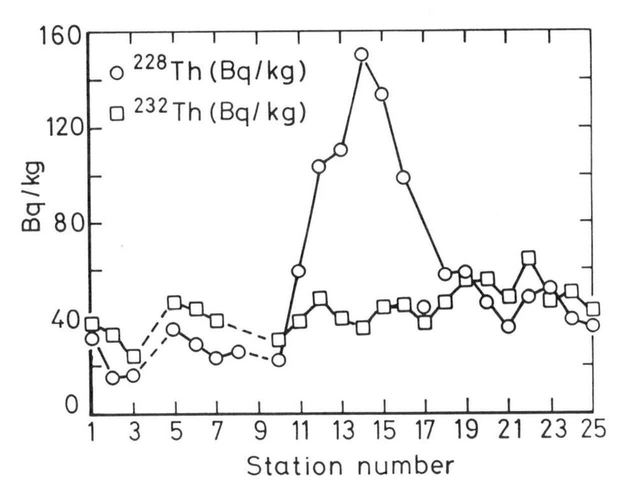

FIGURE 5. ^{223}Th and ^{232}Th*activities (BqKg^{-1}) in Loch Etive surface sediments. *(Determined by INAA).

The ^{228}Th deficiency in the outer basin is almost certainly due to the release of ^{228}Ra from the oxic surface sediments (27,28). The details of the mechanism involved in ^{228}Th enrichment in the inner basin sediments are not fully understood and are being further investigated. Possible contributory processes could be the enhanced residence time of ^{228}Ra in the restricted waters of the inner basin or involvement of ^{228}Th in the depositional and diagenetic recycling chemistry of the manganese. There is in fact a strong linear correlation between ^{228}Th and Mn concentration along the length of the fjord (correlation coefficient r=0.9). This could be the result of a genuine association between these species or could represent the mutual influence of some process or condition in the inner basin.

^{210}Pb, like ^{228}Th, is a highly non conservative radionuclide added to seawater as a result of radioactive decay but the two nuclides do not show any marked similarity of concentration trends in the fjord as seen from the excess ^{210}Pb results (relative to the secular equilibrium value with ^{226}Ra) shown in Figure 4. The excess concentration of ^{210}Pb within the sediments of the fjord is in general higher than the corresponding value in open sea conditions and, as previously indicated for total organic carbon, three distinct minima are observed in the excess ^{210}Pb concentration at stations 10, 17 and 21 (Figure 4) corresponding to the river inlets. The excess ^{210}Pb consequently shows a strong linear correlation with the sediment organic content (correlation coefficient r=0.85) and this could indicate a genuine chemical association with organics or may simply reflect dilution of both species at the same locations as a consequence of enhanced input of terrigenous material. It should be noted that the excess ^{210}Pb does not show any correlation with the Mn distribution. Thus the two non-conservative natural radionuclides ^{210}Pb and ^{228}Th show distinctly different associations in this situation.

Manmade radionuclides are present at readily detectable levels in Loch Etive and the $^{134}Cs/^{137}Cs$ ratios of about 0.03 and $^{238}Pu/^{239,240}Pu$ ratios of about 0.1 indicate that a substantial contribution arises from the Sellafield waste. Both the ^{137}Cs concentration and the $^{134}Cs/^{137}Cs$ ratios in surface sediment show an increase towards the seaward end of the fjord consistent with an increasing marine influence and, interestingly, three minima are observed for ^{137}Cs at stations 10, 17 and 21 analogous to those for excess ^{210}Pb and total organic content. This may reflect the observations made by McKay and Baxter (29), that organic molecules coat clay minerals and trap ^{137}Cs in the exchange sites preventing desorption, or may simply reflect a dilution effect. Finally ^{241}Am tends to increase landward to seaward but there is no obvious relationship between the ^{241}Am and either the Mn concentration or total organic content.

Depth profiles for a number of gravity cores are being derived at present but detailed discussion of these is beyond the scope of this paper and will be discussed elsewhere when complete. The following salient points can so far be summarised:

(1) $^{137}Cs/^{239,240}Pu$ activity ratios for surface sediment in the inner and outer basins have similar values at 22 and 17 respectively

(2) ^{137}Cs and $^{239,240}Pu$ penetrate to similar depths, 35-40cm in both basins, indicating sedimentation rates on the order of $0.5cmy^{-1}$

(3) ^{210}Pb and $^{134}Cs/^{137}Cs$ results suggested limited bioturbative or other mixing of the sediments to depths of about 10cm.

CONCLUSIONS

The following conclusions can be drawn from this initial survey of Loch Etive surface samples.

(1) Different disequilibria systematics are observed in the $^{232}Th \longrightarrow {}^{228}Ra \longrightarrow {}^{228}Th$ decay chain between the two basins of Loch Etive with ^{228}Th deficient in oxic sediments and enhanced in oxygen depleted sediments.

(2) There is a strong linear correlation of excess ^{228}Th and Mn.

(3) There is a linear correlation between excess ^{210}Pb and total organic content.

(4) ^{210}Pb and ^{241}Am do not show any correlation with Mn.

(5) Excess ^{210}Pb and ^{228}Th are present in suitable quantities to study accumulation rates and mixing.

This study is being continued with an investigation of the apparent Pb-C, Cs-C and ^{228}Th-Mn associations identified above, and a comparison of vertical profiles of natural and manmade nuclides in the sediments, interstitial waters and overlying waters.

REFERENCES

1. British Nuclear Fuels Limited, 1974 to 1985. Annual Report on Radioactive Discharges and Monitoring of the Environment.
2. Cambray, R.S., 1982. Annual discharges of certain long lived radionuclides to the sea and to the atmosphere from the Sellafield wastes Cumbria 1957-1981. UKAEA Report AERE-M-3269.
3. MacKenzie, A.B. and Scott, R.D., 1982. Radiocaesium and Plutonium in intertidal deposits from Southern Scotland. Nature, 299, 613-616.
4. Livingston, H.D. and Bowen, V.T., 1979. Plutonium and Caesium-137 in coastal sediments. Earth Planet Sci. Lett., 43, 29-45.

5. Aston, S.R. and Stanners, D.A., 1979. The Determination of Estuarine sedimentation rates by $^{134}Cs/^{137}Cs$ and other artificial Radionuclide Profiles. Estuarine Coastal Mar. Sci., 9, 529-541.
6. Hetherington, J.A., 1976. Environmental Toxicity of Aquatic Radionuclides Models and Mechanisms. Ann Arbor Science Publishers Inc.
7. Nelson, D.M. and Lovett, M.B., 1981. Measurement of the Oxidation State and Concentration of Plutonium in Interstitial Waters of the Irish Sea. Proc.IAEA Symp "Impacts of Radionuclide Releases into the Marine Environment".
8. Day, J.P. and Cross, J.E., 1981. ^{241}Am from the decay of ^{241}Pu in the Irish Sea. Nature, 292, 43-45.
9. Pentreath, R.J., Harvey, B.R. and Lovett, M.B., 1985. Chemical Speciation of Transuranium Nuclides Discharged into the Marine Environment. CEC/NRPB Speciation-85 Seminar.
10. Pentreath, R.J., Lovett, M.B., Jefferies, D.F., Woodhead, D.S., Talbot, J.W. and Mitchell, N.T. Proc. I.A.E.A. Conf. Radioactive Waste Management, 5, 315 (1984).
11. Hamilton, E.I., 1981. α - Particle Radioactivity of Hot Particles from The Esk Estuary. Nature, 290, 690-693.
12. Jefferies, D.F., Preston, A. and Steele, A.K., 1973. Distribution of ^{137}Cs in British Coastal Waters. Marine Pollution Bulletin, 4, 118-122.
13. McKinley, I.G., Baxter, M.S., Ellet, D.J. and Jack, W., 1981. Tracer Applications of Radiocaesium in the Sea of the Hebrides. Estuarine Coastal Shelf Sci, 10, 116-120.
14. Baxter, M.S., McKinley, I.G., MacKenzie, A.B. and Jack, W., 1979. Windscale Radiocaesium in the Clyde Sea Area. Mar Pollution Bulletin, 10, 116-120.
15. Lovett, M.B. and Nelson, D.M., 1978. The Determination of the Oxidation States of Plutonium in Seawater and Associated Particulate Matter. PROC CEGB Conf., Determination of Radionuclides in Environmental and Biological Materials, London, 1978.
16. Murray, C.N., Kautsky, H., Hoppinheit, M. and Domain, N., 1978. Actinide activities in water entering the North Sea. Nature, 276, 225-230.
17. MacKenzie, A.B., and Scott, R.D., 1984. Some aspects of Coastal Marine Disposal of Low-level liquid radioactive waste. Nuclear Engineer, 25, 110-122.
18. Smith, T.J., Parker, W.R. and Kirby, R., 1980. Part 1, Radionuclides in Marine Sediments 1980. IOS Report 110 Unpublished Manuscript.
19. Wood, B.J.B., Tett, P.B. and Edwards, A., 1973. An introduction to the phytoplankton, primary production and relevant hydrography of Loch Etive. J. Ecol, 61, 569-585.
20. Edwards, A. and Edieston, D.J., 1977. Deep Water Renewal of Loch Etive : a three basin Scottish fjord. Estuarine Coastal Mar. Sci., 5, 567.
21. Price, N.B. and Calvert, S.E., 1973. A study of the geochemistry of suspended particulate matter in coastal waters. Mar. Chem., 1, 169-189.
22. Malcolm, S.J. and Price, N.B., 1984. The behaviour of iodine and bromine in estuarine marine sediments. Mar. Chem., 15, 263-271.
23. Ridgeway, I.M. and Price, N.B., 1987. Geochemical Associations and Post Depositional mobility of Heavy Metals in coastal sediments in Loch Etive, Scotland. Mar. Chem. (in press).
24. Ridgeway, I.M., 1984. The behaviour of organic matter and minor elements in Scottish Sea Lochs. Unpubl. Ph.D. thesis, University of Edinburgh.

25. Craib, J.S., 1965. A sampler for taking short undisturbed marine cores. J. Cons Perm. Int. Explor. Med., 1, 34.
26. Elderfield, H. and Greaves, M., 1982. The rare earth elements in seawater. Nature, 296, 214-219.
27. Moore, W.S., 1969. Measurement of ^{228}Ra and ^{228}Th in Sea Water. J. Geophys. Res. 74, 694.
28. Carpenter, R., Peterson, M.L., Bennett and Somayajulu, B.L.K., 1984. Mixing and Cycling of Uranium, Thorium and ^{210}Pb in Puget Sound Sediments. Geochimica et Cosmochimica Acta, 48, 1949 1963.
29. MacKay, W.A. and Baxter, M.S., 1985. The Partioning of Sellafield-derived Radiocaesium in Scottish Sediments. J. Environ Radioactivity, 2, 93-114.

CHEMICAL PARTITIONING OF PLUTONIUM AND AMERICIUM IN SEDIMENTS FROM THE THULE REGION (GREENLAND)

E. Holm, J. Gastaud, B. Oregioni
IAEA
International Laboratory of Marine Radioactivity
Oceanographic Museum
Monaco

and

A. Aarkrog, H. Dahlgaard
Risø National Laboratory
Roskilde
Denmark

and

J.N. Smith
Fisheries and Oceans
Institute of Oceanography
Dartmouth
Canada

ABSTRACT

Chemical partitioning of plutonium and americium was studied by sequential leaching of sediments collected at about 200 m depth in 1968 and 1984 in the vicinity of the point of impact of an accidental loss of a nuclear device near Thule, Greenland, in January 1968.

The fractions separated and determined were exchangeable, bound to carbonates, bound to Fe/Mn oxides, bound to organic matter-sulfides and residual. Radiochemical determinations of total samples showed a considerable inhomogeneity due to the presence of hot particles. These hot particles were associated with the residual fraction. The distribution of plutonium between the different phases did not change significantly between 1968 and 1984. Americium showed a greater tendancy to be associated with the exchangeable, carbonate and Fe/Mn oxide fractions than did plutonium. Americium built up *in situ* from the decay of ^{241}Pu is mainly found in the residual fractions.

INTRODUCTION

Since the accidental loss of a nuclear device near Thule, Greenland, in 1968, several scientific expeditions have taken place in the area for the collection of sediment, water and biota samples, and the results were published (1-5). In 1984, the estimated inventories of $^{239+240}Pu$ and ^{241}Am in the sediments derived from the accident were 1 TBq and 0.1 TBq respectively (4).

Once a chemical species has been incorporated into a sediment, its ultimate fate depends on a number of very complex factors. Some elements can be considered to be irreversibly incorporated into a sedimentary component, whilst others may undergo post-depositional remineralization and take part in various bio-geochemical reactions. In order to predict the fate of transuranics or even make the most primitive evaluations of this kind it is necessary to understand something of the mechanisms by which these elements are incorporated into the various sedimentary phases. It is then not sufficient to carry out chemical analysis on the total samples.

The partitioning of transuranics between the various sedimentary components can be studied by physical or chemical separation and subsequent analysis of individual components. The results will depend not only on the nature of the depositional environment but also on the source terms. The chemical and physical partitioning as well as the isotopic composition of the transuranium elements will differ if the source is in form of liquid release from a nuclear fuel reprocessing plant, global fallout from nuclear weapon tests, a nuclear device accident or fallout from a reactor accident such as the recent Chernobyl disaster.

This paper presents the chemical partitioning of plutonium and americium between exchangeable, carbonate bound, Fe/Mn oxide bound, organic-sulfide bound and residual fractions in sediments collected in 1968 and 1984 contamined by the accidental loss of a nuclear device.

MATERIAL AND METHODS

During the expeditions a HAPS corer (6) was used for sediment collection. The sampling area was 145 cm^2. The cores were cut in 3 cm depth sections to a total depth of 12 cm (or 15 cm when available). Four cores collected close to the point of impact were studied (Table 1). These samples were dried at 100°C, sieved and the fraction <60 µm kept for analysis.

Table 1. Sediments subject to this study

Core No.	Year	Position	Depth (m)	Distance from point of impact (km)
53	1968		203	0.4
84.01184	1984	76°28.8'N, 69°31.8'W	244	7.6
84.01381	1984	76°31.5'N, 69°15.8'W	195	4.5
84.01377	1984	76°31.0'N, 69°18.5'W	200	1.4

The radioanalytical procedure for plutonium and americium has been described elsewhere (7). Cores 53 (1968) and 84.01377 (1984) were selected for the sequential leaching experements. This choice was made on the basis fo their similar $^{239+240}Pu$ activities (~850 mBq g^{-1}) in the surface (0-3 cm) sections. The samples analyzed in this work had a dry weight of 1 g and it has been shown that there is no relation between frequency of hot samples and sample size (4).

The chemical partitioning studies of plutonium and americium were performed by sequential leaching. The fractions separated and determined were:

Fraction 1: Exchangeable (1M CH_3COONH_4, 20 ml, 2h shaking)
Fraction 2: Bound to carbonates (1M CH_3COOH, 20 ml, 4 h shaking)
Fraction 3: Bound to Fe/Mn oxides (0.1 M NH_2OH,HCl in 25% CH_3COOH, 20 ml, overnight shaking)
Fraction 4: Bound to organic matter - sulfides (H_2O_2 pH=2, 10 ml, 85°C, 6h shaking + 1M CH_3COONH_4, 15 ml, 1h shaking)
Fraction 5: Residual (conc. HNO_3, $HClO_4$, HF)

The sequential leaching experiment was done twice on each sample from 1984 and five times on the sample from 1968. The activities were corrected to the date of collection. In this context for the build up of ^{241}Am during sample storage, we used a $^{241}Pu/^{239+240}Pu$ activity ratio of 3.3 ± 0.4 previously determined in 1968 (4).

RESULTS AND DISCUSSION

There are a great variety of chemical techniques which have been designed to establish the distribution of non-residual elements among the components of a sediment (8). The suitability, relative advantages and inconveniences of these methods have been reviewed at the International Laboratory of Marine Radioactivity in Monaco, and a standard method has been adopted after experimentation to check reproducibility. The method used in this work has been used previously to study transuranic geochemical partitioning in deep-sea and near-shore sediments (9-10).

In order to reduce the effect on trace metal or transuranic concentrations which result from grain size difference, it is common practice to exclude the coarser sediment fraction by sieving. In most techniques, the fraction which has a grain size of $< 60\ \mu m$ is used for analysis. Little work has been carried out on the distribution of transuranics in sediments with regard to grain sizes characteristics, but for heavy metals at grain sizes $> 60\ \mu m$, presence of trace metal-rich heavy minerals or large-sized trace metal-poor minerals might affect the results.

When performing this work we encountered several difficulties such as the inhomegeneity of the samples and that the chemical partitioning of fallout plutonium and americium is not known. The fallout background is estimated to be 23 Bq m^{-2} (5) which is very small compared to the levels (~30000 Bq m^{-2}) found in this study. The influence on the results from such plutonium and americium is therefore not likely to be significant.

We have not taken into consideration any alteration of chemical distribution during storage although this might have occured in practice

(11). Americium-241 was partly delivered at the accident, partly grown up from the decay of ^{241}Pu in situ on the sea floor and also during storage in the laboratory. The fact that all results for americium had to be corrected to the date of collection render these results and the interpretation of data more uncertain.

Shortly after the accident detailed nuclear track autography and microscopic studies of melted crust samples were conducted (12). The studies showed the plutonium to be in form of oxide particles with a very wide size distribution with a median diameter of 2 µm. The particles

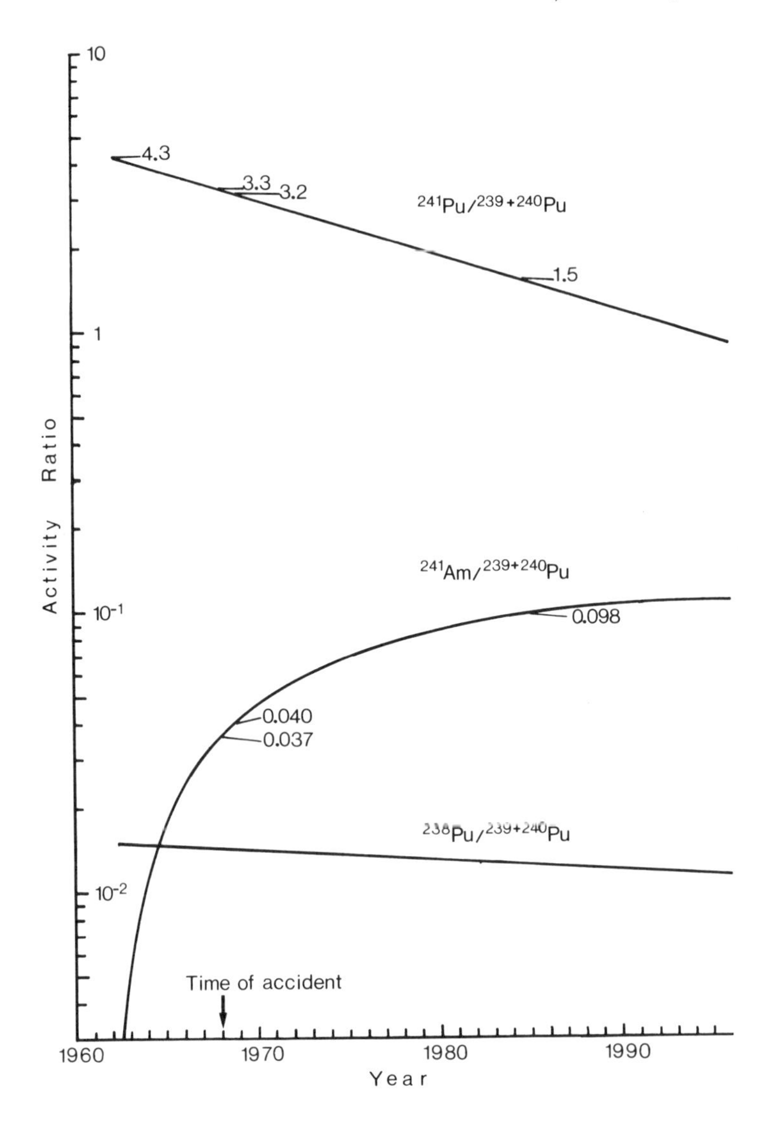

Fig. 1. ^{241}Pu, ^{238}Pu and ^{241}Am related to $^{239+240}$Pu in the accident debris at Thule.

were associated with or adhering to particles and pieces of inert debris of all kinds of material (glass, metal, plastic, rubber, etc.). The subsequent cleaning of the crash site was very effective and it was estimated that only 350 g of plutonium of the 3150 g released were trapped in the ice. Sedimentation studies of melted ice cores showed that 85 to 95% of the debris and associated plutonium oxide sank immediately (12) when contaminated ice was transferred into the sea and broke up in June–July 1968. .

It is obvious that such delivery of plutonium and americium to the sea floor is quite different to other sources such as global fallout, releases from nuclear fuel reprocessing plants, etc. In the Thule case the elements have not undergone any interaction in the atmosphere and a very limited one, if any, in the water column. The interactions and resulting chemical partitioning have taken place in the sediments. Consequently the observed activity ratio sediment/water for plutonium at the point of impacts is in the order of 10^{10} (4), while at areas in a similar marine environment where fallout plutonium has been transferred to the seafloor by particle interaction in the water column this ratio is in the order of $10^4 - 10^5$ (13).

The isotopic ratios $^{238}Pu/^{239+240}Pu$ and $^{241}Pu/^{239+240}Pu$ are different depending on source, being much higher in the releases from nuclear fuel reprocessing facilities and in fallout from the Chernobyl accident than in global fallout from nuclear detonation tests and these, in turn, are higher than those for weapon grade plutonium.

The $^{238}Pu/^{239+240}Pu$ activity ratio was 0.016 ± 0.01 (n=39, 1 SE) and the $^{241}Pu/^{239+240}Pu$ had earlier been determined to 3.3 ± 0.4 (n=6, 1 SE) at the time of the accident (January 1968) (4). The $^{241}Am/^{239+240}Pu$ activity ratio was determined in this study to 0.098 ± 0.005 (n=29, 1SE) in 1984. From the available data we can now calculate the decay of ^{238}Pu and ^{241}Pu relative to $^{239+240}Pu$ and also the build up of ^{241}Am as shown in Fig. 1. We then estimate that the $^{241}Am/^{239+240}Pu$ activity ratio was about 0.037 at the time of the accident. Supposing that there was no ^{241}Am present after the initial plutonium separation we observe that the plutonium was fabricated in 1962 $\pm$ 1 year. Furthermore, we find that when the sample collection took place in August 1968 and August 1984, about 8% and 62% respectively of the americium had been formed *in situ*.

The integrated activity area per unit (Bq m^{-2}) was studied in the three sediment cores collected in 1984 and given in Table 1 as a function of dry mass depth (kg m^{-2}). These cores represent different distances from the point of impact and accordingly different activity levels. The results are given in Fig. 2. We cannot observe any difference in the distribution pattern between the three cores or between americium and plutonium in these total activity analysis. The vertical distribution of americium in 1979 was also similar to that of plutonium, consequently showing unchanged americium/plutonium ratio with depth. A downward displacement of activity as well as a horizontal translocation of activity with time have been observed (4). It was suggested that the presence of plutonium and americium in the deeper layers of sediment is due to bioturbation and that physico-chemical mechanisms are less important (3).

The integrated activity area per unit for the different chemical fractions of sediment core No. 84.01377 is shown in Fig. 3 as a function

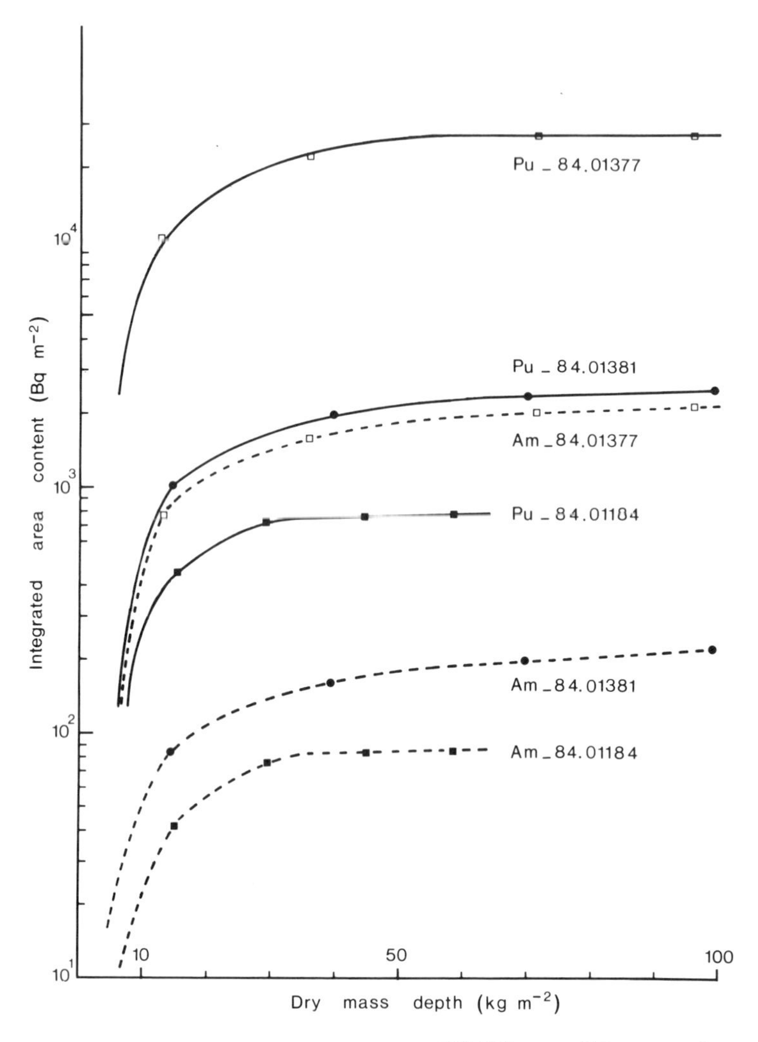

Fig. 2. Accumulated deposition of $^{239+240}$Pu and ^{241}Am (Bq m^{-2}) as a function of mass depth (kg m^{-2}) in 3 different cores from the Thule area, collected in 1984.

of the mass depth (kg m^{-2}). The different fractions show a rather similar shape of the distribution through the core. About 40% of the total activity is contained in the upper 3 cm except for the residual fraction where we here find about 50%.

The most remarkable result is however the excess of americium over plutonium in the "exchangeable" and "carbonate fractions" throughout the sediment column. The americium/plutonium activity ratio in the surface sediment section (0-3 cm) of the 1968 and 1984 cores are compared in Table 2. It is evident that there is also a preferential association of

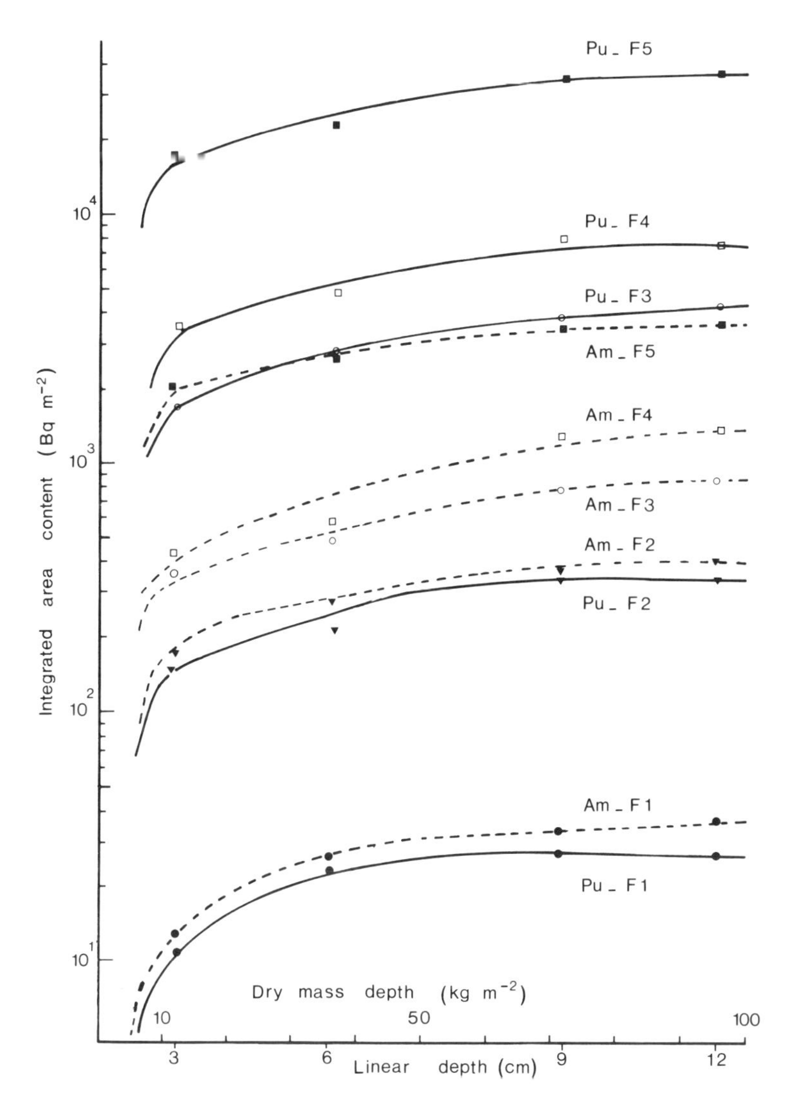

Fig. 3. Integrated area content for different chemical fractions (Bq m^{-2}). F1: exchangeable, F2: carbonate, F3: Fe/Mn oxide, F4: organic-sulfide, F5: residual fraction.

americium relative to plutonium in the "Fe/Mn oxide fraction". The weighted means in Table 2 also include all analysis of total samples. An increased americium/plutonium ratio in the exchangeable fraction was found in the 1984 core as compared with that of 1968.

It is obvious from Figure 3 that the influence on the results from fallout would only be important if all such plutonium (~ 23 Bq m^{-2}) and americium (~ 7 Bq m^{-2}) was associated with the exchangeable fraction. Fallout plutonium is however expected to mainly be associated with the Fe/Mn oxide or residual fractions (14).

Table 2. The $^{241}Am/^{239/240}Pu$ activity ratio for different chemical fractions in surface sediment (0-3 cm) collected in 1968 and 1984.

Fraction	$^{241}Am/^{239+240}Pu$ activity ratio 1968	1984
Exchangeable	0.20 ± 0.04	1.21 ± 0.30
Bound to carbonates	1.43 ± 0.09	1.13 ± 0.10
Bound to Fe/Mn oxides	0.28 ± 0.08	0.20 ± 0.02
Bound to organic matter-sulfides	0.28 ± 0.23	0.12 ± 0.05
Residual	0.026 ± 0.001	0.10 ± 0.03
weighted mean	0.040 ± 0.007 (n=18)	0.098± 0.005 (n=29)

As a distributional diagram the chemical partitioning is shown in Fig. 4. We find about 0.1-1% of plutonium associated with the exchangeable fraction compared to 1-3% for americium in 1968 and 1984. Americium also shows a higher affinity for the carbonate fraction (8-20%) compared to plutonium (0.6-1.5%). About 10% of plutonium and to 25% of americium is associated with the Fe/Mn oxide fraction, compared to only 1% for both plutonium and americium reported for a deep sea sediment (9), although the importance of this fraction (40-90%) has been demonstrated for lake and coastal sediments (15). Plutonium shows a decreasing percentage associated with organic matter with depth. The relative amount of plutonium in the residual fraction increases slightly with depth (69 to 82%) while the opposite is found for americium (55 to 23%). Consequently the americium/plutonium activity ratio decreases with depth in the residual fraction and increases in the organic fraction.

Samples with unusually high activity levels contained hot particles. These hot particles were always found in the residual fraction. The distribution of plutonium between the different phases did not change significantly between 1968 and 1984. This indicates that quasi-equilibrium of the chemical partitioning occured relatively quickly for this element. The americium built up in situ is mainly found in the residual fraction. It has been shown that benthos collected in the area preferentially concentrate americium relative to plutonium (4). Our results showing a higher affinity for the exchangeable and carbonate fractions for americium are suggested to be the explanation for this.

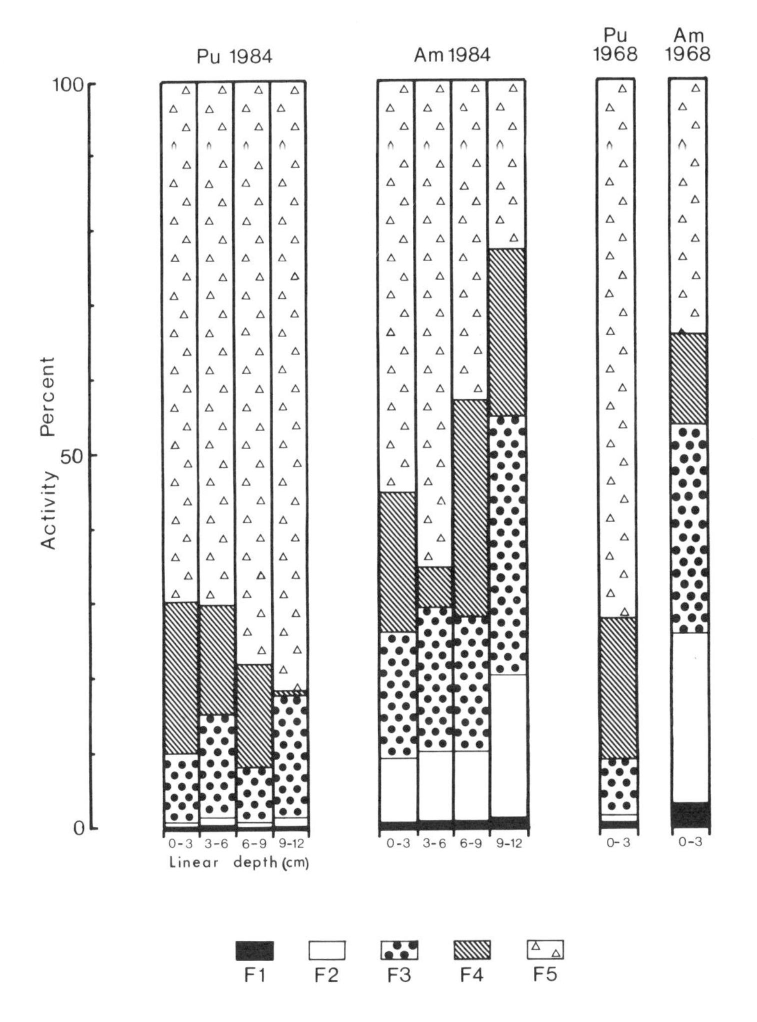

Fig. 4. Chemical partitioning of plutonium and americium in sediments collected in 1968 and 1984 at Thule (Greenland). F1: exchangeable, F2: carbonate, F3: Fe/Mn oxide, F4: organic-sulfide, F5: residual fraction.

AKNOWLEDGEMENT

The International Laboratory of Marine Radioactivity operates under an agreement between the International Atomic Energy Agency and the Government of the Principality of Monaco. The Laboratory gratefully acknowledges this support and that of the Oceanographic Museum at Monaco. The CEC Radiation Protection Programme provided additional financial support. We thank Dr. L.D. Mee for his review of the manuscript.

REFERENCES

1. Aarkrog, A., Radioecological investigations of plutonium in an arctic marine environment. Health Physics, 1971, 20, 31-47.

2. Aarkrog, A., Environmental behaviour of plutonium accidentally released at Thule, Greenland. Health Physics, 1977, 32, 271-84.

3. Aarkrog, A., A re-examination of plutonium at Thule, Greenland, in 1970. In Radioecology applied to the protection of man and his environment, EUR 4800, 1971, pp. 1213-19.

4. Aarkrog, A., Dahlgaard, H., Nilsson, K. and Holm, E., Further studies of plutonium and americium at Thule, Greenland. Health Physic, 1984, 46, 29-44.

5. Aarkrog, A, Boelskifte, S., Buch, E., Christensen, G.C., Dahlgaard, H., Hallstadius, L., Hansen, H. and Holm, E., Environmental Radioactivity in the North Atlantic Region. The Faroe Islands and Greenland included. 1984. Riso National Laboratory, Roskilde, Denmark, Report Riso-R-528, December 1985.

6. Kanneworff, E. and Nicolaisen, W., The HAPS, a Frame-supported Bottom Corer. Ophelia, 1973, 10, 119-29.

7. Ballestra, S. and Fukai, R. An improved radiochemical procedure for low-level measurements of americium in environmental matrices. Talanta, 1983, 30, 45-48.

8. Chester, R. and Aston, S.R., The partitioning of trace metals and transuranics in sediments. In Techniques for Identifying Transuranic Speciation in Aquatic Environments, International Atomic Energy Agency, Vienna, STI/PUB/613, 1981, 173-93.

9. Aston, S.R., Gastaud, J., Oregioni, B. and Parsi, P., Observations on the adsorbtion and geochemical association of technetium, neptunium, plutonium, americium and californium with a deep-sea sediment. In Behaviour of long-lived radionuclides in the marine environment, Commission of the European Communities, EUR 9214, 179-87.

10. Aston, S.R. and Stanners, D.A., Observation on the deposition, mobility and chemical associations of plutonium in intertidal sediments. In Techniques for Identifying Transuranic speciation in Aquatic Environments, International Atomic Energy Agency, Vienna, STI/PUB/613, 1981, 203-17.

11. Thomson, E.A., Luoma, S.N., Cain, D.J. and Johansson, C., The effect of sample storage on the extraction of Cu, Zn, Fe, Mn and organic material from oxidized estuarine sediments. Water, Air and Soil Pollution, 1980, 14, 215-33.

12. Langham, W.H., In Project Crested Ice, eds. G.E. Torres, H.B. Tracy and O.R. Smith, USAF Nuclear Safety, 1970, 65, 36-41.

13. Holm, E., Persson, B.R.R., Hallstadius, L., Aarkrog, A. and Dahlgaard, H., Radiocesium and transuranium elements in the Greenland and Barents Seas, Oceanologica Acta, 1983, 6, 457-62.

14. Edgington, D.N., Characterization of transuranic elements at environmental levels. In Techniques for Identifying Transuranic speciation in Aquatic Environments, International Atomic Energy Agency, Vienna, STI/PUB/613, 1981, 3-25.

15. Edgington, D.N., Alberts, J.J.. Wahlgren, M.A., Karttunen, J.O. and Reeve, C.A. Plutonium and americium in Lake Michigan sediments. In Transuranium Nuclides in the Environment, International Atomic Energy Agency, Vienna, STI/PUB/410, 1976, 493-516.

POLONIUM-210 IN SELECTED CATEGORIES OF MARINE ORGANISMS: INTERPRETATION OF THE DATA ON THE BASIS OF AN UNSTRUCTURED MARINE FOOD WEB MODEL

R.D. Cherry
Department of Physics,
University of Cape Town
Rondebosch, C.P., 7700
SOUTH AFRICA

and

M. Heyraud
International Laboratory of Marine Radioactivity,
Musée Océanographique,
MONACO

ABSTRACT

A substantial body of data for the concentration in marine organisms of the naturally-occurring alpha-radioactive element ^{210}Po is available. Concentrations of ^{210}Po show an extremely large variation between different categories of marine organism, e.g. ^{210}Po concentrations in penaeid shrimp from the genera Gennadas and Bentheogennema are generally more than two orders of magnitude higher than those in euphausiids. We show that the ^{210}Po data for euphausiids, carid shrimp and penaeid shrimp can be fitted to allometric relationships: the slopes for the three categories of organism are similar even though the concentrations vary widely. It is well-established that the dominant source of ^{210}Po in a marine organism is the food eaten by the organism; it should thus be possible to use the ^{210}Po concentrations in marine organisms as natural tracers to obtain information about organism diets. The "unstructured" marine food-web model proposed by Isaacs (1) generated equations which can in principle be used to predict the steady-state concentrations of chemical substances in organisms from various feeding categories. We apply this model to the available ^{210}Po data with modest success. We suggest that the same approach could be used fruitfully with other trace elements: an Isaacs-type model, combined with trace element concentration data, could be an important tool for (a) predicting trace element concentrations in marine organisms and (b) identifying the nature of the food consumed by the organisms.

INTRODUCTION

Over the last decade substantial data concerning the concentration of the naturally-occurring nuclide ^{210}Po in marine organisms have been published. Initially the primary interest in ^{210}Po was from the point of view of its importance as a contributor to the natural radiation dose received by the organisms; subsequently it was suggested that ^{210}Po measurements had considerable potential as an indicator of the nature of the food consumed by a marine organism (2-7). We have particularly extensive ^{210}Po data, essentially all of them already published in the references already cited, for three categories of marine organism, viz. euphausiids, carid shrimp and penaeid shrimp. Initially, we shall concentrate on these three categories. When we started to assemble our data, we had at our disposal not only the ^{210}Po concentrations in the whole animals but also, from our laboratory records, the dry weight per individual of the organism measured. The data are plotted in Fig. 1, and it is immediately obvious that the concentration of ^{210}Po per kg dry weight decreases with increasing dry weight per individual. Each of the three sets of data are fitted to the allometric equation

$$\log Q = a + b \log W \qquad (1)$$

where Q is the ^{210}Po concentration in Bq/kg dry weight and W is the dry weight per animal in kg. The following features of the regression lines should be noted: (i) the correlation coefficients are all significant; (ii) the slopes, b, of the three lines are similar although not identical; (iii) the three intercepts, a, are very different from each other. The clear separation between the three groups of organisms is perhaps the most important feature of the data along with the strikingly wide range in ^{210}Po concentrations, some two orders of magnitude between animals of the same size.

We suggest that this separation reflects different feeding regimes for the three organism categories. From the published data in the references already cited, it seems established that the food eaten is the dominant source of the ^{210}Po in most if not all marine animals. For example, in the euphausiid Meganyctiphanes norvegica there is direct experimental evidence that a decrease in the ^{210}Po content of the animal's diet is followed by a decrease in the ^{210}Po content of the animal, and its hepatopancreas in particular, within a few days (3). Surface adsorption of ^{210}Po from sea water must, moreover, be of minor importance in view of the low ^{210}Po contents of euphausiid and shrimp moults and/or exoskeleta in comparison with high ^{210}Po in internal organs (3 and Heyraud, unpublished data).

If this suggestion is correct, then knowledge of ^{210}Po concentrations in the food eaten by a marine animal, combined with biological parameters (notably the degree of assimilation of ^{210}Po by the animal) should enable the ^{210}Po content of the animal to be calculated. Conversely, knowledge of the ^{210}Po content of the animal could be used to calculate the ^{210}Po content in the food, and thereby give an indication of the nature of the food. Such calculations can be attempted on a piecemeal basis for each organism category; alternatively, intercomparison between different organism categories, such as those in Figure 1, may be better served by relating the ^{210}Po data to a model of the marine food web as a whole. We attempt to do so here using the model proposed by Isaacs (1).

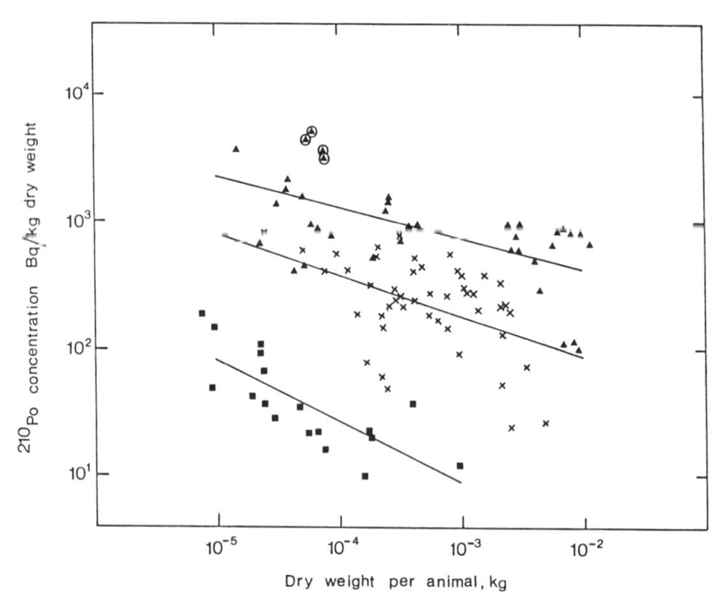

Figure 1. The experimental ^{210}Po concentrations in penaeids, carids and euphausiids plotted as a function of dry weight per animal. Triangles: penaeids; circled triangles: mid-water, mid-Atlantic penaeids; crosses: carids; squares: euphausiids. The results of the allometric fits to equation (1) in the sequence n; b; a; r; P are: Euphausiids 18; -0.489 ± 0.104; -0.535 ± 0.342; -0.76; < 0.001. Carids 47; -0.320 ± 0.093; $+1.304 \pm 0.284$; -0.45; < 0.01. Penaeids 40; -0.239 ± 0.053; $+2.153 \pm 0.167$; -0.59; < 0.001.

THE ISAACS MODEL

Isaacs proposed that unstructured food webs may more closely resemble real marine food webs than do structured models. Most structured models envisage organisms being assigned to definite trophic levels, thus giving rise to a pyramidal structure of the biomass. The unstructured model assumes that, in much of the marine environment, most creatures feed on the total contents of the web, subject to availability, suitability as to size and compatibility with their mode of feeding. To this basic idea Isaacs added the assumptions that the food web is in a steady-state and that any transfer of matter (or energy) between components in food webs can be characterised by three coefficients K_1, K_2 and K_3. The coefficient K_1 is the fraction of the material in food which is converted into living tissue in a single consumer step; K_2 is the fraction which is converted into irrecoverable forms (e.g. by respiration) and K_3 the fraction which is converted into non-living but recoverable forms such as organic detritus. The three fractions must of course add up to unity. Isaacs made the further assumption that the three coefficients are the same for all heterotrophic organisms, recognising that this was a simplification and that it might, in particular, be necessary to assume different coefficients for the first consumer step, that of herbivorous feeding, than for the remainder of the web.

The gross limitations which result from the feeding structures and the habits of organisms define groups in the food web, and Isaacs identified several categories of which the following are the most important for our purposes: Strict herbivores, which feed only on autotrophic material; omnivores, which feed on all material; particle feeders, which feed on autotrophic material and on non-living recoverable material; detrital feeders, which feed on non-living recoverable material; full predators, which feed on heterotrophic living material; non-herbivorous omnivores, which feed on all non-autotrophic material; feeders on a detrital milieu, which feed on non-living recoverable material and on detrital feeders. Isaacs set up equations which predicted the potential biomass in each of the above groups in terms of the three coefficients K_1, K_2 and K_3 and of the basic assumed flux, M_o, of autotrophic material introduced successively into the food web over each time interval.

Isaacs went on to point out that his model could be extended to generate equations which could predict the steady-state concentrations of chemical substances in the various feeding categories. For each particular chemical substance the coefficients K_1, K_2 and K_3 can be defined, and require identification by means of subscripts. Thus K_{1f}, K_{2f} and K_{3f} can represent the coefficients applicable to the organic content of the food, while K_{1c}, K_{2c} and K_{3c} represent the coefficients applicable to a particular trace element. Of course, K_{1c}, K_{2c} and K_{3c} must also add up to unity. Equations expressing the steady state concentrations then involve the six coefficients K and the concentration, C_o, of the trace element with respect to organic material in the primary autotrophic plants.

The Isaacs model attracted considerable attention, and was subsequently refined mathematically (8) and formulated in terms of compartment models (9). As has been pointed out, however (9), the evidence for the Isaacs food web comes mainly from the results of one experiment (10). This experiment involved measurements of caesium in fish, and the data available were very sparse. Here we fit the ^{210}Po data of Figure 1 to the Isaacs model in its original form, without the subsequent refinements and formulation in compartmental models. This restriction will have to be removed when more data are available because the refinements (8-9) allow the assumption that the coefficients K are the same for all heterotrophic organisms to be overridden. Since these coefficients may well in reality vary between organisms consuming plant matter, animal matter and detritus respectively, use of the more refined models will clearly be desirable when the data-base has been expanded.

Before presenting the results of the data fit, two problems are to be noted. Firstly, the Isaacs model does not include, explicitly, any dependence on the dry weight per individual organism. How such a dependence could be incorporated into the Isaacs model is not clear to us, and we shall sidestep the issue by performing our calculations in terms of a particular animal size (*viz.* 25 mg dry, for reasons given below). Our calculations will therefore not throw any light on the slopes of the regression lines in Fig. 1; they will, however, help explain the separation between the three lines. The slopes of the lines are quite possibly related to ingestion rates; thus, in the case of *M. norvegica*, data show that the slope of the allometric relationship between specific ingestion rate and individual animal dry weight is very close to the slope of the ^{210}Po concentration allometric relationship (2,11). A second possibility is that the negative slopes reflect to some extent the fact that ^{210}Po is a radioactive isotope with a half-life of 138 days: higher concentrations of the isotope in the smaller (faster-growing) than in the larger animals might be expected, although the very limited data available (2, 4) suggest that radioactive decay is of minor importance compared with egestion as a mode of elimination of ^{210}Po from a marine organism.

Secondly, there is the problem that the coefficients K_{1f}, K_{2f} and K_{3f} in the Isaacs expressions are not precisely known experimentally, let alone K_{1c}, K_{2c} and K_{3c}. We can, however, make a start, and as a first approach we make use of data which are available for euphausiids.

We can calculate K_{1f} and K_{1c} as follows:

$$K_{1f} = \frac{q_g \rho_g}{q_r \rho_r} \quad (2)$$

$$K_{1c} = \frac{Q_g \rho_g}{Q_r \rho_r} \quad (3)$$

when q_g represents organic matter concentration in the animal in g dry organic matter per g dry animal, q_r represents organic matter concentration in the food the animal eats, Q_g and Q_r are the corresponding ^{210}Po concentrations, and ρ_g and ρ_r are the animal's growth and ingestion rates respectively, both in g dry weight/g dry animal/day. Calculation of K_{3f} and K_{3c} is more problematical, and we shall make an assumption which will certainly be no more than an approximation, viz. we assume that the recoverable fraction K_3 arises entirely from egestion, and that egested material is all recoverable. Some of this fraction will not arise from egestion, and, probably more important, some of the egested material will certainly not be recoverable; hence our caveat about this assumption. We then have:

$$K_{3f} = \frac{q_e \rho_e}{q_r \rho_r} \quad (4)$$

$$K_{3c} = \frac{Q_e \rho_e}{Q_r \rho_r} \quad (5)$$

with the same meanings and units to the symbols as in (2) and (3) but with the additional subscript e representing egested material. Once K_{1f}, K_{3f} and K_{1c} and K_{3c} are known we have of course also calculated K_{2f} and K_{2c} because the three coefficients K add up to unity in each case.

Before inserting the data in equations (2) to (5) we must decide what constitutes the euphausiid food supply. We shall make two alternative assumptions. At the one extreme we assume that the euphausiid is a strict herbivore, feeding on phytoplankton. Such might be the case for the Antarctic Krill, *Euphausia superba*, during the season of abundant phytoplankton (12), although recent data indicate that *E. superba* can feed on particles ranging from nanophytoplankton to macrozooplankton (7, 13). The other, apparently more likely to be the case for most other species of euphausiid (12, 14), assumes that they are omnivorous, feeding on a mixture of phytoplankton, microzooplankton and detritus. Because of the dependence of ^{210}Po concentration on body weight we must choose a particular size of euphausiid, and we have taken an individual of 25 mg dry because it is in this approximate size that data for the biological parameters ρ are available. On this basis, we use the following data. The ρ values (2, 12, 15) are: $\rho_g = 0.015$ g/g dry/day; $\rho_r = 0.200$ g/g dry/day; $\rho_e = 0.038$ g/g dry/

day. Typical q are (14–18): q_r (assuming phytoplankton food) = 0.75 g/g dry; q_r (assuming omnivore food is an equal mixture of phytoplankton, microzooplankton and detritus) = 0.70 g/g dry; q_g = 0.84 g/g dry; q_e = 0.50 g/g dry. The Q values are: Qr (phytoplankton food) = 167 Bq/kg dry (3, 19); Q_r (omnivore food) = 455 Bq/kg dry (2, 3, 19, 20); Q_g = 52 Bq/kg dry (Fig. 1); Q_e = 814 Bq/kg dry (20).

CALCULATION AND DISCUSSION

The Isaacs expressions predicting the concentrations of a trace element in animals from the various categories involve five ratios, designated by the symbols j, p, α, β and δ. In Table 1 we quote the mathematical expressions derived by Isaacs for these ratios and the values of the ratios calculated using the data with the two alternative assumptions for the nature of the euphausiid food. The use of two alternative assumptions is instructive, because it gives a good indication of the sensitivity of the different ratios to changes in the experimental data. Least sensitive is the ratio δ. The values of K_{1f} and jK_{1f} are small, and changes in either the assumptions or the data are unlikely to result in δ differing much from unity. The ratios j, p and α are moderately sensitive. Highly sensitive is the ratio β, which is calculated thus:

$$\beta = \frac{K_{2f}}{K_{2c}} = \frac{1 - K_{1f} - K_{3f}}{1 - K_{1c} - K_{3c}} = \frac{1 - K_{1f} - K_{3f}}{1 - jK_{1f} - pK_{3f}} . \qquad (6)$$

Even modest changes in j, p, K_{1f} and K_{3f} can result in a substantial change in the denominator of this expression, and β becomes extremely sensitive as $jK_{1f} + pK_{3f}$ tends to unity.

In Table 2 we list the Isaacs expressions predicting the concentration C_μ of a particular trace element in organisms from a feeding category μ. It is to be noted that the concentration C_μ, as defined by Isaacs, is the concentration of the trace element with respect to the concentration of organic matter in the organism.

In Table 2 we also list the predicted ^{210}Po concentrations, again with respect to organic matter, as calculated from the Table 1 ratios. It is to be seen that the predictions differ widely between the two assumptions, a fact which is unsurprising since the very sensitive ratio β enters into the Isaacs expression for C_μ for all but the first feeding category. The predictions are nonetheless encouraging because on both assumptions they show a distinct tendency to band into groups, as do the ^{210}Po data in Figure 1. On the first assumption the groups are, in increasing ^{210}Po concentration: (i) the herbivores; (ii) the predators; (iii) the omnivores and the particle feeders; (iv) the non-herbivorous omnivores; and (v) the detrital feeders and the feeders on detrital milieu. On the second assumption the herbivores and the predators are interchanged, but otherwise the sequence remains the same.

The ^{210}Po data with which these predictions are to be compared are taken from Figure 1 for a 25 mg dry animal: euphausiids, 52 Bq/kg dry; carids, 599 Bq/kg dry; penaeids, 1791 Bq/kg dry. For a sub-category of the penaeids, viz. mid-Atlantic mid-water shrimp of the genera *Gennadas* and *Bentheogennema*, we estimate ^{210}Po at about 11840 Bq/kg dry. We have previously (6) measured four mid-Atlantic mid-water (610 to 1500 m) specimens of these genera and found them to contain particularly high ^{210}Po contents; these four specimens are identified in Figure 1, and it is from

Table 1. The Isaacs expressions for the ratios j, p, α, β and δ and the values of these ratios calculated from the experimental data assuming (1) that euphausiids are herbivores eating phytoplankton only, or (2) that euphausiids are omnivores eating an equal mixture of phytoplankton, microzooplankton and detritus. On the first assumption $K_{1f} = 0.084$ and $K_{3f} = 0.127$; on the second assumption $K_{1f} = 0.090$ and $K_{3f} = 0.136$.

Ratio	Isaacs expression	Calculated value Assumption 1	Assumption 2
j	$\frac{K_{1c}}{K_{1f}}$	0.28	0.09
p	$\frac{K_{3c}}{K_{3f}}$	7.3	2.5
α	$\frac{1-K_{2c}}{1-K_{2f}}$	4.5	1.5
β	$\frac{K_{2f}}{K_{2c}}$	16.2	1.2
δ	$\frac{1-jK_{1f}}{1-K_{1f}}$	1.07	1.09

Table 2. The Isaacs expressions predicting the concentration C_μ of a particular trace element in organisms from a particular feeding category, together with the predicted C_μ for ^{210}Po using assumptions (1) and (2) as in Table 1. The concentrations C_μ are the concentrations of the trace element with respect to the concentration of organic matter in the organism. C_0 refers to the concentration of the trace element in the primary autotrophic plants.

Feeding category	C_μ Isaacs expression	Predicted C_μ for ^{210}Po Assumption 1	Assumption 2
Strict herbivores	$C_o j$	$C_o j$	$C_o j$
Full predators	$C_o j^2 \beta$	4.5 $C_o j$	0.11 $C_o j$
Omnivores	$C_o j\ \beta$	16 $C_o j$	1.2 $C_o j$
Particle feeders	$C_o j\ \beta\delta$	17 $C_o j$	1.3 $C_o j$
Non-herbivorous omnivores	$C_o j\ \beta\alpha$	73 $C_o j$	1.8 $C_o j$
Detrital feeders	$C_o j\ p\beta$	118 $C_o j$	3.0 $C_o j$
Feeders on a detrital milieu	$C_o j\ p\beta\delta$	127 $C_o j$	3.3 $C_o j$

them that our estimated ^{210}Po concentration is obtained. The estimate is only approximate, because the four animals measured were all relatively large (56 to 78 mg dry). The predictions for the first assumption, viz. that euphausiids are strict herbivores, can in fact be fitted surprisingly well to the data and what is known about the feeding habits of the different categories of organism. Several references (21-23) have referred to the common occurrence in the guts of Gennadas and Bentheogennema of "olive-green cytoplasm", "greenish-brown debris" and "amorphous green residue", and faecal pellets have been suggested as a possible food source. It seems not unreasonable to place these shrimp in the detrital feeder category, and the prediction that detrital feeders should have a ^{210}Po concentration 118 times higher than that in herbivores is comparable with the factor of 228 observed in the limited data. Further, the data give penaeids as a whole a ^{210}Po concentration 34 times higher than euphausiids, and carids 12 times higher; the concentrations predicted for non-herbivorous omnivores (73 times higher) and omnivores (16 times higher) are roughly consistent with these data. Studies of shrimp feeding (e.g. 23) suggest that they are omnivores. The ^{210}Po data, plus the Isaacs model and the assumption that euphausiids are herbivores, suggest a finer gradation: carid shrimp are omnivores, while penaeid shrimp range from largely non-herbivorous omnivores to detrital feeders.

The predictions for the second assumption, viz. that euphausiids are omnivores, are not satisfactory, because on this assumption there is no category which has predicted ^{210}Po levels more than three times those for omnivores.

We recognise that the data cannot at this stage be considered as giving firm support to the first rather than the second assumption. The situation is too uncertain because (i) the data for the various ρ and Q are sparse and imprecise; (ii) the assumption we made in deriving equations (4) and (5) is no more than an approximation; (iii) modest changes in the ρ and Q data or in the accuracy of the approximation just referred to can give rise to large changes in some of the calculated ratios in Table 1, notably β, and hence to the predicted C_μ in Table 2; and (iv) the Isaacs assumption that the K coefficients are the same for all heterotrophic organisms has been retained, even though different values might apply in reality to the first consumer step and to detrital feeders.

Faced with these uncertainties, it is worth pointing out that the model itself contains inherent constraints. It predicts that the ratio of the trace element concentration in detrital feeders to that in omnivores is equal to p. As Isaacs noted, $jK_{1f} + pK_{3f}$ cannot be greater than unity, and hence p cannot be greater than K_{3f}^{-1}. Isaacs suggested further that 0.1 was a realistic lower limit for K_{3f}; if this is correct, p cannot be larger than 10. However, if euphausiids are omnivores, then the data for the detrital feeding shrimp require p larger than 100. It is accordingly difficult to imagine any change in our estimates of the K values which would enable the data to fit the second assumption. On the other hand, the model predicts that the ratio of the trace element concentration in detrital feeders to that in strict herbivores is equal to $p\beta$. This factor can be larger than 100, as can be seen from Table 2, and it is for this reason that our first assumption came reasonably close to predicting the very wide range in ^{210}Po concentrations which are in fact observed between the different categories of organisms. Although the idea that euphausiids are strict herbivores runs counter to current opinion, we suggest that the unstructured food-web model plus the ^{210}Po data indicate, albeit tentatively, that they are much closer to being herbivores rather than omnivores. A recent paper (24) lends support to this idea. It compared Euphausia pacifica feeding on the diatom Thalassiosira angstii and on Pseudocalanus sp. copepods respectively, and found that the euphausiids obtained a maximum carbon-specific ration of 8.8% per day from the diatom compared with only 0.45% per day from the copepods. Even though euphausiids can, clearly, feed on a wide range of prey, it seems possible that herbivorous feeding may be the optimal mode in terms of their overall metabolic requirements.

CONCLUSION

The present situation is, then, that the Isaacs unstructured food-web model is reasonably successful in explaining the ^{210}Po data. In particular, it predicts the two most important features of the data, viz. the separation between the three organism categories and the overall range of about two orders of magnitude in the ^{210}Po concentrations in animals of the same size. In view of the uncertainties in our estimates of the K coefficients and the sensitivity of certain of the Isaacs expressions to the parameters involved it is probably unreasonable to expect much from the model as a predictive tool at this stage. We are in the process of expanding our ^{210}Po data base in an attempt to establish allometric relationships for organisms from other feeding regimes; this might result in more accurate predictions for the Isaacs ratios. At the same time, in collaboration with other colleagues, we are acquiring data for other trace elements in the three organism categories discussed in this paper. If there are other trace elements which, like ^{210}Po, show a wide overall range in concentration and a clear separation between the concentrations in the different organism-categories, the data for such elements could be used to establish the consistency or otherwise of our diet predictions. Conversely, an Isaacs type model applied to organisms whose diet is known could, in the longer term, be used as a predictor of certain trace element concentrations in these organisms.

ACKNOWLEDGEMENTS

Support from the Foundation for Research Development and from the University of Cape Town is acknowledged. The International Laboratory of Marine Radioactivity operates under an agreement between the International Atomic Energy Agency and the Government of the Principality of Monaco. We repeat our thanks to numerous colleagues in many institutions who generously provided us with sample material for ^{210}Po analysis. We thank M.I. Cherry, A.G. Davies, M.J.R. Fasham, J.G. Field, S.W. Fowler and R.C. Newell for comments.

REFERENCES

1. Isaacs, J.D., Potential trophic biomasses and trace-substance concentrations in unstructured marine food webs. Mar. Biol., 1973, 22. 97-104.

2. Heyraud, M., Fowler, S.W., Beasley, T.M. and Cherry, R.D., Polonium-210 in euphausiids: a detailed study. Mar. Biol., 1976, 34. 127-36.

3. Heyraud, M. and Cherry, R.D., Polonium-210 and lead-210 in marine food-chains. Mar. Biol., 1979, 52. 227-36.

4. Higgo, J.J.W., Cherry, R.D., Heyraud, M., Fowler, S.W. and Beasley, T.M., The vertical oceanic transport of alpha-radioactive nuclides by zooplankton fecal pellets. In The natural radiation environment III, eds. T.F. Gesell and W.M. Lowder, U.S. Department of Energy, CONF-780422, Vol 1, pp. 502-13.

5. Cherry, R.D. and Heyraud, M., Polonium-210 content of marine shrimp: variation with biological and environmental factors. Mar. Biol., 1981, 65. 165-75.

6. Cherry, R.D. and Heyraud, M., Evidence of high natural radiation doses in certain mid-water oceanic organisms. Science, 1982, 218. 54-6.

7. Boyd, C.M., Heyraud, M. and Boyd, Ch.N., Feeding of the Antarctic Krill Euphausia superba. J. Crus. Biol., 1984, 4 (Spec. No. 1). 123-41.

8. Lange, G.D. and Hurley, A.C., A theoretical treatment of unstructured food webs. Fish. Bull., 1975, 73. 378-81.

9. Fasham, M.J.R., Analytic food-web models. In Mathematical models in biological oceanography, eds. T. Platt, K.H. Mann and R.E. Ulanowicz, The Unesco Press, Paris, 1981, 54-65.

10. Young, D., The distribution of cesium, rubidium and potassium in the quasi-marine ecosystem of the Salton Sea. Ph.D. dissertation University of California, San Diego, 1970.

11. Heyraud, M., Food ingestion and digestive transit time in the euphausiid Meganyctiphanes norvegica as a function of animal size. J. Plankton Res., 1979, 1. 301-11.

12. Mauchline, J. and Fisher, L.R., The biology of euphausiids. Adv. Mar. Biol., 1969, 7. 1-454.

13. Laws, R.M., The ecology of the Southern Ocean. American Scientist, 1985, 73. 26-40.

14. Mauchline, J., The biology of mysids and euphausiids. Adv. Mar. Biol., 1980, 18. 1-677.

15. Small, L.F. and Fowler, S.W., Turnover and vertical transport of zinc by the euphausiid Meganyctiphanes norvegica in the Ligurian Sea. Mar. Biol., 1973, 18. 284-90.

16. Bougis, P., Ecologie du Plancton marin, Masson et Cie, Paris, 1974.

17. Childress, J.J. and Nygaard, M., Chemical composition and buoyancy of midwater crustaceans as a function of depth of occurrence off Southern California. Mar. Biol., 1974, 27. 225-38.

18. Fowler, S.W., Benayoun, G. and Small, L.F., Experimental studies on feeding, growth and assimilation in a Mediterranean euphausiid. Thalassia Jugosl., 1971, 7. 35-47.

19. Shannon, L.V., Cherry, R.D. and Orren, M.J., Polonium-210 and lead-210 in the marine environment. Geochim. Cosmochim. Acta, 1970, 34. 701-11.

20. Beasley, T.M., Heyraud, M., Higgo, J.J.W., Cherry, R.D. and Fowler, S.W., ^{210}Po and ^{210}Pb in zooplankton fecal pellets. Mar. Biol., 1978, 44. 325–28.

21. Foxton, P. and Roe, H.S.J., Observations on the nocturnal feeding of some mesopelagic decapod crustacea. Mar. Biol., 1974, 28. 37–49.

22. Hefferman, J.J., Hopkins, T.L., Vertical distribution and feeding of the shrimp genera Gennadas and Bentheogennema (Decapoda: Penaeidea) in eastern Gulf of Mexico. J. Crust. Biol., 1981, 1. 461–73.

23. Roe, N.S.J., The diel migrations and distributions within a mesopelagic community in the North East Atlantic. 2. Vertical migrations and feeding of mysids and decapod crustacea. Prog. Oceanog., 1984, 13. 269–318.

24. Ohman, M.D., Omnivory by Euphausia pacifica: the role of copepod prey. Mar. Ecol. Prog. Ser., 1984, 19. 125–31.

MODELE MATHEMATIQUE DU TRANSPORT DES RADIONUCLIDES SUR LE PLATEAU CONTINENTAL NORD-EUROPEEN

S. DJENIDI et J.C.J. NIHOUL
Mécanique des Fluides Géophysiques
Université de Liège, B5, Sart Tilman
B-4000 LIEGE

et

A. GARNIER
Association EUR/CEA
Institut de protection et de sûreté nucléaires
B.P. 6
F-92260 Fontenay-aux Roses.

RESUME

Le modèle mathématique 3D = 2D + 1D développé par le Laboratoire de Mécanique des Fluides Géophysiques de l'Université de Liège est appliqué à l'étude du déplacement à long terme des masses d'eau sur le plateau Nord-Européen. Les résultats sont exploités pour modéliser le transport et la diffusion des polluants radioactifs issus des sources côtières localisées. Une première comparaison avec les observations permet d'évaluer les performances du modèle et montre le rôle essentiel que peuvent jouer les radiotraceurs dans la validation des modèles hydrodynamiques à grande échelle et la calibration des modèles d'advection-diffusion pour les études de pollution et la gestion de l'environnement marin.

INTRODUCTION

Les études effectuées dans le cadre des programmes de recherche concernant la protection radiologique des personnes et de l'environnement font appel à de multiples disciplines, entre autres l'océanographie. Si la présence de radionuclides pose des problèmes de protection, elle constitue en revanche un outil de travail. Quant aux études faites dans un but de protection, elles peuvent avoir des "retombées" d'autant plus intéressantes qu'elles font appel à des modèles explicatifs très élaborés.

De nombreuses études, mesures et interprétations, ont déjà été effectuées et doivent permettre d'étayer des modèles de dispersion ou d'envisager leur application à des zones bien définies. De nombreux modèles "locaux" existent. On a plutôt envisagé de développer les modèles de circulation "résiduelle" et leur application possible à l'évaluation des concentrations de polluants à longue distance du point de rejet, en tenant compte des paramètres du site d'origine et de la climatologie des régions concernées.

La comparaison des concentrations calculées a posteriori avec les concentrations mesurées au cours des années peut contribuer à apprécier la validité du modèle et à le perfectionner : les rejets radioactifs apparaissent alors comme un outil de travail en océanographie.

MODELES HYDRODYNAMIQUES

Les processus à méso-échelle tels que les marées et les courants transitoires induits par le vent sont de loin les plus intenses dans les mers continentales. Leur temps caractéristique s'étend de quelques heures à quelques jours et les courants qui leur sont associés peuvent largement dépasser des vitesses d'un mètre par seconde.
Cependant, les biologistes et les chimistes sont beaucoup plus intéressés par les phénomènes à macro-échelle qui ont des temps caractéristiques typiques des processus écologiques (de quelques semaines à quelques mois).
L'importance des marées et des tempêtes pour la production de turbulence et pour le mélange des propriétés marines est évidemment considérée comme essentielle par les biologistes, mais beaucoup d'entre eux seraient déjà satisfaits d'une grossière paramétrisation de l'efficacité du mélange turbulent et d'une description générale de ce transport à long terme des masses d'eau qu'on appelle souvent *"circulation résiduelle"*.

Mis à part les courants dus aux différences de densité, que nous n'envisageons pas ici et qui ne sont appréciables que dans les régions où les gradients de température et surtout de salinité sont importants, on peut considérer que les courants résiduels présents dans des bassins continentaux peu profonds tels que la mer du Nord et les mers adjacentes, sont principalement générés et entretenus par trois causes.

Celle qui paraît être prédominante au vu de ses effets à méso-échelle est sans doute la marée, surtout par le truchement de la composante semi-diurne lunaire (M_2) qui joue un rôle essentiel dans la région étudiée. Si on fait abstraction des autres composantes ainsi que des conditions météorologiques, une moyenne sur une période de l'ordre du jour suffit pour éliminer l'effet transitoire de la marée et ne laisser ainsi que le résidu généré par ses interactions non linéaires : *c'est la circulation résiduelle de marée.*

La circulation résiduelle régionale est influencée par la circulation résiduelle globale c'est-à-dire la composante quasi stationnaire des écoulements pénétrant aux frontières.
Les modèles d'inflow-outflow ont montré que les flux aux frontières jouent un rôle important dans le calcul de la dérive à long terme des masses d'eau en ce sens qu'ils arrivent à reproduire les grandes tendances de celle-ci, et qu'à ce titre les conditions aux limites peuvent être considérées comme l'un des trois "forcing" de la circulation résiduelle.

La superposition des effets du vent à ceux de la marée donne lieu à des surcotes et à des modifications courantométriques qui peuvent être si importantes lors de conditions météorologiques sévères qu'il est impensable de négliger l'impact des *forces atmosphériques* sur la dérive à long terme des masses d'eau.
Malheureusement, à l'inverse de la marée qui a un comportement cyclique aisément prévisible, le champ éolien est caractérisé par une irrégularité et une complexité qui rendent délicate la prise en compte de cet effet.
Un circulation résiduelle obtenue par moyenne sur une période de l'ordre de quelques semaines, avec une succession d'événements météorologiques typiques d'une saison semble être la plus significative parce qu'elle présente un caractère "climatique" tout en reflétant la variabilité atmosphérique. Elle reste néanmoins difficile à déterminer à cause de la longueur des périodes à simuler.

Une manière d'éviter ce problème consiste en l'utilisation d'un vent moyen sur deux ou trois cycles de marée. Mais considérer une moyenne sur une période équivalente à celle du résiduel "tidal" n'est fiable que dans le cas d'une situation atmosphérique remarquablement persistance. En effet, si le vent obtenu par moyenne de conditions météorologiques réelles très variables peut donner des résultats valables dans les grandes lignes, il risque cependant de ne pas produire les effets de certains événements intenses qui ont été éliminés ou amoindris par le processus de moyenne, et de "lisser" en quelque sorte la circulation "climatique" associée qu'on veut approcher.

La circulation résiduelle étant implicitement incluse dans les équations dépendantes du temps de la circulation transitoire, il serait tentant de résoudre ces équations et de faire la moyenne de la solution transitoire pour obtenir les courants résiduels.
Sur le plateau continental nord européen, la large prédominance des mouvements à méso-échelle fait en sorte que les courants résiduels sont du même ordre de grandeur que l'erreur sur les courants transitoires calculés par un modèle ou mesurés à l'aide de courantomètres. Pour cette raison, il est plus correct de moyenner, non pas la solution, mais les équations elles-mêmes, sur un temps T suffisamment long pour éliminer les phénomènes transitoires.
Cette moyenne introduit des termes supplémentaires - *les tensions de Reynolds à méso-échelle (TRME)* - dus aux non linéarités des équations, une forme paramétrisée de ces termes constituant la contribution des méso-échelles dans les équations de la circulation résiduelle. Ces équations sont indépendantes du temps à cause de la quasi stationnarité du phénomène (Nihoul, 1985).

Dans cette optique, la résolution des équations stationnaires moyennes qui décrivent la circulation résiduelle exige la prescription, le long des frontières du modèle, de *conditions aux limites* adaptées aux échelles de temps et aux conditions météorologiques considérées.
Les conditions aux limites - élévations ou vitesses résiduelles - sont néanmoins difficiles à obtenir à cause du fait que, d'une part les instruments ne sont pas spécialement conçus pour mesurer des grandeurs résiduelles et que, d'autre part, l'importance des frontières en mer ouverte ne permet pas l'établissement d'un réseau expérimental suffisamment dense; de plus, les séries temporelles de données doivent être de longue durée pour tenir compte des fluctuations saisonnières des phénomènes.

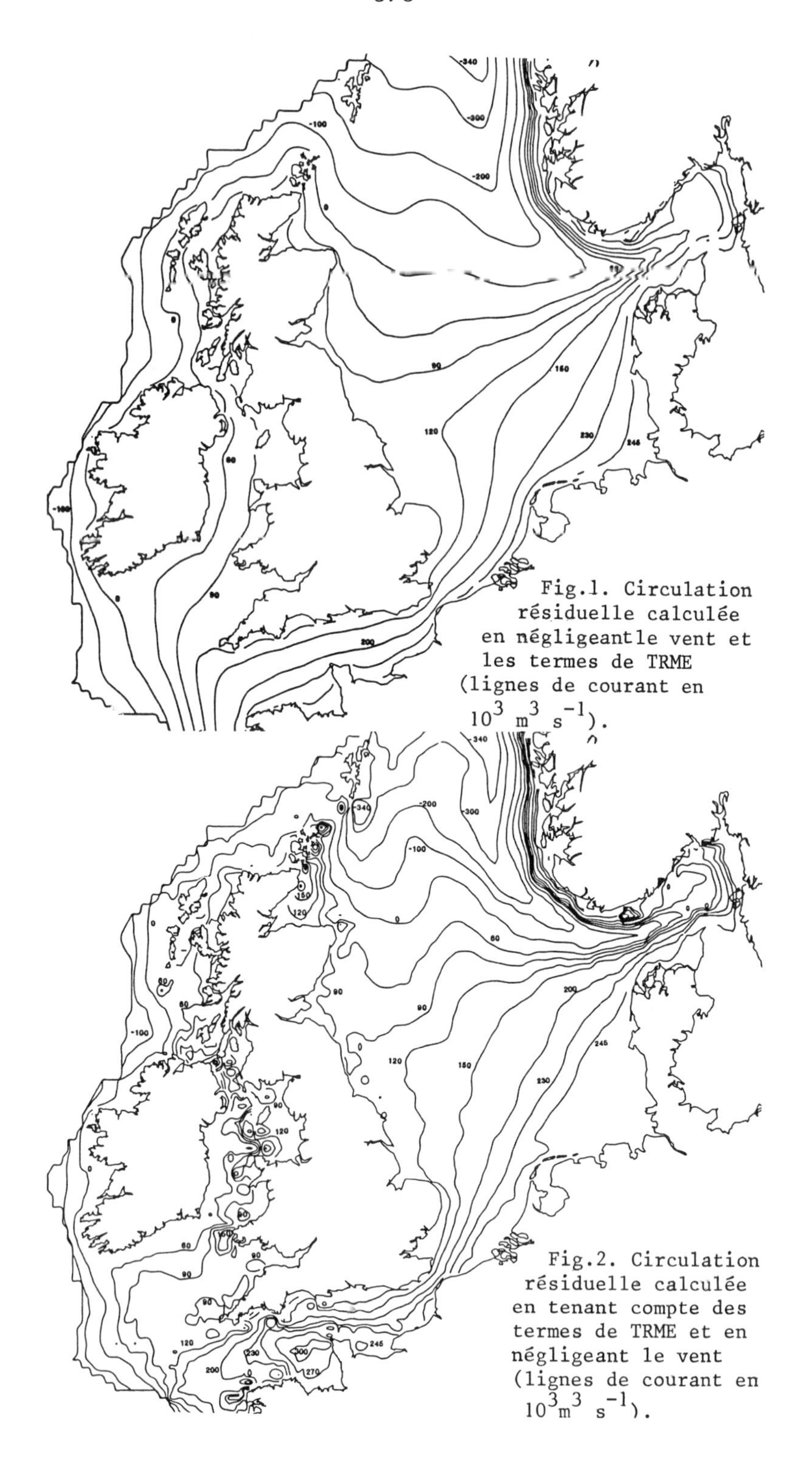

Fig.1. Circulation résiduelle calculée en négligeant le vent et les termes de TRME (lignes de courant en $10^3\ m^3\ s^{-1}$).

Fig.2. Circulation résiduelle calculée en tenant compte des termes de TRME et en négligeant le vent (lignes de courant en $10^3 m^3\ s^{-1}$).

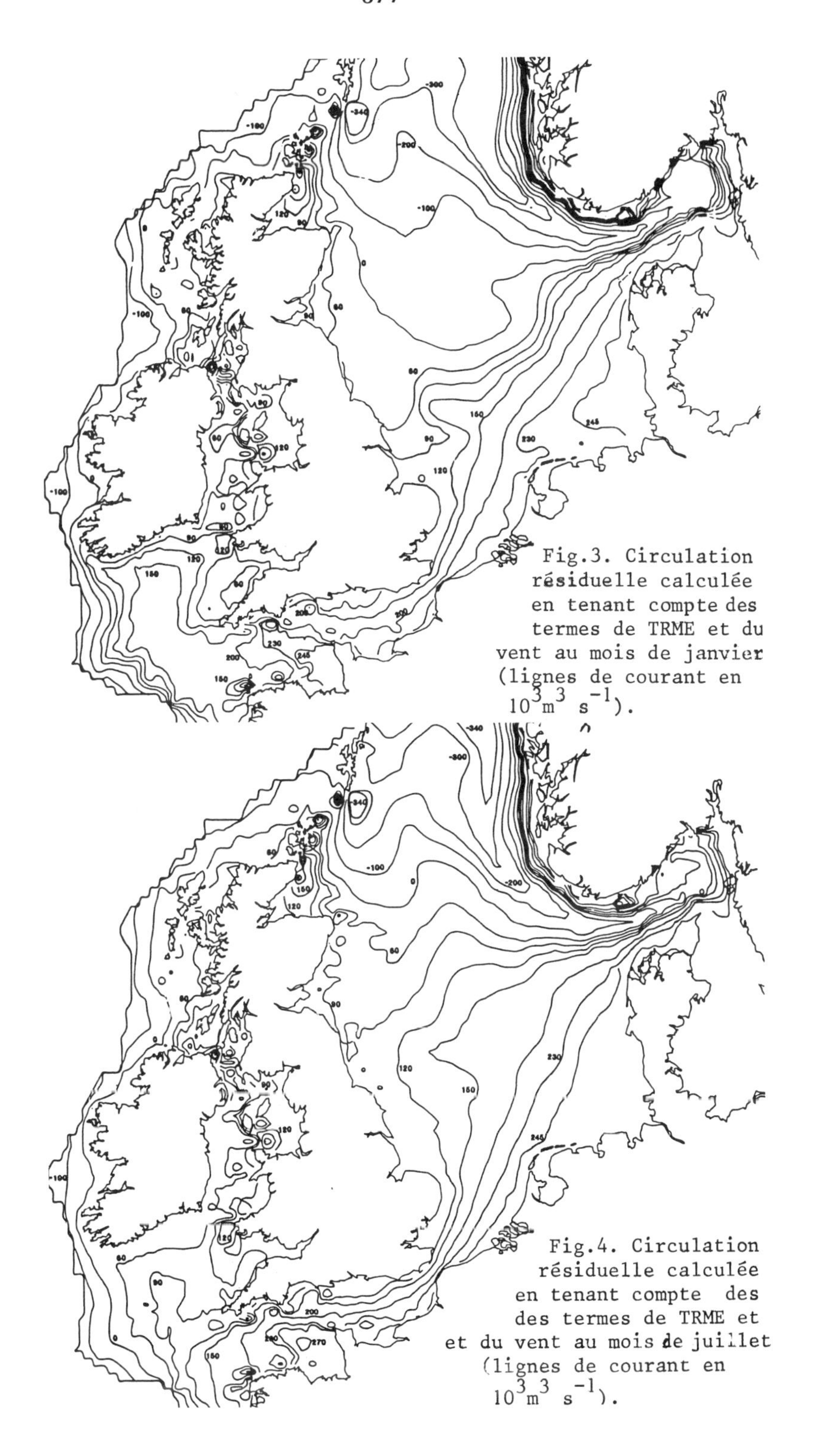

Fig.3. Circulation résiduelle calculée en tenant compte des termes de TRME et du vent au mois de janvier (lignes de courant en $10^3 m^3 s^{-1}$).

Fig.4. Circulation résiduelle calculée en tenant compte des des termes de TRME et et du vent au mois de juillet (lignes de courant en $10^3 m^3 s^{-1}$).

La grille du modèle numérique couvre la mer du Nord, le Skaggerrak, la Manche, la mer Celtique et la mer d'Irlande. La frontière du côté Atlantique est déterminée par l'isobathe de 200 mètres. Pour tenir compte de manière réaliste de la faible largeur de certains détroits tels que le Pas de Calais et le North Channel, une maille de petite dimension est nécessaire. La finesse de la maille (10' x 10') permet non seulement de mieux épouser la géométrie de la côte, mais aussi une meilleure représentation de la topographie du fond

La quasi homogénéité de la colonne d'eau, due aux faibles profondeurs et à l'intensité du brassage par les marées et les tempêtes, autorise l'utilisation de modèles 2D + 1D dont les équations ont été présentées en détail dans des publications antérieures (Nihoul et Djenidi, 1987, Djenidi et al, 1986). Les conditions aux limites ont également été dûment discutées (Djenidi et al, 1985).

Si on néglige les termes de TRME, ce qui revient à ignorer les interactions non linéaires des marées et des tempêtes et à considérer que la circulation résiduelle n'est principalement forcée que par les échanges avec l'Atlantique, on obtient le schéma de circulation de la figure 1. Il représente les grandes lignes de la dérive des masses d'eau induite par les deux branches du courant Nord Atlantique mais ne fait pas apparaître la structure fine et les écoulements secondaires révélés par les observations.

La prise en compte des termes de TRME dans des conditions de vent négligeable montre par contre une circulation beaucoup moins "lisse" associée à des méandres et des gyres, surtout dans les régions à géométrie tourmentée et à forts gradients de profondeurs (Manche, Mer d'Irlande, ...) où la tension de marée joue un rôle important (figure 2).

Pour mettre en évidence les effets du vent, deux situations météorologiques typiques des saisons d'hiver et d'été ont été sélectionnées. Les champs éoliens utilisés sont ceux des mois de janvier et juillet 1976 donnés par le modèle atmosphérique de l'Impérial Collège de Londres. Nous nous sommes intéressés à la circulation résiduelle "climatique" décrite plus haut mais, par souci d'économie de calculs sur ordinateur, nous nous sommes limités à la borne inférieure de l'échelle de temps du phénomène, soit à des périodes d'une semaine.
Les figures 3 et 4 qui représentent les deux cas étudiés montrent comment les conditions météorologiques modifient, en mer du Nord, l'évolution des deux branches du courant nord-atlantique entrant par la Manche et entre la Norvège et l'Ecosse. On peut également constater, d'une situation à l'autre, que la localisation et l'extension des structures tourbillonnaires résiduelles sont nettement perturbées en Mer du Nord, en Manche et en Mer Celtique, alors qu'elles restent peu affectées en Mer d'Irlande. Il apparait aussi que la situation d'été se rapproche sensiblement de celle où les conditions atmosphériques sont négligées.

TRANSPORT DES RADIO-ELEMENTS

D'après leurs mesures des concentrations de ^{137}Cs dans l'eau de mer, effectuées de 1971 à 1978, Kautsky et Murray ont déterminé les trajectoires moyennes du ^{137}Cs en mer du Nord, avec quelques variantes (Kautsky et al., 1981). De même, les mesures de concentrations de ^{90}Sr et ^{137}Cs en mer d'Irlande, à proximité de Sellafield, ainsi qu'à des distances variables du

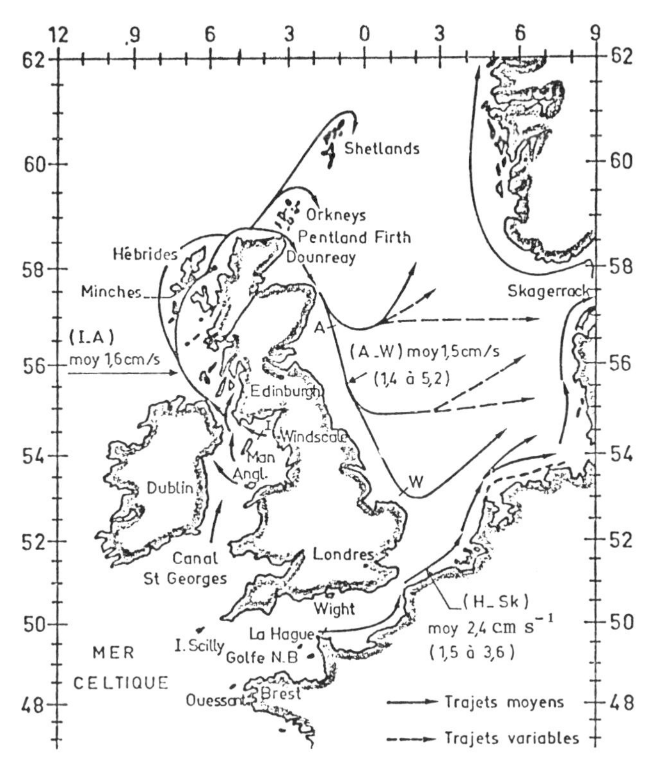

Fig.5. Trajets moyens des radio-isotopes dans l'eau en mer du Nord et en mer d'Irlande (d'après Kautsky et Murray, 1981, et d'après Mauchline, 1980).

Temps de transport	D'après Mauchline	D'après les résultats du modèle
- De LA HAGUE à la Côte Néerlandaise	~ 15 mois	8 à 15 mois
- De LA HAGUE à DOUVRES	~ 3 mois	3 à 10 mois
- De LA HAGUE au SKAGGERAK	18 à 21 mois	16 à 23 mois
- De WINDSCALE au MINCH NORD	0,7 à 1,6 ans	~ 15 mois
- De WINDSCALE au Large d'ABERDEEN en faisant le tour de l'Ecosse (~ 1000 km)	~ 2 ans	~ 21 mois
- De WINDSCALE à La BALTIQUE	4,5 à 5 ans	3 à 7 ans

Tableau 1. Temps de transport d'après Mauchline (1980) et d'après les résultats du modèle.

point de rejet, ont permis de suivre les déplacements de l'effluent, qui (semblent dépendre) des conditions météorologiques et des caractéristiques hydrographiques variables de la mer d'Irlande (Mauchline, 1980); les trajets moyens sont indiqués Fig. 5.
Quelques estimations concernant les temps de transport sont rassemblées dans le tableau 1.

Si les circulations résiduelles déterminées à l'aide du modèle hydrodynamique permettent d'apprécier, aussi bien qualitativement que quantitativement, le déplacement des masses d'eau, elles pourraient également servir à évaluer le temps que mettrait une particule pour aller d'un point à un autre du domaine.
Nous avons essayé d'estimer des temps de transit (de Windscale et de La Hague jusqu'au Skaggerrak) pour un traceur de masses d'eau tel que le ^{137}Cs en supposant que son transport résulte uniquement de l'advection par les courants résiduels. Les trajets considérés correspondant à ceux de la figure 5 suivent en gros des lignes de courant particulières, et les vitesses utilisées pour les différents tronçons du parcours sont extraites des champs de courants résiduels calculés par le modèle.
Ces estimations sont portées également dans le tableau 1 et semblent en assez bon accord avec celles de Mauchline. Les valeurs extrêmes en ce qui concerne les temps de transport à partir de La Hague correspondent aux cas où la particule pénètre d'abord dans le bassin Breton-Normand ou non.

Si la connaissance des temps de transport présente un intérêt certain pour aborder le problème du devenir des éléments radioactifs véhiculés par les masses d'eau, elle reste néanmoins insuffisante pour répondre à la question suivante : quelle est la distribution dans l'espace et dans le temps d'un radio-élément rejeté en mer ?
Aux échelles de temps des phénomènes hydrodynamiques présents en mer, sont associées des phases différentes de la dispersion d'un polluant.
A la micro-échelle correspond un phase typiquement tri-dimensionnelle commandée par les caractéristiques du rejet et concernant des distances de quelques dizaines à quelques centaines de mètres (la phase intrinsèque); cette première dilution dans le champ proche est généralement abordée par des modèles analytiques de jet ou par une équation de diffusion simplifiée, suivant l'importance du débit.
A une distance suffisante de l'émissaire, l'effluent est pris en charge par l'environnement et sa dispersion bidimensionnelle est assurée par l'advection (courants de marée) et la diffusion par effet cisaillant; sont temps caractéristique est de l'ordre du jour (méso-échelle) et elle intéresse des distances équivalentes à l'excursion de la marée (10 à 20 km pour les mers à fortes marées).
Le devenir à long terme (plusieurs semaines à plusieurs mois, voire plus) et à longue distance (~ 100 à 1000 km) est plus délicat à prédire car les mécanismes d'interaction entre les radio-éléments et le milieu marin, notamment avec les sédiments en suspension et ceux du fond, ont une influence marquée et sont complexes. Cette dispersion est due à l'advection par les courants résiduels et une diffusion à grande échelle comportant non seulement la diffusion par effet cisaillant, mais encore une "diffusion" globalisant les phénomènes à méso-échelle non pris en compte.

Si on se limite dans une première étape à la fraction soluble (^{137}Cs par exemple), on peut décrire l'évolution spatio-temporelle d'un élément radio-actif par une équation d'advection-diffusion pour la concentration c en cet élément considéré comme semi-passif; les interactions sont globalement

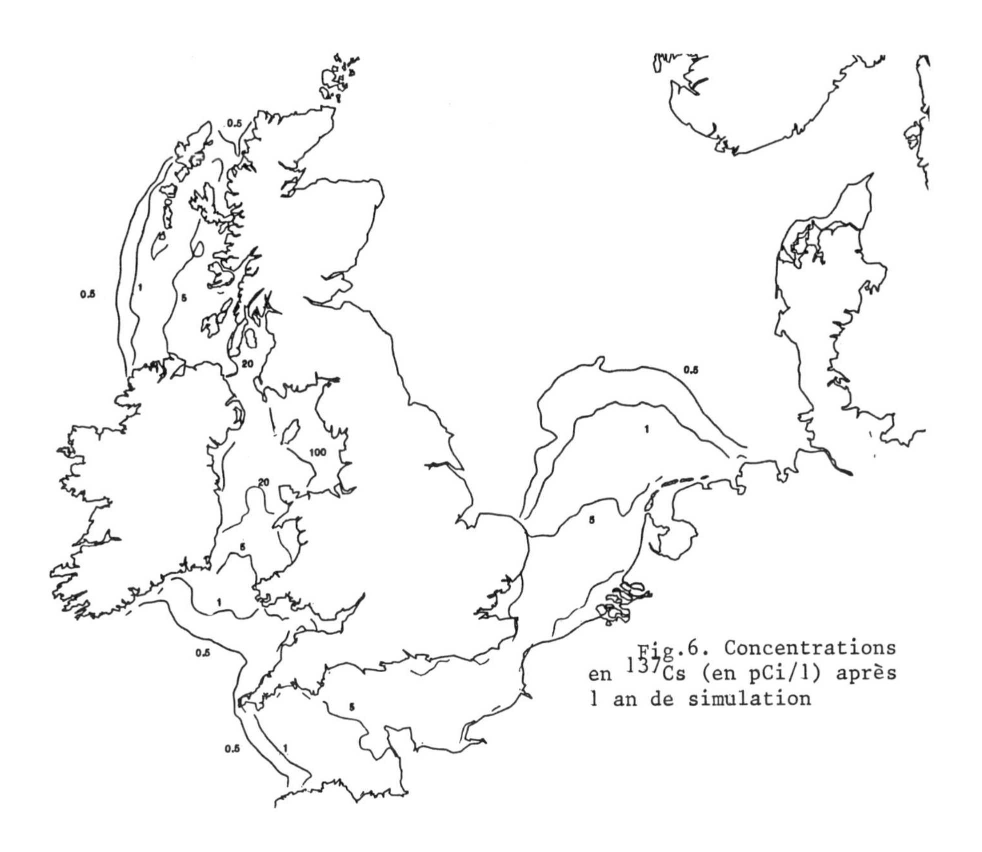

Fig.6. Concentrations en ^{137}Cs (en pCi/l) après 1 an de simulation

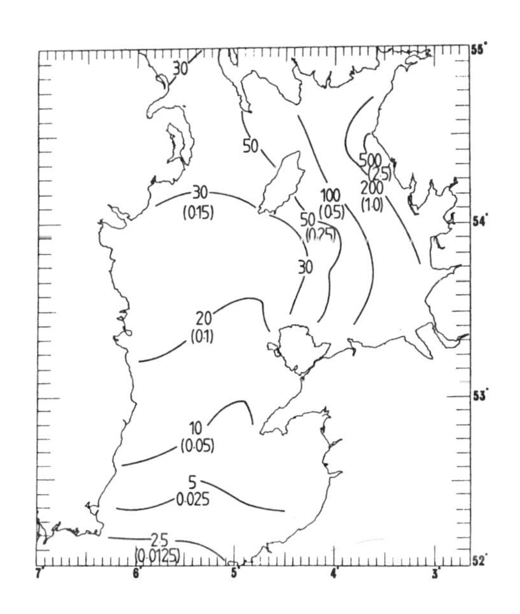

Fig.7. ^{137}Cs en mer d'Irlande Valeurs moyennes de 1970 à 1978 en pCi/l (Jefferies et al, 1982)

représentées par un terme de destruction qui n'est fonction que de la concentration,et faisant intervenir la période de vie de l'élément $Q = -\alpha c$.

$$\frac{\partial c}{\partial t} + u \frac{\partial c}{\partial x} + v \frac{\partial c}{\partial y} = Q + \frac{\partial}{\partial x} (K_x \frac{\partial c}{\partial x}) + \frac{\partial}{\partial y} (K_y \frac{\partial c}{\partial y})$$

où u et v sont les courants résiduels,
K_x et K_y les coefficients de diffusion.

Une première estimation de la dispersion du ^{137}Cs à partir des centrales de retraitement de Windscale et de La Hague a été réalisée moyennant les hypothèses suivantes :

- les concentrations initiales sont nulles dans tout le domaine.
- les concentrations sont nulles aux limites en mer ouverte pendant toute la période simulée (un an).
- les coefficients de diffusion sont constants et égaux, $K_x = K_y = 500\,m^2s^{-1}$; Bowden (1950) proposait $500<K<900\ m^2s^{-1}$ pour la mer d'Irlande.
- les effluents sont rejetés en continu à Windscale (15 000 Curies par mois) et à La Hague (600 Curies par mois); ce sont les valeurs maximum trouvées dans la littérature (Garnier, 1985) pour le ^{137}Cs.

La figure 6 représente les résultats des calculs après un an de simulation (isocourbes de 100, 20, 5, 1, 0.5 pico-curies par litre d'eau de mer) Si la durée de cette simulation est trop courte pour voir le ^{137}Cs "occuper" toute la mer du Nord avec des concentrations significatives, un état "quasi-stationnaire" semble par contre atteint en mer d'Irlande. On est alors tenté d'établir une comparaison avec la moyenne sur une dizaine d'années (1970-1978) des mesures de Jefferies et al (1982), représentée à la figure 7.
Malgré l'intérêt que présentent ces premiers résultats, il était inutile de poursuivre plus longtemps la simulation à cause des limitations imposées par les hypothèses faites. En effet, en plus du fait que les quantités rejetées ne sont pas les quantités réelles, les effluents atteignent les limites géographiques du modèle et il est impératif alors de connaitre les échanges avec l'Atlantique. De plus, les coefficients de diffusion, anisotropes et spatialement variables,doivent être estimés de manière beaucoup plus fine. Si les ordres de grandeur de ceux-ci semblent corrects pour la mer d'Irlande, ils sont largement surestimés pour la mer du Nord (Maier-Reiner, 1979).

REFERENCES

Ancellin J., Guéguéniat P. & Germain P., 1979 - Radioécologie marine. Etude du devenir des radionucléides rejetés en milieu marin et applications à la radioprotection. Paris : Eyrolles.

Bowden K.F., 1950 - Processes affecting the salinity of the Irish Sea. Mon. Not. R. Astron. Soc., Geophys. Suppl., 6, 63-90.

CEE, 1979 - Commission des Communautés Européennes, Méthodologie pour l'évaluation des conséquences radiologiques des rejets d'effluents radioactifs en fonctionnement normal, Rapport CEA-NRPB, Doc V/3865/79-FR - EN.

Djenidi S., Ronday F., Beckers Ph. & Pirlet A., 1985 - Etude de la circulation résiduelle en vue de l'évalution de la dispersion d'effluents radioactifs dans l'environnement marin (Plateau Continental Nord-Européen). Rapport SC-012-B/BIAF/423 - F (SD)/DN01, Commissariat à

l'énergie Atomique, 107 pp.
Djenidi S., Nihoul J.C.J., Ronday F., Garnier A., 1986 - Modèles mathématiques de courants résiduels sur le Plateau Continental Nord-Européen. In : La Baie de Seine, IFREMER. Actes de Colloques n.4, 73-84.
Garnier A., 1985 - Choix de méthodes d'évaluation de l'impact à longue distance de rejets radioactifs en milieu marin. Oceanis, vol. 11, 6, 597-626.
Guéguéniat P., Auffret J.P. & Baron Y., 1979 - Evolution de la radioactivité artificielle gamma dans des sédiments littoraux de la Manche pendant les années 1976-77-78, Oceanologica Acta, 2, 2 , 165-180.
Jefferies D.F., Steele A.K. & Preston A., 1982 - Further studies on the distribution of ^{137}Cs in British coastal waters. I : Irish Sea, Deep Sea Res., 29, 6A , 713-738.
Kautsky H. & Murray C.N., 1981 - Artificial radioactivity in the North Sea. Revue de l'Energie atomique, Suppl. 2, 63-105.
Lochard J. et al., 1981 - L'effet de site sur l'impact des radionucléides rejetés dans le milieu marin. - Coll. Intern. sur l'impact des radionucléides rejetés dans le milieu marin, Vienne, Octobre 1980.
Luyckx F., Broomfield M & Camplin W.C., 1982 - The radiological impact of routine releases from nuclear fuel cycle operations within the European Communities. - Congrès SFRP, Avignon, octobre 1982.
Maier-Reimer E., 1979 - Some effects of the Atlantic Circulation and of River Discharges on the Residual Circulation of the North Sea. Dt. Hydrogr. Z., 32, 126-130.
Mauchline, 1980 - Artificial radioisotopes in the marginal seas of North-Western Europe. In : Elsevier Oceanography Series (F.T. Banner ed.), 24B, 517-542.
Nihoul J.C.J., 1985 - Modelling of Marine Systems, Elsevier Publ. Co., Amsterdam, 272 pp.
Nihoul J.C.J. & Djenidi S., 1987 - Perspective in three-dimensional modelling of the marine system. In : Three-dimensional Models of Marine and Estuarine Dynamics, J.C.J. Nihoul & M. Jamart Editors, Elsevier Publ. Co., Amsterdam, 1-33.

A LAGRANGIAN MODEL FOR LONG TERM TIDALLY INDUCED TRANSPORT AND MIXING. VERIFICATION BY ARTIFICIAL RADIONUCLIDE CONCENTRATIONS

J.C. Salomon
Institut Français de Recherche pour l'Exploitation de la Mer
B.P. 337 - 29273 BREST CEDEX
FRANCE

P. Guegueniat
Commissariat à l'Energie Atomique - Laboratoire de Radioécologie Marine
B.P. 270 - 50107 CHERBOURG

A. Orbi
Faculté des Sciences de Brest
29200 - BREST
FRANCE

Y. Baron
Commissariat à l'Energie Atomique - Laboratoire de Radioécologie Marine
B.P. 270 - 50107 CHERBOURG
FRANCE

ABSTRACT

The paper presents a new way of modelling long term tidally induced water movements in coastal environments. The procedure successively involves a classical Eulerian phase and a transition into barycentric coordinates. Taking the example of the Channel Islands region, calculations are compared with observed ^{125}Sb activities. Both results are in good agreement, confirming the interest of the two techniques and the reciprocal enrichment they can provide to the other.

INTRODUCTION

Long term marine water movement is a natural phenomenom responsible for the transportation of a wide variety of physical, chemical, geological and biological properties. Thus it is of great importance for human and marine life. Unfortunatly, its determination in the coastal zone is often very difficult, especially in tidally dominated regions, and our knowledge of it generally remains poor.

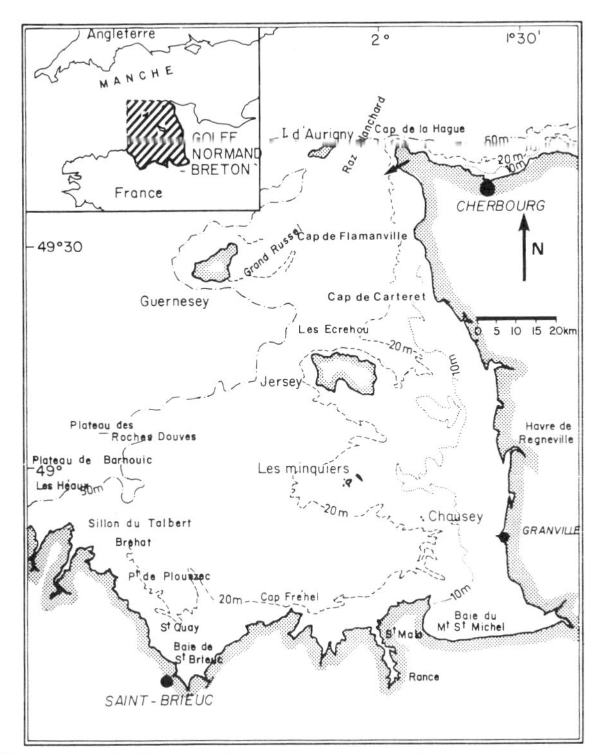

Figure 1. Bathymetric map of the Golfe Normand Breton

Considering the case of the Golfe Normand-Breton (G.N.B.) and taking account both of its size (more than 100 km) and of the usual order of magnitude of the mean currents (about 5 cm s^{-1}), one easily deduces that time scales involved range from one month to about a year.

For such durations usual measuring devices like current meters and even drifters are generally inadequate. Remote sensing has the advantage of giving synoptic insights, but to the authors knowledge there is no detectable parameter conservative at these time scales. Finally, long period radioactive tracers seem to be one of the very few experimental technologies suitable for this problem.

On the other hand, there are theoretical ways, amongst which is numerical modelling. The technique has already proved its efficiency in short term circulation problems but is hardly tractable in the present case because of the very high ratio between strong instantaneous velocities and their weak residual components, and because of the long term simulations which are needed.

Our purpose here is to present an operating numerical technique adapted to perform these calculations and to compare the results obtained in the G.N.B. with in situ radionuclide concentrations.

RADIONUCLIDE DETECTION

La Hague fuel reprocessing plant discharges artificial radionuclides near the headland extremity (fig. 1). Among the various chemical constituants ^{125}Sb (T = 2.7 Y)is considered as a good tracer of the water masses and thus gives the opportunity of describing their movements and mixing conditions (Guégueniat et al., 1988).

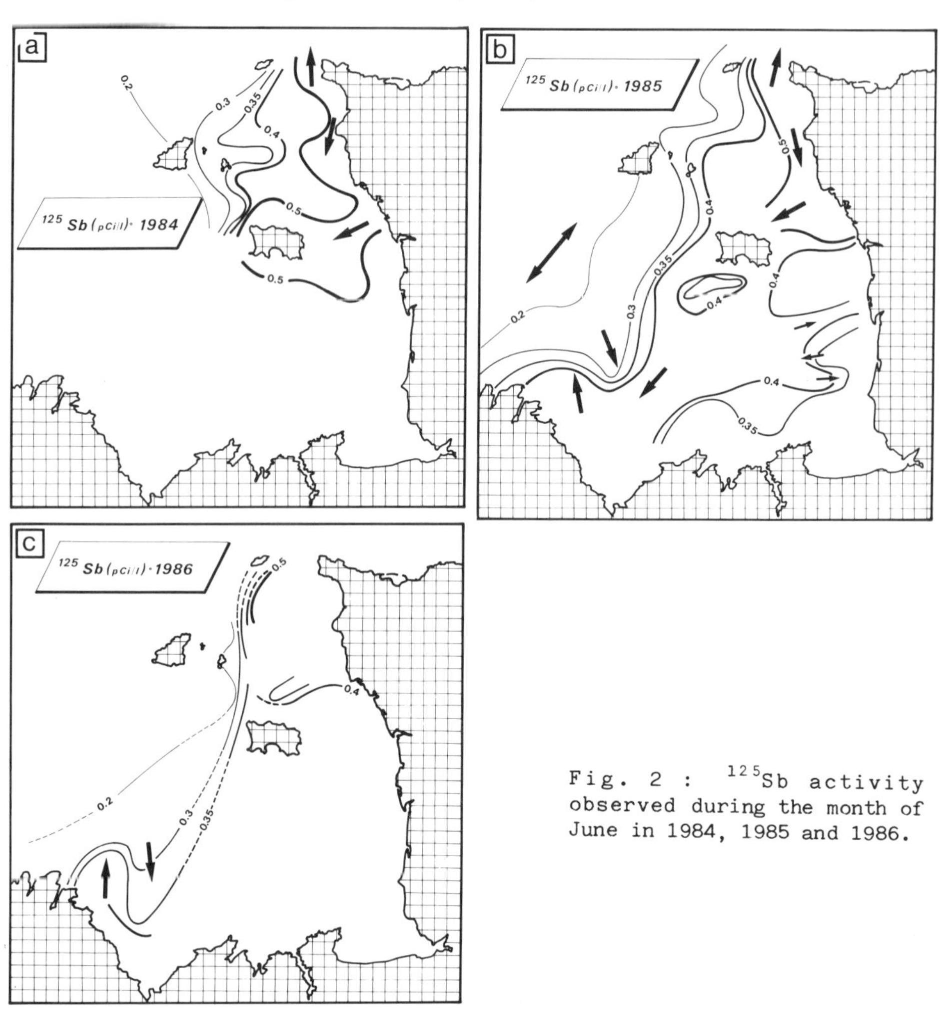

Fig. 2 : ^{125}Sb activity observed during the month of June in 1984, 1985 and 1986.

Fig. 2 presents the results of three annual cruises conducted in 1984, 1985 and 1986 :

1984 cruise was essentially dedicated to the northern part of the G.N.B. Figure 2a shows the sharp decrease of concentrations towards the west, which implies a dominant North-South current, except in front of Cap de Carteret where high concentrations suggest a westward movement.

This outline is confirmed and extended to the centre and the south of the G.N.B. by 1985 measurements (figure 2b).

Isolines indicate :

- A general S.W. - N.E. current in the western part of the gulf.
- A southward water flux along the coast, north of Carteret, then oriented west, and later south-west.
- A probable vortex near Brehat and smaller ones near Les Minquiers.

The last cruise (1986, Figure 2c) also confirms the marked gradient in the region of Alderney and the vortex eastward of Brehat.

NUMERICAL MODELLING OF WATER MOVEMENTS

METHODOLOGY

Tides are the major hydrodynamic phenomenom in the G.N.B. Their range reaches 14 m in extreme spring conditions. On the contrary, fluvial discharge is insignificant. As a result, waters are usually well mixed and currents nearly the same over most of the water depth.

In such a local situation dynamic considerations indicate that long term movements can be induced both by wind action and by tides through non linear terms. Previous measurements and computations (Le Hir et al., 1986) have alreardy shown that wind induced currents are relatively weak and essentially restricted to the shallowest regions, along the coast. Thus, only the tidal action will be considered here.

For simplicity let us first consider the tide to be periodic and simulate the long term tidally induced currents with a two-dimensional model using a classical finite difference formulation (Salomon and Le Hir, 1980 ; Orbi, 1986). Time integration of computed tidal velocities directly results in an Eulerian residual velocity field, but this is far from representing the long term water movements we are looking for : the next step is to convert this into a Lagrangian point of view by creating many labelled water parcel trajectories. As tidal velocities are already known, this can be done in a straight-forward manner by hourly releases at the Eulerian grid point, but the summation of tidal currents has to be done cautiously. In coastal zones, where velocity gradients are high, a few (at least six) points of the initial computational grid seem to be necessary along each trajectory. For a sinusoidal particle velocity with an amplitude Vmax, this simple consideration leads to the following general upper mesh size limit :

$$DX < 2\,000 \,.\, V_{max}$$

Consequently the spatial discretization used here is one mile.

These trajectories are different at the same geographical location (fig. 3), according to the instant of release. This means that the usual first order approximation of Lagrangian velocities, being the sum of Eulerian velocities and Stokes drifts, is not valid here (fig. 4) :

$$\vec{V}_{RL} = \vec{V}_{ER} + \vec{V}_{SD} + \vec{V}_{LD}$$

$\vec{V}_{RL}$: Lagrangian residual

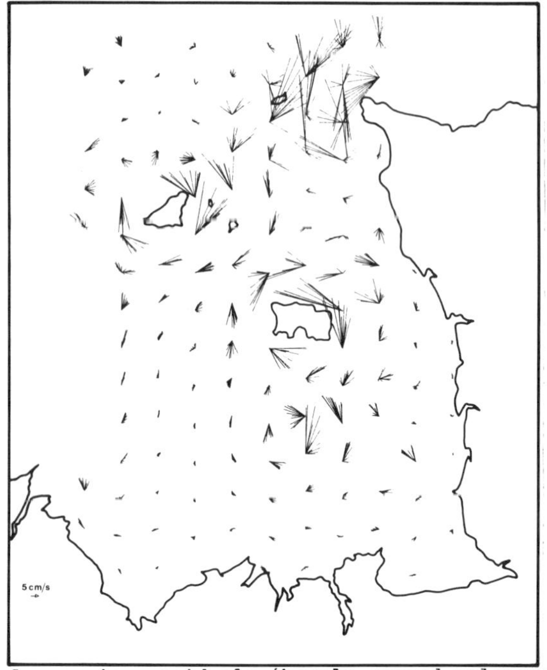

Fig. 3 : Lagrangian residuals (hourly spaced releases)

$\vec{V}_{ER}$: Eulerian residual

$\vec{V}_{SD}$: Stokes drift

$\vec{V}_{LD}$: Lagrangian drift

Analytical expressions of the Lagrangian drift can be obtained in simple theoretical cases (Cheng et al., 1986 ; Feng, 1987). For example Cheng et al. (1986) demonstrated that, in case of a Sverdrup wave, the Lagrangian drift describes an ellipse. But the reality in the G.N.B. is much more complicated and, to cope with this difficulty, we introduce barycentric coordinates :

In these coordinates, the residual velocity vector is placed at the centre of gravity of the trajectory of a water parcel released from a particular point at a particular time (Figure 5, Point A, time t_1). A vector for a parcel released at the same point but at a different time (Point A, time t_2) is placed at a different location in the barycentric coordinate system because it follows a different trajectory. Water parcel released at widely separated locations (Figure 5, point B) may end up close to each other in this system if the time of release (t_3) is such that the trajectories are similar. In this way groups of water parcels released at the same point become widely scattered and the velocity vectors associate with similar ones from widely separated locations (figure 6).

An interpolation procedure is then used to go back to a regularly defined velocity field (fig. 7).

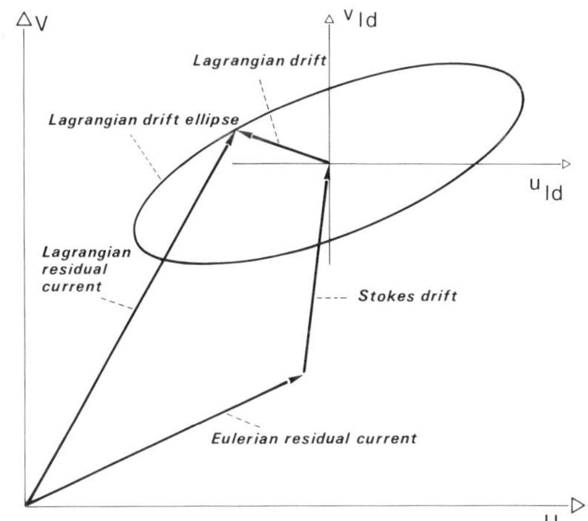

Fig. 4 : Schematic diagram of Lagrangian residual components

This process sometimes fails near sharp headlands and islands, where centers of trajectories may be situated on land. Such trajectories have been eliminated here.

Except for this limited drawback, which it seems quite easy to overcome, the method leads to a unique residual Lagrangian velocity field and preserves most of the information obtained previously. It seems better than a simple averaging of residual Lagrangian drifts (Feng, 1987) which would have eliminated some of the velocity features appearent on fig. 3, such as minima and maxima of bundles of vectors.

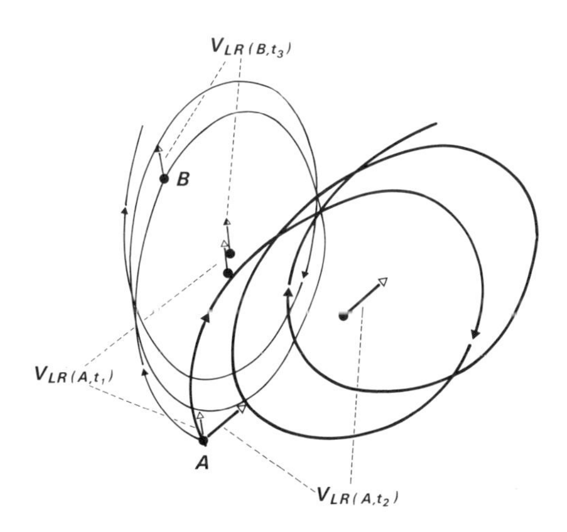

Fig. 5 : Barycentric system

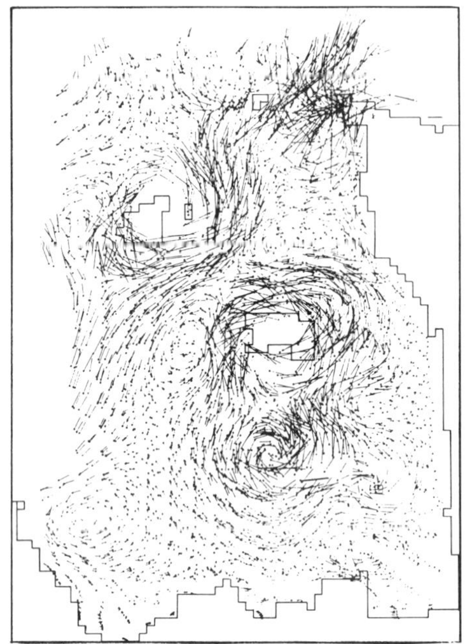

Fig. 6 : Lagrangian residuals in barycentric coordinates

RESULTS

Results presented here are for an average tide.

A complementary presentation of the velocity field (fig. 7) is given fig. 8 by means of trajectories and isotachs. Each representation provides indications on water movements in the G.N.B. :

- strong cyclonic circulations around islands and shallows (Guernesey, Jersey, Minquiers, Chausey),
- sharp vein crossing obliquely across the Gulf from Brehat to "le Raz Blanchard",
- anticyclonic circulations between Flamanville and Carteret,
- double system of coupled vortices, anticyclonic in the bay of St Brieuc and west of Jersey, cyclonic around Jersey and Les Minquiers.

Current structures near the open frontiers of the model, in the north-west part of the G.N.B., are not considered significant, probably arising from boundary problems.

All these features are coherent with artificial radionuclide maps presented above (fig. 2). The Lagrangian model thus explains how things proceed :

The greatest part of discharged material goes North-East and escapes from the Gulf, but a certain part of it is trapped in the eastern branch of Flamanville vortex and flows south to Carteret. Then it turns west as far as a point situated about half way between Jersey and Guernesey where

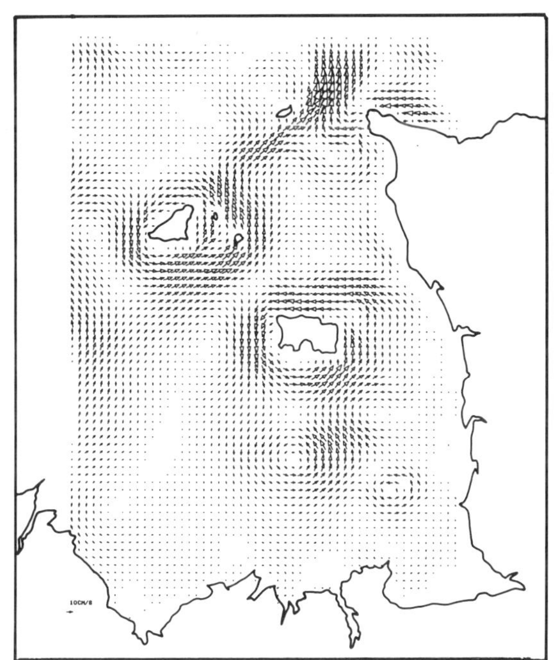

Fig. 7 : Lagrangian residual velocity field
(average tide - barycentric coordinates)

it diverges. Part of it goes north and will either leave the region or be recirculated once more. The remainding part goes south and spreads the southern part of the gulf, by advection through the various existing gyres, and by diffusion.

Another partial confirmation of this residual velocity field has recently been obtained from the observation that when a chemical tracer is released at La Hague, its persistency in neighbouring waters is observed for about six weeks. This is also the time lag predicted from model results, for a particle to describe the anticyclonic vortex situated between Guernesey and Flamanville.

In this scheme, with no wind, non marked water penetrates the southern part of the Gulf between Jersey and Guernesey, and also a small fraction between the two coupled vortices situated north of St Brieuc. This last intrusion is more pronounced during spring tides and, of course, during westerly wind periods.

Travel and residence times, evaluated from fig. 7 and 8, are as follows : movement around Jersey or Guernesey takes about a week ; a traverse of the G.N.B. from Brehat to La Hague needs a little more than one month (5 to 6 weeks) and the same time is required for the anticyclonic gyre in front of Flamanville. Compound circuits in the southern gulf are described more slowly and require 3 to 4 months, whereas the renewing of the more littoral waters through the tidal action alone would need even longer time. As wind action has not been included in computations, this last result has probably no physical significance.

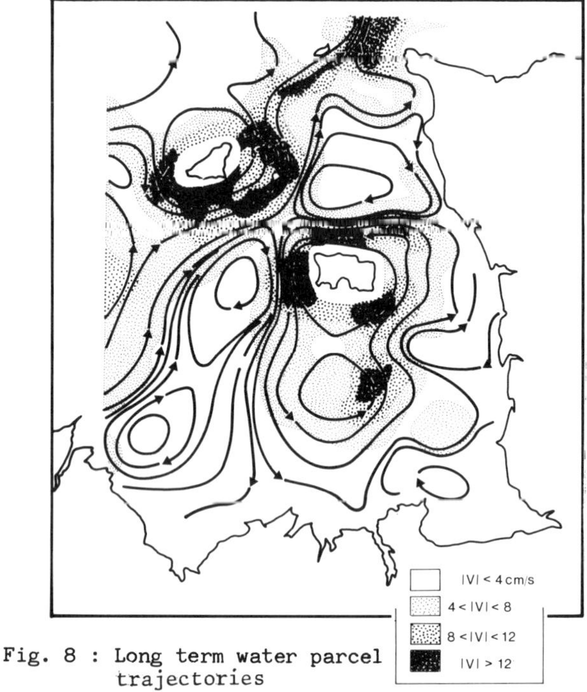

Fig. 8 : Long term water parcel trajectories

Bringing together model results and Antimony concentrations give some additional information on the efficiency of La Hague discharge scenario : Spatial integration of concentrations shown in fig. 2 indicates the total ^{125}Sb activity in the G.N.B. is about 4 TBq. As the total release is 100-150 GBq per year (Guéguéniat et al., 1988), and as a rough estimation of residence time of marked waters may be more than four months (less in the northern part, but much more in the South), one easily deduces that about 5 % of the radionuclides enter the G.N.B. before being evacuated into the eastern Channel.

TRANSPORT - DISPERSION SIMULATION

The Lagrangian residual velocity field can also be used to simulate the long term movement and spreading of dissolved matter in an efficient and economic way.

The advection-dispersion equation, written in Lagrangian form, is as follows :

$$\frac{DC}{Dt} - Kx \frac{\partial^2 C}{\partial x^2} - Ky \frac{\partial^2 C}{\partial y^2} = 0$$

C : concentration of dissolved matter
t : time
x, y : spatial coordinates
Kx, Ky : dispersion coefficients

DC/Dt represents the evolution of concentration along a particle trajectory. On the condition that a slight different meaning of the dispersion coefficients (already poorly known quantities) be accepted, this equation may be solved using the Lagrangian residual velocities, and ignoring the various instantaneous velocity fields. As these velocities are much smaller than tidal ones, standard numerical methods which usually involve some limitations based on the Courant number, will allow considerable increase in integration time step.

An example of results obtained after one year simulation is presented in fig. 9. The simulation is very schematic as a single velocity field has been used (corresponding to a medium tide), as wind action has not been included in calculations, and as no adjustement of dispersion coefficients has been done. All these points could easily be improved in the future. Nevertheless, keeping in mind the high degree of approximation made in drawing fig. 2, simulation and observation results show a surprisingly fair agreement : Highest concentrations along the Normandy coast, north of Port Bail, a sharp frontier in front of Carteret, another frontier crossing diagonally the G.N.B. between the Channel islands, and a plume going south and then spreading in opposite directions are features consistent with observations.

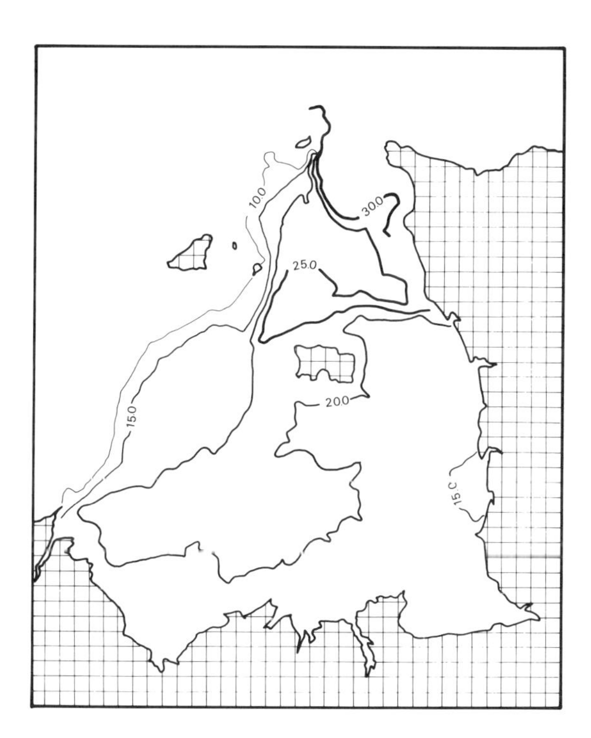

Fig. 9 : Concentrations distribution after one year simulation of hypothetical release at La Hague (arbitrary unit).

CONCLUSION

The main object of this study was to present and test a different way of modelling long term water movements and dissolved matter dispersion in tidally dominated environments. The process successively involves an instantaneous Eulerian hydrodynamic model, then a transition into Lagrangian residuals, the introduction of barycentric coordinates and the resolution of the transport-dispersion equation.

The only data used in the calculations are the bathymetry of the region and the tidal elevations along the open boundaries.

As the process is quite complicated, an assessment of intermediate and final results is needed. This can easily be achieved for the first steps by means of current meters and tide gauges, but with much more difficulty for the following steps which concern long term movements. Radionuclides can play this part efficiently. Results obtained with ^{125}Sb were generally in good agreement with simulations and thus confirmed the modelling technique.

Reciprocally, radionuclides concentrations maps may be difficult to understand in all details and often raise many new questions. Models are then generally the required technique to give the appropriate answers.

The example related here clearly shows that both methods are susceptible of a reciprocal enrichment. Their combined use opens promising perspectives.

REFERENCES

1. Feng, S., A three-dimensional weakly non linear model of tide-induced lagrangian residual current and mass-transport, with an application to the Bohai Sea. In Three dimensional Models of Marine and Estuarine Dynamics, ed. J.C.J. Nihoul and B.M. Jamart, Elsevier Oceanography Series, Amsterdam, 1987, pp. 471-488.

2. Cheng, R.T., Feng, S. and Xi, P., On lagrangian residual ellipse. In lecture notes on Coastal and Estuarine studies, ed. J. Van. de Kreeke, Springer-Verlag, Berlin, 1986, pp. 102-113.

3. Guéguéniat, P., Gandon R., Baron, Y., Salomon, J.C., Pentreath, J., Brylinski, J.M. and Cabioch, L. Utilisation des radionucléides artificiels (^{125}Sb, ^{137}Cs, ^{134}Cs) pour l'observation (1983-86) des déplacements de masse d'eau en Manche. Radioactivity and Oceanography. Elsevier oceanography series, Amsterdam, 1988.

4. Le Hir, Bassoullet P., Erard, E., Blanchard, M., Hamon, D., Jegou A.M., Etude régionale intégrée du Golfe Normand-Breton. Rapport Scient. IFREMER, Brest, 1986, I, 265 p.

5. Orbi, A., Circulation de marée dans le Golfe Normand-Breton. Thèse UBO, Brest, 1986, 225 p.

6. Salomon, J.C. and Le Hir, P., Etude de l'estuaire de la Seine. Modélisation numérique des phénomènes physiques. Rapport Scient. CNEXO/UBO, Brest, 1980, 286 p.

MODELLING THE DISTRIBUTION OF SOLUBLE AND PARTICLE-ADSORBED RADIONUCLIDES IN THE IRISH SEA

P. A. Gurbutt, P. J. Kershaw and J. A. Durance
Ministry of Agriculture, Fisheries and Food
Directorate of Fisheries Research
Fisheries Laboratory
Lowestoft, Suffolk NR33 OHT
England

ABSTRACT

Radionuclides have been discharged, under authorization, into the north-east Irish Sea from the Sellafield reprocessing plant for more than 35 years. Surveys have revealed that whereas most of the fairly soluble nuclides, such as those of caesium, have already left the Irish Sea, a large percentage of those radionuclides which become associated with particulate materials remains within it, incorporated into the sea bed.

This paper describes the difficulties encountered in setting up a simple box model to predict future radionuclide concentrations based on a large amount of data from within the Irish Sea collected by MAFF over the past 15 years. The model includes parameterizations of the water circulation and mixing, resuspension, advection and deposition of particulate material, bioturbation and variation in sediment type. The combined water and sediment model has been run for $^{239+240}$Pu to demonstrate its use for particle associated nuclides.

INTRODUCTION

Alpha-active plutonium radionuclides have been discharged, under authorization, into the north-east Irish Sea from the Sellafield reprocessing plant for more than 35 years, discharges reaching a peak value of 65.1 TBq in 1973, since when they have continually declined[1]. It appears probable that the sea bed is not, as has been predicted in the past, a long-term sink for plutonium, but it is in dynamic equilibrium with the total system; that is, the sea-bed sediment is a short-term sink and a potential future source of plutonium. At present, the complexity of dynamic interactions between the plutonium in the effluent, sea water, suspended load and sea-bed sediment is not fully understood. Clearly a model - or models - based on the information already available, are needed to indicate where effort and resources can best be deployed to improve our understanding of the environmental behaviour of plutonium and americium and our ability to predict the consequences of past, present and future discharges.

There have been many models of all or part of the Irish Sea, the majority of which have been of the water circulation derived from tides, winds and stratification. There have been relatively few which have been used to study the dispersion of contaminants, and even fewer which have attempted to deal with those contaminants which are scavenged from solution by fine particulate matter. Prandle[2] and Onishi and Thompson[3] have both used the discharge of ^{137}Cs from the Sellafield reprocessing plant as a passive, conservative tracer to validate their models. Whereas Prandle's model just includes water circulation with advection and diffusion of the ^{137}Cs, Onishi and Thompson's model includes a sediment transport model and the sorption and desorption of the contaminant by the particles. However, as ^{137}Cs in sea water has a relatively low affinity for particles, this part of the model can have only a small influence on the results, but, given adequate data about sheer stress of erosion and of deposition of bottom sediment, it could be used to simulate the dispersion of actinides.

Unlike the above, this paper is concerned with the variation in concentration of plutonium and americium in the whole of the Irish Sea, especially the eastern part, in order to predict annual doses resulting from discharged activity for periods from one year to 100 years.

In many modelling problems there is a paucity of data against which to validate model predictions, but fortunately, in this case, there have been many measurements of man-made and natural radionuclides in the marine environment over a number of years. However, the construction of a model and the parameterization of the processes included often reveal areas of ignorance especially about the rates at which various exchanges occur.

A box model approach was adopted for a number of reasons: the timescales of interest in the output are long; it is easier to integrate different parameterizations of sediment scavenging into a box model, bearing in mind the lack of current information about cohesive sediments in the Irish Sea; it is easier to investigate gross perturbations to the system to answer hypothetical questions about the future behaviour of the activity in the sea bed; the model resolution can be refined as more and better data become available; the long lead time to develop a hydrodynamic circulation model is not necessary; and finally computational times for simulations of 100 years are not too long.

MODEL DESCRIPTION

Surveys of plutonium and americium in sea water and sediments of the Irish Sea indicate that there is a concentration gradient away from the discharge point of the Sellafield pipeline[4], with marked tendency for isopleths to run parallel to the English coast. Plutonium has a high K_d in sea water and is, in any case, discharged in association with an iron floc[5]. A map of the bottom sediment type, contoured in terms of the percentage of fine particles less than 62.5 μm[4] reveals that there is a mud patch covering approximately, but not precisely, the same area as the highest concentrations in plutonium. There is a second mud patch to the west of the Isle of Man which does not noticeably affect the contours of plutonium concentration. However, ^{241}Am which is discharged and grows in from ^{241}Pu (half life 14.4 years) has a higher K_d and shows some influence of the second mud patch. As a result of examining these data, the boxes in the model were designed to resolve the gradients around the pipeline, with coarser resolution further away (Figure 1). The mud patch in the western Irish Sea has been divided into two boxes because of the ^{241}Am distribution in the sea bed.

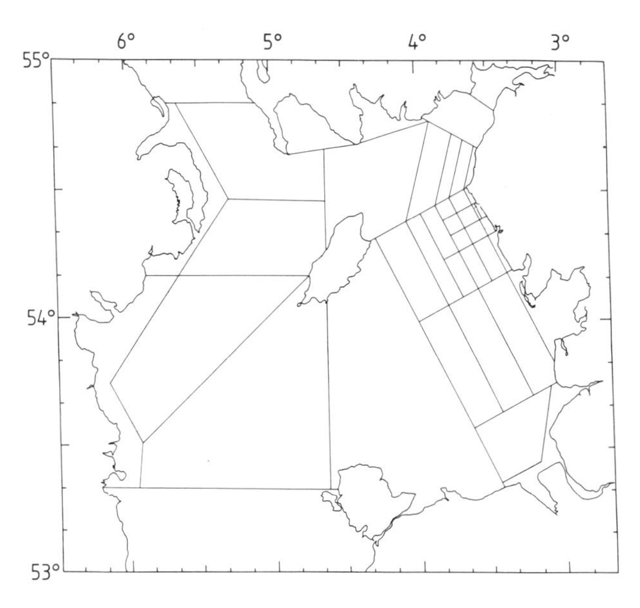

Figure 1. Layout of the numbered boxes for the Irish Sea model.

As one of the main aims of the model is to examine the behaviour of plutonium and americium in the Irish Sea, a model of the movement of the water, while necessary, is not sufficient. The water carries a suspended particulate load which exchanges with the sea bed. Also, the sea bed contains biota which mix the sediment down to as much as 1.5 m in some places in the Irish Sea[6]. Both of these factors influence the distribution of particle associative nuclides.

To reflect this, each geographical area, as seen in Figure 1, has been given the vertical structure as represented in Figure 2. There is a water box occupying the bulk of the water column and what has been labelled an interface box between this and the settled sediments of the sea bed. In the water box, suspended particulates move with the water. Originally the concentration was assumed to be the same in each water box, but this has been changed to take account of the different sea-bed types and the observed decreases in the concentration of suspended particulates away from the coast. In order to conserve particle mass and fixed particle concentrations in the boxes, where there is a convergence

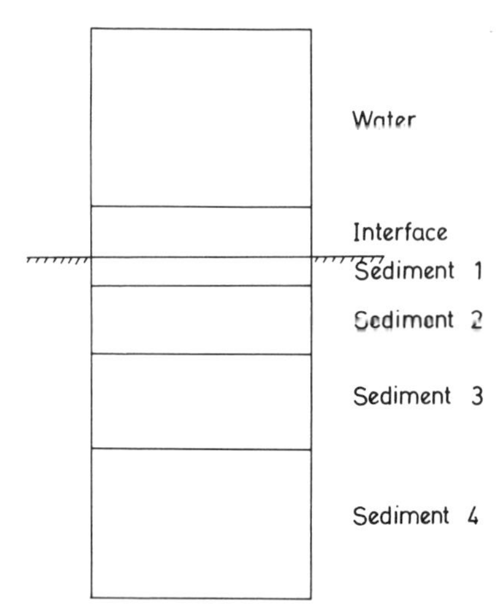

Figure 2. Vertical structure of the model, which can be fitted into each numbered box region shown in Figure 1.

of particles, the excess is deposited, and where there is a divergence, part of the sea bed is resuspended to supply the necessary material. However, in the interface box of thickness 10 cm, a higher suspended load is assumed in an attempt to represent the resuspension and deposition of material which takes place near the bed. The interface box is a means of trying to parameterize the near bed exchanges which some authors[7] model as a diffusive layer acting as a semi-barrier between the water and the sea bed.

The sea bed has been divided into up to four layers in order to represent the gradients in the observed concentrations of radionuclides. Box thicknesses of 20 cm, 20 cm and 1.6 m give a total mixed sea-bed depth of 2 m with a sink for activity below that if there is net deposition of sediment. However, mixing in all places in the Irish Sea does not extend to this depth and so not all regions need to have all the sediment boxes. In general, the muddy areas have all the boxes whereas the sandy areas have only one, plus the sink box.

THE WATER CIRCULATION

There are several ways in which the circulation pattern can be determined for use in this model: from a numerical hydrodynamic model of the Irish Sea, driven by the appropriate wind, heating and cooling and fresh water inputs; from measurements of the currents from the Irish Sea; by inference from the distribution of natural and man-made tracers. For radiological assessment purposes, the shortest period of interest in the model output is one year, so initially it has been assumed that there is an annual mean circulation pattern. This could have been derived in any of the above methods but the tracer method was chosen because the model is trying to predict radionuclide distributions, so the tracer fields should contain all of the averaging appropriate to the dispersion of the activity; the use of current meter measurements requires a comprehensive set of observations from the Irish Sea for a lengthy period (greater than one year), which is not available. However, MAFF is undertaking long-term current measurements in the eastern Irish Sea and has 4.9 meter years of good data from the first year of deployments; circulation models are a major research area and, even if the code of a model were to be obtained, there might be problems producing a suitably averaged flow field because non-linearities in the system mean that a flow generated by an average wind is not necessarily the same as the averaged flow from a fluctuating wind[8]. This approach can only be used when there is an adequate number of data sets. A model set up in this way is not readily transferable to a different geographical region, unlike a numerical circulation model.

^{137}Cs is almost conservative in sea water and, until Chernobyl, there has been only one major source (Sellafield) in the Irish Sea. For these reasons, and because MAFF has a long time series of observations of the distribution of ^{137}Cs in the water, it was selected as the tracer from which to deduce the flow field.

Figure 3 displays the annual discharge and monthly discharge rates (PBq a^{-1}) from Sellafield from 1964 to 1985; longer time scales are given by Pentreath[1]. It is also clear that the amount released varied quite significantly from month to month, with most of the annual discharge occurring over a few months.

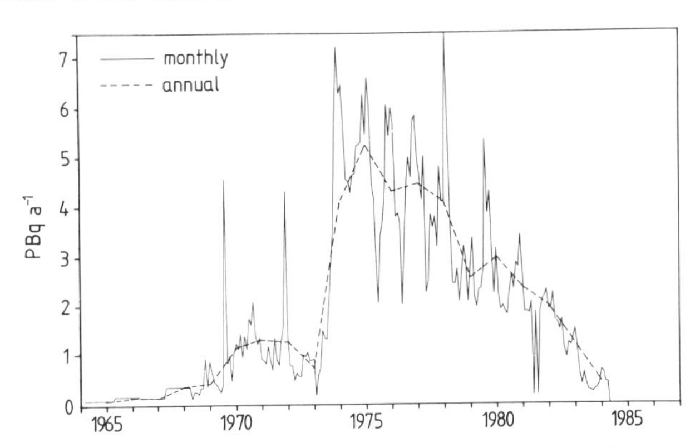

Figure 3. Discharge rates (PBq a^{-1}) of ^{137}Cs from Sellafield for the period 1964 to 1984.

MAFF has regularly measured concentrations of ^{137}Cs in surface, mid and near-bottom waters at many locations in the Irish Sea during the period from 1972 to the present as published in the annual monitoring reports, the latest of which is that of Hunt[9]. These results show that the overall concentrations varied in relation to the annual discharges from Sellafield and that activity tended to leave the Irish Sea through the North Channel, passing to the north of the Isle of Man.

Using the survey and discharge data, the exchanges between the boxes were derived by trial and error to obtain a 'good' fit to the observations, as Camplin et al.[10] did with a simpler box model of the European continental shelf. The gross features of the annual mean circulation are included: the northward flow in the western Irish sea which exits through the North Channel; a branch of this northward flow passing anticlockwise round the Isle of Man; a narrow southerly current along the Irish coast and a southerly drift down the Cumbrian coast towards Liverpool Bay. Mixing between boxes was calculated from an eddy diffusivity of 10^6 cm^2 s^{-1}, the cross-sectional area of the interface between boxes and a mixing length.

The concentrations calculated from this data set fit the observed values reasonably well in the Eastern and Northern Irish Sea, but calculated concentrations to the southwest of the Isle of Man are underestimated by a factor of about five. While this is encouraging it hides two problems: those of spatial and of temporal averaging of the measurements. The model has been set up to use an annual mean circulation pattern so that the results are relevant to periods from one to 100 years. The resolution of the model, even near the Sellafield pipeline, is still fairly coarse, so the predicted concentrations of activity in the water and sea bed are averaged over a large area. However, measurements are taken at one location at one time.

For comparison purposes, the measured concentrations within a model box have been arithmetically averaged, and the mean value assumed to be representative of the concentration in the region throughout the year. Even with this crude approximation, good visual agreement can be seen between plots of the model output and the data, when boxes are block coloured according to their concentration. However, the discrepancies are noticeable, particularly close to the discharge point. Figure 4 shows the predicted concentration in the water of ^{137}Cs in box 5, just

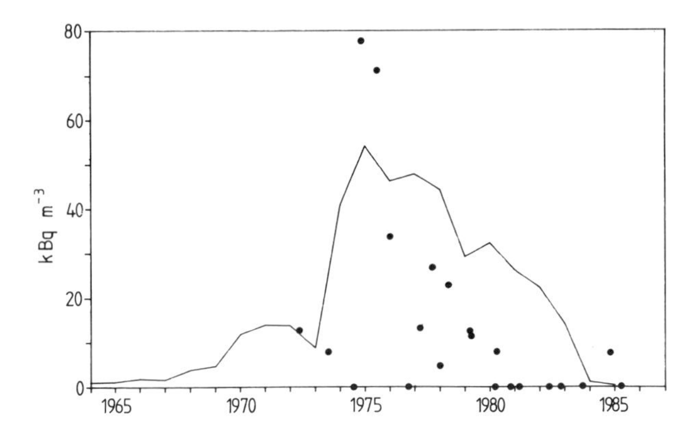

Figure 4. Comparison of model prediction (-) for ^{137}Cs with observations (●).

beyond the end of the pipeline, and the spot measurements taken within the box. Not surprisingly, the concentration in the box closely follows the annual discharges. However, the observations are widely scattered about this line, again not surprising, as the monthly discharge rates fluctuate widely. Indeed, if the concentrations derived from the monthly discharges (Figure 3) are compared with the observations, the agreement is rather better. This example does provide an indication of the variability in the system and provides a measure of the accuracy of the model.

THE SEDIMENT MODEL

Not all radionuclides behave almost conservatively in water, like ^{137}Cs; instead, many are more strongly sorbed by particles. The depth to which activity penetrates into the sea bed depends on the rate of turn-over of the sea bed by biota, the half-life of the nuclide of interest, and the rates of resuspension, deposition, and burial of particles. Their distribution through the Irish Sea also requires a knowledge of how and where suspended particles move, and the amount of activity they carry. Whereas, Kershaw *et al.*[11] deals more fully with the data upon which the model is based, this paper examines some of the limitations and strengths of the approach adopted.

In general, the mixing of the sea bed, and redistribution of radio-nuclides, by biota is frequently modelled as a bio-diffusion process, which pre-supposes certain conditions on spatial homogeneity of the mix-ing[12]. Indeed, this is the approach adopted here, even though the presence in the Irish Sea of a species of Echiurian worm, *Maxmulleria lankesteri* and a species of Thalassinid shrimp, *Callianassa subterranea*, may call this approximation into question. However, from the many ^{234}Th profiles ($t_{\frac{1}{2}}$ = 24 d) in the top 5-6 cm, it is possible to obtain a bio-mixing rate of 42 ± 15 cm^2 a^{-1}[11]. The results from ^{210}Pb ($t_{\frac{1}{2}}$ = 22.2 a) profiles are not so convincing, producing variations of two orders of magnitude, and the profiles not showing exponential decreases with depth, but instead revealing subsurface maxima. These may be related to the effects of burrowing animals and work is in progress at Lowestoft based on that of Boudreau[13] and Smith *et al.*[14] to interpret many of the ^{210}Pb profiles from cores collected in the eastern Irish Sea[11]. Even with this local variation within cores, the bio-diffusion approximation may be appropriate when applied to the large areas defined by the box model.

To check the magnitude of the likely errors in the solution of the diffusion equation with a three-box representation compared with a more accurate spatial discretisation, the one-dimensional equation was solved with different spatial resolution. From this study it can be concluded that the Irish Sea model with just three layers in the sea bed will tend to over estimate the amount of activity in the sea bed by 10% compared with that observed. As there are sampling problems involved in estimat-ing bed inventories from cores of greater than this, it was decided that three layers would be sufficient for the full Irish Sea box model.

Whilst using a model similar to this to find the best fit for the vertical exchanges based on bio-diffusion, it was noticed that results could be improved by reducing the transfer across the sediment-water interface. Nyffeler *et al.*[7] also found a 'diffusive barrier' at the sediment-water interface necessary when interpreting movement of radionuclides in laboratory experiments.

Radionuclides are most rapidly scavenged from solution by fine par-ticulate material, so in order to follow the cycle of uptake and recycling of material in the water and the surface of the sea bed it is necessary to understand the resuspension, deposition and consolidation of cohesive sediments. Whereas the concentration of most material in suspension away from the coast is of the order of 1 mg l^{-1}[15], large amounts of fine material can be resuspended during storms. MAFF has a field programme of measurement of bed shear stress so that the processes can be understood and modelled in detail over short periods. However, the Irish Sea model requires a parameterization of the general effects of resuspension and deposition averaged over longer time periods, from a month to a year.

(15) To this end, a suspended sediment load which varied geographically was added to all the water boxes, with a higher suspended load in the thin interface box, representing a higher bed load. The artificial net deposition of the order of 1 mm a^{-1}, resulting from conservation of suspended particulate material parameterization, is of the order of other likely errors in the model, but because it is always in the same areas the distribution of particle reactive contaminants becomes dominated by these spurious erosions and depositions. This is currently investigated, along with the variation of the percentage of silt, sand and gravel in different regions of the Irish Sea.

If spiked water is shaken up with uncontaminated sediment in the laboratory, the concentration of the radionuclide in water and on sediment gradually approaches an equilibrium. If this equilibrium is reached sufficiently rapidly then the usual K_d approximation can be applied to the partition of activity between water and particles. A simple two-box model of water and sea bed, using a rate on - rate off representation of the kinetics can be used to demonstrate that this equilibrium is reached three times more quickly in the water than the top sediments, but these periods are sufficiently short compared with the timescales of interest in this model for the detailed kinetics to be ignored and the K_d approximation to be used. The K_ds used in these calculations are taken from an IAEA Technical Report[16].

TRIAL SIMULATIONS OF THE PLUTONIUM DISTRIBUTIONS

Thus far, we have described how the model was set up and some of the sensitivity analysis which has already been performed in order to test the assumptions about the parameterizations. The sub-models have been assembled to examine whether there are any additional features which need to be included to describe the behaviour of plutonium and americium. Even though the model has been used to predict concentrations of nuclides in a decay chain, these results are not described here, as there are no additional parameterization problems.

The data for $^{239+240}Pu$ can be averaged onto the box grid in the same manner as the ^{137}Cs and used for comparison in the model output. Even though the annual $^{239+240}Pu$ discharges have varied in a similar manner to ^{137}Cs[1], there is likely to be less variation in the bed inventory of $^{239+240}Pu$ than in the spot readings of ^{137}Cs near the pipeline because the sea bed accumulates the activity. This should provide a more suitable set of measurements for validating the model.

Figure 5a shows the observed 1978 distribution of $^{239+240}Pu$ per square metre of the sea bed (in approximately the top 30 cm) averaged over the model boxes from the data[4]. As expected, there is a high concentration around the Sellafield pipeline, and in the area of the mud patch which runs almost parallel to the English coast. There are lower concentrations in the south-western part of the Irish Sea, where there are fewer fine particulates and there is an inflow of water. There is a distinct movement to the north around St Bees Head and also the activity moves to the north of the Isle of Man, via the mud patch in the western Irish Sea towards the North Channel.

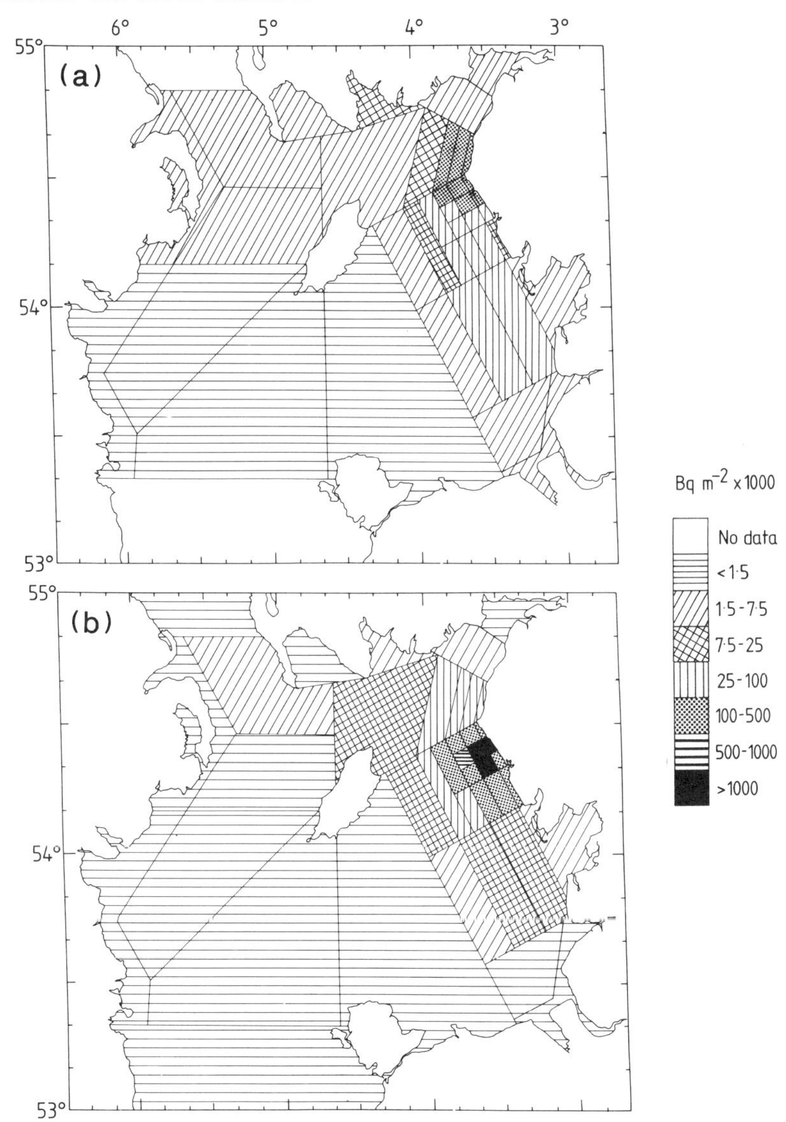

Figure 5. Sea bed inventories (Bq m^{-2}) averaged onto the box model grid: (a) amount of $^{239+240}Pu$ in top 30 cm of bed for 1978; (b) model output of the total amount of $^{239+240}Pu$ in the sea bed for the year 1978.

The model has been run using the circulation pattern and vertical exchange rates described above. Figure 5b is the model output of the in-bed inventory (in Bq m^{-2}) for 1978 of $^{239+240}Pu$ with the same shading as Figure 5a. The patterns look different. There are higher amounts in the sea bed along the English coast and to the north of the Isle of Man and the distinctive tail round St Bees Head is not so pronounced. The former may be the effect of using a coarse resolution of the vertical structure in the sea bed coupled with the making of a comparison between the upper 30 cm of the sea bed (observed data) and the top 2 m (model results) and the use of only one type of bed sediment rather than a mud/sand mixture. The latter may be the result of ignoring, in the model, the effects of the practice of discharging activity in freshwater on the ebb tide which moves north along the coast.

In the water, the concentration of $^{239+240}Pu$ follows the discharge pattern as the ^{137}Cs does but the bed inventory is different. Figure 6 shows the in-bed amounts in regions 2, 1, 5, 12, 21, 22 and 27, a line running from Sellafield normal to the coast. Box 1, the pipeline region, shows an increase and a decrease, almost in line with the discharges and

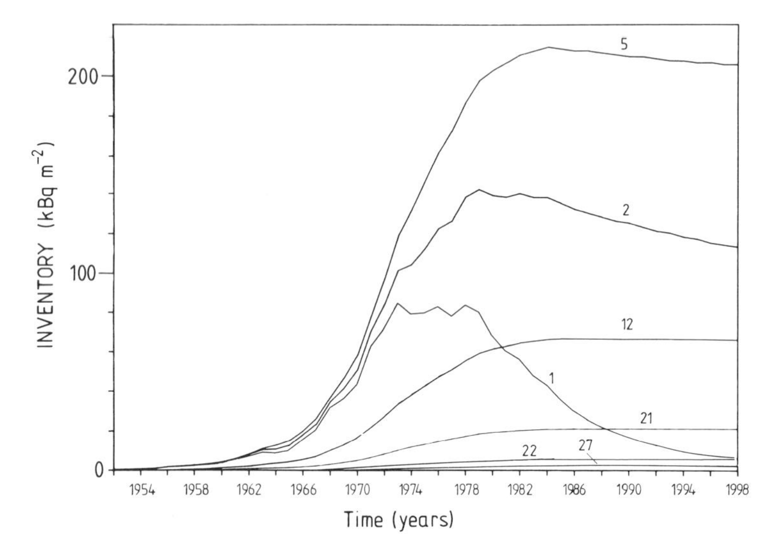

Figure 6. Time development of the amounts in the sea-bed boxes (Bq m^{-2}) under regions 2, 1, 5, 12, 21, 22 and 27. These boxes fall on a line from Sellafield normal to the coast.

water concentrations. However, all the other boxes show an increase with time but only a slow decrease when the discharges have been greatly reduced. In some regions, further from the pipeline, for example box 21, the levels in the sea bed are still increasing. This indicates that activity is being removed from the sea bed in regions of high concentration near the discharge point and is migrating, via the water, to regions of lower sea-bed inventory. Further model improvements and more validation runs are in progress to improve on the uncertainty.

CONCLUSIONS

The setting up of a simple box model, to be used for predicting the distribution of plutonium and americium in the Irish Sea over the next century, has many problems, even though there is a large data set of field observations upon which to base and validate the model. However, some conclusions can be drawn from this study: natural variability in the environment and fluctuations in the source of activity makes it difficult to compare in detail observations with the results of a long time-averaged model; the parameterization of the suspended particulate movement and sea-bed exchanges play a crucial role in determining the distribution of particle-sorbed radionuclides; equilibrium kinetics are an adequate representation of the particle sorption/desorption of radionuclides; a conveyor-belt description of mixing may be necessary to explain individual sediment profiles of radionuclides, but the bio-diffusion description is probably adequate for the calculation of long time average, coarse resolution estimates of sea-bed inventories; a better parameterization of the asymmetries in the release of actinides from Sellafield is needed to explain the high sea-bed concentrations north of St Bees Head.

The model described here is still under development, and it is now being used to examine a number of scenarios of likely (and unlikely) events which may affect the distribution of actinides.

REFERENCES

1. Pentreath, R. J., Sources of artificial radionuclides in the marine environment. International Symposium on Radioactivity and Oceanography: Radionuclides - a Tool for Oceanography, Cherbourg, France, June 1987, (this volume).

2. Prandle, D., A modelling study of the mixing of ^{137}Cs in the seas of the European continental shelf. Phil. Trans. Roy. Soc., Lond., 1984, A310, 407-436.

3. Onishi, Y. and Thompson, F. L., Mathematical simulation of sediment and radionuclide transport in coastal waters. Volume 1: Testing of the sediment/radionuclide transport model, FETRA Report NUREG/CR-2424, Vol. 1, PNL-5088-1. Pacific Northwest Laboratory, Richland, Washington, USA, 1984, 115 pp.

4. Pentreath, R. J., Radioactive discharges from Sellafield (UK). Annex 1. In Behaviour of Radionuclides Released into Coastal Waters. ed. R. J. Pentreath, IAEA-TECDOC-329, IAEA, Vienna, 1985, 183 pp.

5. Pentreath, R. J., Harvey, B. R. and Lovett, M. B., Chemical speciation of transuranium nuclides discharged into the marine environment. In Speciation of Fission and Activation Products in the Environment, ed. R. A. Bulman and J. R. Cooper, Elsevier, London and New York, 1986, pp. 312-332.

6. Kershaw, P. J., Swift, D. J., Pentreath, R. J. and Lovett, M. B., The incorporation of plutonium, americium and curium into the Irish Sea sea bed by biological activity. Sci. Total Environ., 1984, 40, 61-81.

7. Nyffeler, U. P., Santschi, P. N. and Li, Y-H., The relevance of scavenging kinetics to modelling of sediment-water interactions in natural waters. Limnol. Oceanogr., 1986, 31, 277-292.

8. Hainbucher, D., Pohlmann, T. and Backhaus, J., Transport of conservative passive tracers in the North Sea: First results of a circulation and transport model. Submitted to Continent. Shelf Res., in press.

9. Hunt, G. J., 1986. Radioactivity in surface and coastal waters of the British Isles, 1985. Aquat. Environ. Monit. Rep., MAFF Direct. Fish. Res., Lowestoft, 1986, 14, 48 pp.

10. Camplin, W. C., Durance, J. A. and Jefferies, D. F., A marine compartment model for collective dose assessment of liquid radioactive effluents. Sizewell Inquiry Ser. MAFF Direct. Fish. Res., Lowestoft, 1982, 4, 7 pp.

11. Kershaw, P. J., Gurbutt, P. A., Young, A. K. and Allington, D. J., Scavenging and bioturbation in the Irish Sea from measurements of $^{234}Th/^{238}U$ and $^{210}Pb/^{226}Ra$ disequilibria. International Symposium on Radioactivity and Oceanography: Radionuclides - a Tool for Oceanography, Cherbourg, France, June 1987 (this volume).

12. Boudreau, B. P., Mathematics of tracer mixing in sediments: I. Spatially-dependent diffusive mixing. Am. J. Sci., 1986, 286, 161-198.

13. Boudreau, B. P., Mathematics of tracer mixing in sediments: II. Nonlocal mixing and biological conveyor-belt phenomena. Am. J. Sci., 1986, 286, 199-238.

14. Smith, J. N., Boudreau, B. P. and Noshkin, V., Plutonium and ^{210}Pb distributions in northeast Atlantic sediments: subsurface anomalies caused by non-local mixing. Earth Planet. Sci. Lett., 1986, 81, 15-28.

15. Kershaw, P. J. and Young, A. K. Scavenging of ^{234}Th in the eastern Irish Sea. J. Environ. Radioact., in press.

16. IAEA, Sediment K_ds and concentration factors for radionuclides in the marine environment. Tech. Rep., int. atom. En. Ag., Vienna, 1985, 247, 73 pp.

HIGH-LEVEL RADIOACTIVE WASTE DISPOSAL INTO THE SEABED : INVESTIGATION OF THE SEDIMENT-BARRIER EFFICIENCY BY MEANS OF ANALYTICAL MODELS

BOUST, D.
IPSN-DERS-SERE - Laboratoire de Radioécologie Marine
C.E.A. Centre de la Hague, BP 508, F50105 CHERBOURG

POULIN, M.
Ecole des Mines de Paris
35, rue Saint-Honoré, F77305 FONTAINEBLEAU

CHESSELET, R.
Centre des Faibles Radioactivités, C.N.R.S.
Avenue de la Terrasse, BP 1, F91190 GIF-SUR-YVETTE

ABSTRACT

In order to assess the feasibility of the concept of burying high-level radioactive waste into deep-sea sediments, sorption-diffusion models have been set up. The analytical models presented here, are derived from the equations of heat conduction in solids ; they are more simple to handle than numerical models, but can only deal with homogeneous sediment column. They yield total fluxes of radionuclides at the sediment-seawater interface and radionuclide activities at any point of the storage field. The results are presented in terms of fluxes and concentrations versus time, for $^{241-243}Am$, ^{237}Np (^{241}Am daughter element) and ^{99}Tc. The sensitivity of the models to some parameters (distribution coefficients and burial depth) is also discussed. This allows an estimation of the efficiency of the sediment-barrier with respect to these radionuclides, in standard conditions (50 m burial depth, no porewater advection, no faults, perfect hole-closure).

INTRODUCTION

Among the radioactive wastes generated by the nuclear industry, are the high-level radioactive wastes, so called because they consist of long-lived and toxic radionuclides. The high-level wastes considered here are vitrified reprocessed wastes. Many specifications concerning them can be found in the PAGIS inventory (1). They are made of both β - γ emitters fission products and actinides. Their composition at the time of vitrification is summarized on fig. 1 in which initial activity per ton of heavy metal (= uranium) of each radionuclide is plotted versus its radioactive half-life. It is a very simple way to identify potentially critical radionuclides but many other parameters must be taken into account.

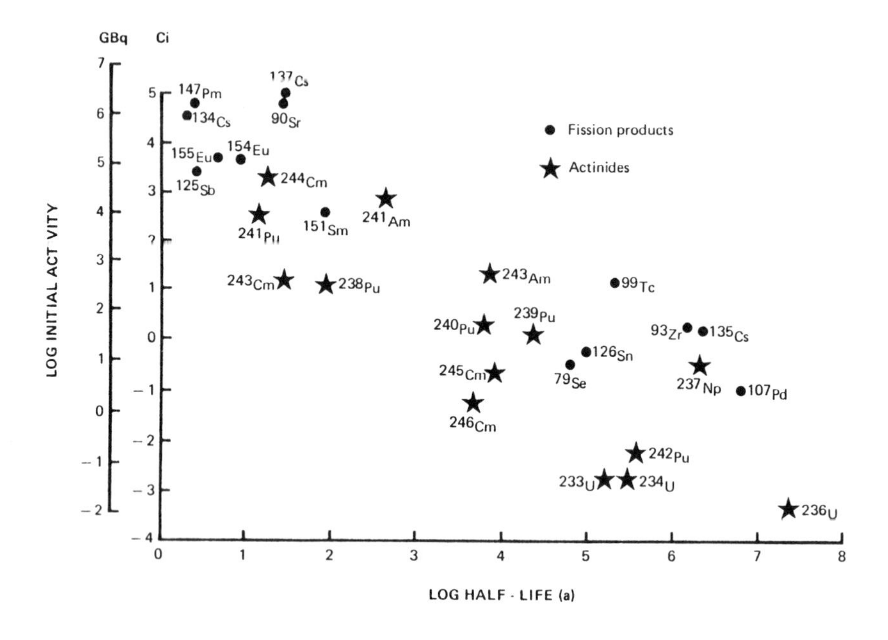

Figure 1. Composition of the high-level radioactive waste per ton of heavy metal.

The volume of waste to be disposed would be of 11000 m^3 and would result of the consumption of ~ 10^5 tons of uranium (with a 3.5 % ^{235}U enrichment). Consequently, all the initial activities mentionned on fig. 1 must be multiplied by a factor of 10^5. One of the possibilities for the long-term geological disposal is the SEABED option. In this hypothesis, high level radioactive wastes are buried into deep-sea sediments of abyssal plains. The retention is obtained by three successive barriers : (a) the glass itself ; (b) the canister ; (c) the sediment-column. Because both glass and canister are not supposed to resist corrosion for much more than 1000 years, the sediments will be the final barrier liàble to delay the long-term migration of the radionuclides up to seawater. As a consequence, we will only consider here the so-called sediment-barrier, and show what can be expected from simple analytical models about its efficiency.

BASIC CONSIDERATIONS

The computation of radionuclide fluxes versus time through the sediment-seawater interface from an internal source enable us to estimate the sediment-barrier efficiency, and to identify radionuclides or parameters (diffusion coefficients, distribution coefficients between particulate and soluble phases, burial depth...) which must be taken into account for predictive impact studies.

Within the sediment column, radionuclides can diffuse as free ions, complexes, in dissolved or colloidal state, interact with sediment particles when undergoing many processes of adsorption ; they can also be transported by porewater advection.

The general equation taking into account these 3 basic processes can be written as follows :

$$\omega D \frac{\partial^2 C_n}{\partial x^2} - U_D \frac{\partial C_n}{\partial x} = \omega_{an} \left(\frac{\partial C_n}{\partial t} + \lambda_n C_n \right) - \omega_{an-1} \lambda_{n-1} C_{n-1} \qquad (1)$$

DIFFUSION ADVECTION SOLID-LIQUID INTERACTION + radioactive decay

in which :

- C_n is the concentration of the radioelement n and λ_n its decay constant.
  The index n-1 refers to its parent.
- ω is the porosity.
- D is the molecular diffusion coefficient corrected for tortuosity.
- U_D is the Darcy porewater velocity.
- ω_a is the apparent porosity ;

$$\omega_a = \omega + (1 - \omega)\, \rho_s\, Kd$$

If both porosity and Kd are assumed to be constant throughout the sediment column, this equation has an analytical solution. Analytical solutions have been used for numerical model verification (2-3) and are used here for sensitivity studies.

The integration leads to the expression of the total flux (F) of an element emitted by a point source in a semi-infinite medium limited by a zero concentration surface, and assuming no advection :

$$F = \frac{N \, . \, h \, . \, e^{-\left(\lambda t \frac{L^2}{4 D_e t}\right)}}{t \, (4 \pi D_e t)^{\frac{1}{2}}} \qquad (2)$$

in which :

- h is the distance between the point source and the sediment-seawater interface (= burial depth).
- N is the initial number of atoms.
- D_e is the effective diffusion coefficient defined as :

$$D_e = \omega D / \omega_a$$

- t is the time after burial.

This analytical solution is related to a homogeneous medium and one element. For multi-element decay chain and heterogeneous medium, analytical solutions are quite complex. However, some of them have been derived for sensitivity studies, for chain migration and two-layered systems (4-5). Let us look at the example of the ^{243}Am decay chain. Radioactive half-lives and distribution coefficients recommended for deep-sea sediments have been plotted on fig. 2 for the eight first daughter elements. This shows how contrasting are those two parameters for the radioelements of the decay chain, thus leading to highly constrasting behaviours.

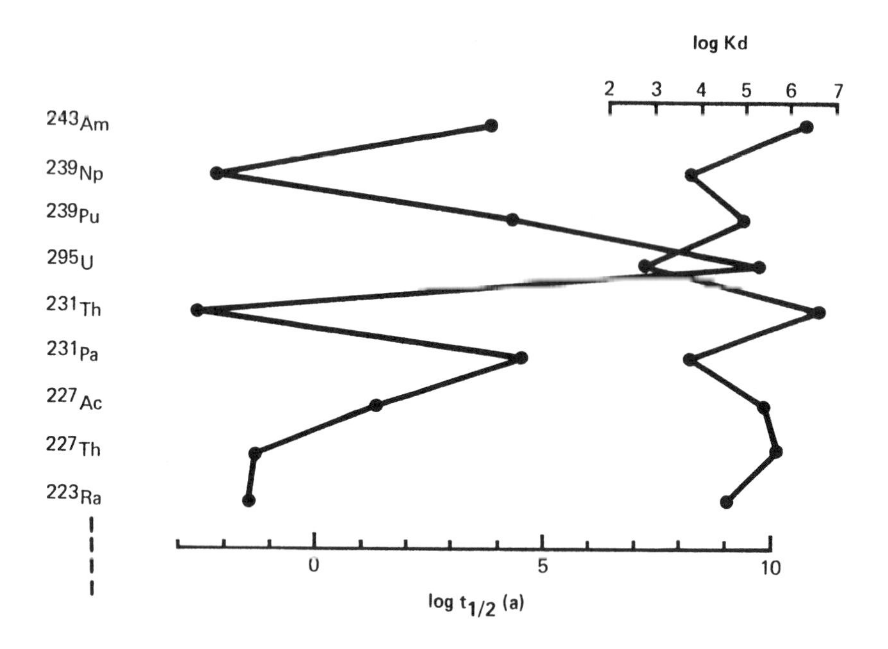

Figure 2. Recommended Kd values (in ml g^{-1}; IAEA Technical Report Series - 247) and radioactive half-lives ($t_{\frac{1}{2}}$ in years) for the eight first daughter elements of ^{243}Am.

FLUXES OF $^{241\text{-}243}$Am, ^{237}Np AND ^{99}Tc AT THE SEDIMENT-SEAWATER INTERFACE

We will now give some results obtained with the simple analytical solution described above, for some of the most critical radioelements for storage conditions : americium, neptunium and technetium.

For both $^{241\text{-}243}$Am isotopes (initial amounts of which are 5 x 10^{28} and 3 x 10^{28} atoms, respectively), the fluxes calculated at the sediment-seawater interface are not significant (many orders of magnitude lower than 1 atom per thousand years) at any time after burial, even for distribution coefficients as low as 10^3. Consequently, the sediment-barrier can be assumed to be total with respect to these radionuclides.

The fluxes of ^{237}Np at the sediment-seawater interface versus time in standard conditions (50 m burial depth) are plotted on fig. 3, for various Kd values.

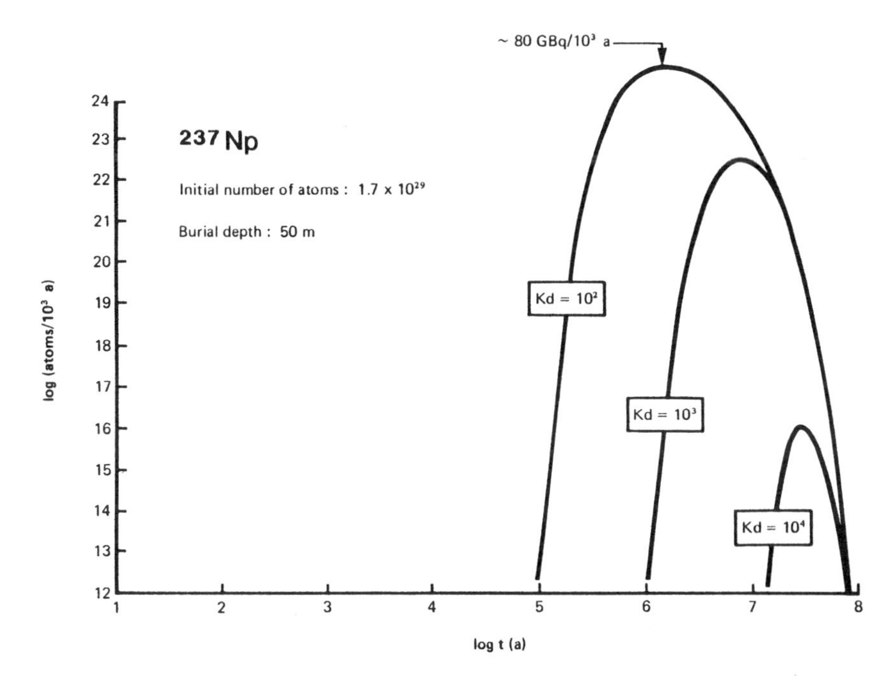

Figure 3. Total ^{237}Np fluxes at the sediment-water interface versus time.

The results show that, for a Kd value of 100 ml g^{-1}, peak fluxes of 6.3 x 10^{24} atoms per thousand year are predicted, that is to say ~ 10 moles/10^3 a, which is 80 GBq/10^3 a, after a 1 or 2 millions years long period. For higher Kd values, peak values of flux get lower and are delayed (3 x 10^{22} atoms/10^3 a after 8 x 10^6 years, for Kd = 10^3).

In the case of ^{99}Tc, recommended Kd values range between 0 and 10^3 ml g^{-1}. This dramatically reduces the time of peak release. For a zero Kd value, peak value of the flux occurs after 10000 years and reaches 5 x 10^{27} atoms/10^3 a (Fig. 4). In this case, an increase of the Kd value of two orders of magnitude from 1 to 100 decreases peak flux of 4 orders of magnitude, and delay it from some tens of thousands years to about 1 million years.

SENSITIVITY OF THE MODEL TO BURIAL DEPTH

The sensitivity of this simple model can also be tested for burial depth (Fig. 5). As it could be expected, burial depth affects peak values of the fluxes much more efficiently for higher Kd values. It could be, also shown that an increase of the burial depth delays the period of peak values of fluxes at sediment seawater interface. As a result of both sensitivity studies and penetrator experiments, the burial depth has been fixed at 50 m in standard conditions. For ^{99}Tc (Kd = 10), one can estimate that if 10 % of the penetrators reach a burial depth of 25 m instead of 50 m, they could induce fluxes of the same order of magnitude as the 90 % remaining penetrators, after a period of time of 50000 years instead of 200 000 years.

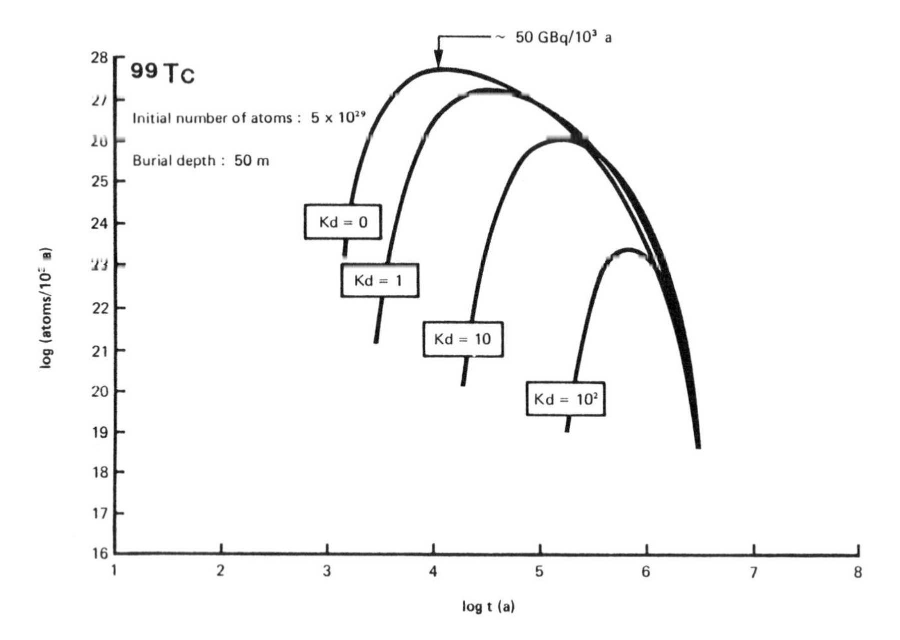

Figure 4. Total ^{99}Tc fluxes at the sediment-water interface versus time.

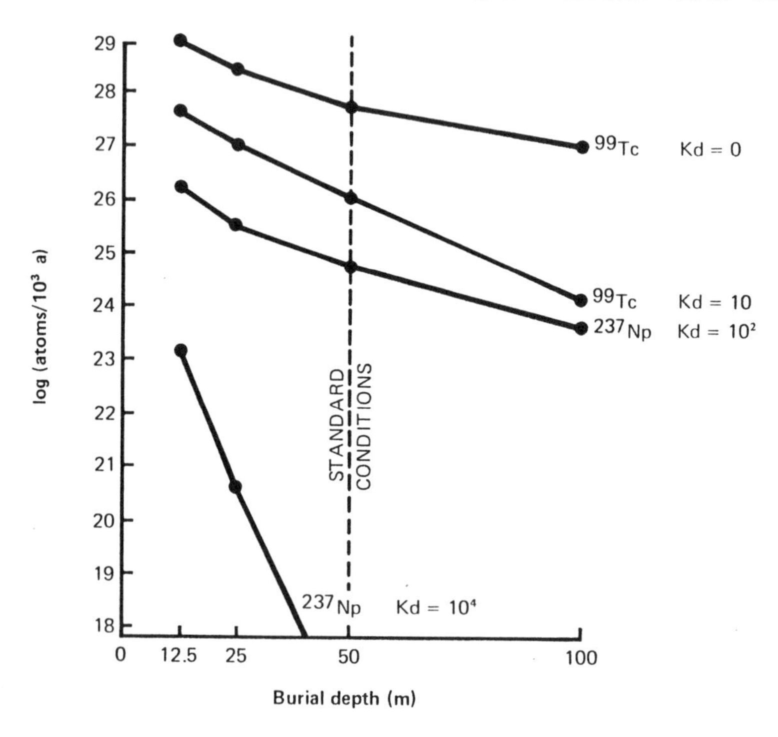

Figure 5. Peak fluxes of ^{237}Np (Kd = 10^2-10^4) and ^{99}Tc (Kd = 0-10) at the sediment-seawater interface versus burial depth.

RADIONUCLIDE CONCENTRATIONS NEAR SEDIMENT-SEAWATER INTERFACE

Radionuclide concentrations generated by each individual penetrator (loaded with 5 canisters) can be derived at any point of the sediment-barrier from the following formula, assuming a zero concentration at sediment-seawater interface as shown in Fig. 6 :

$$C = \frac{M}{(4\pi D_e t)^{3/2}} \cdot e^{-\left(\frac{x^2+y^2}{4D_e t}\right)} \cdot \left[e^{-\frac{(z+h)^2}{4D_e t}} - e^{-\frac{(z-h)^2}{4D_e t}} \right] \cdot e^{-\lambda t} \quad (3)$$

in which :

- x, y are the horizontal coordinates.
- z is the vertical coordinate.
- h is the burial depth.
- other symbols are the same as in equation (2).

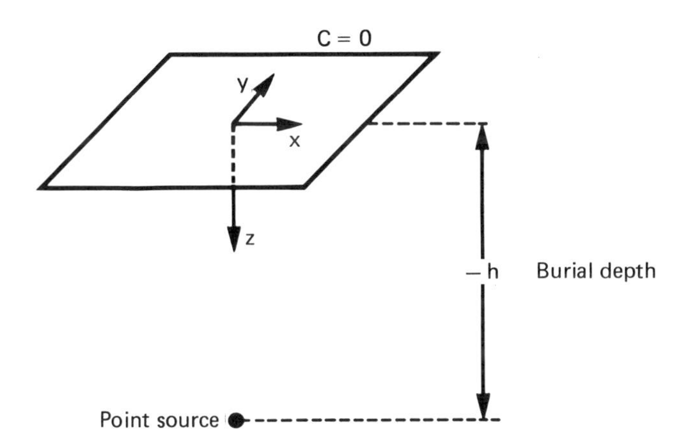

Figure 6. Definition of axes for equation (3).

As an example, we calculated the concentrations of ^{237}Np and ^{99}Tc at a 10 cm depth below sediment-seawater interface, induced by one penetrator (point source) at a 50 m burial depth. Some of the results are summarized in Table 1.

Table 1. Concentrations of ^{237}Np and ^{99}Tc at 10 cm depth below sediment-seawater interface, induced by one penetrator at 50 m burial depth (initial numbers of atoms are 1.14×10^{25} and 3.34×10^{25} for ^{237}Np and ^{99}Tc, respectively).

	Kd	Time of maximum concentration	Maximum concentration Bq kg^{-1}
^{237}Np	100	4×10^{6} a	10^{-2}
^{99}Tc	100	9×10^{5} a	10^{-4}
	0	9×10^{4} a	0.5

By comparison, ^{237}Np concentrations measured in the sediments around the reprocessing plant of La Hague (France) are in the range 10^{-3} - 10^{-2} Bq kg^{-1} (GERMAIN, pers.com.).

For ^{99}Tc, very few data are available ; sediment concentrations measured in the vicinity of Sellafield reprocessing plant range between 1.2 and 4 Bq kg^{-1} (6-7-8).

CONCLUSION

We have shown how useful for sensitivity studies. can be simple analytical models yielding radionuclide fluxes and concentrations at or near the sediment-seawater interface, At a standard burial depth of 50 m, the distribution coefficient between soluble and particulate phases (Kd) appears to play a prominent role on the long term behaviour of the radionuclides within the sediment-column. One-order-of-magnitude variation of Kd can affect peak fluxes by several orders of magnitude, and time of peak release. This enhances the fact that more accurate and realistic Kd values must be obtained from batch experiments, and when possible, from in situ and/or stable analogs measurements as fully examplified in reference 9.

REFERENCES

1. Cadelli, N., Cottone, G., Bertozzi, G. and Girardi, F., Performance assessment of geological isolation systems ; common methodological approach based in European data and models ; summary report of phase 1. Nuclear Science and Technology Series, 1984, E.U.R. 9220, VI, 156 p.p.

2. Poulin, M., Construction d'un modèle complet d'analyse des systèmes du SEABED. Rapport CEA-Ecole des Mines, 1984, LHM RD 8460.

3. Poulin, M., Programme français SEABED ; analyse des systèmes du SEABED. Rapport CEA-Ecole des Mines, 1985, LHM RD 8578.

4. Klett, R.D., Hertel, E.S. and Garner, J.W., Systems models for nuclide transport in saturated geologic media. Report Sand 83-0768, 1986, Sandia National Laboratories, Albuquerque, New Mexico, 87185.

5. Nakayama, S., Takagi, I., Nakai, K. and Higashi, K., Migration of radionuclide trough two-layered geologic media. Journal of Nuclear Science and Technology, 1984, 21, p.p. 139-47.
6. Hunt, G.J., Radioactivity in surface and coastal water of the British Isles, 1978. Aquat. Environ. Monit. Rep. n°4, 1981, MAFF, Lowestoft.

7. Hunt, G.J., Radioactivity in surface and coastal water of the British Isles, 1979. Aquat. Environ. Monit. Rep. n°6, 1981, MAFF, Lowestoft.

8. Hunt, G.J., Radioactivity in surface and coastal water of the British Isles, 1980. Aquat. Environ. Monit. Rep. n°8, 1982, MAFF, Lowestoft.

9. Boust, D., Les terres rares au cours de la diagénèse des sédiments abyssaux ; analogies avec un transuranien : l'américium. Thèse Univ. Caen, 1987, 297 pp.

DESCRIPTION DES RESULTATS OBTENUS EN MATIERE D'OCEANOGRAPHIE DANS LE CADRE DU PROGRAMME RADIOPROTECTION DE LA CCE

G. DESMET
et
C. MYTTENAERE

Programme de Radioprotection
Commission des Communautés Européennes
Bruxelles
Belgique

Résumé

Dans le cadre du Programme d'Action Recherche (PAR) Santé et Sécurité, la Commission a présenté au Conseil des Ministres un nouveau Programme Radioprotection pour la période du 1.1.85 au 31.12.89. Ce programme fait suite aux programmes exécutés par la Commission depuis 1959. De nombreuses Recherches en Radioécologie ont été effectuées depuis cette date et les écosystèmes estuariens et marins ont fait l'objet d'une attention toute particulière.

D'autres travaux incombant au Programme de la Direction "Santé et Sécurité" (DGV, Lg) sont également brièvement décrits dans cette note; l'ensemble de ces résultats montre qu'il est nécessaire de disposer de modèles plus précis permettant d'évaluer de manière plus exacte les doses à l'homme et aux populations. Tchernobyl nous a indiqué quelques voies nouvelles d'investigation et les recherches ultérieures seront en partie orientées dans ce sens.

1. Introduction

La Commission a défini sept thèmes prioritaires dans le "Programme Cadre" qui répondent aux exigences générales des stratégies scientifique et technique de la Commission des Communautés Européennes pour l'avenir. Un des buts consiste en "l'amélioration des conditions de vie et de travail" comprenant les deux objectifs "Amélioration de la Sécurité et de la Protection Sanitaire" et "Protection de l'Environnement et Prévention des Risques". Le Programme d'Action Recherche (PAR) "Santé et Sécurité" constitue l'élément moteur principal pour la mise en oeuvre du premier objectif.

Dans le cadre de ce PAR, la Commission a présenté au Conseil de Ministres un nouveau sous-programme "Radioprotection" pour la période du 1.1.1985 au 31.12.1989. Ce Programme fait suite aux programmes exécutés par la Commission depuis 1959.

Les priorités scientifiques ont été choisies pour répondre aux conséquences de la situation radiobiologique présente et future sur une variété de thèmes formant un programme très intégré et pertinent pour la protection de l'homme, incluant l'amélioration future des bases scientifiques pour les "Normes de base de sécurité pour la protection de la santé de la population et des travailleurs contre les dangers des radiations ionisantes". Par commodité, bien que un peu arbitrairement, le programme est divisé en six secteurs importants :

- Dosimétrie des rayonnements et son interprétation ;
- Comportement et contrôle des radionucléides dans l'environnement ;
- Effets non stochastiques des rayonnements ionisants ;
- Radiocancérogénèse ;
- Effets génétiques des rayonnements ionisants ;
- Evaluation des risques d'irradiation et optimisation de la protection.

En poursuivant ces efforts, la Commission entend :

- coordonner et complémentariser les efforts nationaux, sans les reproduire ;
- réorienter la recherche suivant les besoins, par le développement de la science et de la technologie ;
- poursuivre la diffusion des résultats du programme, aussi largement que possible ;
- coopérer avec les organisations internationales agréées ainsi qu'avec les programmes des pays n'appartenant pas à la Communauté.

Le programme est exécuté par le biais de contrats à frais partagés conclus avec des institutions nationales et universités travaillant différents points particuliers du programme. Ceci permet une coordination constructive et un emploi plus économique du potentiel scientifique et des compétences des Etats membres tout en stimulant la recherche actuelle dans une voie compétitive.

Le principal objectif des études radioécologiques est l'application et, si nécessaire, l'élaboration de modèles de transfert local, régional et global dans des conditions de rejet essentiellement statiques et aussi dynamiques.

Les modèles actuels qui décrivent le transfert de radionucléides dans l'environnement suite à un rejet en fonctionnement normal donnent généralement des valeurs pessimistes et ne se prêtent pas très bien à la mise en oeuvre des procédés d'optimisation. Par ailleurs, il y a lieu de considérer les données nécessaires pour effectuer les calculs d'optimisation du stockage des déchets dans différents milieux.

Des modèles dynamiques sont nécessaires pour étudier les évènements anormaux dans les centrales nucléaires (rejets accidentels de radioactivité). Il faut les améliorer en approfondissant les connaissances de la cinétique des processus écologiques responsables du transfert des radionucléides à l'homme.

L'environnement aquatique marin a toujours fait l'objet d'une attention particulière dans le volet "Comportement et contrôle des radionucléides dans l'environnement".

Cela est particulièrement vrai pour certains aspects tels que les situations d'estuaire, le dépôt dans les grands fonds et l'influence à long terme des nucléides qui, en principe du moins, peuvent entraîner une exposition de l'homme sur de nombreuses générations.

En outre, les progrès de notre connaissance du comportement des matières radioactives dans l'environnement nous aident souvent à mieux comprendre l'écologie marine et l'océanographie dans son ensemble et ils présentent par conséquent de l'intérêt pour la solution d'autres problèmes tels que ceux posés par les polluants non radioactifs.

2. Principaux résultats du programme de 1980-1984

Il est difficile de faire une synthèse des résultats obtenus dans le cadre du nouveau programme 1985 - 1989. Les contrats ont été signés en fin 1985 et l'accident de Tchernobyl a retenu l'attention des chercheurs. Nous nous proposons donc de faire un bref compte rendu des résultats obtenus durant le programme précédent.

2.1. Les milieux estuariens

Une attention particulière a été accordée à trois estuaires : la zone deltaïque hollandaise, l'estuaire de Ravenglass et l'embouchure du Rhône.

La zone deltaïque hollandaise qui associe les estuaires du Rhin, de l'Escaut et de la Meuse a été étudiée notamment par le Delta Institute de Yerseke. Bien que les concentrations de Pu dans les sédiments ne différaient guère en valeur absolue de celles qui pouvaient être dues aux retombées, la normalisation par rapport à la teneur en Al a révélé un excès de Pu qui va en augmentant vers la mer à l'intérieur de l'estuaire et le long de la côte en direction de la Manche ; l'examen de la composition isotopique du plutonium a confirmé que celui-ci ne pouvait être attribué exclusivement aux retombées. Il semble que les matières proviennent des installations de retraitement du Cap de la Hague et/ou de Sellafield ; les rejets de cette dernière ont été sensiblement plus élevés que ceux de la Hague, mais une petite fraction seulement, non encore quantifiée, sort de la mer d'Irlande par le Sud et pénètre dans la Manche, d'où elle pourrait être entraînée vers la côte hollandaise.
Les travaux réalisés par le NERC de Plymouth dans l'estuaire de Ravenglass, à plus ou moins 10 km au Sud de l'installation de retraitement de Sellafield, ont permis de décrire qualitativement les processus de redistribution estuarienne affectant la radioactivité introduite par les rejets de Sellafield dans la mer d'Irlande.

La remobilisation des actinides dans les sédiments a aussi été étudiée par l'AERE de Harwell dans ce même estuaire. Il a été constaté que les courants de marée provoquaient des pertes nettes d'Am et de Pu dans les sédiments en suspension, bien qu'il n'y ait pas de transfert net de masse. Il faut pour cela postuler un mécanisme qui permet de remobiliser l'activité qui s'est déposée antérieurement sur les sédiments. Plusieurs mécanismes de ce genre ont été proposés à titre d'hypothèses à vérifier.

Le CEA de Cadarache a étudié les effets des variations de salinité sur les coefficients de distribution pour divers nucléides autres que les actinides. Le mélange eau de mer/eau douce induit une variation des valeurs du Kd pour differents radionucléides : Fe-59, Mo-99, Tc-99, Cr-51, Cs-137, Mn-54, Zn-65, Na-22, Co-60.

Les effets du vieillissement sur la spéciation de l'Am dans les eaux saumâtres ont été étudiés par le Biologische Anstalt d'Holgoland. Ces études ont révélé un mouvement important d'Am(III) de l'état dissout à l'état de macroparticules et détecté la formation d'un complexe américohumique en présence d'acides humiques.

En plus des particules chaudes contenant de l'uranium et provenant sans doute directement de Sellafield, le NERC de Plymouth a identifié des particules chaudes associées au débris de moules. Un aspect particulier des travaux concernait les répartitions et la dynamique de l'absorption de radioactivité par les moules.

Ces études montrent que les modèles de transfert de l'eau vers les sédiments et vice-versa doivent faire intervenir d'autres mécanismes que les processus physicochmiques mis en évidence.

2.2. Milieux côtiers

Les nombreuses croisières d'échantillonage effectuées par le laboratoire de Risø ont permis de suivre les courants sur plus de 7000 km depuis le littoral européen jusqu'aux zones du Groenland au moyen du Cs et du Tc provenant des rejets de l'industrie nucléaire. Un facteur de dilution de 10 a été trouvé entre la mer du Nord et l'Océan Arctique puis un facteur de 300-500 à l'arrivée dans le courant du Groenland Oriental.
Des travaux plus théoriques comme ceux du MAFF de Lowestoft ont montré qu'à l'écart des zones immédiates de rejet, le Np et le Pu sont tous les deux principalement à l'état de valence V tandis que l'Am et le Cm se présentent tous deux principalement dans les états de valence réduite, III et IV. Il a été constaté que la bioturbation provoquait un important mélange des sédiments. A Cadarache et Cherbourg, le CEA a étudié la grande amplitude du coefficient de diffusion lié à Eu, Ce et Am et il a conclu que l'Eu constitute un bon homologue de l'Am. En outre, l'absorption d'Am et de Pu par les espèces benthiques s'est révélée faible, même si certaines espèces absorbent l'Am non seulement à partir de l'eau interstitielle mais aussi diretement à partir des sédiments.

Des études de l'ENEA de Santa Teresa et de l'IRSNB de Bruxelles sur le technétium ont révélé la persistance du pertechnetate en solution dans un large éventail de conditions de pH et d'oxydo-réduction sauf en présence de sédiments à grains fins riches en matières organiques Des études effectuées sur les algues et un Flavobacterium ont révélé la réduction du pertechnetate et la formation de complexes organiques. On déduit d'un modèle à compartiments de la répartition du Tc que la contamination des chaînes alimentaires marines par le Tc ne constitue probablement pas une voie importante d'exposition pour l'homme.

Le laboratoire national de Risø a comparé Fucus vesiculosus et Mytilus edulis comme bioindicateurs en fonction de la température et de la salinité. Dans la plupart des cas, Fucus possède un facteur de transfert plus élevé, au point même que pour les lanthanides et les actinides les vitesses d'absorption et de rejet sont si élevées qu'elles conduisent à préférer Mytilus comme bioindicateur.

Les études pratiques du Trinity College de Dublin ont montré que pour l'Irlande le Cs-137 est le seul nucléide important dans la voie de consommation de poissons associée aux prises effectuées en mer d'Irlande ; et même pour ce nucléide, l'exposition du groupe critique n'est que de l'ordre de 1 $mrem.a^{-1}$.

2.3. Milieux des grands fonds

Ces travaux visent à améliorer notre compréhension de la radioécologie fondamentale de ces milieux.

En conséquence, le NIOZ de Den Burg a élaboré une description systématique de la microflore et de la macrofaune benthiques du point de vue de leur composition, de leur densité, de leur biomasse et de leur répartition verticale. Le MAFF de Lowestoft a pour sa part construit un modèle de sédimentation tenant compte des effets de bioturbation. L'importance de ces effets est démontrée par la présence de Pb-210 à des profondeurs de sédiments allant jusqu'à 8 cm et elle est confirmée par une étude des profils de C-14 ; les données relatives à ces deux nucléides ont servi de base pour la modélisation. Le MAFF est passé maintenant à la conception de modèles permettant d'estimer les doses à la faune.

2.4. Autre sujets

L'AERE de Harwell et le CEA de Cadarache ont longuement étudié le transfert de radioactivité à l'atmosphère sous la forme d'aérosols. Dans le cadre de ces travaux, ils ont tous deux utilisé des barboteurs placés juste sous la surface de l'eau pour provoquer la formation d'aérosols et confirmer ainsi l'enrichissement en Pu et en Am des aérosols par rapport aux concentrations dans l'eau. Le phénomène est attribué à l'absorption de macroparticules par les aérosols et revêt une telle ampleur que les facteurs d'enrichissement de 30 à 3000 ont été mesurés dans les conditions expérimentales.

Dans le domaine de la biochimie cellulaire, le NERC de Plymouth et l'Université de Nantes ont étudié les transuraniens dans les moules, les coques, les bigorneaux et les astéries. D'autres travaux ont porté sur le technétium dans les astéries.

3. Travaux exécutés dans le programme 1985-89

Les nucléides à longue période, en particulier le Pu, l'Am et le Tc, occupent à nouveau une place de choix et les milieux à étudier vont des situations estuariennes jusqu'aux zones de grandes profondeurs en passant par les conditions côtières. Du point de vue géographique, les études vont des eaux du Groenland à la Méditerranée en passant par l'Atlantique Nord (et la mer Baltique).

Les recherches sur la dispersion océanique, les mécanismes et les vitesses de transfert de la radioactivité en direction et en provenance des sédiments, la remobilisation des sédiments, la formation d'aérosols marins et les processus biologiques sont poursuivis.

L'impact des rejets de Tchernobyl sur l'environnement marin a préoccupé la plupart des contractants durant l'année 1986. Cette contamination accidentelle a modifié les teneurs en Cs-137 des écosystèmes marins étudiés et plus particulièrement de l'Atlantique-Nord.

D'autre part l'entrée de l'Espagne et du Portugal dans la Communauté permettra de mettre l'accent sur la contamination des écosytèmes subtropicaux et un premier contrat signé avec le CIEMAT, Madrid, Espagne a pour but d'étudier en détail la contamination de la région de Palomarès (teneurs du Pu-239/240 et Am-41 de la région côtière et des fonds marins ainsi que des mécanismes responsables de la fixation et de la remobilisation de ces éléments).

4. Etudes théoriques supplémentaires

4.1. Projet MARINA

En 1979, la Commission a publié une "Méthodologie pour l'évaluation des conséquences radiologiques des rejets d'effluents radioactifs en fonctionnement normal".

En 1985, il a été décidé de procéder à une révision approfondie du modèle à la lumière des nouvelles connaissances acquises sur les processus écologiques et des discussions ont été entamées à cette fin avec les contractants initiaux, à savoir le NRPB de Harwell et le CEA de Fontenay-aux-Roses. Toutefois, étant donné les développements de nos connaissances des processus marins et l'intérêt accrû de la population et des spécialistes à l'égard des voies d'exposition qui s'y rattachent pour l'homme, un groupe de travail spécial a été créé auquel tous les Etats membres de la Communauté sont invités à participer et qui est chargé de modéliser le transfert à l'homme par l'intermédiaire de l'Atlantique du Nord-Est et des mers qui communiquent avec lui. Ces travaux, dénommés projet MARINA, devront évidemment être menés en liaison étroite avec ceux qui concernent la révision générale de la méthodologie.

La mission du groupe MARINA consiste à rédiger un rapport comprenant les points suivants :

- un inventaire des déchets radioactifs solides et liquides jetés dans l'Atlantique du Nord-Est (y compris la mer d'Irlande, la mer du Nord et la Manche) ;
- une analyse des modèles correspondants de dispersion marine en vue de faire un choix parmi eux ou d'en tirer un modèle approprié. Dans le cadre de cette analyse, les prévisions obtenues à partir des modèles devront être comparées avec les niveaux mesurés de radioactivité ;
- une enquête sur les quantités et l'utilisation des produits marins concernés ;
- l'estimation, sur base des travaux ci-dessus, des doses individuelles et collectives aux populations des Etats membres.

Quatre sous-groupes ont été constitués pour s'occuper respectivement d'un inventaire des rejets, du mesurage dans l'environnement et des doses aux groupes critiques, des produits marins et enfin, de la modélisation.

4.2. Données concernant les rejets

Dans le passé, la Commission a publié tous les deux ans les données disponibles sur les rejets d'effluents liquides et gazeux de toutes les centrales et usines de retraitement nucléaire de la Communauté européenne, avec un rappel des principales doses qui en résultent pour les membres de la population. Pour différentes raisons, un seul rapport de ce genre a été publié ces dernières années, en 1983 il porte sur les rejets effectués au cours de la période 1976-1980. La Commission a l'intention de relancer cette série de rapports.

5. Conclusions

Depuis de nombreuses années, la Commission Européenne apporte son soutien aux recherches sur la pollution radioactive marine et octroît pour ces travaux des aides financières considérables. Les études entreprises vont de la collecte de données sur les rejets jusqu'à la construction de modèles pour l'estimation des doses à l'homme en passant par les phénomènes de transport et de transfert d'origine physique, chimique ou biologique.

En dépit des difficultés économiques actuelles, des dispositions ont déjà été prises afin de poursuivre ce soutien jusqu'à la fin de la décennie.

Les résultats obtenus ont montré qu'il était nécessaire de disposer de modèles plus précis permettant d'évaluer aussi bien que possible les doses à l'homme et aux populations. Les conséquences des récents évènements plaident en faveur d'une estimation aussi précise que possible des doses par le biais de modèles radioécologiques dynamiques. D'autre part les problèmes posés plaident également en faveur d'une meilleure formation de nos radioécologistes, d'une collaboration plus étroite entre modélistes et biologistes et aussi, faut-il le souligner, d'une meilleure information du public. Cette tâche sera la nôtre demain!

REFERENCES

MYTTENAERE,C et FRASER, G., C.E.C. Studies relating to radioactive releases from land-based sources, PARCOM/NEA Workshop on marine radioactivity research and monitoring in the Paris convention area, OCDE, Paris, 18-20 Paris, 18-20 February, 1986

SCOPPA,P. et MYTTENAERE, C., Behaviour of radionuclides in the marine environment : present state of knowledge and future needs, Environmental Inorganic Chemistry, p.299-306 (1985).

XXX., Proposition pour une décision du conseil arrêtant le programme pluriannuel de recherche et de formation pour la Communauté Européene de l'Energie Atomique dans le domaine de la Radioprotection (1985-1989), COM(83) final (1983.)

CEC, EURATOM, Programme Radioprotection 1980-1984, I, II, EUR 9733 DE/EN/FR 1985.

^{10}Be IN THE WATER COLUMN OF THE EQUATORIAL ATLANTIC

M. Segl* and A. Mangini**
G. Bonani*** , H.J. Hofmann*** , M. Suter*** , W. Wölfli***

* Fachbereich Geowissenschaften, Universität Bremen, 28 Bremen 33, FRG
** Heidelberger Akademie der Wissenschaften c/o. Institut für Umweltphysik Im Neuenheimer Feld 366, 69 Heidelberg, FRG
*** Institut für Mittelenergiephysik der ETH Zürich, Hönggerberg, CH-8093 ZURICH

On a transect across the Atlantic from Panama to Nepals, 5 water profiles for ^{10}Be measurements were taken. The ^{10}Be concentration was measured at the AMS facility of the ETH Zürich.
^{10}Be is produced by cosmic rays in the high atmosphere. It gets attached to particles within a few hours after production and is rained out on the ocean surface.
Earlier measurements in the Central Pacific and in the Atlantic yielded. Concentrations of 2200 - 6000 atoms/g. From this a residence time of some 100 to > 1000 years could be calculated.
From the calculation of the ^{10}Be-flux to sediments in upwelling areas of West-Africa, a lower residence time in this area followed. This is due to a scavenging effect of the settling particles. The consequence is a concentration gradient in the water column between the coastal areas and the open ocean.
The ^{10}Be concentrations near the coast are about 1500 atoms/g (25°W), the open-ocean concentrations near the Mid-Atlantic Ridge are > 2000 atoms/g. This yields a residence time for the open ocean of about 1000 years and a few 100 years for the coastal areas.

The profile near the Mid-Atlantic Ridge (45°W) shows a maximum of the ^{10}Be concentration at water depths of 1000-2000 m. This may correspond to a water mass from the Labrador Sea.

VERTICAL FLUX OF RADIONUCLIDES IN MARINE SYSTEMS MEDIATED BY BIOGENIC DEBRIS

N. S. Fisher*, J. K. Cochran**, S. Krishnaswami***

* Oceanographic Sciences Division, Brookhaven National Laboratory, Upton, New York 11973 U.S.A.
** Marine Sciences Research Center, State University of New York, Stony Brook, New York 11794 U.S.A
*** Physical Research Laboratory, Ahmedabad, INDIA

The degree to which radionuclide flux in the sea can be attributed to sinking biogenic debris deriving principally from phytoplankton can be calculated for specific water columns in the Atlantic and Pacific Oceans. Radionuclide flux predictions are based on experimentally determined concentration factors in phytoplankton, on dissolved radionuclide concentrations in surface waters, and on new production estimates or organic carbon flux for specific ocean regions. The flux of organic matter is assumed to be directly related to new production, *sensu* Eppley and Peterson. The concentration factors in phytoplankton are determined using radiotracer methodology and assess the degree to which the radionuclides fractionate between dissolved and particulate phases at equilibrium. Laboratory-derived values generally compare favorably with estimates from field samples. Predictions of radionuclide fluxes are compared with actual sediment trap measurements of radionuclide fluxes at different depths in these waters. It can be shown that in oceanic regions the downward flux from surface waters of many of the particle-reactive radionuclides is governed principally, if not entirely, by sinking biogenic debris.

^{10}Be FLUXES FROM SEDIMENT TRAPS

G.M. Raisbeck*, F. Yiou*, S. Honjo**, J. Klein*** and R. Middleton***

* Laboratoire René Bernas, 91406 Orsay, FRANCE

** Woods Hole Oceanographic Institution, Woods Hole, Ma 02543, U.S.A

*** Department of Physics, University of Pennsylvania, Philadelphia, Pa 19014, U.S.A.

Exploitation of the various potential applications of ^{10}Be in the ocean, discussed elsewhere at this meeting, requires an improved understanding of the marine geochemistry of this isotope. As part of such a study ^{10}Be has been measured in material collected in sediment traps exposed at various depths and at three locations (Panama Basin, Demerara Abyssal Plain, and East Hawaii Abyssal Plain), and in surface sediments at or near these locations. The results demonstrate the important role of particle scavenging in the vertical transport of ^{10}Be in the ocean. The data also suggest a significant amount of horizontal transport of ^{10}Be at intermediate depths in the ocean (either in "soluble" form, or associated with very small particles), before being definitively incorporated into sediments.

NATURAL RADIONUCLIDES IN TWO ANOXIC MARINE ENVIRONMENTS

J. F. Todd*, R. J. Elsinger** and W. S. Moore*

* Department of Geology, University of South Carolina
Columbia, South Carolina 29208

** Chevron U.S.A., Inc., 935 Gravier Street,
New Orleans, Louisiana 70112

Anoxic basins are unique environments for the examination of interactions between dissolved and particulate species over various oxidation/reduction conditions. The wide range of redox conditions found at the interface of oxic and anoxic waters can significantly influence the distribution of particulate manganese and iron oxides which are effective carrier phases for many radionuclides. The influence of iron and manganese particulate phases on the distribution of natural radionuclides in anoxic environments has only recently been investigated.

Vertical profiles of naturally-occuring U-238, U-234, Ra-226, Ra-228 and Th-228 have been obtained from two anoxic marine systems : Framvaren Fjord, Norway (August 1981) and Saanich Inlet, British Columbia (May 1981). While Framvaren Fjord is permanently anoxic below a depth of 18 m (oxygen-hydrogen sulfide interface), Saanich Inlet is intermittently replenished with oxygenated water on a yearly timescale. Maximum hydrogen sulfide concentrations in Framvaren Fjord are more than two orders of magnitude higher than those found in Saanich Inlet.

Concentrations of U-238 and Ra-228 were fairly constant throughout the water column of Saanich Inlet. A slight gradient was observed for Ra-226, which increased from 11 dpm/100 L in the surface waters to 17 dpm/100 L at 195 m. A pronounced Th-228 maximum was observed at 160 m, howewer, just below the particulate Mn maximum. In Framvaren Fjord, Ra-226 concentrations increased linearly from 9 dpm/100 L (8 m) to 70 dpm/100 L (153 m), with the exception of a maximum located near the oxygen-hydrogen sulfide and dissolved Mn maxima. Sharp increases were also observed in the profiles of U-238 and Th-228 at this depth, suggesting that Mn cycling may, in part, control the distribution of uranium and thorium near the oxic-anoxic interface. A similar mechanism has been invoked to explain the behavior of Pb-210 and Ra-226 at the oxic-anoxic interface of Orca Basin, Gulf of Mexico.

PROCESSES INVOLVED IN LONG-TERM BEHAVIOUR OF AMERICIUM IN DEEP-SEA SEDIMENTS INVESTIGATED BY RARE EARTH ELEMENTS ANALOGY

D. Boust*, C. Lambert**,
R. Chesselet**, J.L. Joron***

* IPSN-DERS-SERE - Laboratoire de Radioécologie Marine
C.E.A. Centre de la Hague, BP 270, 50107 CHERBOURG

** Centre des Faibles Radioactivités, C.N.R.S.
Avenue de la Terrasse, BP 1, 91190 GIF-SUR-YVETTE

*** Laboratoire "Pierre Süe", Groupe des Sciences de la Terre
CEN Saclay, 91191 GIF-SUR-YVETTE Cedex

The interest of studying behaviour of geochemical analogs of radionuclides which could be burried into abyssal sediments has been early pointed out (International Subseabed Disposal Program). In this respect, rare earth elements (REE) are often cited as analogs of trivalent transuranics, especially Eu as analog of Am. Although this concept is now well accepted, little work has been done in this way, partly because of analytical problems set by the detection of stable elements at ultra-low level concentrations. As a contribution to the understanding of long-term behaviour of Am isotopes in deep-sea sediments, a study of REE reactivity during oxic to suboxic diagenesis is presented.

REE profiles were obtained by neutronic activation analysis and subsequent chemical extraction in interstitial waters of a 13 m long core collected on the Cape Verde Abyssal Plateau (4850 m water depth). Dissolved REE concentrations (Eu, Tb, Yb, Lu) range between $2\text{-}200\times10^{-12}$ M. In the suboxic part of the core, they are 3 (Eu) to 17 (Yb) times higher than in the oxic layer, thus suggesting a strong dependance of REE solid-liquid equilibria upon redox conditions. Dissolved profiles display significant positive gradients in both oxic and suboxic layers; they are interpreted as resulting of sorption-diffusion processes and discussed in the wider context of element recycling in the deep ocean.

In addition, sequential leaching experiments yield to the identification of REE host-phases in the sediment column. Mn-Fe rich mineral coatings are shown to play a prominent in the REE behaviour and therefore could be efficient sinks for Am isotopes liable to reach the abyssal environnement.

$^{10}Be/^{9}Be$ AS A POTENTIAL PALEOCEANOGRAPHIC TRACER
(Poster)

D. Bourlès, G.M. Raisbeck and F. Yiou

Laboratoire René Bernas, 91406 Orsay, FRANCE.

Most of the ^{10}Be introduced to the ocean is formed by cosmic ray interactions in the atmosphere, and is input in soluble form, mainly by precipitation. ^{9}Be is introduced to the ocean by erosion, in both soluble and insoluble form. Using a selective leaching procedure, we have found that the $^{10}Be/^{9}Be$ ratio associated with the authigenic fractions of marine sediments (presumably representing the "soluble" or easily exchangeable portion of these two isotopes) varies considerably with location. These variations are likely due to both the location of the sediment with respect to the sources of the two isotopes, and the circulation patterns of the overlying water. If this hypothesis can be confirmed, and the present dependence established, measurement of $^{10}Be/^{9}Be$ in sediment cores might be able to serve as an indicator of paleoceanographic circulation patterns for the past ~ 15 My.

MIXING RATES ON THE CONTINENTAL SLOPE DERIVED FROM RADIUM MEASUREMENTS
(Poster)

P.A Gurbutt

Ministry of Agriculture, Fisheries and Food
Directorate of Fisheries Research
Fisheries Laboratory, Lowestoft
Suffolk NR33 OHT, ENGLAND

Large volume water samples were collected in June 1982 for measurements of ^{228}Ra and ^{226}Ra along a section running from the continental shelf west of Ireland down the continental slope at 51°42'N, near Porcupine Bank. Whereas previous ^{228}Ra measurements across the shelf edge have been from surface samples, the data described here show how the concentration varies with depth. The highest concentrations of ^{228}Ra are near the sea bed on the shelf, and in a tongue of water near the sea surface extending over deeper water. The ^{226}Ra concentrations increase with water depth, with a few intermediate higher concentrations close to the slope. These data are related to the hydrography of the area. Where simple one dimensional diffusion models can be applied to concentration profiles, they reveal mixing rates which are consistent with deep ocean values. However, a two dimensional model is needed to describe most of the data.

LE KRYPTON-85 OCEANIQUE : TECHNIQUE DE MESURE ET INTERET EN OCEANOGRAPHIE
(Poster)

P. Jean Baptiste, J. Radwan et J-F. Ternon

Laboratoire de Géochimie Isotopique - LODYC (UA CNRS 1206)
CEA/IRDI/DESICP - Département de Physico-Chimie,
91191 GIF-SUR-YVETTE Cedex, FRANCE

L'intérêt des traceurs radioactifs et/ou transitoires tels que ^{14}C, ^{3}H pour les études de circulation océanique n'est plus à démontrer. L'utilisation de nouveaux traceurs transitoires tels que les fréons ou le krypton-85, dont les conditions aux limites sont différentes de celles des précédents traceurs et dont les fonctions d'entrée dans l'océan sont comparables à celle du CO_2 anthropogénique, doit permettre de mieux contraindre les modèles de circulation et de progresser dans l'étude du rôle de l'océan mondial dans le cycle du CO_2.

Une technique de mesure du Krypton-85 est présentée :
Vu la faible activité du ^{85}Kr dans l'eau de mer ($\sim$ 2dpm/m^3 en surface), cette technique nécessite le traitement de 200 litres d'eau par échantillon. Une unité embarquable d'extraction et de stockage des gaz dissous a été développée. Les échantillons prélevés au moyen de bouteilles General Oceanics de 100 litres sont pulvérisés à travers un gicleur dans une cuve d'extraction maintenue sous vide. Les gaz ainsi extraits sont transférés et comprimés par des pompes à membranes dans un container étanche qui sera ramené au laboratoire.

Le taux de récupération des gaz a été évalué pour l'oxygène, l'hélium ainsi que pour l'ensemble des gaz dissous. Il se situe dans la gamme 97% - 99,5% suivant les conditions expérimentales (température de l'eau, etc ...).

Le krypton présent dans le mélange gazeux ainsi récupéré est séparé au laboratoire en plusieurs étapes par un système de colonnes de chromatographie (carbosieve) et de piégeage/dépiégeage sur charbons actifs. La quantité totale de krypton (typiquement 14 microlitres NTP) est mesurée en fin de cycle par un catharomètre. La quantité de krypton perdue pendant la séparation est inférieure à 4%.

L'échantillon de krypton est ensuite introduit dans un compteur proportionnel de petite taille (1,5 cc) et l'activité en ^{85}Kr est mesurée par comptage β bas-niveau. Cette dernière partie est encore au stade du développement dans notre laboratoire.

RADIOLUCLIDES IN SUBMARINE HYDROTHERMAL SYSTEMS (Poster)

D. Kadko
Oregon State University, U.S.A

Naturally occurring radionuclides in the submarine hydrothermal systems which are associated with seafloor spreading centers can yield information on the rates of processes that occur both within the crust and within the dispersing hydrothermal effluent. Measurements of radionuclides in the hot vent water which emanates directly from the seafloor, and in the sulfide deposits associated with these vents, can be used to constrain the crustal residence time of the hydrothermal fluid. On the Endeavour segment of the Juan de Fuca Ridge, $^{210}Pb/Pb$ ratios in boths sulfide and vent water samples collected by DSRV ALVIN were in the range ~ 0.3-0.6 dpm $^{210}Pb/\mu g$ Pb, which is comparable to the $^{238}U/Pb$ ratio of basalts from the area. This suggests that the residence time of the fluid from the onset of basalt alteration is much less than the mean life of ^{210}Pb (32 years) which would be required to produce additional ^{210}Pb from the large measured excess ^{222}Rn in the system (~ 500 dpm/l). Thus, if the residence time was long, the $^{210}Pb/Pb$ ratio of the venting fluid and sulfides would be much greater than the $^{238}U/Pb$ ratio of the basalt. A limited number of samples were analyzed for $^{228}Ra/^{226}Ra$ and these values are consistent with a residence time of only a few years.

From the site of venting, the hydrothermal effluent is carried away in plumes which have been detected 100-300 m above the seafloor. These plumes contain large excesses of ^{3}He, CH_4, Mn, particles, and other constituents. In of itself, this plume is a dynamic chemical system. Numerous reactions occur that involve precipitation, oxidation and scavenging processes and many may be microbially mediated. Such a plume over the Endeavour Ridge has been studied and excesses of both ^{222}Rn ($t_{1/2}$ = 3.85 d) and ^{234}Th ($t_{1/2}$ = 24.1 d) exceeding 250 dpm/100 l have been measured within it. These nuclides may act as "clocks" to constrain both the rate of dispersion of the plume, and the kinetics of the reactions that occur within it.

RADIOCARBON DATING OF IRISH SEA SEDIMENTS (Poster)

P.J. Kershaw

Ministry of Agriculture, Fisheries and Food
Fisheries Laboratory, Lowestoft, Suffolk NR33 OHT

The fate of long-lived, particle-reactive radionuclides discharged from the Sellafield nuclear fuel reprocessing plant is inextricably linked to the fate of the sediments in which a significant proportion of these artificial radionuclides presently reside. Radiocarbon dating of Irish Sea sediments was undertaken to provide a better understanding of the sedimentary regime into which the waste was being discharged. Initial ^{14}C measurements of the bulk sediment carbonate content revealed near-constant ages to depths of ~ 150 cm - implying complete turnover by bioturbation to that depth. More recent ^{14}C dates of <u>Turritella communis</u> shells indicate linearly decreasing ^{14}C ages with depth, in contrast to a more variable ^{14}C age of the surrounding sediment. It appears that <u>T. communis</u> shells may be a more sensitive indicator of the recent sedimentary climate and that bioturbation acts to selectively avoid disturbance of the shell fragments.

MAJOR IONS AS MASTER PARAMETERS IN RADIONUCLIDE PARTITION BETWEEN SEDIMENTS AND SEAWATER (Poster)

A.M. Hansen*, L.D. Mee*, and J.O. Leckie**

* Instituto de Ciencias del Mar y Limnologia
Universidad Nacional Autonoma de Mexico
Apartado Postal 70-305, Mexico 04510 D.F.
MEXICO.

** Environmental Engineering and Science
Department of Civil Engineering
Stanford University. Stanford, CA 94305
U.S.A.

The chemical speciation of trace radionuclides in aqueous systems is determined by a variety of competitive reactions, including the formation of surface species. The sediments act as traps for many radionuclides, forming complexes and otherwise adsorbed surface compounds. In this way, the dissolved ions exist in dynamic equilibrium with bed sediment and suspended particles. The reactions between dissolved radionuclides and functional surface groups can be described as coordination reactions, and the surface oxides, as multiple ligand systems.

Many factors inherent to both sediment and solution can influence the concentration of radionuclides and its species. These factors include the pH, type of mineral, surface area and/or amount of adsorbing surface sites, complexing ligands, and competing ions. The relative importance of these factors was studied using an isotope dilution technique. A variety of coastal sediments from the Gulf of Mexico were used as the solid matrixes.

Seawater contain a variety of dissolved metal ions that compete for different surface sites on the sediments. Most studies of competitive adsorption have involved trace metals in the presence of a large excess of calcium and magnesium ions. Under these conditions, magnesium ions have a greater effect on trace metal adsorption than do calcium ions. Magnesium ions tend to suppress trace metal adsorption in the ocean by decreasing the number of available binding sites on the surface of sediments. These ions may also be responsable for the decrease in adsorption velocity of radionuclides onto sediments in seawater with respect to river water, and for the desorption of adsorbed trace elements in rivers and estuaries upon contact with seawater.

The information obtained in this investigation helps predict the solution concentration of radionuclides and allows extrapolation of the results to other situations.

PARTICLE ASSOCIATED ^{10}Be IN THE OCEAN (Poster)

F. Yiou*, G.M. Raisbeck*, D. Bourlès*, M. Bacon**

* Laboratoire René Bernas, 91406 Orsay, FRANCE

** Woods Hole Oceanographic Institute, Woods Hole, Massachusetts 02543, U.S.A.

Present evidence suggests that the adsorption on particles plays an important role in the transport of ^{10}Be in the ocean. As a pilot study for a possible larger study, we have therefore carried out an experiment in which both the "soluble" and particle associated ^{10}Be have been measured in a 271 liter surface water sample taken near Bermuda. The results show that < 2% of the ^{10}Be in this sample is retained by the 1 μm filter. This is about an order of magnitude smaller retained fraction than for Th isotopes under similar conditions, thus suggesting that Be is somewhat less reactive than Th in the ocean.

TENEURS EN ^{99}Tc CHEZ FUCUS SP. ET DANS LES EAUX DE MER SUR LE LITTORAL DE LA MANCHE : SYNTHESE (Poster)

M. Masson* F. Patti**, D. Calmet***

* CEA, IPSN, SERE, Laboratoire de Radioécologie Marine de La Hague
B.P. 270 - 50107 CHERBOURG - FRANCE

** CEA, IPSN, DPS, SEAPS, Laboratoire de Radiotoxicologie et Biochimie
CEN/FAR, B.P. 6 - 92265 FONTENAY-AUX-ROSES Cedex - FRANCE

*** CEA - Station marine de Toulon. IFREMER/CT
B.P. 330 - 83507 LA SEYNE-SUR-MER Cedex - FRANCE

Le poster synthétise les résultats d'analyses de ^{99}Tc chez l'indicateur biologique <u>Fucus sp.</u> et dans l'eau de mer sur le littoral de la Manche.

- La mise au point de techniques d'analyses du ^{99}Tc (Patti et al., 1982 ; Patti et al., 1983) a permis de montrer que ce radioélément artificiel était quantitativement le plus important chez l'algue <u>Fucus sp.</u> sur le littoral de la Manche et que le FC était d'environ 50 000 (Jeanmaire et al., 1981 ; Masson et al., 1981 ; Patti et al., 1985).

- L'étude de l'évolution dans le temps des teneurs en ^{99}Tc chez Fucus, à Goury, près de l'émissaire des rejets de l'usine de La Hague, montre une augmentation exponentielle des rejets de 1974 à 1983, puis l'amorce d'un palier de 1983 à 1986 (environ 2 900 Bq kg^{-1} sec) (Patti et al., 1985 et résultats récents non publiés).

- Les analyses effectuées en 1979/80 (Masson et al., 1983) en 1983 (Patti et al., 1985) et 1983/84 (Calmet et al., sous presse) montrent une décroissance dyssymétrique des teneurs en ^{99}Tc chez Fucus en s'éloignant de part et d'autre de l'émissaire, confirmant une dérive principale des eaux vers l'est.

- L'observation d'un rythme saisonnier des teneurs en ^{99}Tc chez Fucus aux stations de Wimereux à l'est et de Roscoff à l'ouest en 1983 et 1984 (Calmet et al., sous presse), suggère deux hypothèses :

1. observation d'un temps de latence de l'impact des rejets dû au temps de transit des eaux depuis La Hague,

2. cycle saisonnier des teneurs en ^{99}Tc chez le Fucus selon un rythme physiologique.

THE MARINE DISPERSION OF RADIONUCLIDES FROM WINFRITH (Poster)

S. E. Nowell

United Kingdom Atomic Energy Authority
AEE Winfrith, Dorchester, Dorset
ENGLAND DT2 8DH

The nuclear power plant at Winfrith, on the south coast of England, discharges to sea a variety of radionuclides, subject to the regulations imposed by the Department of the Environment. Those easily detectable in the marine environment include Co-60, Zn-65 and Mn-54.

In developing a computer model to simulate the transport and distribution of particular nuclides, local tidal and geological characteristics and concentration factors have to be determined. The measured concentrations of Co-60 in seawater (averaged over a tidal period) are compared with the discharge levels and with measurements made on various sediments and seaweed.
An estimate is made of the delay in the adsorption of the nuclide onto sampling media, to allow the concentration factors to be calculated. Further measurements on suspended load are discussed.

The posters summarise the development of the model and the monitoring work undertaken. The concentrations in seawater predicted by the model are compared with actual measurements. The long term aim is to be able to predict the effects on the marine environment of changes in the quantity or content of the nuclides discharged.

PROCESSUS SEDIMENTAIRE ET NIVEAUX D'ACTIVITE DU ^{137}Cs DANS LES SEDIMENTS DE MEDITERRANEE NORD OCCIDENTALE

(Poster)

D. Calmet et J.M. Fernandez

CEA/IPSN/SMT, IFREMER-CT, BP 330
83507 LA-SEYNE-SUR-MER, FRANCE

Depuis 1977, le Service d'Etude et de Recherche sur l'Environnement du Commissariat à l'Energie Atomique développe à la Station Marine de Toulon des études au sein d'un programme général d'évaluation des processus marins intervenant dans la dispersion des radioéléments naturels ou anthropogéniques en Méditerranée (Programme RADMED pour RADioactivité et MEDiterranée).

Le ^{137}Cs issu des retombées atmosphériques à la suite des essais d'armes nucléaires dans l'atmosphère de l'hémisphère nord a marqué les différentes composantes de l'environnement marin méditerranéen. Aussi dans le cadre du programme RADMED, les distributions spatiales horizontale et verticale du ^{137}Cs au sein des sédiments ont été étudiées sur l'ensemble des zones côtières françaises et de la plaine abyssale du bassin occidental méditerranéen en liaison avec les principaux processus sédimentaires et hydrodynamiques.

Au total les campagnes annuelles RADMED regroupent 777 échantillons de sédiments côtiers, hors zone d'extension du prodelta rhodanien et 58 échantillons prélevés dans des étangs et lagunes en bordure du littoral. 40 autres échantillons proviennent de sites choisis le long des pentes continentales des bassins Algéro-provençal et d'Alboran et ce jusqu'à plus de 2700 mètres de profondeur. L'ensemble de ces résultats a permis de dresser une première cartographie de la distribution du ^{137}Cs des dépôts récents.

En dehors de la zone littorale directement soumise à l'influence rhodanienne, les niveaux en ^{137}Cs apparaissent relativement homogènes, fluctuant entre 0.5 et 35 $Bq.kg^{-1}$ de sédiments secs. Cette distribution est à relier avec l'affinité de ce radioélément pour le matériel pélitique d'un diamètre inférieur à 2 µm. Les variations horizontales s'expliquent par l'action des phénomènes de transport et d'accumulation des pellites sous l'effet de l'hydrodynamisme local et de l'existence d'un néphéloide benthique. Les zones plus profondes sont alimentées soit par ce dernier, soit par du matériel arraché sur le plateau continental ou encore par sédimentation directe des poussières atmosphériques.

Les profils verticaux en ^{137}Cs permettent également de cerner l'influence des processus biologiques de bioturbation sur la distribution de cet élément dans les premiers centimètres de la colonne sédimentaire.

DETERMINATION OF RECENT SEDIMENTATION RATE IN THE TYRRHENIAN SEA USING 239, 240-Pu AND 210-Pb (Poster)

R. Delfanti*, C. Papucci* AND PAGANIN**

* ENEA - CREA, C.P. 316 La Spezia (ITALY)

** ENEA - CRE CASACCIA, C.P. 2400 Roma (ITALY)

Two exemples are presented on the use of natural (210-Pb) and man-made (239, 240-Pu) radionuclides for the calculation of the sedimentation rate in coastal marine environments (Gaeta and La Spezia Gulfs).

In both cases the calculated average sedimentation rate is in the order of 0.5 cm/y.

The identification of the subsurface maximum of 239, 240-Pu in sediment cores, corresponding to the peak fallout in 1963, has been found effective to calculate sedimentation rates greater than 1 cm/y. Radionuclide profiles at sites of lower-sedimentation rate are often altered because of sediment mixing and the 1963 fallout peak loses definition.

However, the Plutonium profile from the La Spezia core shows a well defined subsurface maximum suggesting that the mixing rate at this station was low and the sedimentation rate calculated by this profile, 0.52 cm/y, is in good agreement with that calculated by 210-Pb.

For the Gaeta Gulf the mixing rate seems to be higher and the Plutonium subsurface maximum is still evident, but has lost definition. In this case, a range for the sedimentation rate was estimated to be 0.4 - 0.6, which was confirmed by 210-Pb depth profile.

DISTRIBUTIONS OF 239, 240-Pu AND 14-C IN NORTH-EAST ATLANTIC SEDIMENTS (WASTE DISPOSAL AND NOAMP AREAS) (Poster)

G. Papucci, R. Delfanti

ENEA - CREA, C.P. 316 La Spezia (ITALY)

Analyses of 239, 240-Pu and 14-C in sediment cores collected from the Atlantic low-level radioactive disposal site and from a reference area (NOAMP) are presented.

Pu inventories in the two areas are compared and mixing characteristics of the sediments are discussed, using Pu and 14-C depth distributions.

Radionuclide concentration levels in sediments from the dumpsite are substantially indistinguishable from fallout contamination levels in deep-sea sediments at this latitude. The average thickness of the mixed layer in both areas is 7 - 10 cm. In most cases a subsurface maximum in the 3 - 4 cm depth interval was observed in Pu vertical profile.

In only one core, collected from the NOAMP area, Pu vertical distribution could be described by a classical diffusion equation and a mixing coefficient of 0.2 cm^2 y^{-1} was calculated.

SELLAFIELD WASTE RADIONUCLIDES IN THE INTERTIDAL SEDIMENTS OF SKYREBURN BAY, SOUTH WEST SCOTLAND (Poster)

A.B. MacKenzie and R.D. Scott

Scottish Universities Research and Reactor Centre East Kilbride, Glasgow G75 OQU, SCOTLAND

Skyreburn Bay is a shallow bay containing extensive areas of intertidal sediments lying in the north shore of the Solway Firth some 60 km from the BNFL nuclear fuel reprocessing plant at Sellafield. The sediments of the bay are typical of many parts of the north Solway coastal area and consist of a heterogeneous mixture of silt, sand and shell fragments. Concentrations of manmade radionuclides in the sediments of the bay are dominated by waste from the Sellafield plant and a time sequence study of the distributions of radionuclides in the sediments has been carried out over a seven year period in order to define the supply mechanisms affecting the area. Variations in radionuclide depth profiles as a function of time do not reflect variations in the Sellafield discharge and indicate that, while general levels of contamination do vary as a function of the discharge, radionuclide supply to the bay is subject to more complex variations. Nuclide and isotope ratios for the sediment also reveal that there is not a simple relationship between the Sellafield discharge and the resulting radionuclides deposition in the Solway sediments but suggest rather that there is considerable mixing of the waste nuclides with previously discharged waste. This high degree of mixing is also suggested by ^{210}Pb measurements. The observed nuclide ratios and the preferential association of the radionuclides with the finer grained material suggest that the transport of contaminated particulate material from a mixed reservoir in the Irish Sea is the dominant mode of radionuclide supply to the area.

PATHWAYS OF REMOVAL OF CHERNOBYL-INDUCED RADIONUCLIDES FROM NORTHWESTERN MEDITERRANEAN WATERS

S.W. Fowler*, P. Buat-Menard*, S. Ballestra*,

Y. Yokoyama**, H.V. Nguyen**

* International Laboratory of Marine Radioactivity, IAEA, C/O Musée Océanographique, MC 98000, MONACO

** Centre des Faibles Radioactivités, Laboratoire Mixte CNRS-CEA, BP 1, F 91190, GIF-SUR-YVETTE, FRANCE

Vertical flux of materials in the northwestern Mediterranean was examined by a sediment trap array moored in a 2000 m water column during spring 1986. A single sediment trap was placed at 200 m water depth off the north coast of Corsica in April 1986. The trap collectors were set to sample sequentially every 6.25 days. During this period an input of radioactivity from the Chernobyl incident was recorded in the northwestern Mediterranean. Increases of radioactive fallout were recorded in Monaco during the first day of May. Particulate material collected between 26 April and 2 May in the sediment trap registered a slight increase in radioactivity during this period. Maximum radioactivity of ^{137}Cs, ^{141}Ce and ^{144} Ce was noted in particulate samples trapped between 8-15 May, after which time activity decreased substantially in sediment trap samples for ^{141}Ce and ^{144}Ce. Considering the time approximated for the input of these artificial radionuclides to the sea surface and their arrival at 200 m depth, it is calculated that a short transit time on the order of a few days was necessary for the radionuclides to reach 200 m. This clearly confirms the rapid vertical transport rates of artificial radionuclides associated with sinking particulate material which have been hypothesized from experimental studies. Analyzes of zooplankton and their fecal pellets collected during this period confirmed the importance of sinking biogenic particulates in effecting the vertical flux of these radionuclides. Continual monitoring at this site should give insight into the processes involved in removing artificial radionuclides from the upper water column and transporting them to depths.

SELLAFIELD TRACERS IN THE NORWEGIAN SEA AND ARCTIC OCEAN

J.N. Smith, K.M. Elis, E.P. Jones

Fisheries and Oceans
Bedford Institute of Oceanography
P.O. Box 1006
Darmouth, N.S. B2Y 4A2
CANADA

The activities of the radionuclides, Cs-134 and Cs-137, derived from the Sellafield (Windscale) fuel reprocessing plant on the west coast of England were measured in surface water collected in the Norwegian Sea in 1982. Correlation of the spatial distribution of these tracers with their input function from the reprocessing plant and with measurements made in other years, provide information on water circulation processes in the North Atlantic Ocean. A Cs-137 water depth profile, measured in 1981 at the FRAM III ice station, located several hundred kilometers north west of Svalbard, provide further information on water circulation in the Norwegian Sea. Elevated levels of Cs-137 (up to 11.0 mBq/l) were measured in the surface water at this site and are ascribed to Sellafield-labelled water from the Norwegian Sea. Below the surface layer, Cs-137 activities decrease to values of 0.33 mBq/l in the Arctic Bottom Water. An anomalously high Cs-137 activity measured at a water depth of 1500 m, may be the result of penetration of Sellafield-labelled surface water to this depth as a result of convection processes in the Norwegian Sea. However, the measurements of Cs-134 and Cs-137 at the CESAR ice camp in the central Arctic Ocean over the Alpha Ridge in 1983 and at the Canadian Ice Island, just north of Ellesmere Island, in 1985 and 1986 exhibit no evidence for the transport of Sellafield-labelled water into the Arctic Ocean.

RADIOACTIVE TRACERS FROM THE EUROPEAN NUCLEAR INDUSTRY IN FRAM STRAIT AND GREENLAND SEA OCEANOGRAPHY

H. Dahlgaard*, A. Aarkrog*, and E. Holm**

* Risø National Laboratory, DK-4000 Roskilde, DENMARK

** Lund University, Lasarettet, S-221 85 Lund, SWEDEN

Radionuclides released from nuclear fuel reprocessing facilities in Western Europe have been detectable in the East Greenland Current at least since 1977. Radionuclide data across the Fram Strait at 78-80°N from 1983, 1984 and 1985 are presented. The distribution of ^{137}Cs, ^{134}Cs, ^{90}Sr and ^{99}Tc in the Fram Strait and the Greenland Sea indicates that the surface waters of the East Greenland Current receives a significant amount of tracer-rich North Atlantic water south of 79°N in the Fram Strait.

This is now explained on the basis of recently published oceanographic studies in the area.

The transfer factors for transport of radionuclides from Europe to Greenland waters will be summed up, and the implications for modelling conventional pollutant transfer will be discussed.

L'UTILISATION DES TRACEURS RADIOACTIFS ARTIFICIELS ET DES CAPTEURS RADIOMETRIQUES EN SEDIMENTOLOGIE DYNAMIQUE

A. Caillot

Commissariat à l'Energie Atomique,
Service d'Applications des Radioéléments

Les problèmes que posent la défense des côtes, le maintien des profondeurs dans les chenaux de navigation envasés, le devenir des rejets urbains ou industriels sous forme particulaire dans le milieu marin sont parmi les plus difficiles à résoudre pour les ingénieurs maritimes.

Cela résulte de la puissance de la mer et du nombre de paramètres à prendre en considération mais aussi de la mauvaise connaissance théorique et pratique des transports sédimentaires en suspension et par charriage sous les actions simultanées des courants, des houles et des vents.

L'immersion volontaire puis le suivi dans l'espace et le temps de sédiments marqués par des radionucléides artificiels émetteurs gamma et de périodes courtes ou moyennes (de quelques heures à quelques mois) ont le grand avantage de conduire à des résultats quantitatifs obtenus en nature dans des conditions expérimentales choisies et hydrométéorologiques connues. Ainsi sont étudiés :

- le "régime des côtes" pour lutter contre leur érosion,

- le rejet et le devenir des produits de dragage,

- le transfert des rejets urbains dans les milieux estuariens et marins depuis la station de traitement jusqu'à leur accumulation dans les organismes vivants.

L'emploi de jauges nucléaires permet de localiser l'interface mal définie eau-vase déposée dans les chenaux de navigation puis de mesurer en temps réel le profil vertical de concentration en sédiment à l'intérieur du dépôt. Cet instrument à diffusion de photons gamma (source de césium 137) est récemment devenu une aide à la décision pour gérer les chantiers de dragage et pour assurer le passage des navires à grand tirant d'eau dans les chenaux envasés.

Ces procédés et ces techniques nucléaires hier encore moyen de recherche sont devenus aujourd'hui des outils au service des ingénieurs. Une illustration en est donnée par l'emploi de radionucléides pour mesurer la dispersion en fonction du temps et de la distance des sédiments fins remis en suspension par le rejet des produits de dragage ou de polluants sous forme particulaire.

ORIGIN AND FATE OF ARTIFICIAL RADIONUCLIDES IN THE RHONE DELTA, FRANCE

J.M. Martin and A. Thomas

Institut de Biogéochimie marine, Ecole Normale Supérieure,
1, rue Maurice Arnoux, 92120 MONTROUGE

The Rhone is the most important river entering the Mediterranean sea since the daming of the Nile river. It represents the major source of river water dissolved salts, nutrients and sediments to the Golfe of Lion, and has a major effect upon the whole area.

In order to specify the actual significance of the Rhone river upon the Mediterranean sea, it is important to trace the material discharged to this area. To gain more insight into these problems, it has been decided to use artificial radionuclides discharged to the Rhone by different nuclear power plants and mainly by the reprocessing plant of Marcoule.

The aim of this paper is :

- to provide the first assessment of the total dissolved and particulate concentration of more than ten radionuclides as well as their partitioning between different particulate phases.

- to present a preliminary survey of the fate of these different artificial radionuclides in the mixing zone between river water and sea water.

ACCUMULATION DU ^{137}Cs DANS LES SEDIMENTS ESTUARIENS ET LITTORAUX DE LA MANCHE

P. Walker*, P. Gueguenïat**, J.P. Auffret***, Y. Baron**

* Université de Caen, Laboratoire de Géologie Marine
Adresse actuelle : CREO - Allée des Tamaris LA ROCHELLE

** Laboratoire de Radioécologie Marine, C.E.A. Centre de La Hague

*** Laboratoire de Géologie Marine, Université de CAEN

Des profils verticaux de l'activité du ^{137}Cs ont été établis en analysant des sédiments prélevés par carottage dans des environnements intertidaux des côtes françaises de la Manche entre la baie de Saint-Brieuc et la baie de Somme.

La composition minéralogique des échantillons est assez homogène mais les variations granulométriques doivent être prises en compte pour l'interprétation des courbes de radioactivité du ^{137}Cs. L'activité du ^{40}K est liée à la teneur en éléments fins de l'échantillon et un lissage des profils verticaux de radioactivité artificielle a été obtenu en utilisant cet élément naturel.

Les courbes d'activité du ^{137}Cs, ainsi corrigées des variations granulométriques s'avèrent alors représentatives de l'évolution de la concentration du radionucléide dans la masse d'eau et comparables à des profils d'accumulation obtenus par modélisation des apports liés aux retombées atmosphériques ou aux rejets des usines de retraitement.

La chronologie des dépôts est ensuite établie et des vitesses de sédimentation caractéristiques des différents environnements étudiés sont proposées.

Des taux de sédimentation de quelques centimètres par an (1 à 4) ont été observés sur les schorres des havres du Cotentin ou sur les slikkes vaseuses du Nord-Cotentin et des petits estuaires du sud de la baie de Seine. La vitesse de dépôt des sédiments est beaucoup plus élevée en certains points de la baie du Mont-Saint-Michel ou de l'estuaire de la Seine et peut alors atteindre 8 à 10 cm/an.

AUTHOR INDEX

AARKROG A. 294, 351, 446
AGNEDAL P.O. 240
ALLINGTON D.J. 131
ANDRIE C. 35, 45
ARNOLD M. 64
AUFFRET J.P. 449
BACON M. 437
BALLESTRA S. 444
BARD E. 64
BARNES C. 162
BARON Y. 217, 260, 312, 384, 449
BATTAGLIA A. 143
BAXTER M.S. 183, 281
BONANI G. 427
BOURLES D. 55, 432, 437
BOUST D. 407, 431
BRADLEY P.E. 281
BRADSHAW K. 321
BRANICA M. 171
BRYLINSKI J.M. 260
BUAT-MENARD P. 121, 444
BUESSELER K.O. 204
CABIOCH L. 260
CAILLOT A. 447
CALMET D. 217, 312, 438, 440
CHERRY R.D. 362
CHESSELET R. 1, 431
COCHRAN J.K. 162, 428
DAHLGAARD H. 294, 351, 446
DELFANTI R. 441, 442
DESMET G. 416
DJENIDI S. 373
DJOGIC R. 171
DUNIEC S. 294
DUPLESSY J.C. 64
DURANCE J.A. 395
ECONOMIDES B.E. 281
ELLETT D.J. 281
ELLIS K.M. 445
ELSINGER R.J. 430
FERNANDEZ J.M. 217, 440
FISHER N.S. 428
FOWLER S.W. 121, 444
GANDON R. 260
GARNIER A. 383
GASTAUD J. 351
GERMAIN P. 312
GERSHEY R.M. 92
GREEN D.R. 92
GUEGUENIAT P. 260, 294, 384, 449
GURBUTT P.A. 131, 395, 432
HAMILTON-TAYLOR J. 321
HANSEN A.M. 436
HEUSSNER S. 121
HEYRAUD M. 362
HOFMANN H.J. 427
HOLM E. 294, 351, 446
HONJO S. 429
HUH Chih-An 84
IZDAR E. 204
JEAN-BAPTISTE P. 45, 433
JONES E.P. 445
JORON J.L. 431
KADKO D. 434
KAUTSKY H. 271
KELLY M. 321
KERSHAW P.J. 131, 395, 435
KLEIN J. 429
KNIEWALD G. 171
KONUK T. 204
KRISHNASWAMI S. 428
LALOU C. 1
LAMBERT C. 431
LA ROSA J. 121
LECKIE J.O. 436
LELU M. 45
LIVINGSTON H.D. 204
MacAULAY I.R. 304
MacKENZIE A.B. 341, 443
MANGINI A. 427
MARTIN J.M. 448
MARTINOTTI W. 143

MASSON M. 312, 438
MAUNIER P. 217
MAURICE P. 64
MEE L.D. 436
de MEIJER R.J. 195
MERLIVAT L. 35
MIDDLETON R. 429
MOORE R. M. 111
MOORE W.S. 430
MUDGE S. 321
MYTTENAERE C. 416
NEVISSI A.E. 153
NGUYEN H.V. 121, 444
NIES H. 227, 250
NIHOUL J.C.J. 373
NIVEN S.E.H. 111
NOWELL S.E. 439
ORBI A. 384
OREGIONI B. 351
PAGANIN G. 441
PAPUCCI C. 441, 442
PATTI F. 438
PENTREATH R.J. 12, 260
POLLARD D. 304
POULIN M. 407
PRICE N.B. 101, 341
PUT L.W. 195
QUEIRAZZA G. 143
RADWAN J. 433
RAISBECK G.M. 55, 429, 437
de REUS J. 195
REYSS J.L. 121
RHEIN M. 74
RIDGWAY I.M. 341
ROETHER W. 74
SALOMON J.C. 260, 384
SCHMIDT S. 121
SCHUILING R.D. 195
SCOTT R.D. 341, 443
SEGL M. 427
SHIMMIELD G.B. 101
SMITH J.N. 351, 445
SUTER M. 427
TERNON J.F. 433
THOMAS A. 448
THOMSON J. 183
TODD J.F. 430
TOOLE J. 183
WALKER P. 449
WEDEKIND C. 227
WIERSMA J. 195
WILLIAMS T.M. 341
WOLFLI W. 427
WOODHEAD D.S. 331
YIOU F. 55, 429, 437
YOKOYAMA Y. 121, 444
YOUNG A.K. 131

LIST OF PARTICIPANTS

Dr A. AARKROG
Riso National Laboratory
Post Box 49
DK-4000 ROSKILDE

Dr P.O. AGNEDAL
Studsvik Energiteknik AB
S-61182 NYKOPING

Melle G.APROSI
EDF-DER
6 Quai Watier
F-78400 CHATOU

Mr J. AVOINE
Université de Caen
Laboratoire de Géologie Marine
F-14032 CAEN CEDEX

Mr J.P. BABLET
Service Mixte de Contrôle Biologique
BP 16
F-91310 MONTLHERY

Mr J. BAILLY
CEA/IPSN/DERS
F-13108 ST PAUL LEZ DURANCE

Mr Y. BARON
GEA-EAMEA
F-50115 CHERBOURG NAVAL

Dr A. BATTAGLIA
CISE SpA
Via Reggio Emilia 39
I-20090 SEGRATE (MILAN)

Dr M. BAUMANN
Universitat Bremen
D-2800 BREMEN

Dr M.S. BAXTER
Scottish Universities Research and Reactor Centre
East Kilbride
UK-GLASGOW G75 OQU

Mr Y. BELOT
CEA/IPSN/DERS/SERE/SRTCM
F-92265 FONTENAY AUX ROSES

Dr D. BENTLEY
INTECHMER
BP 324
F-50103 CHERBOURG CEDEX

Mme C. BONNOT
CNRS
Laboratoire de Geomorphologie
F-35800 DINARD

Mr D. BOURLES
CNRS
49 rue de la Sablière
F-91120 PALAISEAU

Mme J. BOUSSARD
IFP
127 rue du Mal Foch
F-95150 TAVERNY

Mr R. BOUSSARD
SFEN
127 rue du Mal Foch
F-85150 TAVERNY

Dr D. BOUST
CEA/IPSN/DERS/SERE
Laboratoire de Radioécologie Marine
BP 508
F-50105 CHERBOURG CEDEX

Dr P. BRADLEY
Scottish Univ. Research and Reactor Centre
East Kilbride
UK-GLASGOW G75 OQU

Dr. P. BUAT-MENARD
CNRS Centre des Faibles Radioactivités
BP 1
F-91198 GIF SUR YVETTE CEDEX

Dr K.O. BUESSELER
Woods Hole Oceanographic Institution
Chemistry Dept. Redfield 330
Woods Hole MA 02543 (U.S.A.)

Mr J. BUSSAC
CEN FAR - DRSN
F-92265 FONTENAY AUX ROSES

Dr L. CABIOCH
CNRS
Station Biologique
F-29211 ROSCOFF

Dr A. CAILLOT
CEA/SAR CEN SACLAY
BP 21
F-91191 GIF SUR YVETTE

Dr D. CALMET
CEA-SMT/IFREMER-CT
BP 330
F-83507 LA SEYNE SUR MER CEDEX

Dr M.S.N. CARPENTER
24 rue Paul Le Flem
F-35200 RENNES

Dr F.D.P. CARVALHO
LNETI/DPSR
Estrada nacional 10
P-2685 SACAVEM

Dr G. CHABERT D'HIERES
I.M.G.
BP 53 X
F-38330 ST ESMER

Melle S. CHARMASSON
CEN FAR
BP 6
F-92265 FONTENAY AUX ROSES

Mr M. CHARTIER
CE N FAR - DPS
BP 6
F-92265 FONTENAY AUX ROSES

Dr C. CHASSARD-BOUCHAUD
Université P. et M. Curie
Biologie et Physiologie des Organismes Marins
4 pl. Jussieu
F-75230 PARIS CEDEX 05

Mr M. CHEVALIER
EDF/GDF
76 Bd Mendès France
F-50100 CHERBOURG

Dr R.D. CHERRY
Physics Department
University of Cape Town
CAPE TOWN CP 7700 (AFRIQUE DU SUD)

Dr G. CICERI
CISE-SPA
PO BOX 12081
I-20134 MILAN

Dr A.A. CIGNA
ENEA
I-13040 SALUGGIA VC

Mr A. CLECH
COGEMA LA HAGUE
BP 508
F-50105 CHERBOURG CEDEX

Mr. J. CLUCHET
CEA/DAM/DQS
31 Rue de la Fédération
F-75752 PARIS CEDEX 15

Dr J.K. COCHRAN
Marine Sciences Research Centre
State University of New York
STONY BROOK, NEW YORK 11743 (U.S.A.)

Dr F. COGNE
CEN/FAR
BP6
F-92265 FONTENAY AUX ROSES

Dr.J.R. COOPER
National Radiological Protection Board
CHILTON
UK-DIDCOT, OXON

Dr. D. COSSA
IFREMER Centre de Nantes
BP1049
F-44037 NANTES CEDEX

Mr J. COUTURE
SFEN
48 rue de la Procession
F-75724 PARIS

Dr H. DAHLGAARD
Riso National Laboratory
DK-4000 ROSKILDE

Dr DELAPIERRE
4 rue de la Chaintre
F-45740 LAILLY EN VAL

Mr H. DELAUNAY
COGEMA LA HAGUE
BP 508
F-50105 CHERBOURG CEDEX

Dr R. DELFANTI
ENEA-CREA
CP 316
I-LA SPEZIA

Dr G. DESMET
CCE
Rue de la Loi, 200
B-1049 BRUXELLES

Dr S. DJENIDI
Université de Liège
Mécanique des Fluides géophysiques
B5, Sart Tilman
B-4000 LIEGE

Dr R. DJOGIC
Center for Marine Research
Rudjer Boskovic Institute
POB 1016
Y-41001 ZAGREB

Mme M. DOUCHET
EAMEA
Bd de la Bretonnière
F-50115 CHERBOURG NAVAL

Mr M. DUBOUCHET
SODERN
Av Descartes
BP 23
F-94451 LIMEIL-BREVANNES CEDEX

Mr P.C.C. DUCOUSSO
Service Mixte de Contrôle Biologique
BP 16
F-91310 MONTLHERY

Dr J.C. DUPLESSY
CNRS Centre des Faibles Radioactivités
BP 1
F-91198 GIF SUR YVETTE CEDEX

Mr B. DUPOUYET
EDF/GDF
76 Bd Mendès France
F-50100 CHERBOURG

Mr C. DUSAOT
EDF/GDF
76 Bd Mendès-France
F-50100 CHERBOURG

Dr. B. ECONOMIDES
Scottish Universities Research and
Reactor Center
East Kilbride
UK-GLASGOW G75 OQU

Mr Y. EYMERY
EDF/GDF
76 Bd Mendès France
F-50100 CHERBOURG

Mr FALCOZ-VIGNE
COGEMA
9 Av de la Gare
F-84150 JONQUIERES

Mr FARGES
CEN/FAR
F-92265 FONTENAY AUX ROSES

Mr J.M. FERNANDEZ
CEA-SMT/IFREMER-CT
BP 330
F-83507 LA SEYNE SUR MER

Dr N.S. FISHER
Brookhaven National Laboratory
Oceanographic Sciences Division,
UPTON, NEW YORK (U.S.A.)

Mr L. FITOUSSI
CEN/FAR/IPSN/DRSN
BP 6
F-92265 FONTENAY AUX ROSES

Dr A. FRAIZIER
CEA/IPSN/DERS/SERE
Laboratoire de Radioécologie Marine
BP 508
F-50105 CHERBOURG CEDEX

Dr C. FREJACQUES
CNRS
15 Quai Anatole France
F-75700 PARIS

Dr R. GANDON
CEA/IPSN/DERS/SERE
Laboratoire de Radioécologie Marine
F-84150 JONQUIERES BP 508
F-50105 CHERBOURG CEDEX

Dr A. GARNIER
CEA/IPSN/DPS/SEGP
BP 6
F-92265 FONTENAY AUX ROSES

Dr J. GASTAUD
IAEA/ILMR
Musée Océanographique
MC-98000 PRINCIPAUTE DE MONACO

Mr J. GAUSSENS
SFEN
48 Rue de la Procession
F-75231 PARIS CEDEX 05

Dr P. GERMAIN
CEA/IPSN/DERS/SERE
Laboratoire de Radioécologie Marine
BP 508
F-50105 CHERBOURG CEDEX

Dr E. GILAT
PO BOX 3743
ISRAEL-HAIFA 34731

Dr D. GREEN
SEAKEM OCEANOGRAPHY LTD
PO BOX 696
C-DARTMOUTH, NS, B2Y 3Y9

Dr J.C. GUARY
INTECHMER
BP 324
F-50103 CHERBOURG CEDEX

Dr P. GUEGUENIAT
CEA/IPSN/DERS/SERE
Laboratoire de Radioécologie Marine
BP 508
F-50105 CHERBOURG CEDEX

Dr P.A. GURBUTT
Directorate of Fisheries Research
MAFF
Fisheries Laboratory
UK-LOWESTOFT , SUFFOLK NR33 OHT

Dr A. HANSEN
Instituto de Sciencias del Mar y Limnologia
Univ. Nacional Autonoma de Mexico
AP 70-305
M-04510 DF MEXICO

M. HENRY
COGEMA
BP 4
F-78141 VELIZY-VILLACOUBLAY CEDEX

Dr M. HEYRAUD
IAEA/ILMR
Musée Océanographique
MC-98000 PRINCIPAUTE DE MONACO

Dr E. HOLM
IAEA/ILMR
Musée Océanographique
MC-98000 PRINCIPAUTE DE MONACO

Dr J. HOWORTH
Enrion. and Medical Sciences Division
AERE HARWELL
GB-DIDCOTT, OXON

Dr Chih-An HUH
College of Oceanography
Oregon State University
CORVALLIS, OREGON (U.S.A.)

Dr T.S. JACOBSEN
The Marine Pollution Laboratory
Jaegersbord Alle 1 B
DK-2920 CHARLOTTENLUND

Dr P. JEAN-BAPTISTE
CEN/SACLAY
CEA/DPC/LGI
F-91191 GIF SUR YVETTE

Mr G. JOLY
EDF/GDF
76 Bd Mendès France
F-50100 CHERBOURG

Dr D. KADKO
College of Oceanography
Oregon State University
CORVALLIS , OREGON 97330 (U.S.A.)

Dr A.D. KARPF
Dornier System, NTK
Postfach 1360
D-7990 FRIEDRICHSHAFFEN

Dr H. KAUTSKY
VOGTWELLSSTRASSE 24a
D-2000 HAMBURG 54

Dr P.J. KERSHAW
Directorate of Fisheries Research
MAFF
Fisheries Laboratory
UK-LOWESTOFT, SUFFOLK NR33 OHT

Dr C. LALOU
CNRS-Centre des Faibles Radioactivités
BP 1
F-91198 GIF SUR YVETTE CEDEX

Dr C. LAMBERT
CNRS-Centre des Faibles Radioactivités
BP 1
F-91198 GIF SUR YVETTE CEDEX

Dr C. LARSONNEUR
Université de Caen
Laboratoire de Géologie Marine
F - 14032 CAEN

Dr R. LEBORGNE
EDF/GDF
76 Bd Mendès France
F - 50100 CHERBOURG

Dr I.R. McAULAY
Trinity College
University of Dublin
IRL - DUBLIN

Dr P. McDONALD
Scottish Universities Research and Reactor Centre
East Kilbride
UK-GLASGOW G75 OQU

Dr J. MacKENZIE
Safety and Reliability Directorate
UKAEA
UK-WARRINGTON CHESHIRE

Dr A.B. MacKENZIE
Scottish Universities Research
and Reactor Centre
East Kilbride
UK - GLASGOW G75 OQU

Mr F. MADELAIN
IFREMER
66 Av d'Iéna
F - 75116 PARIS

Dr. J.M. MARTIN
Ecole Normale Supérieure
Laboratoire de Biogéochimie Marine
1 rue Maurice Arnoux
F-92120 MONTROUGE

Dr W. MARTINOTTI
ENEL - CRTN
Via Rubattino 54
I-20134 MILANO

Dr M. MASSON
CEA/IPSN/DERS/SERE
Laboratoire de Radioécologie Marine
BP 508
F-50105 CHERBOURG CEDEX

Dr J.L. MAUVAIS
IFREMER - Centre de Brest
BP 70
F-29263 PLOUZANE

Dr Y. MEAR
INTECHMER
BP 324
F-50103 CHERBOURG CEDEX

Dr L. MEE
IAEA/ILMR
Musée Océanographique
MC-98000 PEINCIPAUTE DE MONACO

Dr L. MERLIVAT
CNRS/LODYC
Université P. et M. Curie
4 Place Jussieu
F-75252 PARIS

Dr J. MEYER
Service Mixte de Contrôle Biologique
BP 16
F-91310 MONTLHERY

Dr R.J. de MEIJER
Kernfysich Versneller Instituut
Zernikelaan 25
NL-9747 AA GRONINGEN

Dr J.C. MILLIES-LACROIX
SMSR
BP 16
F-91310 MONTLHERY

Dr P. MIRAMAND
INTECHMER
BP 324
F - 50103 CHERBOURG CEDEX

Dr R.M. MOORE
Dept. of Oceanography
Dalhousie University
C - HALIFAX NS - B3H 4J1

Mr C. MORGANT
I.S.O. SNC
125 bis Rue Emile Zola
F - 50100 CHERBOURG

Dr S. MUDGE
University of Lancaster
Dept. of Environmental Science
UK - LANCASTER

Dr A. MURAT
INTECHMER
BP 324
F - 50103 CHERBOURG CEDEX

Dr C. MYTTENAERE
C.C.E.
Rue de la Loi, 200
B-1049 BRUXELLES

Dr A.E. NEVISSI
Laboratory of Radiation Ecology
WH-10, University of Washington
SEATTLE, WASHINGTON 98195 (U.S.A.)

Dr H. NIES
Deutsches Hydrographisches Institut
Bernhard-Nocht-Str. 78
D-2000 HAMBURG 4

Dr J. NIHOUL
Université de Liège
Mécanique des Fluides Géophysiques
B5, Sart Tilman
B-4000 LIEGE

Dr S.E.H. NIVEN
Dalhousie University
Dept. of Oceanography
C-HALIFAX, NS B3H 4J1

Dr S.E. NOWELL
UK ATOMIC ENERGY AUTHORITY
210, A40, AEE WINFRITH
UK-DORCHESTER, DORSET DT 2 8DH

Dr C. PAPUCCI
ENEA-CREA
I-LA SPEZIA

Dr F. PATTI
CEA/IPSN/DPS/SEAPS
CEN - FAR
F-92265 FONTENAY AUX ROSES

Dr R.J. PENTREATH
Directorate of Fisheries Research
MAFF
UK-LOWESTOFT, SUFFOLK, NR33 OHT

Mr PITIE
GEA - Marine Nationale
BP 19
F-50115 CHERBOURG NAVAL

Dr A. PRESTON
Directorate of Fisheries Research
MAFF
Fisheries Laboratory
UK-LOWESTOFT, SUFFOLK, NR33 OHT

Dr N.B. PRICE
Edinburgh University
Grant Institute of Geology
UK-EDINBURGH

Dr G.M. RAISBECK
CNRS
Laboratoire René Bernas
F-91406 ORSAY

Dr J. RANCHER
CEA/SMSR
BP 10
F - 91310 MONTLHERY

Dr M. RHEIN
Institut fur Umweltphysik
D - 6900 HEIDELBERG

Mr R. RIEHL
EDF/GDF
76 Bd Mendès France
F - 50100 CHERBOURG

Dr L. ROMERO GONZALEZ
CIEMAT-JEN
Avda. Complutense, 22
S - MADRID

Dr J.C. SALOMON
IFREMER Centre de Brest
BP 70
F-29263 PLOUZANE

Dr J. SAUREL
CNAM
292 Rue St Martin
F - 75141 PARIS CEDEX 03

Dr E.H. SCHULTE
ENEA - EURATOM
Centro Ricerche Energia Ambiente
I - 19100 LA SPEZIA

Dr R. SCOTT
Scottish Universities Research and Reactor Centre
East Kilbride
UK - GLASGOW G75 OQU

Dr. M. SEGL
Universitat Bremen
FB5
D-2800 BREMEN

Dr G. SHIMMIELD
Edinburgh University
Grant Institute of Geology
UK-EDINBURGH

Dr J.N. SMITH
Bedford Institute of Oceanography
PO BOX 1006
C- DARTMOUTH, NS, B2Y 4A2

Mr L. SOMBRE
CCE
Rue de la Loi, 200
B-1049 BRUXELLES

Dr P. SOUDET
SFEN - Hopital Pasteur
Av du Val de Saire
F-50100 CHERBOURG

Dr STAUB
CEA/IPSN/DAS/SASC
CEN - FAR
BP 6
F-92265 FONTENAY AUX ROSES

Dr TEILLAC
CEA
31 rue de la Fédération
F-75752 PARIS CEDEX 15

Dr THOMSON
Institute of Oceanography Sciences
Brook Road
UK-GODALMING WORMLEY GU8 5UB

Dr J.F. TODD
University of South Carolina
Department of Geology
COLUMBIA, SOUTH CAROLINA 29208 (U.S.A.)

Dr J. TOOLE
Scottish Universities Research and Reactor Centre
East Kilbride
UK- GLASGOW G75 OQU

Mr H . VERNET
EGF-SFEN
Les Fontinelles - Orsau
F - 30200 BAGNOLS / CEZE

Mr S. VITTRAUT
EDF/GDF
76 Av Mendès France
F - 50100 CHERBOURG

DR P. WALKER
Compagnie de Recherches et d'Etudes
Océanographiques
Allée des Tamaris
F - 17000 LA ROCHELLE

Dr A. WALTON
IAEA/ILMR
Musée Océanographique
MC - 98000 PRINCIPAUTE DE MONACO

Dr M. WARTEL
Université Lille I
11 Rue Thiers
F - 59655 VILLENEUVE D'ASCQ CEDEX

Dr T. WILLIAMS
Scottish Universities Research and Reactor Centre
East Kilbride
UK - GLASGOW G75 OQU

Dr. D.S. WOODHEAD
Directorate of Fisheries Research
MAFF
Fisheries Laboratory
UK - LOWESTOFT, SUFFOLK NR33 OHT

Dr F. YIOU
CNRS
Laboratoire René Bernas
F-91400 ORSAY